Cotton Precision Breeding

Mehboob-ur-Rahman • Yusuf Zafar
Tianzhen Zhang
Editors

Cotton Precision Breeding

Springer

Editors
Mehboob-ur-Rahman
Agricultural Biotechnology Division
National Institute for Biotechnology &
Genetic Engineering (NIBGE) College
Faisalabad, Pakistan

Pakistan Institute for Engineering and
Applied Science (PIEAS)
Islamabad, Pakistan

Yusuf Zafar
Pakistan Agriculture Research Council
Islamabad, Pakistan

Tianzhen Zhang
College of Agriculture and Biotechnology
Zhejiang University
Hanzhou, China

ISBN 978-3-030-64503-8 ISBN 978-3-030-64504-5 (eBook)
https://doi.org/10.1007/978-3-030-64504-5

This Springer imprint is published by the registered company Springer Nature Switzerland AG
The registered company address is: Gewerbestrasse 11, 6330 Cham, Switzerland

Preface

Cotton is cultivated for harvesting natural lint fiber that sustains textile industry worldwide, generating revenue of around US$ 899 billion. Economies of few countries are heavily dependent on cotton production. In the present scenario, harvesting sustained cotton yield has become a major issue owing to changing environments, high pressure of expanding human population, and high cost of inputs including pesticides, fertilizers, water, etc. The availability of manmade fibers at low price also competes out the natural fiber in the international market. Resultantly, it depresses the cotton farm profitability that may cause significant reduction in cotton growing area. This phenomenon has been witnessed in many regions, including Pakistan, Argentina, Brazil, and Paraguay.

The sustainable enhancement in cotton production can be achieved largely by developing cotton varieties that can mitigate the fast evolving resistance in insect pests and pathogens and instable prevailing environments. Cotton cultivars with brilliant genetics can be bred in the shortest possible time by the application of genomic knowledge as well as genetic transformation assay. The revolution in genomic science together with parallel amazing innovations in bioinformatics and transformation technologies particularly witnessed after 2010s has made us to initiate "Precision Breeding in Cotton."

In this particular book, efforts are made to summarize the key issues depressing cotton production as well as means to combat these challenges for awareness and better utilization of natural fiber cotton. Five different research areas including cotton germplasm, genomic resources and their evolution, genetically modified (GE) cotton developed through transformation assay, genome editing, skeptical views about GE cotton and cotton production beyond 2030 have been discussed.

Breeding for improved cotton cultivars through avoiding the repetitive use of few highly adaptive germplasm—stagnating cotton production worldwide, has got the top priority for bringing new genes under cultivation. Diverse untapped genetic resource (under-exploited cotton species) is not only an asset for breeders to enrich the cultivated gene pool with new genes and/or alleles but also for molecular biologist for unraveling various genetic mechanisms.

In this book, extraordinary natural diversity exists in *Gossypium* for various traits, including fiber morphology and stress tolerance, and also to agronomic traits have been described. It has been explained that high resilience and yield potential in *G. hirsutum* than that of *G. barbadense* is the outcome of divergent evolution of several genes that was the major driving force for imparting this species a wider adaptability. Similarly, within a species, several genetically distinct landraces, accessions, etc. exist. However, within cultivated cotton varieties, the genetic diversity is very narrow as indicated in several studies. It makes the cotton improvement very challenging.

At the moment, cumulatively 70,000 accessions, cultivars, landraces, etc. are available in nine gene banks of different cotton-growing countries, particularly at Fort Collins, USDA, and CIRAD, Montpellier, France. These germplasm resources can be utilized for tapping genes for adding resilience to abiotic and biotic stresses. However, it is expected that several accessions are duplicated, which can be avoided by characterizing the available germplasm resources phenotypically as well as genotypically. For bringing diversity in cotton varieties, exchange of cotton genetic resource among the cotton breeders of different countries is the key to success. Such activities would add synergy to cotton improvement programs. Possible means to explore such kind of untapped genetic resources have been discussed.

The potential of genetic diversity occurs in germplasm and has not been fully utilized owing to several inherent issues, including linkage drags of undesirable traits, requirement of several rounds of backcrosses for recovering the genetic background of recurrent parent genotype, and incompatibility barriers. Generation of enormous amount of genomic data has made it possible to use such genetic resources in breeding programs through designing DNA markers around the genes followed by making selection using these markers in succeeding generation, which can reduce sufficient years for releasing a cotton variety. For example, availability of gap-less genome assemblies of all cultivated cotton species and their progenitor species and high-density genetic maps developed using interspecific and intraspecific populations are available, which can be used in warranting marker-assisted selection. Several articles on cotton genome sequencing appeared from 2012 till date. This information was also helpful in knowing the evolutionary course of the genus *Gossypium*, landscape of cotton genome, and also the function of various genes involved in conferring various traits. Also the use of mutagens for inducing mutations by exposing cottonseed with mutagens like EMS and gamma radiations would be instrumental in expanding the allelic diversity at various loci that can buffer the potential epidemics of insect pests and diseases. Utilization of genetic diversity and means to enhance the genetic diversity for improving resistance to diseases as well as to water-limited conditions have been explained in this book

Improvement in fiber traits has remained the top priority area of several cotton-breeding programs. Multiple genes have been identified which confer high fiber traits. These are found in abundance in *G. barbadense* as elucidated by undertaking genome-wide association analysis (GWAS) and re-sequencing of cotton germplasm. Genetic mechanism for fiber traits development has also been explained. Similarly, genetics of colored lint fiber has been discussed in one of the chapters and also the scope of breeding for developing cotton varieties producing colored natural fiber.

Another major section of this book is related to the development of GE cotton and its cultivation in the field. First generation of GE cotton has offered protection from bollworms and herbicides, which together depreciate the cost of production across all the cotton-growing countries. Limited number of genes of Cry series and genes conferring resistance to glyphosate have been commercialized in cotton; highlighting the need for introducing new genes especially derived from plant sources, if possible, would have high chances of acceptance among the community. Thus, the possibilities of exploring new technologies and genes for building more walls of protection to stresses have also been discussed. This gives a birth of new generation of GE cotton. For example, pyramiding of Bt genes with RNA interference (RNAi) has been used to delay the evolution of resistance by chewing insect pests of cotton. Similarly, RNAi method was deployed to develop to generate resilience to whitefly and diseases including Verticillium wilt. Also the role of RNAi in improving agronomic traits and resistance to other stresses has been discussed in this book. Scope of third generation of GE cotton, aiming for producing huge quantity of industrial products cheaply, has also been explained. Acquiring high expression of pharmaceuticals products can be achieved by introducing genes in plastid. Also the use of CRISPR technology for developing third generation GE cotton has several advantages, including development of transgene-free cotton and gossypol-free cottonseed by silencing the genes conferring gossypols in seed. Major advantage of this technology is that the new cultivars can be evolved without introgression of foreign gene; hence, the technology will be acceptable to countries having skeptical views about the GE technology. The success of second and third generation of GE cotton has not yet been demonstrated in the field, however, with the given knowledge and high-tech genomic assays, in near future, when second and third generation of GE cotton will be cultivating at farmer's field. In the second last chapter, skeptical views as well as counter arguments pertaining to GE cotton have been compiled. Discussion on pros and cons of using GE cotton products and by-products will help end-user to get benefit of the modern technologies aiming at improving the quality of life.

Hence, the scope of harvesting sustained cotton production beyond 2030 has been discussed. Many well-known scientists from the cotton research community have shared wisdom, knowledge, views, and application of the genomic knowledge that would lay a firm foundation for initiating precision breeding in cotton. Thus, the book comprehensively covering such kind of information is the need of time.

Faisalabad, Pakistan — Mehboob-ur-Rahman
Islamabad, Pakistan — Yusuf Zafar
Hanzhou, China — Tianzhen Zhang

Acknowledgments

The idea for writing this book was conceived upon a scientific visit to China through a collaborative project "Mining of Elite Genes for High Yield, Super Fiber Qualities and Heat Tolerance and Their Usages in Pakistan and China Cotton Cultivars"—Project No. PSF/BSFC-AGR/P-NIBGE (12).

Contents

Part I
Cotton Breeding and Genomics

Chapter 1
Historical Perspectives: From Conventional to Precision Breeding in Cotton

Mehboob-ur-Rahman, Sana Zulfiqar, Abid Mahmood, Yusuf Zafar, and Tianzhen Zhang

1.1 Historical Perspective

Use of natural fibers for making fabrics is very prehistoric that is dated back to 36,000 BP (Balter 2009; Kvavadze et al. 2009). Natural fibers, made of cellulosic components, can be derived from several plant species including hemp, cotton, jute, flax, sisal, etc., while the animal fibers (wool and silk) are made of proteins (Hansen and Bjorkman 1998). Synthetic fiber, another major type, is made from petroleum products, which contributes around 60% of the total fiber production worldwide. These fibers are rayon, polyester, nylon, acetate, acrylic, spandex, lyocell, etc. (Fig. 1.1). At present, synthetic fibers are considered durable and economically cheaper. Though the debate on the superiority as well as benefit of both types of fibers is going on over the last many years, natural fibers are still preferred by the masses because these are produced naturally and thus considered much safer.

Mehboob-ur-Rahman (✉)
Agricultural Biotechnology Division, National Institute for Biotechnology & Genetic Engineering (NIBGE) College, Faisalabad, Pakistan

Pakistan Institute for Engineering and Applied Science (PIEAS), Islamabad, Pakistan
e-mail: mehboob@nibge.org

S. Zulfiqar
Plant Genomics and Mol. Breeding Lab, National Institute for Biotechnology and Genetic Engineering (NIBGE), Faisalabad, Pakistan

A. Mahmood
Ayub Agricultural Research Institute (AARI), Faisalabad, Pakistan

Y. Zafar
Pakistan Agriculture Research Council, Islamabad, Pakistan

T. Zhang
College of Agriculture and Biotechnology, Zhejiang University, Hanzhou, Zhejiang Province, China

Mehboob-ur-Rahman et al. (eds.), *Cotton Precision Breeding*,
https://doi.org/10.1007/978-3-030-64504-5_1

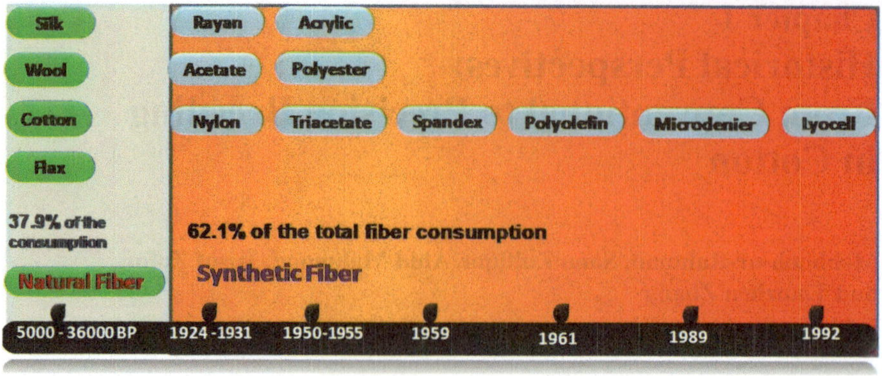

Fig. 1.1 Fibers: an important part of human civilization

Among natural fibers, cotton fiber has been used extensively for making fabrics especially after the onset of industrial revolution in the eighteenth and nineteenth centuries. Later on, semisynthetic fiber, viscose, and synthetic fiber, nylon, were commercialized in 1905 and 1930s, respectively. These fibers particularly nylon captured a major market share of natural fibers. For overcoming this issue, efforts were made for developing cotton varieties with enhanced yield and better lint qualities which can easily be amenable to new textile machinery. All these efforts resulted in the development of improved cotton cultivars by incorporating genes from unadapted cotton germplasm in the mid-twentieth century. Thereafter, discoveries in genomic science paved the way of transferring genes from alien sources into cotton. First genetically engineered (GE) cotton containing *Cry 1Ac* gene was commercialized in 1995 (Rahman et al. 2015). However, over the last many years and even in the present time, the prices of petroleum have been depreciated significantly that may further depress the market share of natural fiber owing to the influx of low-cost synthetic fiber.

1.2 Challenges to Cotton Production

1.2.1 Insect Pest and Diseases

Infestation of insect pests and infection by diseases are the main causes of low yield in cotton. Protection to chewing pests was largely achieved by incorporating *Cry* genes (Rahman et al. 2012). This protection was weakened; resultantly chemicals are sprayed for killing the chewing insect particularly pink bollworm. The potential of minor pests for becoming major pests can be the future challenge, for example, infestation of mealybug and dusky bug in Pakistan. This scenario may also arise in other countries. Similarly, evolution of new strains of pathogens especially new virus strains that cause leaf curl disease, verticillium wilt—also called cotton cancer—and bacterial blight are the major potential threats to cotton production. Cotton leaf curl can spread to cotton-growing countries where whitefly is prevalent

(Rahman et al. 2017). If these two factors are not properly addressed, it may lead to the elimination of cotton cultivation from several cotton-growing regions.

1.2.2 Changing Climate

The gradual increasing trends of heat, drought, unexpected heavy rains and other environmental extremities are negatively impacting the agricultural productivity (Ray et al. 2013; Mills et al. 2018). These fluctuations in environments if occurring together may reduce crop productivity very drastically. Breeding cotton cultivars has been carried out in high-input environments (Rahman 2016); thus the genetic potential of these varieties will not be able to sustain under the adverse environments.

Like many other crops, global warming is a detrimental threat to cotton production worldwide, and its impact partly has been witnessed in the form of high temperature, extensive rainfall, unpredictable climatic adversities, etc. It has been reported that temperature prevailing the crops will continue to increase in the twenty-first century. High temperature retards cotton growth that leads to shedding of small fruiting bodies (square) and reduced boll size; together these factors lessen the lint yields. High temperature also reduces the effectiveness of pesticides. This phenomenon can be more detrimental to those cotton-growing regions where temperature often exceeds 40 °C.

1.2.3 Competitive Ability with Synthetic Fiber

If the present declining trend of petroleum price persists, synthetic fiber can capture the market share of natural fiber in the future. These factors may lead to the depression of the price of natural fiber in the international market; ultimately the profitability of farmers will be reduced. Thus the cotton-growing community can think cultivation of other crops for increasing their farm's profitability. For example, in Pakistan, during the last three years, area under cotton has significantly been reduced by ~30%. The major factor is the high cost of production as well as infestation of pink bollworm.

1.2.4 Yield Losses and Contamination

Yield losses during harvesting time and trash contamination in seed cotton are two major issues which are supposed to be addressed for deriving maximum benefit from cotton. For example, in developing countries cotton is picked manually. In the future, it will be very difficult to harvest cotton crop well in time owing to the mass migration of families from rural to urban areas. Secondly, hesitation of new generation to work in the field is also aggravating the situation. Hence, for harvesting sustained cotton production, it is important to shift on mechanical picking system like many other advanced countries including USA and Australia which have adopted this system. There is a need to improve the efficiency of mechanical picking system for avoiding the yield losses (at least 10% during picking). Govt department should provide mechanical pickers to farmers at subsidized rates so that harvesting can be carried out well in time. Similarly, contamination issues will be more serious in those cotton-growing countries where varieties have trichomes on various organs including leaves. At maturity when farmers spray defoliants before harvesting, the falling leaves usually stick with seed cotton—thus adding contamination.

1.2.5 Yield Stagnation

Cotton varieties have achieved maximum yield potential by utilizing conventional as well as transgenic approaches. Almost every breeding program is working with very limited genetic diversity (Rahman et al. 2002, 2008; Hu et al. 2019); thus using present breeding approaches, cotton production will not help in breaking yield barriers. Future improvements can be made if some novel genetic resources will be available. These can help in developing cotton varieties with outstanding genetics for combating the challenging environments. Thus improving resilience together with best management practices can ensure harvesting of maximum sustained yield potential with improved lint quality.

1.3 Roadmap for Achieving Sustainable Cotton Production

1.3.1 Genetic Resources

The valuable genetic diversity is present in the available germplasm in various forms including landraces, genotypes, accessions, etc. These germplasm resources can be utilized for tapping genes required for combating abiotic and biotic stresses. Cotton fiber is contributed by four major cultivated cotton species including *G. arboreum*, *G herbaceum*, *G. hirsutum* and *G. barbadense.* One of the major breeding challenges is to broaden the narrow genetic base of the cultivated cotton varieties for adding resilience to environmental factors. For this, new alleles or genes can be introduced from alien species. For the introduction of new alleles or genes from the genus *Gossypium*, it is important to have well-characterized genetic resources. About 70,000 accessions are available in several gene banks of 9 cotton-growing countries (USA, Uzbekistan, India, China, Australia, France, Russia, Brazil, and Pakistan; Fig. 1.2). While in CottonGen database, a total of 20,253 accessions-genotypes-cultivars representing almost all *Gossypium* species have been reported (https://www.cottongen.org/find/germplasm).

The major issue with these gene banks is the preservation of similar accessions. For avoiding these duplicated accessions, it is important to characterize all

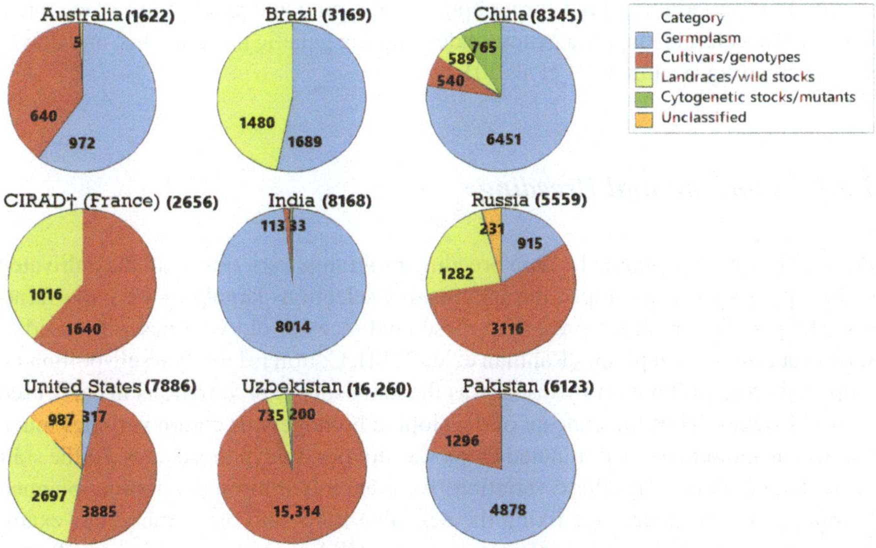

Fig. 1.2 Genetic resources of the genus *Gossypium* in different gene banks across the world

the available germplasm resources phenotypically as well as genotypically (Iqbal et al. 2015). Only the selected accessions (by omitting the duplicated accessions) should be characterized by conducting multilocation trials in several other cotton-growing countries. In this regard, exchange of germplasm among breeders across the cotton-growing countries will remain the most important tool for studying the response of cotton accessions in diverse environments. For example, a big activity of germplasm transfer from the USA to Pakistan (sponsored by the USDA through cotton productivity enhancement project, ID-1198) resulted in the identification of several cotton accessions conferring resistance to cotton leaf curl disease (Rahman and Zafar 2018). Among these, Mac-07 was extensively used in developing cotton cultivars with improved tolerance to the disease. Similarly, another joint activity was initiated [(Pak-China project entitled mining genes for high yield, super fiber, and heat tolerance, PSF/NSFC-AGR/P-NIBGE (12)] for screening germplasm to heat as well as studying fiber traits in various environments. Through this project, heat-tolerant cotton genotypes have been identified for using in future cotton breeding programs. Such activities would add synergy to cotton improvement programs.

For scoring traits of cotton germplasm, it is important to devise a uniform scale for rating or scoring each character so that everyone should speak the same language. It is also important to select leading cultivated cotton varieties and genotypes/accessions with unique traits for doing re-sequencing. It will help in identification of polymorphisms which can be exploited in fabricating unique universal SNP chip as well as issuing passport to each accession and/or cultivar. The SNP chip can be used for initiating targeted breeding, and thus the time for developing new cotton varieties can be reduced. Hence, a substantial amount of funds is required for undertaking high-throughput phenotyping and genotyping procedures, these will accelerate mapping genes, understanding genetic mechanisms, and downstream breeding (Dempewolf et al. 2014).

1.3.2 Conventional Breeding

Prior to the onset of planned cotton breeding programs, varieties of all the cultivated cotton species were developed through making selections largely by the early farmers. The genetic variations present in these land races or old varieties were used to select superior cotton plants (Rahman et al. 2014). Cotton is largely a self-pollinated crop, and cross pollination by insect was the main source of variations in early bred cotton varieties. Thus the concept of developing homogenous cotton varieties much like wheat and other self-pollinated crops was not possible in those days. At the start of the twentieth century, these variations were exploited through adopting planned hybridization procedures for fixing the desirable traits into one variety. For example, escape from boll weevil infestation in the USA was managed by developing early maturing cotton varieties, thus replaced gradually the cultivated late maturing cotton varieties. The Deltapine cotton varieties were developed by attempting several selections and crossing procedures in 1911 (Poehlman 1987). Wilt resistant cotton plant (cocker 100 wilt), Stoneville type, storm resistant types, etc. were selected from 'Lone Star' type. Similarly, several cotton varieties were developed by selecting plants from the cotton field of a small village 'Acala', Mexico in 1906–1907. Similarly, Pima cotton (*G. barbadense*) varieties were developed in the USA by selecting best cotton plants when grown in Arizona and Southern California in 1903. Heterogeneity was desirable for having persistent hybrid vigor in a variety by bulking seed of cotton progenies showing sufficient uniformity for morphological characteristics, disease and insect resistance, and lint quality (Poehlman 1987).

Like self-pollinated crop varieties, planned hybridization followed by pedigree selection procedure was deployed for developing cotton varieties (Rahman et al. 2014). Similarly, backcross breeding procedure was adopted for bringing new alleles conferring resistance to diseases from untapped genetic resources. Backcrosses were also attempted for bringing male-sterile genes or fertility restoring genes into a variety. Chromosomes can also be transferred into sterile cytoplasms by attempting several rounds of backcrosses.

Other option for enhancing the yield could be the exploitation of hybrid vigor. For instance, hybrid vigor resulted in multifold increase in corn production worldwide. Developing cotton hybrids showing 30% increase in lint production over the standard OP cotton variety may convince farmers to cultivate hybrid cotton. This task is difficult but achievable through exploring the best combiners. Hybrid breeding cotton is handicapped due to the non-availability of reliable genetic resources, lack of mechanical procedures for emasculation and also the efficacy of gametocides. Various genes encoding sterility and restoring fertility were identified, and efforts were made to produce hybrid cottonseed as successfully demonstrated in corn by adopting the ABR system (Poehlman 1987). However, these genes in cotton were found to be temperature sensitive and thus the potential benefit of this system was not harvested as witnessed in corn. Work on understanding the genetics of hybrid vigor can be the best choice which will help in targeting important genes for amending their expression, thus hybrid vigor can be fixed.

Development of resilience in cotton through conventional breeding can also be helpful for addressing the issues of changing environment. One of the best strategies to combat escalating heat issue is to develop cotton varieties which can tolerate 50 °C day temperature and 30 °C night temperature. In this regard, few traits, for example, boll retention at high temperature is considered as the most appropriate selection criterion. However, such varieties are usually high input demanding. When inputs are scarce, such varieties could not show their yield potential. Thus selection for varieties which can keep on growing even by sacrificing few bolls on first sympodial branches would be the desirable feature for encountering scarce water resources as well as high temperature. Secondly, breeding for small-to-medium sized leaves would allow sun light to reach lower leaves and opened bolls—thus boll rottening can be avoided in rainy season. Canopy of such cotton varieties also facilitates the uniform application of pesticides on even lower leaves of cotton plants.

Sustained cotton production can be achieved by studying the response of genetic resources (parent genotypes and their progenies) under low-input environment. For instance, landraces evolved in low-input regions can be utilized in breeding programs. Deployment of genomic-based selection procedures would be the best choice for transferring genes into the domesticated high-yielding varieties. For this purpose, extensive screening of germplasm would be required for grabbing genotypes exhibiting traits such as resistance to various insect pests and diseases, better nutrient economy, delayed leaf senescence, and yield consistency across the environments. Germplasm originated in different cotton-growing regions can be screened in those countries which are facing the impact of climate change. Segregating populations can also be screened through joint collaboration. Resulting newly developed candidate lines can be tested in different environments for studying their adaptability. This approach is straightforward, and does not require enough resources. Thus, exchange of cotton germplasm will remain the most instrumental tool. DNA markers too can be very instrumental since the trait is not as complex as drought tolerance.

1.3.3 Introgression Breeding

Introgression breeding, transferring of alleles or genes from unadapted genetic resources to adapted genotypes (Hernandez et al. 2020), has played gigantic role in transferring useful genes from different closely related species into the cultivated species including cotton (Rahmat et al. 2019). Before the domestication of tetraploids, exchange of gene transfer between *G. hirsutum* and *G. barbadense* was reported (Hu et al. 2019). These introgressions facilitated the adoption of a specie in a challenging environment by expanding its genetic diversity. This concept was translated by cotton breeders to transfer genes from other closely related cotton species to the cultivated cotton species at the National Key Laboratory of the Crop Genetic and Germplasm Enhancement, College of Agriculture, Nanjing Agricultural University, Nanjing China under the leadership of Professor Tianzhen Zhang. Pima cotton, *G. barbadense*, is easily crossable with upland cotton *G. hirsutum*. Similarly, diploid cotton species including *G. arboreum*, *G. herbaceum*, *G. raimondii*, etc. have also been used for introducing new genes into the *G. hirsutum* background.

For improving the lint quality, genes related to superior lint quality were introduced in old cotton varieties of *G. hirsutum* through interspecific crosses. The interspecific population may have several transgressive segregants at both extremes for quantitative traits including plant height, leaf shape, number of bolls, boll size, trichome density, lint quality, etc. (Zhang 1993). This phenomenon needs further explorations. There are several stable lines with superior fiber developed after doing extensive backcrossing followed by selections of desirable plants (Ma and Liu 1982; Zhang 1993; Cantrell and Davis 2000; Percy et al. 2005; Gore et al. 2012; Liu et al. 2005).

Some desirable genes from Pima cotton were transferred to Acala cotton for improving its lint quality but success was very limited. Similarly, high-quality lines Pee Dee were developed through interspecific crossing (May 2001). Later on, these lines were used in introducing genes from Acala and Pee Dee lines into cultivated cotton varieties (Bowman and Gutierrez 2003).

Several other genes with qualitative impact were transferred from Pima cotton to upland cotton. These alleles or genes are related to traits such as glandless (Yuan et al. 2000), sub-okra leaf shape (Zhang 1993, 2011) and resistance to bacterial blight (Percy and Kohel 1999), verticillium wilt (Wilhelm et al. 1985; Zhang et al. 2012), thrips (Zhang et al. 2013a, b), and spider mite (Zhang et al. 1992, 1993). Introgressed lines conferring tolerance to drought and salinity were also developed (Zhang and Hughs 2012; Tiwari et al. 2013).

The genetic diversity found in wild tetraploids was also utilized for enriching the genome of cultivated cotton species. Crossing between tetraploid species was easy owing to the equal number of chromosomes. The brown fiber producing tetraploid *G. tomentosum* ($2n = 4\times = 52$) exhibited resistance to leaf hopper and thrips (Jayaraj and Palaniswamy 2005). The tetraploid *G. mustelinum* produced terpenoid aldehydes in leaves that conferred resistance to insect pests (Altaf et al. 1997).

Similarly, genes from diploid cotton species were transferred for improving resilience (Rahmat et al. 2019) as well as fiber traits. For instance, *G. longicalyx* ($2n = 2\times = 26$, F-genome), native to Africa, is resistant to reniform nematode (Yik and Birchfield 1984), and also has genes for fiber quality (Weaver et al. 2013). Similarly, *G. armourianum* ($2n = 2\times = 26$, D_{2-1} genome) is resistant to whitefly (Jayaraj and Palaniswamy 2005). The *G. herbaceum* and *G. arboreum* (both are diploid cultivated species) are resistant to cotton leaf curl disease (Rahman et al. 2005). Several methods were used to hybridize these two species. In few studies, chromosomes of these diploid species were doubled ($2n = 4\times$) followed by crossing with *G. hirsutum* to have F_1 for making backcrosses with *G. hirsutum* (Nazeer et al. 2014; Rahmat et al. 2019). Development of synthetic tri-species hybrids have been suggested to introgress genes from diploid species to cultivated tetraploids (Mergeai 2006; Robinson et al. 2007).

The process of introgression breeding has claimed limited success in improving cotton owing to linkage drag of unwanted genes, hybrid breakdown, instability and selective elimination of desirable genes during selfing (Zhang et al. 2014). In the future, DNA markers can help in overcoming the issues of linkage drags as well as lengthy breeding procedures in developing stable lines. Since the genome of many important cotton species have been sequenced, thus designing new DNA markers around the loci involved in conferring high lint quality trait will help in reviving the introgression breeding work.

1.3.4 Mutation Breeding

Use of mutagens for expanding the genetic diversity of cultivated cotton resulted in the development of useful germplasm as well as crop varieties. Mutagens can create new traits or enhance the expression of already present genes. In total, ~3308 mutant crop varieties of 200 crop species have been developed worldwide (https://mvd.iaea.org/). Majority of these mutant varieties were released in China, India, and Japan. First mutant cotton variety was released in the early 1960s, and later on, 48 mutant cotton cultivars were released. Out of these, almost 25% were released in Pakistan by exposing the genetic material with gamma radiations.

By the 1990s, mutation detection assays existed to enable the development of reverse-genetics with point mutations, a process known as targeted induced localized lesion in genomes (TILLING). In the present time, next-generation sequencing assays have made it possible to discover several mutant alleles in one experiment. By developing large M_2 populations, one can identify thousands of mutant alleles in each population. Thereafter, only the selected mutant plants containing maximum number of desirable mutations can be retained for performing phenotypic characterization. Thus the available genetic information can further help in reducing the number of desirable mutants which can be used in breeding programs. The typical mutants can be traced in segregating generations using DNA markers by synthesizing primers around the gene or allele of interest.

Complex traits can also be improved to some extent by exposing the genetic material with chemical mutagens (ethyl-methane sulfonate, *N*-ethyl-*N*-nitrosourea, etc.). However, the role of physical mutagens is marginal for improving the complex traits. For example, in multiple studies, significant improvements in ginning outturn have been reported (Patel et al. 2014, 2016; Aslam et al. 2017; Witt et al. 2018). Key to success is to grow maximum number of mutants in M_2 population for identifying very few mutants containing the desirable traits.

1.3.5 Genomic Resources

The genomic science is heavily dependent upon the availability of genetic resources followed by associating phenotypic diversity with the sequence diversity. Extensive re-sequencing will be required for identifying the natural variations occurred in a particular species. Once the genes are identified, these can be used for introducing in cultivated cotton varieties using transgenic approaches or can be directly mutated for altering their expression using targeted breeding approaches like RNAi and CRISPR. Before deploying these technologies, consensus for cracking the genome of cotton specie was made by a group of scientists for facilitating the process of improving and understanding the genetic circuits of various traits.

The idea for sequencing the most simple cotton genome (*G. raimondii*) was discussed in 2006 by a group of cotton scientists led by Prof. Andrew H Paterson, University of Georgia, USA (Chen et al. 2007). Finally, two papers on D-genome sequencing were published in 2012 by two different research groups (Paterson et al. 2012; Wang et al. 2012). Two years later, one study on sequencing *G. arboreum* (A-genome) was published (Li et al. 2014). Thereafter, two papers on sequencing genome of *G. hirsutum* (known for high yield) were published independently by two different research groups in 2015 (Li et al. 2015; Zhang et al. 2015). In the same year, two studies on sequencing the genome of *G. barbadense* (known for high-quality lint) were published by two different consortia (Liu et al. 2015; Yuan et al. 2016). However, the previous genome assemblies of tetraploid cottons were not contiguous; for example, intergenic regions were represented poorly—rendering a lot of information inaccessible to the researchers. Construction of a contagious genome assembly was really a mammoth task because of their large genome size (~2.6 Gb, near to the human genome size).

More recently, genomes of *G. hirsutum* and *G. barbadense* were sequenced for reducing gaps in the reported genome assemblies by involving several assays including single-molecule real-time sequencing, BioNano optical mapping, and high-throughput chromosome conformation capture. Significant improvement in contiguity was achieved particularly in regions containing a lot of repeats, for example, centromeric regions. Massive genome rearrangements particularly inversions in para-/pericentric regions of 14 chromosomes occurred after polyploidization have been reported. In total, 13 QTLs conferring high fiber quality trait have been identified. These regions can be exploited in future cotton improvement programs (Wang et al. 2019).

Same year, Zhang and colleagues (Hu et al. 2019) adopted a series of procedures and approaches for generating a huge number of sequence reads (800 Gb, >330×)—sufficient to differentiate errors from the actual sequence, which were used to construct assemblies of *G. hirsutum* and *G. barbadense* by merging the corrected scaffolds (long stretch of DNA, often reaching to the length of a chromosome). Order and orientation of the scaffolds were verified by three-dimensional proximity information and ultradense genetic map—constructed by placing ~6.1 million SNPs. Hence, several long-awaited puzzles including evolutionary dynamics of cotton genome and functional dynamics of genes after combing in one nucleus and getting genomic insight of the traits, were resolved. In total, 2.22 gigabases—91.4% of the total estimated genome size of *G. barbadense* var. Hai7124—was assembled that showed 47- and 90-fold increased contiguity over the two previously poorly assembled genomes (Liu et al. 2015; Yuan et al. 2016), and major chunk (98.2%) was assigned into 26 chromosomes. Similarly, 2.30 Gb of *G. hirsutum* var. TM-1 was assembled, of this 97.4% was organized into 26 chromosomes, and impressively the contiguity of the assembly was increased over 10-fold to one sequence assembly (Zhang et al. 2015) and 20-fold over the second published genome assembly (Li et al. 2015). These assemblies can be used as reference for comparing the sequences of other genotypes of these species as well as with some other species of the genus *Gossypium* or even distantly related plant species (Hu et al. 2019).

Centromere of each chromosome composed of several hundreds of kilobases of repetitive elements—keep on expanding/contracting or giving birth of new sequences at remarkably high rate. This feature makes the mapping of centromere very difficult even if the resolution of genome assembly is very high. This issue was resolved by developing a contiguous assembly of each species, and the distorted position of centromeres on each chromosome reported earlier was painted very precisely. For example, the average length of centromeric regions was narrowed down to 270 Kb in Hai7124 and 385 kb in TM-1. These centromeres happened to be localized in the same corresponding regions of the corresponding chromosomes, further authenticating the high quality of the newly assembled genomes (Hu et al. 2019).

Relatively more number of genes were predicted for *G. barbadense* var. Hai7124 than that of the *G. hirsutum* var. TM-1 (75,071 versus 72,761), and most of these showed a clear demarcation of exon-intron boundaries. Majority of these genes were found in duplicated copy due to their ploidy nature. A huge portion of both sequenced genomes comprised of transposable elements (TE) (1460.46 Mb in TM-1 versus 1374.61 Mb in Hai7124), which is doubled in A subgenomes compared to the D subgenome. These were inserted during the three whole genome duplications occurred each around 15, 26, and 48 MYA, well before the synthesis of allotetraploids (Zhang et al. 2015). Some empirical evidences, for example, high colinearity and conserved gene order between TM-1 and Hai7124, suggested that both allotetraploids diverged from a common parent around 0.4–0.6 MYA. The subgenomes of allotetraploid evolved much faster than their diploid progenitors owing to the frequent exchanges of chromosome blocks occurred between the two subgenomes. Relatively a large number of lost, disrupted, and positively selected genes were found in A subgenome, showing the signatures of strong selection pressure exerted during the evolution of these allotetraploids. Deep sequencing of several accessions of *G. barbadense* and *G. hirsutum* suggested that both species evolved very rapidly, and introgression of large genomic regions was found in all *G. barbadense* accessions particularly expressing extra-long staple, confirming that *G. barbadense* was originated in the Northwestern Peru/SW Ecuador region (Westengen et al. 2005). These introgressions helped in the adaptation of domesticated cottons to different environments.

Several other genomic factors including genes with structure variations (10,633 in number) which are responsible for creating novel traits in a particular species have been characterized, e.g., 2-bp deletion in *WLIM1a* of Hai7124. Then authors probed several presence/absence variations (PAVs) which were found unevenly throughout the genomes of TM-1 and Hai7124. Such variations led to generate several species-specific traits, for example, extra-long staple in *G. barbadense* and high yield potential in *G. hirsutum*; both are remarkable features that facilitated their adaptation and domestication (Hu et al. 2019).

Producing quality lint/fiber together with high productivity has remained the central part of cotton breeding programs; however, both traits are difficult to bring into one cultivar due to the complex genetics of long fiber (ESL). More than 50% of the genes (~45,000) contribute in shaping the fiber phenotype. It has also been shown that sucrose transporter (*GbTST1*), Na^+/H^+ antiporter (*GbNHX1*), aluminum-activated malate transporter (*GbALMT*), and vacuole-localized vacuolar invertase are responsible for ELS in Hai7124. Secondly preferential expansion of few gene families in Hai7124, for example, ADP ribosylation factor (ARF) GTPase family, also plays a role in conferring long but strong fiber in the ELS cotton. Similarly, high resilience and yield potential of TM-1 are the outcomes of divergent evolution (after duplication) experienced by several genes after domestication, thus facilitating the adaptation in new environments (Hu et al. 2019). This information can be used for initiating targeted breeding programs for improving high-quality traits in cotton.

More recently, Chen and colleagues (Chen et al. 2020) reported high-quality genome assemblies of five allotetraploid species including *G. hirsutum* $(AD)_1$, *G. barbadense* $(AD)_2$, *G. tomentosum* $(AD)_3$, *G. mustelinum* $(AD)_4$, and *G. darwinii* $(AD)_5$. Relatively large genome size (2.305 Gb) was reported for *G. hirsutum* than the previous studies. High-quality genome assemblies of *G. tomentosum* $(AD)_3$, *G. mustelinum* $(AD)_4$, and *G. darwinii* $(AD)_5$ with 74,699, 78,338, and 78,303 genes, respectively, were reported. Almost, 99% for *G. mustelinum*, 99.2% for *G. tomentosum*, and 99.1% for *G. darwinii* genome were placed in all chromosomes. However, the genome coverage in chromosomes for *G. hirsutum* (98.9%) and *G. barbadense* (97.0%) was less than that of the three wild tetraploid species (Chen et al. 2020). These findings are useful for undertaking future research on designing new markers as well as editing the genome with upcoming new technologies.

1.3.6 New Breeding Procedures

During the last few years, new breeding procedures have been developed to expedite the breeding process.

1.3.6.1 Marker-Assisted Breeding

Cotton varieties with outstanding genetics together with the application of best management practices can ensure harvesting of maximum yield with improved lint quality. Future yield enhancement with improved lint quality is expected to derive from cultivating the resilient cotton varieties. Thus the role of genetics of a cotton variety will be more important particularly for helping the resource-poor cotton-farming communities.

Several DNA markers including restriction fragment length polymorphism (RFLP), amplified fragment length polymorphism (AFLP), random amplified polymorphic DNA (RAPD), simple sequence repeats (SSRs), single nucleotide polymorphisms (SNPs), etc. have been used extensively in cotton (Shaheen et al. 2006, 2016; Rahman et al. 2009). However, SSRs and SNPs are easy to use, abundantly present, amenable to automation, and codominant in expression; all these qualities make them the markers of choice. There are several SNP chips available which can be used for the identification of SNPs in cotton. For instance, a CottonSNP63 (Hulse-Kemp et al. 2015), CottonSNP80K (Cai et al. 2017), etc. have been fabricated for developing high-density genetic maps.

DNA markers have been used to detect QTLs for several agronomic traits (Kalivas et al. 2011; Li et al. 2016), tolerance to drought and salinity (Saeed et al. 2011; Shaheen et al. 2013; Jia et al. 2014; Zhao et al. 2016), and yield and lint quality (Zhang et al. 2011; Cai et al. 2014; Nie et al. 2016; Iqbal and Rahman 2017; Wang et al. 2020). The generated information is being used in the identification of DNA markers and their utility in selection of cotton plants from segregating populations.

The lint-to-seed ratio beyond 50% is very much important to improve per unit yield. Natural genetic variations are present in a few accessions of *G. hirsutum*. A few genotypes exceed 50% ginning outturn which can be used to improve the lint-to-seed ratio. In most countries lint recovery ranges from 30 to 45%. This trait is very complex, but can be tailored using DNA markers otherwise difficult to improve using conventional breeding approaches (Iqbal and Rahman 2017; Feng et al. 2019). These markers can be deployed in interspecific and intraspecific populations for screening plants containing the desirable loci. For developing such populations, one adapted cotton cultivar which can be crossed with the obsolete varieties, accessions, and even wild species.

Deployment of genomic-based selection procedures for making selections (foreground and background) would be the best choice for transferring genes into the domesticated high-yielding varieties. Genomic selection procedure can be opted for improving resistance to insect pests and diseases, better nutrient economy, delayed leaf senescence, and yield consistency across the environments. This approach is straightforward, and does not require enough resources.

1.3.6.2 Development of GM Cotton

Resistance to insect pests in cotton genome has been developed using *cry* genes excised from a soil bacterium—success has been demonstrated (Rahman et al. 2012). Now the resistance conferred by these genes has been broken down in different parts of the world (Seetharaman 2018). The potential of converting minor pests into major pests is another threat to cotton sustainability. For example, before the cultivation of *Bt* cotton in Pakistan, mealybug and dusky bug remained unnoticed on cotton—indirectly controlled by the application of insecticides used to kill lepidopteron pests. Now farmers are supposed to kill these newly emerged insect pests by the application of specific insecticide for this particular pest. Such a scenario can emerge in other cotton-growing countries. Improving genetics of cotton by adding a range of novel genes derived from other alien sources under the tissue-specific promoters would be very handy in combating the insect pests more effectively. For example, efforts toward the identification of new genes from other sources such as *Hvt* gene derived from a spider were tested in cotton for studying its response to chewing insect pests, and were found encouraging (Khan et al. 2006). Similarly, several other genes encoding phytohormones (such as jasmonates) which can add in defense to insect herbivory have recently been characterized, which have the potential for combating bollworms in the future.

The challenge of changing climate can also be addressed using transgenic approaches. For example, tolerance to drought can also be enhanced by deploying genes from other organisms including distantly related plants growing under harsh climatic conditions. Efforts are underway; for example, drought-tolerant genes including *TsVP* and H^{+}*-PPase*-coding gene derived from *Thellungiella halophile* were introduced in cotton that resulted in improved shoot and root growth than that of their wild type (Lv et al. 2009). Also genes and/or their transcription factors (DREBs, ERFs, ZIP, WRKY, etc.) derived from other plant species (Dou et al. 2014; Zhou et al. 2015; Wang et al. 2016a, b) can be characterized followed by their introduction in cotton for improving resilience to abiotic stresses.

Cotton production can be enhanced by introducing genes using transgenic approaches. For example, *ScALDH21* taken from *Syntrichia caninervis* exhibited greater plant height, larger bolls, and greater fiber yield in cotton (Zhang et al. 2013a, b; Khan et al. 2018). For improving complex traits substantially, for instance cotton yield, it is important to pyramid several such genes (even from other species).

Cotton has been remained the target for improving its resilience against biotic and abiotic stress and also to the herbicide. Efforts are also being made to reduce anti-nutrient contents of cotton such as gossypol. First generation of GM cotton delivered protection to insect pests as well as herbicide; however, the insertion of transgene was random. Secondly, resistance generated by the insertion of *Bt* genes was weakened. Later on, pyramiding of Bt genes with RNA interference (RNAi) against *Helicoverpa armigera* has been used to delay the evolution of resistance by chewing insect pests of cotton (Ni et al. 2017), often referred to as second generation of GM cotton. Resistance to whitefly, another destructive pest in Indian subcontinent, has been demonstrated by targeting specific genes of whitefly using RNAi assays (Malik et al. 2016; Raza et al. 2016). Similarly, resistance to cotton diseases including verticillium wilt has been expressed using RNAi and virus-induced gene silencing (VIGS) (Wang et al. 2017).

Similarly, RNAi was used to silent genes which controls gossypol formation in seed while retaining these gossypols in all other plant organs for avoiding predation (Wedegaertner and Rathore 2015). For expressing the Bt protein in all plant tissues except seed, green tissue-specific promoter PNZIP was used in specific organs, and it has shown the expression of *Cry9C* gene in all vegetative parts except 100-time lower in reproductive organs including pollen, petals, and developing cotton seed (Wang et al. 2016a, b). RNAi was also used in improving several agronomic traits including fiber as well as resistance to stresses (Wang et al. 2016a, b) and response to stress (Zahid et al. 2016). However, these results are not demonstrated yet at farmer's field. Genes for improving fiber quality can be introduced by excising from *Calotropis procera* that produce hallow trichomes much longer than cotton (Cheema et al. 2010). For using such genes, comprehensive understanding about the development of cotton fiber is required so that the genetic circuits can be changed or new genes can be added which can supplement the existing mechanisms.

The scope of CRISPR technology for developing third generation GM cotton has several advantages including development of transgene-free systems (Zaidi et al. 2019). For example, gossypol-free cottonseed can be produced by silencing the genes conferring gossypols in seed. Major advantages of this assay are that the function of gene can be characterized and new cultivars can be evolved without introgressing foreign gene; hence, the technology will be acceptable to countries having skeptical views about the GM technology.

1.4 Future Prospects

Further enhancement in cotton production is possible by the integration of genetic resources as well as new genetic assays including DNA markers and transformation technologies. For using germplasm resources, a consortium of cotton scientists across all the cotton-growing countries should be involved in phenotypic as well as genotypic characterization. Expression studies at various developmental stages will also add synergy to characterization work. Use of robotic technologies can be very instrumental in collecting the huge quantity of data (particularly physiological) in least possible time from several hundreds of accessions in one experiment with limited hands. Earlier much emphasis was given to explore the upper parts of cotton plant. Diversion of focus is required to study the root traits that will help in identifying genes conferring high root biomass. These studies would also help in studying the interaction of microbes with roots—may lessen the dependency on chemical fertilizer. The generated data can be analyzed quickly using bioinformatic tools. Effective communication among the scientific community is required for dissemination of data for using in breeding programs.

Once these resources have been phenotypically characterized, efforts toward pangenome of the genus *Gossypium* will help in elucidating genetic circuits of important complex traits in cotton as well designing new DNA markers for the traits of interest. This will lead to the initiation of targeted breeding in cotton. Thus a required genetics of a cotton plant can be designed in a computer. For achieving this task, collaboration among cotton experts including molecular biologists, breeders, agronomists, physiologists, and extension workers is the need of time for making the cotton crop more competitive and profitable.

Cotton transformation assay for developing a transgenic plant is very lengthy as well as one or two genotypes respond to regeneration system. In the future, understanding the mechanism for converting callus to plant may lead to the shortening of the current timeline for developing resilience in transgenic cotton.

There are three genomes in a plant cell; however, success story of genetic transformation came from the nuclear genome transformation. The insertion of foreign gene is random. Secondly, the expression is either low or complete silencing of the introduced gene. While insertion of transgene in chloroplast is site specific and also results in very high expression (>70% of total soluble proteins), the genome of chloroplast does not travel through pollen, particularly in field crops, resultantly transgene can be contained. Till now, chloroplast transformation could not bear any fruit owing to laborious and lengthy tissue culture-based protocols for recovering transplastomic plants. Any successful effort will lead to the overexpression of the transgenes; thus protection to insect pest and harvesting high cotton yield with improved quality can be achieved.

Thus, it is summarized that the adoption of high-tech management practices, utilization of untapped genetic resources in breeding, cultivation of cotton varieties with excellent genetics, monitoring of risk and efficacy of transgene in ecosystem, and continued search for new genetic resources would help in sustaining cotton production.

Acknowledgments The idea for writing this chapter and the book entitled Cotton Precision BreedingEditors: Rahman, Mehboob Ur, Zafar, Yusuf, Zhang, Tianzhen (Eds.) was conceived when M.R visited China through a collaborative project "Mining of Elite Genes for High Yield, Super Fiber Qualities and Heat Tolerance and Their Usages in Pakistan and China Cotton Cultivars"—Project No. PSF/BSFC-AGR/P-NIBGE (12). Funds for this visit were provided by the Pakistan Science Foundation.

References

Altaf M, Stewart J, Murphy J (1997) Survey of cotton germplasm for terpenoid aldehydes important in host plant resistance. In: Oosterhuis D, Stewart JM (eds) Special reports—University of Arkansas Agricultural Experiment Station, pp 153–155

Aslam M, Ahmad K, Akhtar MA, Maqbool MA (2017) Salinity stress in crop plants: effects of stress, tolerance mechanisms and breeding strategies for improvement. *J Agric Basic Sci* 2:70–85

Balter M (2009) Clothes make the (hu) man. *Sci* 325(5946):1329

Bowman DT, Gutierrez OA (2003) Sources of fiber strength in the US upland cotton crop from 1980 to 2000. *J Cotton Sci* 7:164–169

Cai C, Ye W, Zhang T, Guo W (2014) Association analysis of fiber quality traits and exploration of elite alleles in upland cotton cultivars/accessions (*G. hirsutum* L.). *J Integr Plant Biol* 56:51–62

Cai C, Zhu G, Zhang T, Guo W (2017) High-density 80 K SNP array is a powerful tool for genotyping *G. hirsutum* accessions and genome analysis. *BMC Genomics* 18:654

Cantrell RG, Davis DD (2000) Registration of NM24016, an interspecific-derived cotton genetic stock. *Crop Sci* 40:1208

Cheema HMN, Bashir A, Khatoon A, Iqbal N, Zafar Y, Malik KA (2010) Molecular characterization and transcriptome profiling of expansin genes isolated from *Calotropis procera* fibers. *Electron J Biotechnol* 13(6):10–11

Chen ZJ, Scheffler BE, Dennis E, Triplett BA, Zhang T, Guo W, Chen X, Stelly DM, Rabinowicz PD, Town CD, Arioli T, Brubaker C, Cantrell RG, Lacape JM, Ulloa M, Chee P, Gingle AR, Haigler CH, Percy R, Saha S, Wilkins T, Wright RJ, Deynze AV, Zhu Y, Yu S, Abdurakhmonov I, Katageri I, Kumar PA, Rahman M, Zafar Y, Yu JZ, Kohel RJ, Wendel JF, Paterson AH (2007) Toward sequencing cotton (*Gossypium*) genomes. *Plant Physiol* 145(4):1303–1310

Chen ZJ, Sreedasyam A, Ando A, Song Q, Santiago LMD, Hulse-Kemp AH, Ding M, Ye W, Kirkbride RC, Jenkins J, Plott C, Lovell J, Lin YM, Vaughn R, Liu B, Simpson S, Scheffler BE, Wen L, Saski CA, Grover CE, Hu G, Conover JL, Carlson JW, Shu S, Boston LB, Williams M, Peterson DG, McGee K, Jones DC, Wendel JF, Stelly DM, Grimwood J, Schmutz J (2020) Genomic diversifications of five *Gossypium* allopolyploid species and their impact on cotton improvement. *Nat Genet* 52:525–533

Dempewolf H, Eastwood RJ, Guarino L, Khoury CK, Muller JV, Toll J (2014) Adapting agriculture to climate change: a global initiative to collect, conserve, and use crop wild relatives. *Agroecol Sustain Food Syst* 38:369–377

Dou L, Zhang X, Pang C, Song M, Wei H, Fan S, Yu S (2014) Genome-wide analysis of the WRKY gene family in cotton. *Mol Genet Genomics* 289:1103–1121

Feng L, Zhang S, Xing L, Yang B, Gao X, Xie X, Zhou B (2019) QTL analysis for yield and fibre quality traits using three sets of introgression lines developed from three *Gossypium hirsutum* race stocks. *Mol Genet Genomics* 294:789

Gore M, Percy RG, Zhang JF, Fang D (2012) Registration of NM24016/TM-1 RIL, a population of recombinant inbred lines of cotton (*Gossypium hirsutum* L.). *J Plant Regist* 6:124–127

Hansen CM, Bjorkman A (1998) The ultrastructure of wood from a solubility parameter point of view. *Holzforschung* 52:335–344

Hernandez J, Meints B, Hayes P (2020) Introgression breeding in barley: perspectives and case studies. *Front Plant Sci* 11:761

Hu Y, Chen J, Fang L, Zhang Z, Ma W, Niu Y, Ju L, Deng J, Zhao T, Lian J, Baruch K, Fang D, Liu X, Ruan Y-L, Rahman M, Han J, Wang K, Wang Q, Wu H, Mei G, Zang Y, Han Z, Xu C, Shen W, Yang D, Si Z, Dai F, Zou L, Huang F, Bai Y, Zhang Y, Brodt A, Ben-Hamo H, Zhu X, Zhou B, Guan X, Zhu S, Chen X, Zhang T (2019) *Gossypium barbadense* and *Gossypium hirsutum* genomes provide insights into the origin and evolution of allotetraploid cotton. *Nat Genet* 51:739–748

Hulse-Kemp AM, Lemm J, Plieske J, Ashrafi H, Buyyarapu R, Fang DD, Frelichowski J (2015) Development of a 63K SNP array for cotton and high density mapping of intraspecific and interspecific populations of *Gossypium* spp. *Genes Genomes Genet* 5(6):1187–1209

Iqbal MA, Rahman M (2017) Identification of marker-trait associations for lint traits in cotton. *Front Plant Sci* 8:86

Iqbal MA, Abbas M, Zafar Y, Rahman M (2015) Characterization of *Gossypium arboreum* L. genotypes for various fiber quality traits. *Pak J Bot* 47(6):2347–2354

Jayaraj S, Palaniswamy P (2005) Host plant resistance in cotton to major insect pests: perspectives and progress. In: Ignacimuthu S, Jayaraj S (eds) Sustainable insect pest management. Alpha Science Int'l Ltd, New Delhi

Jia Y, Sun J, Wang X, Zhou Z, Pan Z, He S, Pang B, Wang L, Du X (2014) Molecular diversity and association analysis of drought and salt tolerance in *G. hirsutum* L. germplasm. *J Integr Agric* 13:1845–1853

Kalivas A, Xanthopoulos F, Kehagia O, Tsaftaris AS (2011) Agronomic characterization, genetic diversity and association analysis of cotton cultivars using simple sequence repeat molecular markers. *Genet Mol Res* 10:208–217

Khan SA, Zafar Y, Briddon RW, Malik KA, Mukhtar Z (2006) Spider venom toxin protects plants from insect attack. *Transgenic Res* 15:349–357

Khan A, Pan X, Najeeb U, Tan DKU, Fahad S, Zahoor R, Luo Z (2018) Coping with drought: stress and adaptive mechanisms, and management through cultural and molecular alternatives in cotton as vital constituents for plant stress resilience and fitness. *Biol Res* 51:47

Kvavadze E, Bar-Yosef O, Belfer-Cohen A, Boaretto E, Jakeli N, Matskevich Z, Meshveliani T (2009) 30,000-year-old wild flax fibers. *Science* 325(5946):1359

Li F, Fan G, Wang K, Sun F, Yuan Y, Song G, Li Q, Ma Z, Lu C, Zou C, Chen W, Liang X, Shang H, Liu W, Shi C, Xiao G, Gou C, Ye W, Xu X, Zhang X, Wei H, Li Z, Zhang G, Wang J, Liu K, Kohel RJ, Percy RG, Yu JZ, Zhu YX, Wang J, Yu S (2014) Genome sequence of the cultivated cotton *Gossypium arboreum*. *Nat Genet* 46:567–572

Li F, Fan G, Lu C, Xiao G, Zou C, Kohel RJ, Ma Z, Shang H, Ma X, Wu J, Liang X, Huang G, Percy RG, Liu K, Yang W, Chen W, Du X, Shi C, Yuan Y, Ye W, Liu X, Zhang X, Liu W, Wei H, Wei S, Huang G, Zhang X, Zhu S, Zhang H, Sun F, Wang X, Liang J, Wang J, He Q, Huang L, Wang J, Cui J, Song G, Wang K, Xu X, Yu JZ, Zhu Y, Yu S (2015) Genome sequence of cultivated upland cotton (*Gossypium hirsutum* TM-1) provides insight into genome evolution. *Nat Biotechnol* 33:524–530

Li CQ, Xu XJ, Dong N, Ai NJ, Wang QL, Jenkins J (2016) Association mapping identifies markers related to major early-maturating traits in upland cotton (*Gossypium hirsutum* L.). *Plant Breed* 135:483–491

Liu JS, Yang BX, Gao ZY, Yang SJ, Zheng F, Yan Y, Ban ZJ (2005) High fiber strength medium and long staple cottons developed by hybridization between upland and sea-island cotton. *China Cottons* 32:12–14

Liu X, Zhao B, Zheng HJ, Hu Y, Lu G, Yang CQ, Chen JD, Chen JJ, Chen DY, Zhang L, Zhou Y, Wang LJ, Guo WZ, Bai YL, Ruan JX, Shangguan XX, Mao YB, Shan C, Jiang JP, Zhu YQ, Jin L, Kang H, Chen ST, He X, Wang R, Wang YZ, Chen J, Wang LJ, Yu ST, Wang BY, Wei J, Song SC, Lu XY, Gao ZC, Gu WY, Deng X, Ma D, Wang S, Liang WH, Fang L, Cai CP, Zhu XF, Zhou BL, Chen ZJ, Xu SH, Zhang YG, Wang SY, Zhang TZ, Zhao GP, Chen XY (2015) *Gossypium barbadense* genome sequence provides insight into the evolution of extra-long staple fiber and specialized metabolites. *Sci Rep* 5:14139

Lv SL, Lian LJ, Tao PL, Li ZX, Zhang KW, Zhang JR (2009) Overexpression of *thellungiella halophila* H + -PPase (TsVP) in cotton enhances drought stress resistance of plants. *Planta* 229:899–910

Ma FZ, Liu JS (1982) Genetic advances in economic traits of interspecific progenies between *Gossypium hirsutum* and *G. barbadense*. *J China Agric Univ* 8:23–32

Malik HJ, Raza A, Amin I, Scheffler JA, Scheffler BE, Brown JK, Mansoor S (2016) RNAi-mediated mortality of the whitefly through transgenic expression of double-stranded RNA homologous to acetylcholinesterase and ecdysone receptor in tobacco plants. *Sci Rep* 6:38469

May OL (2001) Registration of PD 94045 germplasm line of upland cotton. *Crop Sci* 41:279–280

Mergeai G (2006) Introgressions interspécifiques chez le cotonnier. *CAH Agric* 15:135–143

Mills G, Sharps K, Simpson D, Pleijel H, Frei M, Burkey K, Emberson L, Uddling J, Broberg M, Feng Z, Kobayashi K, Agrawal M (2018) Closing the global ozone yield gap: quantification and cobenefits for multistress tolerance. *Glob Chang Biol* 24:4869–4893

Nazeer W, Ahmad S, Mahmood K, Tipu A, Mahmood A, Zhou B (2014) Introgression of genes for cotton leaf curl virus resistance and increased fiber strength from *Gossypium stocksii* into upland cotton (*Gossypium hirsutum*). *Genet Mol Res* 13:1133–1143

Ni M, Ma W, Wang X, Gao M, Dai Y, Wei X, Zhang L, Peng Y, Chen S, Ding L, Tian Y, Li J, Wang H, Wang X, Xu G, Guo W, Yang Y, Wu Y, Heuberger S, Tabashnik BE, Zhang T, Zhu Z (2017) Next-generation transgenic cotton: pyramiding RNAi and Bt counters insect resistance. *Plant Biotechnol J* 15:1204–1213

Nie XH, Huang C, You CY, Li W, Zhao WX, Shen C, Zhang B, Wang H, Yan Z, Dai B, Wang M, Zhang X, Lin Z (2016) Genome-wide SSR-based association mapping for fiber quality in nation-wide upland cotton inbreed cultivars in China. *BMC Genomics* 17:352

Patel JD, Wright RJ, Auld D, Chandnani R, Goff VH, Ingles J, Pierce GJ, Torres MJ, Paterson AH (2014) Alleles conferring improved fiber quality from EMS mutagenesis of elite cotton genotypes. *Theor Appl Genet* 127:821–830

Patel JD, Wright RJ, Chandnani R, Goff VH, Ingles J, Paterson AH (2016) EMS-mutated cotton populations suggest overlapping genetic control of trichome and lint fiber variation. *Euphytica* 208:597–608

Paterson AH, Wendel JF, Gundlach H, Guo H, Jenkins J, Jin D, Llewellyn D, Showmaker KC, Shu S, Udall J (2012) Repeated polyploidization of *Gossypium* genomes and the evolution of spinnable cotton fibres. *Nature* 492:423–427

Percy RG, Kohel RJ (1999) Qualitative genetics. In: Smith CW, Cothren JT (eds) Cotton: origin, history, technology and production. Wiley, New York, pp 319–360

Percy RG, Zhang JF, Cantrell RG (2005) Genetic variation for agronomic and fiber properties in an introgressed recombinant inbred population of cotton. *Crop Sci* 46:1311–1317

Poehlman JM (1987) Breeding field crops, 1st edn. Springer, Netherlands

Rahman M (2016) Cotton improvement for economically stressed economies. ICAC *Recorder* 1:12–13

Rahman M, Zafar Y (2018) Introductory chapter: updates on achieving sustainable cotton production. In: Rahman M, Zafar Y (eds) Past, present and future trends in cotton breeding. Intech Publisher, pp 3–10

Rahman M, Hussain D, Zafar Y (2002) Estimation of genetic divergence among elite cotton (*Gossypium hirsutum* L.) cultivars/genotypes by DNA fingerprinting technology. *Crop Sci* 42:2137–2144

Rahman M, Hussain D, Malik T, Zafar Y (2005) Genetics of resistance to cotton leaf curl disease in *Gossypium hirsutum*. *Plant Pathol* 54:764–772

Rahman M, Ullah I, Ahsraf M, Stewart JM, Zafar Y (2008) Genotypic variation for drought tolerance in cotton. *Agron Sustain Dev* 28:439–447

Rahman M, Zafar Y, Paterson AH (2009) *Gossypium* DNA markers types, number and uses. In: Paterson AH (ed) Genomics of cotton. Springer, Dordrecht

Rahman M, Shaheen T, Tabbasam N, Iqbal MA, Ashraf M, Zafar Y, Paterson AH (2012) Cotton genetic resources: a review. *Agron Sustain Dev* 32:419–432

Rahman M, Rahmat Z, Mahmood A, Abdullah K, Zafar Y (2014) Cotton germplasm of Pakistan world cotton germplasm resources, Ibrokhim Y. Abdurakhmonov, IntechOpen. https://doi.org/10.5772/58620

Rahman M, Zaman M, Shaheen T, Irem S, Zafar Y (2015) Safe use of Cry genes in genetically modified crops. *Environ Chem Lett* 13:239–249

Rahman M, Khan AQ, Rahmat Z, Iqbal MA, Zafar Y (2017) Genetics and genomics of cotton leaf curl disease, its viral causal agents and whitefly vector: a way forward to sustain world cotton fiber security. Frontier in Plant Sciences (Virology). https://doi.org/10.3389/fpls.2017.01157

Rahmat Z, Ahmad S, Zia Z, Rahman M (2019) Genetic monitoring of introgressed alleles from G. arboreum into G. hirsutum using SSR markers: a potential approach for bringing new alleles under cultivation. *Pak J Bot* 51(2):479–486

Ray DK, Mueller ND, West PC, Foley JA (2013) Yield trends are insufficient to double global crop production by 2050. *PLoS One* 8:e66428

Raza A, Malik HJ, Shafiq M, Amin I, Scheffler JA, Scheffler BE, Mansoor S (2016) RNA interference based approach to down regulate osmoregulators of whitefly (*Bemisia tabaci*): potential technology for the control of whitefly. *PLoS One* 11:e0153883

Robinson A, Bell A, Dighe N, Menz M, Nichols R, Stelly D (2007) Introgression of resistance to nematode into upland cotton (*Gossypium hirsutum*) from *Gossypium longicalyx*. *Crop Sci* 47:1865–1877

Saeed M, Guo W, Ullah I, Tabbasam N, Zafar Y, Rahman M, Zhang T (2011) QTL mapping for physiology, yield and plant architecture traits in cotton (*Gossypium hirsutum* L.) grown under well-watered versus water-stress conditions. *Electron J Biotechnol* 14(3)

Seetharaman G (2018) These two issues could put the brakes on the Bt cotton story. The Economic Times. https://economictimes.indiatimes.com/news/economy/agriculture/the-brakes-are-applied-on-the-bt-cotton-story/articleshow/62583116.cms. Accessed 21 Jan 2018

Shaheen T, Rahman M, Zafar Y (2006) Chloroplast RPS8 gene of cotton reveals the conserved nature through out taxa. *Pak J Bot*. 38(5):1467–1476

Shaheen T, Zafar Y, Rahman M (2013) QTL mapping of some productivity and fiber traits in Gossypium arboreum. *Turk J Bot* 37:802–810

Shaheen T, Zafar Y, Rahman M (2016) Phylogenetic analysis of cotton species (diploid genomes) using Single Nucleotide Polymorphisms (SNPs) markers. *Pak J Agric Sci* 53:283–290

Tiwari R, Picchioni G, Steiner R, Jones D, Hughs SE, Zhang JF (2013) Genetic variation in salt tolerance during seed germination in a backcross inbred line population and advanced breeding lines derived from *G. hirsutum* × *G. barbadense*. *Crop Sci* 53:1974–1982

Wang K, Wang Z, Li F, Ye W, Wang J, Song G, Yue Z, Cong L, Shang H, Zhu S, Zou C, Li Q, Yuan Y, Lu C, Wei H, Gou C, Zheng Z, Yin Y, Zhang X, Liu K, Wang B, Song C, Shi S, Kohel R, Percy RG, Yu JZ, Zhu UX, Wang J, Yu S (2012) The draft genome of adiploid cotton *Gossypium raimondii*. *Nat Genet* 44:1098–1103

Wang C, Lu W, He X, Wang F, Zhou Y, Guo X, Guo X (2016a) The cotton mitogen-activated protein kinase kinase 3 functions in drought tolerance by regulating stomatal responses and root growth. *Plant Cell Physiol* 57:1629–1642

Wang H, Wang H, Shao H, Tang X (2016b) Recent advances in utilizing transcription factors to improve plant abiotic stress tolerance by transgenic technology. *Front Plant Sci* 7:67

Wang L, Wu SM, Zhu Y, Fan Q, Zhang ZN, Hu G, Peng QZ, Wu JH (2017) Functional characterization of a novel jasmonate ZIM-domain interactor (NINJA) from upland cotton (*Gossypium hirsutum* L.). *Plant Physiol Biochem* 112:152–160

Wang M, Tu L, Yuan D, Zhu D, Shen C, Li J, Liu F, Pei L, Wang P, Zhao G, Ye Z, Huang H, Yan F, Ma Y, Zhang L, Liu M, You J, Yang Y, Liu Z, Huang F, Li B, Qiu P, Zhang Q, Zhu L, Jin S, Yang X, Min L, Li G, Chen LL, Zheng H, Lindsey K, Lin Z, Udall JA, Zhang X (2019) Reference genome sequences of two cultivated allotetraploid cottons, *Gossypium hirsutum* and *Gossypium barbadense*. *Nat Genet* 51:224–229

Wang F, Zhang J, Chen Y, Zhang C, Gong J, Song Z, Zhou J, Wang J, Zhao C, Jiao M, Liu A (2020) Identification of candidate genes for key fibre-related QTL s and derivation of favorable alleles in *Gossypium hirsutum* recombinant inbred lines with *G. barbadense* introgressions. *Plant Biotechnol J* 18(3):707–720

Weaver DB, Sikkens RB, Lawrence KS, Sürmelioğlu Ç, van Santen E, Nichols RL (2013) RENlon and its effects on agronomic and fiber quality traits in upland cotton. *Crop Sci* 53:913–920

Wedegaertner T, Rathore K (2015) Elimination of gossypol in cottonseed will improve its utilization. *Procedia Environ Sci* 29:124–125

Westengen OT, Huaman Z, Heun M (2005) Genetic diversity and geographic pattern in early south American cotton domestication. *Theor Appl Genet* 110:392–402

Wilhelm S, Sagen JE, Tietz H (1985) Phenotype modification in cotton for control of verticillium wilt through dense plant population culture. *Plant Dis* 69:283–288

Witt TW, Ulloa M, Pelletier MG, Mendu V, Ritchie GL (2018) Irrigation's effect and applied selection on the fiber quality of Ethyl Methane Sulfonate (EMS) treated upland cotton (*Gossypium hirsutum* L.). *J Cotton Res* 1:17

Yik CP, Birchfield W (1984) Resistant germplasm in *Gossypium* species and related plants to *Rotylenchulus reniformis*. *J Nematol* 16:146–153

Yuan YL, Chen YH, Tang CM, Jing SR, Liu SL, Pan JJ, Kohel RJ, Zhang TZ (2000) Effects of the dominant glandless gene Gl^e_2on agronomic and fibre characters of upland cotton. *Plant Breed* 119:59–64

Yuan D, Tang Z, Wang M, Gao W, Tu L, Jin X, Chen L, He Y, Zhang L, Li Y, Liang Q, Lin Z, Yang X, Liu N, Jin S, Lei Y, Ding Y, Li G, Ruan X, Ruan Y, Zhang X (2016) The genome sequence of sea-island cotton (*Gossypium barbadense*) provides insights into the allopolyploidization and development of superior spinnable fibres. *Sci Rep* 5:17662

Zahid KR, Ali F, Shah F, Younas M, Shah T, Shahwar D, Hassan W, Ahmad Z, Qi C, Lu Y, Iqbal A, Wu W (2016) Response and tolerance mechanism of cotton *Gossypium hirsutum* L. to elevated temperature stress: a review. *Front Plant Sci* 7:937

Zaidi SS, Vanderschuren H, Qaim M, Mahfouz MM, Kohli A, Mansoor S, Tester M (2019) New plant breeding technologies for food security. *Science* 363:1390–1391

Zhang JF (1993) Transferring and utilization of genes from *Gossypium barbadense* L. to *G. hirsutum* L. Agron Doctor Degree Diss, Huazhong Agric, Univ, Wuhan

Zhang JF (2011) Twenty-five years of introgression breeding through interspecific hybridization between *Gossypium hirsutum* and *G. barbadense*. In: Proc beltwide cotton conf, pp 711–716

Zhang JF, Hughs SE (2012) Field screening for drought tolerance in cotton. In: Proc beltwide cotton conf, pp 713–718

Zhang JF, Sun JZ, Liu JL, Wu ZB (1992) Genetic analysis of cotton resistance to spider mites. *J Huazhong Agric Univ* 11:127–133

Zhang JF, Sun JZ, Liu JL, Wu ZB (1993) Identification of cotton varieties resistant to carmine spider mites and exploration of resistance mechanism. *Acta Phytophylacica Sin* 20:155–161

Zhang W, Liu F, Li SH, Wang W, Wang CY, Zhang XD (2011) QTL analysis on yield and its components in recombinant inbred lines of upland cotton. *Acta Agron Sin* 37:433–442

Zhang JF, Sanogo S, Flynn R, Baral JB, Bajaj S, Hughs SE (2012) Germplasm evaluation and transfer of verticillium wilt resistance from pima (*Gossypium barbadense*) to upland cotton (*G. hirsutum*). *Euphytica* 187:147–160

Zhang JF, Fang H, Zhou HP, Hughs SE, Jones DC (2013a) Inheritance and transfer of thrips resistance from pima cotton to upland cotton. *J Cotton Sci* 17:163–169

Zhang Y, Zhang F, Li X, Baller JA, Qi Y, Starker CG, Bogdanove AJ, Voytas DF (2013b) Transcription activator-like effector nucleases enable efficient plant genome engineering. *Plant Physiol* 161:20–27

Zhang J, Percy RG, McCarty JC (2014) Introgression genetics and breeding between upland and pima cotton: a review. *Euphytica* 198:1–12

Zhang T, Hu Y, Jiang W, Fang L, Guan X, Chen J, Zhang J, Saski CA, Scheffler BE, Stelly DM, Hulse-Kemp AM, Wan Q, Liu B, Liu C, Wang S, Pan M, Wang Y, Wang D, Ye W, Chang L, Zhang W, Song Q, Kirkbride RC, Chen X, Dennis E, Llewellyn DJ, Peterson DJ, Thaxton P, Jones DC, Wang Q, Xu X, Zhang H, Wu H, Zhou L, Mei G, Chen S, Tian Y, Xiang D, Li X, Ding J, Zuo Q, Tao L, Liu Y, Li J, Lin Y, Hui Y, Cao Z, Cai C, Zhu X, Jiang Z, Zhou B, Guo W, Li R, Chen ZJ (2015) Sequencing of allotetraploid cotton (*Gossypium hirsutum* L. acc. TM-1) provides a resource for fiber improvement. *Nat Biotechnol* 33:531–537

Zhao YL, Wang HM, Shao BX, Chen W, Guo ZJ, Gong HY, Sang XH, Wang JJ, Ye WW (2016) SSR-based association mapping of salt tolerance in cotton (*Gossypium hirsutum* L.). *Genet Mol Res* 15:1–10

Zhou L, Wang NN, Gong SY, Lu R, Li Y, Li XB (2015) Overexpression of a cotton (*Gossypium hirsutum*) WRKY gene, GhWRKY34, in arabidopsis enhances salt-tolerance of the transgenic plants. *Plant Physiol Biochem* 96:311–320

Chapter 2
Evolution and Diversity of the Cotton Genome

Guanjing Hu, Corrinne E. Grover, Josef Jareczek, Daojun Yuan, Yating Dong, Emma Miller, Justin L. Conover, and Jonathan F. Wendel

2.1 Introduction

The cotton genus includes agronomically important species as well as many others that serve as examples of the evolution of biodiversity. Worldwide, cotton is most famous for the epidermal seed trichomes, or "fiber," of the cultivated species, the production of which comprises a multibillion dollar industry employing millions of people. Biologically, this crop is represented by four independently domesticated species at two different ploidy levels, generating additional interest as a naturally replicated evolutionary experiment. Accordingly, considerable attention has been given to the evolutionary relationships among species, the consequences of polyploidy, the domestication process, and the underlying biology that makes cotton a valuable crop species. In addition, recent technological advances continue to accelerate our understanding of cotton biology and evolution. Here we explore our understanding of cotton evolution and diversity, drawing attention both to the extraordinary evolutionary history of the genus and the importance of this diversity for agronomic improvement.

The original version of this chapter was revised. The correction to this chapter is available at https://doi.org/10.1007/978-3-030-64504-5_18

G. Hu
State Key Laboratory of Cotton Biology, Institute of Cotton Research, Chinese Academy of Agricultural Sciences, Anyang, China

Agricultural Genomics Institute at Shenzhen, Chinese Academy of Agricultural Sciences, Shenzhen, China

C. E. Grover · J. Jareczek · D. Yuan · Y. Dong · E. Miller · J. L. Conover · J. F. Wendel (✉)
Department of Ecology, Evolution, and Organismal Biology, Iowa State University, Ames, IA, USA
e-mail: jfw@iastate.edu

Mehboob-ur-Rahman et al. (eds.), *Cotton Precision Breeding*,
https://doi.org/10.1007/978-3-030-64504-5_2

2.2 Origin of the Cotton Genus (*Gossypium*)

Gossypium belongs to a small taxonomic tribe Gossypieae, both of which are characterized by the punctae or lysigenous cavities ("gossypol glands") that differentiate these taxa from other members of the family Malvaceae. The tribe Gossypieae is monophyletic (LaDuke and Doebley 1995; Seelanan et al. 1999; Wendel et al. 2002) and is divided into 9 small- to modestly sized genera which collectively include approximately 120 species (Fryxell 1968, 1979; Phuphathanaphong 2006). Five of these genera either are monotypic or contain fewer than four extant species, all of which have restricted geographic distributions: *Lebronnecia* (monotypic; Marquesas Islands), *Cephalohibiscus* (monotypic; New Guinea and the Solomon Islands), *Thepparatia* (monotypic; Thailand), *Gossypioides* (two species; East Africa and Madagascar), and *Kokia* (2–3 extant species, one extinct; Hawaii). The remaining genera are moderately sized with broader geographic ranges, i.e., *Hampea* (21 neotropical species), *Cienfuegosia* (25 species; neotropics and parts of Africa), *Thespesia* (17 tropical species), and *Gossypium* (Fig. 2.1).

Gossypium is the largest and most widely distributed genus in the tribe (Fryxell 1992), with over 55 recognized species whose naturally occurring ranges encompass the tropics and subtropics worldwide (Table 2.1). Despite their extensive distribution and extraordinary morphological and cytogenetic diversity, molecular phylogenetic analyses have confirmed the monophyletic origin of the cotton genus (Seelanan et al. 1999; Cronn et al. 2002) and identified its closest relative, i.e., the sister lineages *Kokia* and *Gossypioides*. This observation is somewhat surprising given the distant, restricted ranges of *Kokia* (Hawaii) and *Gossypioides* (East Africa, Madagascar) and their reduced chromosome number ($n = 12$), which is distinct from diploid species in *Gossypium* and most of the remainder of the cotton tribe ($n = 13$). That two genera now separated by over 17,500 km of open ocean are each other's closest relatives implies that transoceanic dispersal was involved in the evolution of one or both genera, representing one of the many examples of long-distance, salt-water dispersal in the tribe (Stephens 1958, 1966; Fryxell 1979; Wendel and Percival 1990; Percy and Wendel 1990; Wendel and Albert 1992; DeJoode and Wendel 1992). The distribution of the tribe itself includes many distant islands (e.g., Hawaii, Wake Island, Solomon Islands), whose phylogenetic distribution serves to underscore the importance of oceanic dispersal during the evolution of the tribe (Seelanan et al. 1999).

While the center for divergence of *Gossypium* consequently remains unclear (Grover et al. 2017a), the evolutionary timetable now is generally agreed upon (Senchina et al. 2003; Wendel and Grover 2015; Chen et al. 2016, 2017c, 2020). Divergence of *Gossypium* from its sister lineage, i.e., *Gossypioides-Kokia*, was initially estimated at approximately 10–15 million years ago (mya) based on relatively few chloroplast and nuclear genes (Seelanan et al. 1999; Cronn et al. 2002). This estimate was more recently upheld using de novo whole-genome sequences and thousands of genes representing the entire genome (Grover et al. 2017a). Interestingly, divergence between *Kokia* and *Gossypioides* was estimated at approximately 3–5

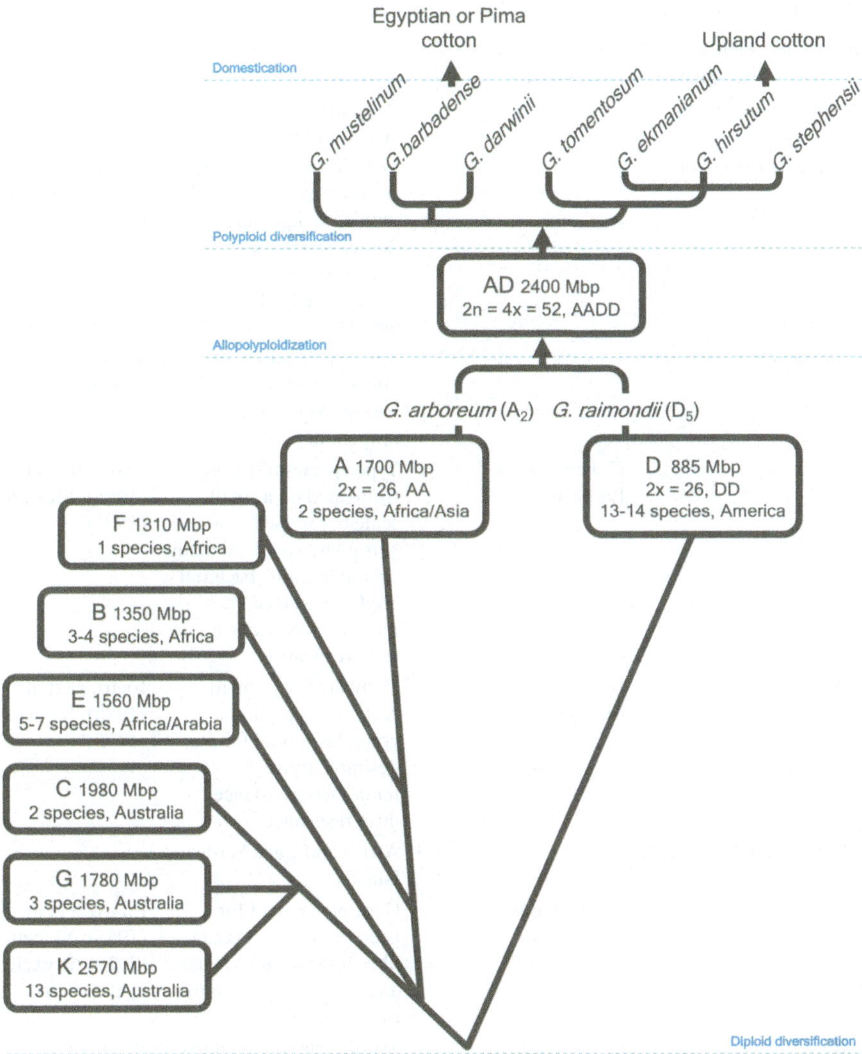

Fig. 2.1 Evolutionary history of *Gossypium.* Following the genus origin 5–10 Mya, diploid *Gossypium* rapidly diversified into three major lineages of eight monophyletic genome groups: the New World clade (D), the African-Asian clade (A, B, E and F), and the Australian clade (C. G and K). This global radiation involved several trans-oceanic dispersal events and was accompanied by morphological, ecological, and genome size differentiation. During the Pleistocene 1–2 mya, allopolyploid cottons formed following trans-oceanic dispersal of an A-genome diploid to the Americas, where the immigrant underwent hybridization, as female, with a native D-genome diploid similar to modern *G. raimondii*. Among the seven modern allopolyploid species, *G hirsutum* and *G. barbadense* were independently domesticated for fiber production

Table 2.1 Taxonomy of *Gossypium* species and notable features

Gossypium L.	Genome designations	Species presently recognized		Comments	Reference
Subgenus *Gossypium*				African, Arabian, and Asian diploid species	
Section *Gossypium*				Wild forms known only from Southern Africa, but with an indigenous domesticated range encompassing parts of Africa, the Middle East, and Asia	Fryxell (1979, 1992); Vollesen (1987)
Subsection *Gossypium*			2	The female parent of the allopolyploid formation was an A-genome progenitor	Wendel et al. (1989)
	A_1	*G. herbaceum* Linnaeus		Type species of the genus. Cultivated on a small scale but is used as a germplasm pool for many desirable traits: bacterial blight resistance, *R. reniformis* resistance, thrip resistance	Endrizzi et al. (1985); Stewart (1995)
	A_2	*G. arboreum* Linnaeus		Cultivated on a small scale but is used as a germplasm pool for many desirable traits: *R. reniformis* resistance, thrip resistance	Endrizzi et al. (1985); Stewart (1995)
Subsection *Anomala* Todaro			3–4	Africa and Cape Verde Islands	
	B_1	*G. anomalum* Wawra and Peyritsch		Germplasm pool for bacterial blight resistance, fiber improvement (finer and stronger), and cotton rust resistance	Endrizzi et al. (1985); Mehetre (2010); Fryxell et al. (1984)
	B_2	*G. triphyllum* (Harvey and Sonder) Hochreutiner		Trifoliolate leaf with fine tomentum that limits water loss from transpiration. Adapted to an extreme desert environment. Molecular data suggests affinities are with other B-genome species	Wendel and Albert (1992); Seelanan et al. (1997); Fryxell (1986)
	B_3	*G. capitis-viridis* Mauer		From Cape Verde Islands	Vollesen (1987); Fryxell et al. (1984)
	B*	(*G. trifurcatum*)			Vollesen (1987)

(continued)

Table 2.1 (continued)

Gossypium L.	Genome designations	Species presently recognized		Comments	Reference
Subsection *Pseudopambak* (Prokhanov) Fryxell			5–7	Poorly understood native ranges of many species, but collectively in parts of the Arabian Peninsula, Northeast Africa, and Southwest Asia	Vollesen (1987); Fryxell (1992)
E_1		*G. stocksii* Masters in Hooker		Potential source of cotton leaf curl resistance and improved fiber strength. Demonstrates increase resistance to *R. reniformis*	Nazeer et al. (2014); Yik and Birchfield (1984)
E_2		*G. somalense* (Gurke) Hutchinson		Demonstrates increase resistance to *R. reniformis*. Potential complementary lethal factor when bred with *G. hirsutum*	Yik and Birchfield (1984)
E_3		*G. areysianum* Deflers		Foliage smells putrid	
E_4		*G. incanum* (Schwartz) Hillcoat		Tends to grow in dry beds that flash flood	Fryxell (1986)
E*		(*G. benadirense* Mattei) (*G. bricchettii* (Ulbrich) Vollesen) (*G. vollesenii* Fryxell)		Reinstated from limited materials thus incompletely understood	Vollesen (1987); Fryxell (1992)
Subsection *Longiloba*			1		
F_1		*G. longicalyx* Hutchinson and Lee		Endemic to East Africa. Cytologically unique and unusually adapted for mesic environments preferring shade and higher rainfall. Useful in breeding schemes that transfer desirable wild traits to cultivated accessions. Like the A-genome, it has long fibers. Completely immune to *R. reniformis*	Phillips and Strickland (1966); Fryxell (1979, 1986); Wendel et al. (2010); Bell et al. (2014, 2015); Wendel and Grover (2015); Yik and Birchfield (1984)

(continued)

Table 2.1 (continued)

Gossypium L.	Genome designations	Species presently recognized		Comments	Reference
Subgenus *Houzingenia* (Fryxell) Fryxell			13–14	New-World diploids; large shrubs and small trees in the South and Middle America (primarily Mexico), with range extensions into Peru, the Galapagos Islands, and Southern Arizona	
Section *Houzingenia*				Large shrubs and small trees found primarily in Mexico	Fryxell 1(992)
Subsection *Houzingenia*				The two species in this subsection are morphologically similar and interfertile	Fryxell (1965, 1968, 1979)
D_1		*G. thurberi* Todaro		Northernmost species adapted to temperate climate into the mountains of Arizona, which tolerates mild frost via defoliation and becomes fully dormant during the winter. D-genome species employed by J.O. Beasley to create the triple hybrid that was used to introgress high fiber strength into *G. hirsutum.* Tends to grow in dry beds that flash flood. Resistance against thirps and silverleaf whitefly which carries disease such as cotton leaf crumple. Lacks seed fiber	Fryxell (1979, 1986); Fryxell et al. (1984); Walker and Natwick (2006)
D_8		*G. trilobum* (DC.) Skovsted		Fully tropical and susceptible to freezing temperature; grown in higher elevation than others (up to 2600 m) and perhaps the least xerophytic in distribution. Source of male sterile cytoplasm and restorer factor resistance to thrips	Fryxell (1965, 1968); Fryxell et al. (1984); Stewart (1992)

(continued)

Table 2.1 (continued)

Gossypium L.	Genome designations	Species presently recognized		Comments	Reference
Subsection *Integrifolia* (Todaro) Todaro				Interspecific hybrids between either species in this subsection and several other species except B, C, and G are embryo lethal	Phillips (1977); Lee (1981, 1986)
	D_{3d}	*G. davidsonii* Kellogg		Source of salt tolerance. Broad ecological range in Baja California, with primary distribution in the Cape Region up to 2000 ft in elevation from seaside; seaside locations tend to be back from the leading edge of salt spray zone	Wendel and Percival (1990); Zhang et al. (2016); J Nason (personal observation)
	D_{3k}	*G. klotzschianum* Andersson		A derivative species of *G. davidsonii* from Baja California to the Galapagos Islands through long-distance dispersal. Source of salt tolerance	Wendel and Percival (1990); Wei et al. (2017)
Subsection *Caducibracteolata* Mauer				These species are calciphiles, typically found in arid habitats around the Gulf of California. They have reduced leaves with thick cuticles and a double palisade layer and the largest seeds of the diploid species. The floral bracts are caduceus, abscising well before anthesis in *G. armourianum*, and shortly before to just after anthesis in the other species. They are subjected to high temperatures, high isolation, and low rainfall almost year-round	Phillips and Clement (1967); Fryxell (1986, 1992)

(continued)

Table 2.1 (continued)

Gossypium L.	Genome designations	Species presently recognized		Comments	Reference
D_{2-1}		*G. armourianum* Kearney		Germplasm pool for bacterial blight and *R. reniformis* resistance. Large seeds 8–10 mm. Adapted to extreme water loss with leaves that have a double layer of palisade cells and a thick cuticle	Endrizzi et al. (1985); Fryxell et al. (1984)
D_{2-2}		*G. harknessii* Brandegee		Source of cytoplasmic male sterility and restorer factors. Large seeds 8–10 mm. Adapted to extreme water loss with leaves that have a double layer of palisade cells and a thick cuticle	Meyer (1975); Fryxell et al. (1984)
D_{10}		*G. turneri* Fryxell		Sister species to or derivative from *G. harknessii*. Adapted to extreme water loss with leaves that have a double layer of palisade cells and a thick cuticle	Fryxell et al. (1984); Fryxell (1986)
Section *Erioxylum* (Rose and Standley) Prokhanov					
Subsection *Erioxylum*				This group of species has a unique flowering phenology adapted to wet season-dry season cycle. At the height of the dry season, while leafless, the plants flower and fruit. After the fruits mature, the plants remain dormant until returning rains to stimulate new vegetative growth	Fryxell et al. (1984)
D_4		*G. aridum* (Rose and Standley) Skovsted		Widely distributed and abundant in western Mexico. There is evidence of cytoplasmic introgression from subsection *Integrifolia* into populations of *G. aridum* from the State of Colima, Mexico; other populations are normal in this respect. Source of salt tolerance	Wendel et al. (1995a, b); Fryxell (1979); Fan et al. (2015)

(continued)

Table 2.1 (continued)

Gossypium L.	Genome designations	Species presently recognized		Comments	Reference
D_7		*G. lobatum* Gentry		Restricted distribution while locally abundant in Michoacán, Mexico. Leaves nearly distichous	Fryxell (1979, 1992)
D_9		*G. laxum* Phillips		Restricted distribution in Guerrero, Mexico, while locally abundant as subdominant species in the vegetation of Cañón del Zopilote	DeJoode and Wendel (1992); Wendel and Albert (1992)
D_{11}		*G. schwendimanii* Fryxell and Koch		The most recently described species among the New-World diploids	Fryxell (1979)
D_{12}		(*G. sp.nov.*)		A newly collected accession, US-72, genetically distant from other species in the subsection	Feng et al. (2011)
Subsection *Selera* (Ulbrich) Fryxell					
D_6		*G. gossypioides* (Ulbrich) Standley		The only diploid species that shows evidence of the original A×D hybridization that gave rise to the allotetraploids. This species may have arisen via introgressive speciation. Lacks foliar nectaries	Fryxell et al. (1987); Fryxell et al. (1984)
Subsection *Austroamericana* Fryxell					
D_5		*G. raimondii* Ulbrich		A relatively recent immigrant to Peru, model of the D-genome parent of allopolyploid cotton. Tends to grow in dry beds that flash flood	Endrizzi et al. (1985); Wendel et al. (1995a, b); Fryxell (1986)
Subgenus *Sturtia* (R. Brown) Todaro				All the indigenous Australian species	Fryxell (1992)
Section *Sturtia*			2	Central and Western Australia. Mauve flower. Neither of these deposit gossypol. Useful in studying the regulation and biosynthesis of gossypol production. Breeding schemes have already introduced the "glandless-seed and glanded plant" phenotype to *G. hirsutum*	Fryxell (1992); Brubaker et al. (1996); Mammadov et al. (2018); Zhu et al. (2005); Liu et al. (2015b)

(continued)

Table 2.1 (continued)

Gossypium L.	Genome designations	Species presently recognized		Comments	Reference
C_1		*G. sturtianum* Willis		"Sturt's Desert Rose," the floral emblem of the Northern Territory, is distributed widely across the Australian continent in the temperate arid zone. Limited cold resistant, it can withstand a few degrees below freezing when in full leaf. Waxy leaves prevent water loss and can fold leaves inward when exposed to water stress. Small seeds 4–5 mm	Fryxell (1979, 1986); Fryxell et al. (1984); Craven et al. (1994)
C_2		*G. robinsonii* Mueller		Potentially basal in the Australian *Gossypium* lineage. Prefers to grow in intermittent water beds to get direct access to moisture for a brief period	Craven et al. (1994); Fryxell (1986)
Section *Grandicalyx* Fryxell			13	Northwest Australia (especially the Kimberley region), Cobourg Peninsula, and Northern Territory, Australia. In contrast to large shrubs of other Australian species, these are sub-shrubby and produce short-lived stems from a perennial rootstock, as an adaption to the dry-season fires of the monsoon zone. White flowers with strongly contrasting red petal spots, occasional pink flowers occur; seeds rely on ant dispersal. Additionally, this section has the largest genome size in all of *Gossypium*	Fryxell (1992); Fryxell (1992); Craven et al. (1994); Stewart (1995); Wendel et al. (2010)
K_8		*G. costulatum* Todaro		One of the first Australian *Gossypium* species to be collected, along with *G. cunninghamii* and *G. populifolium*, by Alan Cunningham between 1818 and 1820 where each occurs near coastal waters in northwest Australia accessible by ship	Craven et al. (1994)

(continued)

Table 2.1 (continued)

Gossypium L.	Genome designations	Species presently recognized		Comments	Reference
K_9		*G. cunninghamii* Todaro		May have originated from a hybridization event. The only sessile or subsessile species in *Gossypium*. This species may have originated from an ancient hybridization in which one parent (maternal) was a species similar to present-day *G. sturtianum*. A similar cytoplasm is also found in *G. bickii*. The paternal parent, however, is located in the Northern Territory	Wendel and Albert (1992)
K_1		*G. exiguum* Fryxell, Craven and Stewart		More widely distributed than other species in this section. *G. exiguum*, *G. rotundifolium*, and *G. pilosum* may be difficult to distinguish in the field and may have imprecise taxonomic descriptions	Fryxell (1992); Stewart, Craven, Brubaker and Wendel (personal observations)
K_2		*G. rotundifolium* Fryxell, Craven and Stewart			
K_4		*G. pilosum* Fryxell			
K_3		*G. populifolium* (Bentham) Mueller ex Todaro		These suberect and erect species are more narrowly distributed than the former three species, with some being known from only a few populations	Fryxell (1992); Stewart, Craven, Brubaker and Wendel (personal observations)
K_5		*G. marchantii* Fryxell, Craven and Stewart			
K_6		*G. londonderriense* Fryxell, Craven and Stewart			
K_7		*G. enthyle* Fryxell, Craven and Stewart			
K_{10}		*G. pulchellum* (Gardner) Fryxell			
K_{11}		*G. nobile* Fryxell, Craven and Stewart			
K_{12}		*G. anapoides* Stewart, Craven and *Wendel*			

(continued)

Table 2.1 (continued)

Gossypium L.	Genome designations	Species presently recognized		Comments	Reference
Section *Hibiscoidea* Todaro			3	Central Australia. Like *Sturtia* they do not deposit gossypol in seeds. Recurved pedicels which could prevent boll rot and rain damage to lint	Brubaker et al. (1996); Brown and Ware (1958)
G_1		*G. bickii* Prokhanov		Created by hybridization of a maternal *G. sturtianum*-like plant and a paternal *G. australe-/ nelsonii*-like plant	Wendel et al. (1992)
G_2		*G. australe* Mueller		Wind dispersed. Small seeds 4–5 mm. Fibers are straight	Wendel et al. (2010); Fryxell et al. (1984)
G_3		*G. nelsonii* Fryxell		Wind dispersed. Fibers are straight	Wendel et al. (2010); Fryxell et al. (1984)
Subgenus *Karpas*			7	The allopolyploid cottons were probably formed during the Pleistocene by hybridization of diploids from the A and D genomes	Wendel et al. (1989); Wendel and Albert (1992); Seelanan et al. (1997)
AD_1		*G. hirsutum* Linnaeus		Large amount of phenotypic diversity which provides many agronomically desirable traits. Originally distributed in Central America	Bell (1984); Meredith (1991); Brubaker et al. (1993); Brubaker and Wendel (1994)
AD_2		*G. barbadense* Linnaeus		Distributed and domesticated in South America. Modern cultivars are highly introgressed with *G. hirsutum*. Some accessions have increased resistance to thrips and increased fiber quality	Shepherd (1974); Meredith (1991); Percy and Wendel (1990); Wang et al. (1995); Zhang et al. (2013)
AD_3		*G. tomentosum* Nuttall ex Seemann		Endemic to the Hawaiian Islands and lacks foliar nectaries. Resistance to thrips and verticillium. Important source of salt tolerance	Zhang et al. (2013); Meyer and Meyer (1961); DeJoode and Wendel et al. (1992); Zhang et al. (2011); Oluoch et al. (2016)

(continued)

Table 2.1 (continued)

Gossypium L.	Genome designations	Species presently recognized		Comments	Reference
AD_4		*G. mustelinum* Miers ex Watt		Widely scattered populations in NE Brazil; resistance to thrips	Wendel et al. (1994); Bowman and McCarty (1997)
AD_5		*G. darwinii* Watt		Closely related to *G. barbadense* and found on the Galapagos Islands. Source of resistance to *Fusarium* and *Verticillium*	Bell (1984); Wendel and Percival (1990); Liu et al. (2016)
AD_6		*G. ekmanianum* Wittmack		Suggested as a separate species in 1928 but not confirmed until 2014	Grover et al. (2014)
AD_7		*G. stephensii* J. Gallagher, C. Grover and Wendel		Inhabits Wake Atoll	Gallagher et al. (2017)

Citations in this table: Meyer and Meyer (1961); Fryxell (1965, 1979, 1986, 1992); Phillips and Strickland (1966); Phillips and Clement (1967); Shepherd (1974); Meyer (1975); Phillips (1977); Bell (1984); Fryxell et al. (1984, 1987); Yik and Birchfield (1984); Endrizzi et al. (1985); Vollesen (1987); Wendel et al. (1989); Wendel et al. (1989, 1994, 1995a, 2010); Wendel and Percival (1990); Percy and Wendel (1990); Meredith (1991); Stewart (1992, 1995); Wendel and Albert (1992); DeJoode and Wendel (1992); Brubaker et al. (1993, 1996); Brubaker and Wendel (1994); Craven et al. (1994); Wang et al. (1995); Bowman and McCarty (1997); Seelanan et al. (1997); Zhu et al. (2005); Walker and Natwick (2006); Mehetre (2010); Feng et al. (2011); Zhang et al. (2013, 2016); Nazeer et al. (2014); Bell et al. (2014, 2015); Grover et al. (2015c); Wendel and Grover (2015); Fan et al. (2015); Liu et al. (2015b); Oluoch et al. (2016); Gallagher et al. (2017); Wei et al. (2017); Mammadov et al. (2018)

mya (Grover et al. 2017a), which is approximately the same time during which the major lineages of *Gossypium* became established and began to diversify (Cronn et al. 2002; Grover et al. 2019a). During this time, the *Kokia-Gossypioides* lineages experienced a shared reduction in chromosome number (Seelanan et al. 1999; Udall et al. 2019) and subsequently experienced remarkable genome downsizing, including massive, differential gene loss (Grover et al. 2017a). This insight into the *Kokia-Gossypioides* lineage provides an essential context in using these genera as phylogenetic out-groups, having consequences for understanding genome evolution, as well as evolutionary patterns and processes within *Gossypium*.

2.3 Diversification of *Gossypium* Diploid Species

Diploid members of *Gossypium* are divided into eight monophyletic genome groups (A through G and K; Table 2.1; Fig. 2.1), as determined by interspecific meiotic pairing behavior (Beasley 1940, 1942; Endrizzi et al. 1985; Zhang and Endrizzi 2015). Classification within and among these groups reflects decades of accumulated understanding that emerged from basic plant exploration, as well as taxonomic and evolutionary analysis (Watt 1907; Hutchinson et al. 1947; Saunders 1961; Fryxell 1979, 1992; Wendel et al. 2009; Wendel and Grover 2015; Wang et al. 2018a). According to the most modern and widely followed taxonomic classification of Fryxell (1979, 1992), species are grouped into four subgenera and eight sections primarily based on morphological, cytogenetic, and geographical evidence. Subsequent molecular phylogenetic analyses (Wendel and Albert 1992; Cronn et al. 1996; Seelanan et al. 1999; Feng et al. 2011; Grover et al. 2015b, c, 2019a; Chen et al. 2016; Gallagher et al. 2017) have confirmed that the genealogical lineages of *Gossypium* species are largely congruent with genome designations and geographic distributions. That is, each genome group is monophyletic, representing a single natural lineage, and in most cases, these lineages are also geographically cohesive. Corresponding to continental division, the three major lineages of diploid species include the Australian subgenus *Sturtia* (C-, G-, K-genomes), the American subgenus *Houzingenia* (D-genome) of the New World, and the African, Arabian, and Asian subgenus *Gossypium* (A-, B-, E-, and F-genomes) of the Old World. Phylogenetic analyses place the earliest divergence between the New World (D-genome) lineage and the ancestor of all African/Australian taxa, followed shortly by divergence of the Australian cottons (i.e., C-, G-, and K-genomes), although the internodes in this basalmost split are short and hence uncertain. Diversification continued with the divergence of the African-Arabian E-genome species from the African A-, B-, and F-genome cottons, and, finally, B-genome divergence from the sister clades composed of the A- and F-genomes, the latter having a single species, *G. longicalyx*. The observation that *G. longicalyx* is sister to the A-genome species is particularly important in that *G. longicalyx* represents the wild ancestor in the evolution of spinnable fiber, which is unique (among diploids) to the A-genome.

Morphologically, the genus *Gossypium* is quite diverse (Fig. 2.2). Cotton species have collectively achieved a nearly worldwide distribution, with species-rich regions

Fig. 2.2 Striking morphological diversity of *Gossypium*. Reprinted (adapted) with permission from Cotton Incorporated

in the arid or seasonally arid tropics and subtropics over all major continents (see Table 2.1 for geographic distribution of species). Consequently, and in response to ecological and environmental demands, plant habits range from the fire-adapted, herbaceous perennials in NW Australia to small trees in SW Mexico that escape the dry season by dropping their leaves. Corolla colors are equally variable, spanning a rainbow that includes blues/purples (e.g., *G. triphyllum*), mauves/pinks (e.g., "Sturt's Desert Rose," *G. sturtianum*), whites/pale yellows (multiple species from NW Australia, Mexico, Africa-Arabia), and even a deep sulfur-yellow (i.e., *G. tomentosum* from Hawaii). Seed coverings range from nearly glabrous (e.g., *G. klotzschianum* and *G. davidsonii*) to short stiff, dense, brown hairs that aid in wind dispersal (*G. australe* and *G. nelsonii*), to the long, fine white fibers that characterize highly improved forms of the four cultivated species. There are even seeds that produce fat bodies to facilitate ant dispersal (section *Grandicalyx* from NW Australia; Seelanan et al. 1999). At the other end of the ant coevolution spectrum is *G. tomentosum* from the Hawaiian Islands, which lost the foliar and extrafloral nectaries that are common in other *Gossypium* species, presumably in response to the absence of native ants. Much of this morphological diversity is described in detail by Fryxell (1979).

In addition to the impressive morphological and ecological diversification, extensive chromosomal evolution has been documented for the genus (Endrizzi et al. 1985; Konan et al. 2009; Soltis et al. 2009). As mentioned above, the genome groups were originally designated based on chromosome size and the meiotic behavior of chromosomes during interspecific crosses. That is, those within the same genome group generally produce hybrids exhibiting bivalent chromosome formation (Beasley 1940, 1942; Endrizzi et al. 1985; Zhang and Endrizzi 2015), and increased occurrence of univalent formation generally reflects structural differentiation between more distant species (Gerstel 1953; Phillips 1966). Notably, although all diploid species share $n = 13$, there is more than threefold variation in DNA content per cell (Hendrix and Stewart 2005) that is evident even from gross chromosome morphology (Stephens 1947; Katterman and Ergle 1970; Abdul Kadir 1976). Genome sizes range from less than 1 Gbp in the American D-genome cottons (average 900 Mbp) to over 2.5 Gbp in the Australian K-genome species (Wendel et al. 2002; Hendrix and Stewart 2005), with most diploid cottons ranging between 1300 and 2000 Mbp. This extraordinary variation in genome size is largely attributed to changes in repetitive DNA content, particularly LTR transposable elements (TEs) (Hawkins et al. 2006; Grover et al. 2008, 2017a, 2019a; Renny-Byfield et al. 2016). Comparisons among closely related species of similar genome size have also revealed a cryptic and dynamic scenario of genome size gain and loss. Recent comparative studies within the A- (Renny Byfield et al. 2016) and D-genome groups (Grover et al. 2019a), as well as between the outgroup genera *Kokia* and *Gossypioides* (Grover et al. 2017a) whose genome size appears static at 520 Mbp, found that relative stasis in size belies a much more complicated scenario of opposing actions, i.e., TE proliferation, TE loss, and small-scale insertions and deletions. Recent estimates of gene numbers also suggest more variability in gene content than previously considered (Page et al. 2013; Grover et al. 2019a), although further research with high-quality genomes is required to characterize gene evolution in diploid cotton.

2.4 Origin and Diversification of Polyploids

Allopolyploid cotton is the fortuitous result of one of the transoceanic dispersals that characterize the Gossypieae. Over half a century ago, a rich body of classic cytogenetic evidence followed by numerous experimental studies established that cotton species with $n = 26$ are allopolyploids composed of two co-resident genomes (aka subgenomes), one from an African A-genome species and the other from a Mesoamerican D-genome species (reviewed in detail by Endrizzi et al. (1985) and Wendel and Cronn (2003)). Since the native distribution of most polyploid species is primarily Mesoamerican, the formation of the AD-genome allopolyploid species most likely occurred somewhere in Mesoamerica following the long-distance transoceanic dispersal of an African A-genome species to the New World. Given the presumed rarity of such transoceanic dispersals, formation of the allopolyploid (i.e., A-/D-genome hybridization and subsequent doubling) is inferred to have occurred only once, in contrast to the prevalence of multiple polyploid origins in other plant species (Soltis et al. 1993, 2004; Soltis and Soltis 1999; Tate et al. 2006). Indeed a monophyletic origin for the polyploid clade is further supported by molecular evidence (Grover et al. 2012a, 2015a), most recently using whole-genome sequences for five allopolyploid species and both model diploid progenitors (Chen et al. 2020). Although hypotheses for the emergence of the polyploid clade originally ranged from a very ancient Cretaceous origin, perhaps 60–100 mya (prior to the separation of the South American and African continents; Saunders 1961; Endrizzi et al. 1985, 1989), to a very recent human-mediated origin (circa 6000 years ago via intentional intercontinental transfer; Hutchinson et al. 1947; Hutchinson 1959; Johnson 1975), a mid-Pleistocene origin (1–2 mya) was generally supported by botanists based on cytogenetic evidence (Phillips 1964) and ecological considerations (Fryxell 1965) and subsequently confirmed by numerous sources of molecular and DNA sequencing data (Wendel and Cronn 2003; Wendel and Grover 2015).

Numerous sources of molecular and sequence data have also supported the identification of the closest model progenitors to the actual parents of the polyploid clade. *Gossypium raimondii* has long been recognized as the closest extant relative of the actual D-genome parent of allopolyploid cotton, from the earliest efforts comparing leaf development between diverse synthetic hybrids and the natural allopolyploids to modern sequencing data (Stephens 1944; Endrizzi et al. 1985; Wendel 1989; Wendel and Cronn 2003; Wendel et al. 2012; Yu-xiang et al. 2013; Li et al. 2014; Grover et al. 2015a). The A-genome parent was historically less clear, although some have considered *G. herbaceum* to be a better candidate than *G. arboreum* based on cytogenetics (Endrizzi et al. 1985). Subsequent molecular and phylogenetic evidence, however, supports sister status for both extant A-genome species, which, together, are equally divergent from the actual A-genome parent of the allopolyploids (Wendel et al. 1989; Wendel and Albert 1992; Cronn et al. 1996; Liu et al. 2001b). Initial evidence based on restriction site data and Southern hybridization analysis of cytoplasmic DNA (Wendel 1989; Galau and Wilkins 1989; Small and Wendel 1999) identified the A-genome as the cytoplasmic donor, i.e., maternal parent. This was later confirmed by numerous analyses of complete chloroplast (Xu et al. 2012; Chen et al. 2017b) and mitochondrial (Tang et al. 2015; Chen et al. 2017c) genomes. An additional unexpected consequence of the numerous phylogenetic analyses devoted to understanding the origin of the polyploid species is the realization that the extant A-genome species better represent the actual A-genome parent than *G. raimondii* which represents the D-genome parent, by a factor of two.

Following their initial origin, allopolyploid cottons rapidly radiated into three major lineages (Wendel 1989; Percy and Wendel 1990; DeJoode and Wendel 1992; Stanton et al. 1994; Wendel et al. 1995a; Cronn et al. 1996; Small and Wendel 1999; Grover et al. 2012a, 2015a; Gallagher et al. 2017) while spreading throughout the coastal tropical and subtropical regions in the Caribbean, northern South America, and Central America, with some very distant extensions into the Pacific Ocean (Brubaker and Wendel 1994; Wendel et al. 2010). At present, the allopolyploid AD genome cottons contain seven recognized species (Wendel and Grover 2015; Wang et al. 2018a): $(AD)_1$ *G. hirsutum*; $(AD)_2$ *G. barbadense*; $(AD)_3$ *G. tomentosum*; $(AD)_4$ *G. mustelinum*; $(AD)_5$ *G. darwinii*; $(AD)_6$ *G. ekmanianum*; and $(AD)_7$ *G. stephensii*. The latter two comprise recently described additions that were both newly collected and cryptically archived in germplasm collections as variants of *G. hirsutum* (Krapovickas et al. 2008; Grover et al. 2015c; Gallagher et al. 2017).

Relationships among the polyploid species are relatively well understood, despite the challenges inherent in characterizing rapid radiations. The earliest divergence (~0.63 mya estimated by Chen et al. 2020) separates the lineage of *G. mustelinum* from the rest of the polyploids, which subsequently split into the *G. barbadense-G. darwinii* clade and the *G. hirsutum-G. tomentosum* complex, the latter of which also includes the two newly assigned species *G. ekmanianum* and *G. stephensii*. It is ecologically noteworthy that most of the species are island endemics (with the exception of *G. hirsutum* and *G. barbadense*) which originated following long-distance dispersal events. The two newly described species *G. ekmanianum* and *G. stephensii*, which are sister species to *G. hirsutum*, are island endemics from Hispaniola (Grover et al. 2015a) and Wake Atoll (Gallagher et al. 2017), respectively. *Gossypium darwinii* is native to the Galápagos Islands, where it may form large and continuous populations in some areas (Percy and Wendel 1990). *Gossypium tomentosum*, on the other hand, is native to the Hawaiian Islands, where it has a more diffuse population structure, occurring mostly as scattered individuals and small populations on several islands (DeJoode and Wendel 1992). Even the earliest diverging allopolyploid, *G. mustelinum*, has an island-like distribution in the sense that it is an uncommon species restricted to a relatively small region of northeast Brazil (Wendel et al. 1994). Notably, it is only the two cultivated species, *G. hirsutum* and *G. barbadense*, that have large indigenous ranges. *Gossypium hirsutum* is widely distributed in Central and South America and the Caribbean and even reaches distant islands in the Pacific (Solomon Islands, Marquesas), whereas *G. barbadense* has a more southerly indigenous range, centered in the northern third of South America but with a large range of region overlap with *G. hirsutum* in the Caribbean. These species also encompass a wealth of morphological forms that span the wild-to-domesticated continuum (Fryxell 1979; Wendel et al. 1992; Brubaker and Wendel 1994;

Brubaker et al. 1999a). Both species possess truly wild forms, although limited in *G. barbadense*, which are important in research and to our understanding the process of domestication in these two species.

2.5 Polyploidy and Evolutionary Genomics of Cotton

Polyploidy is a phenomenon that is common among plants and which can fundamentally alter the evolutionary trajectory of the incipient species. The biological, ecological, and evolutionary consequences of polyploidy are potentially numerous and lineage dependent (Madlung 2013; Soltis and Soltis 2016), and the underlying molecular changes can be remarkably complicated. These molecular changes include those that immediately operate upon genome doubling (or shortly thereafter), as well as those that operate to return polyploid genome to a modified, diploid-like state. The realization that all flowering plants are paleopolyploid (Jiao et al. 2011) indicates that these processes are repeated and cyclical (Wendel 2015).

As early as the 1930s, not only were allopolyploid cotton species identified as such, but diploid cotton itself was suggested as paleopolyploid based on the observation of secondary associations during meiotic metaphase (Lawrence 1931; Davie and Hugh Davie 1933; Skovsted 1933, 1937). Classical cytogenetic data and molecular experiments led to the speculation that $n = 7$ was ancestral to the tribe, possibly even to the Malvaceae (Davie and Hugh Davie 1933; Abraham et al. 1940; Saunders 1961; Seelanan et al. 1999; Muravenko et al. 1998). This supposition suggested a minimum of one polyploidization for diploid cotton ($n = 13$). This observation remained unconfirmed until the publication of the *Gossypium raimondii* genome (Paterson et al. 2012), which revealed a surprisingly complex structure of five- to sixfold ploidy increase approximately 60 mya after the divergence of the cotton family and its many allies (Malvadendrina, including the traditionally recognized Malvaceae (Conover et al. 2019) from chocolate (*Theobroma cacao*, Malvaceae subfamily Byttnerioideae). Subsequent analyses indicate that this multiple polyploidy history resulted not from a single event but through multiple successive events (Wang et al. 2016b), although pinpointing the lineages involved in the paleopolyploidy of cotton remains challenging due to the complicated evolutionary history of genome doubling in the Malvaceae (Conover et al. 2019). Extensive gene loss (c. 70%) following the ancient whole-genome multiplication event(s) is inferred, given the current gene count for the modern *G. raimondii* genome. It seems likely that this gene loss was nonrandom with respect to the progenitor genome contributions; that is, more gene losses were observed in the duplicated regions having lower level of gene expression and higher density of transposable elements (Renny-Byfield et al. 2015). This phenomenon of biased fractionation is in concordance with those originally identified in maize (Schnable et al. 2009, 2011; Woodhouse et al. 2010) and *Brassica* (Wang et al. 2011c; Tang et al. 2012; Cheng et al. 2012) .

For contemporary allopolyploid species, the molecular consequences of genome doubling are more readily apparent. Previous research has characterized the suite of changes accompanying polyploidy in *Gossypium*, including gene loss and silencing, mobilization of transposable elements, epigenetic modifications, and massive genome-wide transcriptomic responses (previously reviewed by Adams et al. (2009), Wendel et al. (2012) and Wendel and Grover (2015)). While many of these changes are more subtle than has been reported for other polyploid species, as noted by Wendel et al. (2012), both the redundant nature of polyploidy and the changes that have accompanied allopolyploidization in cotton are factors in the evolution and domestication of the species. The latter is of particular interest, given that one of the parents to the polyploid does not possess spinnable fiber (D-genome), yet appears to have consequences for the domesticated fiber phenotype (Jiang et al. 1998; Rong et al. 2007; Said et al. 2015).

2.5.1 Genome-Wide Structure Variation Upon Polyploidy

In contrast to the prevalence of chromosomal rearrangements found in other model plant allopolyploids, most notably wheat (Feldman et al. 1997; Liu et al. 1998a, b; Ozkan et al. 2001; Shaked et al. 2001) and *Brassica* (Song et al. 1995; Lukens et al. 2006), structural variation in allopolyploid *Gossypium* does not appear prominent (Gerstel 1953; Menzel and Brown 1954; Brubaker et al. 1999b; Paterson et al. 2000; Liu et al. 2001a). Perhaps as a consequence of the twofold difference in genome (and, correspondingly, chromosome) size between diploid progenitors, large-scale recombination between allopolyploid subgenomes is rare (Salmon et al. 2010; Flagel et al. 2012; Grover et al. 2017b). Although the allopolyploid genome size is not quite additive and the A subgenome has slightly smaller chromosomes than the diploid A-genome species (Davie and Hugh Davie 1933; Endrizzi et al. 1985), the twofold size difference is largely preserved between subgenomes (Skovsted 1933; Endrizzi et al. 1985; Hendrix and Stewart 2005). More recently, high-quality genome sequences have become available for most of the tetraploid cottons (Zhang et al. 2015; Chen et al. 2020 in preparation; Wang et al. 2019) as well as the model diploid progenitors (Paterson et al. 2012; Du et al. 2018). These sequences were inferred with the integration of long-range scaffolding technologies (e.g., BioNano optical mapping, high-throughput chromosome conformation capture data (Hi-C), etc.), which provide a foundation for detailed comparative genomics and characterization of genome-wide structure variation. While broad-scale colinearity is largely preserved among diploid and polyploid cottons, comparisons among genome sequences have identified smaller-scale structural differences that were previously uncharacterized. For example, comparisons between the *G. hirsutum* and *G. barbadense* genomes found extensive accumulation of pericentric and paracentric inversions (Zhang et al. 2015; Yuan et al. 2015; Wang et al. 2019) in one relative to the other. With the release of additional polyploid cotton genomes, the phylogenetic placement and polarization of these events becomes possible. Such analyses will also facilitate understanding of the evolutionary and functional consequences of these smaller-scale structure variations.

2.5.2 Dynamics of Repetitive Elements

The dynamic genomic response observed in polyploid genomes is frequently associated with activities and alterations in the transposable element (TE) fraction of the nascent allopolyploid genome. Activation of TEs during hybridization and polyploidization is a long observed phenomenon (McClintock 1984; Sarilar et al. 2011; Parisod and Senerchia 2012; Piednoël et al. 2013; Senerchia et al. 2014; An et al. 2014; Vicient and Casacuberta 2017) that is linked to decreased repression and/or increased transposition (Kashkush et al. 2003; Madlung et al. 2005; Kawakami et al. 2010; Lopes et al. 2013; Ågren et al. 2016; Springer et al. 2016). Consequences for the genome include altered gene expression levels and genomic rearrangements (reviewed by Parisod and Senerchia (2012) and Vicient and Casacuberta (2017)), both of which are common to polyploid species and the latter of which may contributed to fractionation and/or diploidization (Vicient et al. 1999; Freeling and Thomas 2006; Bruggmann et al. 2006; Lim et al. 2007; Woodhouse et al. 2010, 2014; Vicient and Casacuberta 2017).

Some polyploids, such as allopolyploid cotton, exhibit relative quiescence in TEs post polyploidization (Liu et al. 2001a; Ben-David et al. 2013; Sarilar et al. 2013). As in many other plant species, genome size variation among diploid cottons is largely attributable to specific families and classes of dispersed repetitive elements (e.g., *gypsy* and *copia* LTR retrotransposons), which have differentially proliferated in different *Gossypium* lineages (Hawkins et al. 2006, 2008; Renny-Byfield et al. 2016; Grover et al. 2017a, 2019a). Following allopolyploid formation, however, there is no phylogenetic evidence to support any quantitatively significant proliferation of TEs (Hu et al. 2010). Also, whole-genome sequencing suggests that the TE composition of the diploid progenitors is mainly retained in the allopolyploids, using *G. hirsutum* as a model (Zhang et al. 2015). Despite the absence of massive bursts of transposition, evidence of post-polyploidization TE activity has been found using fluorescent in situ hybridization (Zhao et al. 1998; Hanson et al. 1998, 2000). That is, both Hanson et al. (1998) and Zhao et al. (1998) found that A-genome-specific repetitive elements have spread to the D subgenome following allopolyploid formation. More recently, de novo genome sequences of multiple allopolyploid cotton species (Chen et al. 2020 in preparation) suggest a decrease in the repetitive content in the A subgenome relative to its progenitor and, conversely, a higher repetitive fraction in the D subgenome relative to its diploid progenitor. This is consistently observed in all five sequenced allopolyploid genomes and, together with the observation that the allopolyploid genome size is slightly less than the sum of the two diploid progenitors, may suggest cryptic turnover in TE elements that includes one or more episodes of directionally biased, inter-subgenome TE transpositions and/or biased genomic removal of TE sequence in spite of an overall downsizing.

2.5.3 Fates of Duplicated Genes by Polyploidy

Polyploidy is predicted to reduce the selective pressures on individual genes compared to their orthologs in diploid progenitors, resulting in the opportunity for three classically recognized flavors of functional evolution. A gene can gain a new function relative to its orthologs in the diploid progenitors (neofunctionalization); homoeologous gene pairs can split their ancestral function (subfunctionalization) either at the biochemical level (e.g., enzymes with two active sites for two different substrates can split each function between the two homoeologous proteins) or at the gene expression level (e.g., homoeologs are differentially expressed across different developmental times and/or tissues); or one of the two gene copies can be lost (non-functionalization). Among the earliest plant examples of subfunctionalization were those in cotton (Adams et al. 2003), where A- and D-homoeologous copies (thereafter, At and Dt) of *adhA* were shown to be reciprocally silenced in different whorls of the same flower and many other genes showed strongly biased homoeologous expression (Adams et al. 2004; Liu and Adams 2007; Flagel et al. 2008; Chaudhary et al. 2009; Adams and Wendel 2013).

Although the molecular evidence is still rather scarce regarding functional evolution of At and Dt homoeologs (e.g., Zhao et al. 2018), we have known a great deal about the various outcomes of their sequence evolution from a phylogenetic perspective. Specifically, if homoeologs evolve independently following allopolyploid formation, they should be each phylogenetically sister to the orthologous copy from their diploid progenitors, rather than to each other; falsification of this null hypothesis is indicative of interactions between homoeologs, such as gene conversion, gene loss, and unequal evolutionary rates between homoeologs and their orthologs in the progenitor diploids. This topic has been reviewed (Wendel and Cronn 2003; Adams et al. 2009; Wendel et al. 2012), but only prior to the recent explosion of genomic data; here we provide an updated synopsis.

Homoeolog sequences may evolve in a "concerted" fashion, as mediated by gene conversion or other mechanisms of homoeolog exchange. Following an early report of bidirectional inter-subgenomic homogenization of ribosomal genes (Wendel et al. 1995b), Cronn et al. (1999) studied the homoeologous sequences for 16 nuclear genes and concluded that in principle, single- or low-copy number genes have evolved independently between A and D subgenomes. In the last decade, various sources of large-scale sequence data, from expression sequence tags (Salmon et al. 2010; Flagel et al. 2012) to genomic sequencing (Guo et al. 2014; Li et al. 2015; Page et al. 2016), have been used in an effort to assess the extent of nonindependent homoeolog evolution in allopolyploid *Gossypium*. Noting the methodological caveats and analytic artifacts that could lead to overestimation of homoeolog gene conversion, it is likely that the proportion of the allopolyploid genome that has experienced homoeolog conversion is low, most likely well below 5% (Page et al. 2016).

When homoeologous loci appear to evolve independently, one key question regarding genic evolution is whether evolutionary forces and rates have been equivalent for duplicated genes within the same nucleus, and whether this is true in comparison with the orthologous copies from the progenitor diploids. Using small numbers of nuclear genes, early work first demonstrated a potential enhancement of evolutionary rate in both subgenomes of allopolyploid *Gossypium* relative to its diploid progenitors (Cronn et al. 1999; Senchina et al. 2003). This speculation was supported by subsequent molecular evolutionary analyses at much larger scales, based on transcriptome (Flagel et al. 2012) and whole-genome sequencing data (Zhang et al. 2015; Hu et al. 2019). In the latter studies, Zhang et al. (2015) and Hu et al. (2019) also found that the A subgenome appears to evolve faster than the D subgenome, according to the distribution of synonymous and nonsynonymous substitution rates (Ks and Ka, respectively) estimated from over 20,000 homoeologous and orthologous gene sets. Interestingly, the opposite conclusions were reached by Li et al. (2015) from a separate assembly of the *G. hirsutum* genome, finding that the diploid A and D genomes are evolving faster than their orthologs in *G. hirsutum* and that the Dt subgenome is evolving faster than the At subgenome. While these contradicting results are likely rooted in the methodology used for determining evolutionary rates, comparing the evolutionary rates between multiple allopolyploid species will be useful in determining the generalities of these conclusions.

Since allotetraploid formation, the majority of At and Dt homoeologs have been retained, and this duplicate gene retention is widely thought to provide raw materials for the origin of the morphological, physiological, and ecological novelty of the allopolyploids. Despite the observation that gene loss in the allopolyploid cotton genome is rare, the *G. hirsutum* genome sequences indicate that gene loss is asymmetric between A and D subgenomes; that is, more genes have been lost in the A subgenome than in the D subgenome (228 vs 141 losses in Zhang et al. (2015) and 523 vs 461 losses in Li et al. (2015), respectively). These observations suggest that gene loss is most likely an ongoing, biased process in allopolyploid cotton. More recently, based on de novo genome sequences of multiple allopolyploid cotton species (Chen et al. 2020 in preparation), the net At gene loss is threefold the Dt loss prior to the diversification of allopolyploids, suggesting that biased gene loss is immediate following the formation of allopolyploid cotton. Whether this biased loss is stochastic or destined by "genomic legacy" of the parental diploid genomes is a matter for further investigation. Similarly, the relevance of these patterns of gene loss and retention is of particular interest to advance our understanding of phenotypic novelty accompanying allopolyploidy and domestication.

2.5.4 Evolution of Duplicate Gene Expression

In addition to sequence evolution, homoeolog genes can exhibit divergent expression patterns, a phenomenon that is widely recognized as "homoeolog expression bias" in allopolyploids (Grover et al. 2012a; Yoo et al. 2014). Following the initial report of this phenomenon (Adams et al. 2003), many aspects of this observation have been examined in allopolyploid *Gossypium* (Adams et al. 2009, 2012; Adams and Wendel 2013), including the scope and scale of biased homoeolog expression, its context specificity (e.g., tissue-specific and stress-related), the genome-wide bias, the associated phenomenon of expression-level dominance (Rapp et al. 2009), revamping of coexpression networks (Gallagher et al. 2016; Hu et al. 2016), and the temporal scale at which alterations in homoeolog expression evolve. These studies demonstrate that homoeolog bias is the rule rather than the exception in cotton, and likely most other allopolyploids. In spite of the sweeping nature of this process, virtually nothing is known about its physiological and evolutionary relevance, although clues are beginning to emerge.

At the transcriptome level, 20–55% of homoeologous gene pairs exhibit A- or D-biased expression across multiple tissues and developmental conditions (Yoo and Wendel 2014; Hovav et al. 2015; Zhang et al. 2015). This dynamic range is in concordance with early findings on the tissue and organ specificity of homoeolog expression changes (Adams et al. 2004; Liu and Adams 2007; Chaudhary et al. 2009). Regarding the genome-wide pattern, neither A or D subgenome appears globally dominant in terms of biased homoeolog expression in all tissues. That is, the numbers of A- and D-biased genes often are more or less balanced, as during fiber development (Yoo and Wendel 2014). For example, more A-biased genes were found in ovules at the stage of fiber initiation (i.e., 3 days pre- and post-anthesis) (Yang et al. 2006), but the reverse was observed in seeds (Hovav et al. 2008). In addition, both balanced pattern (Rambani et al. 2014) and unbalanced bias favoring the D subgenome (Flagel et al. 2008) have reported for the petal tissue. The most sweeping study to date, in this respect, was by Zhang et al. (2015), who demonstrated a preferential utilization of the D subgenome in 31 of the 35 tissue/stage samples in *G. hirsutum.* Although some have speculated that genome-wide bias may have been selected to facilitate tissue-specific function and morphology that is more similar to progenitor of the favored subgenome, this doesn't seem to be the case in *Gossypium* given the fact that D-genome diploid does not produce spinnable fibers.

Whereas the mechanistic basis of homoeolog expression bias remains to be elucidated, experimental work does reveal that it can be attributed to two temporal stages, the changes that ensue immediately upon hybridization and those arising later from polyploidy and subsequent evolution (Flagel et al. 2008, 2012; Yoo et al. 2013; Rambani et al. 2014). One consensus observation from these studies, as well as in other allopolyploid plants (Xu et al. 2014; Wang et al. 2016c; Sun et al. 2017), is that a considerable proportion of the expression biases that accompany hybridization become maintained during allopolyploid evolution, and the initial bias often becomes amplified on an evolutionary timescale. Differential expression of homoeologs accompanying hybridization of diverged diploid genomes is diagnostic of *cis*-regulatory divergence between the parental genes, because the common *trans* effect alone should lead to equal homoeolog expression in the absence of *cis* differences (Wittkopp et al. 2004). Hu and Wendel (2019) recently pointed out the opportunities for understanding allopolyploid gene expression and hence phenotypes, through extending the classic *cis-trans* model to explicitly incorporate genome doubling.

Distinct from homoeolog-specific expression, a series of other expression patterns have also been studied for the aggregated expression of a given homoeologous pair, such as additive and nonadditive expression, expression-level dominance, and transgressive expression, as previously reviewed (Grover et al. 2012a; Yoo et al. 2014; Hu and Wendel 2019). About 23–61% of homoeologous pairs exhibited nonadditive expression among variable cotton tissues (Flagel and Wendel 2010; Yoo et al. 2013; Rambani et al. 2014). The phenomenon of expression-level dominance was originally discovered and elaborated in cotton (Rapp et al. 2009; Flagel and Wendel 2010), defined as the state where total expression of a homoeolog pair mimics the expression level of one of the two diploid parents of an allopolyploid; this parent is referred to as "dominant." At the genome-wide scale, Rapp et al. (2009)

demonstrated that in the synthetic AD allopolyploid (2(A_2D_1)), more than 10,000 genes are D-dominant, significantly higher than ~800 A-dominant genes. At present, the functional significance is unknown, but it's scale and scope suggest that it is important from many perspectives, including crop plant improvement. Both homoeolog expression bias and expression-level dominance also have been reported at the protein level (Hu et al. 2011, 2013, 2015).

Beyond the gene-centric characterization of expression changes, coexpression network analysis in polyploids has the potential to facilitate a better understanding of the mechanistic underpinnings of phenotypic and ecological traits, and also may provide novel insight into interactions among duplicated genes and genomes (Gallagher et al. 2016). Hu et al. (2016) conducted coexpression network analysis of developing seeds from diploid and allopolyploid cotton species and found that the network topology in polyploids is to some extent a modular combination of that of its progenitor genomes. Interestingly, the allopolyploid network resembles the network of the A-genome diploid more than that of the D-genome parent, despite its D-like phenotype in oil content. Expression modifications for entire modules of gene expression include analogies to phenomena described for single pairs of duplicated genes, i.e., coexpression-level dominance and transgressive expression.

2.5.5 Epigenetic Modifications

Epigenetic modifications are another consequence of polyploidization, particularly for allopolyploids, whose formation includes hybridization of divergent species. These modifications include DNA methylation, histone modifications, and chromatin remodeling, among others, whose patterns of change under polyploidization are reviewed elsewhere (Chen 2007; Paun et al. 2007; Song and Chen 2015; Vicient and Casacuberta 2017; Ding and Chen 2018).

While the epigenetic consequences of allopolyploidization in cotton are understudied, results from methylation surveys suggest that the allopolyploid cotton genome is epigenetically stable (Liu et al. 2001a). This observation was reiterated in later surveys, which also found that homoeologous methylation changes correlated with expression bias (Song et al. 2015, 2017). Furthermore, homoeologous demethylation of the *COL2D* gene in cultivated cottons was associated with photoperiod insensitivity (Song et al. 2017), demonstrating the potential of evolutionarily relevant epigenetic modifications in allopolyploid cotton. The relevance of methylation patterns to agronomic traits in allopolyploid cotton has been considered for nearly 50 years, when demethylation was associated with *Verticillium* stress in *G. hirsutum* (Guseinov et al. 1975). Subsequent surveys have noted that demethylation is also associated with salt tolerance (Zhao et al. 2010; Wang et al. 2016a) and the response to cold stress (Fan et al. 2013) and that methylation patterns can influence fiber growth (Jin et al. 2013). Given that methylation diversity in *G. hirsutum* is higher than genetic diversity (Keyte et al. 2006; Osabe et al. 2014), and that methylation can be significantly different among tissues (Osabe et al. 2014), further understanding of the patterns of methylation among species, accessions, and tissues can provide insight into biased gene expression patterns (Adams et al. 2003; Flagel and Wendel 2010; Grover et al. 2012a; Song et al. 2017) and, ultimately, phenotypes. Although other forms of epigenetic modification are far less understood for cotton, initial research into histone modifications (Zheng et al. 2016), microRNAs (Guan et al. 2014), and long-noncoding RNAs (Wang et al. 2015a) suggests that understanding these may also provide insight into how the homoeologous genomes of allopolyploid cotton are used and/or interact, with consequences for fiber (Guan et al. 2014) and other phenotypic traits.

2.6 Genomes of the Four Domesticated Species

2.6.1 Diversity and Origin

A remarkable feature of the cotton genus is that domestication has occurred not once but independently four times, involving two allopolyploid species from the Americas, *G. hirsutum* and *G. barbadense*, and two diploids from Africa-Asia, *G. arboreum* and *G. herbaceum*. In each of these four cases, aboriginal people discovered several thousand years ago the unique properties of cotton fibers and made them useful for ropes, textiles, and other applications. As a consequence, all four species are currently grown worldwide, occupying 30–36 million hectares globally in 30 countries (Kranthi 2018) and with smaller shares in more than 70 additional countries (Chee et al. 2016). Cotton fiber production comprises a multibillion dollar industry responsible for hundreds of millions of jobs annually. Accordingly, the history and impact of domestication on these four species is of substantial interest.

2.6.2 Domesticated Diploids

As noted above, both extant species of the A-genome (i.e., *G. herbaceum* and *G. arboreum*) have been domesticated for fiber production. Collectively known as the "short staple cottons" (Khadi et al. 2010), both diploid species produce comparably short and coarse fibers that are primarily used for handloom textiles, hosiery, fillers (e.g., mattress stuffing), and absorbent materials (Kranthi 2018). Although these are the least grown cotton species, they are of regional importance due to pest and/or environmental resistance (Basu 1996; Rajendran et al. 2005). Commercial production of these species is ongoing in India, Pakistan, Iran, Myanmar, and Thailand; *G. arboreum* alone is 22% of the cotton in Myanmar and over 50% in Thailand, and *G. herbaceum* occupies up to 0.6 million hectares in India (Kranthi 2018). Because the importance of these species as cultivars is limited to a few countries, our understanding of the extant diversity is generally limited.

Of the two diploids, only *G. herbaceum* has a known wild form, native to the savannahs of Southern Africa (Vollesen 1987; Wendel et al. 1989; Khadi et al. 2010). While the center of domestication of *G. herbaceum* is unclear, it likely was in Northern Africa or the Near East as pointed out by Fryxell (1979). Range expansions extended the initial growing range through the Persian Gulf states and the Indian subcontinent (Kranthi 2018). Genetic diversity in *G. herbaceum* relative to its sister species is unclear, with some reporting lower diversity in *G. herbaceum* (Wendel et al. 1989), whereas others report higher diversity (Jena et al. 2012). These conflicting observations are attributable to differences in germplasm evaluated, as well as markers (i.e., allozymes versus AFLP markers). Morphological diversity has also been characterized (Wendel et al. 1989), which, together with geographical factors, has partitioned the species into five infraspecific races (Stanton et al. 1994; Khadi et al. 2010); however, the utility of these races is unclear, and the partitioning of new accessions into a particular race is highly subjective.

Diversity in *G. arboreum* is similarly understudied, although the few estimates also indicate that diversity is low (Page et al. 2013; Fang et al. 2017a). Colloquially referred to as "Desi cotton" in the Indian subcontinent, *G. arboreum* has no true wild forms and was once considered a possible derivative of *G. herbaceum* that became separated as the result of a reciprocal translocation (Gerstel 1953). While recent research suggest that *G. arboreum* underwent speciation from its sister species, *G. herbaceum*, long prior to domestication (Renny-Byfield et al. 2016; Du et al. 2018), there is little known about the wild ancestors of this cultivated species and/or its domestication and diversification into the modern cultivated forms. The Indus Valley has historically been suggested as the origin for the species, as it represents a center of diversity and has archaeological evidence (Gulati and Turner 1928; Wendel et al. 2010). This, however, may actually represent a secondary center of diffusion post-domestication (Hutchinson 1954; Wendel et al. 2010). More recent analyses of resequencing data place the origin of Chinese accessions in South China with subsequent radiation to the Yangtze and Yellow River regions (Du et al. 2018). Traditionally, *G. arboreum* production has been limited to Asia, ranging from India to Korea (Wendel et al. 1989; Basu 1996; Guo et al. 2006; Khadi et al. 2010). While accounting for only 1–3% of the total cotton-growing area in India, *G. arboreum* occupies far more of the growing area in Myanmar and Thailand (Kranthi 2018). As with *G. herbaceum*, *G. arboreum* has been traditionally partitioned into five botanical races, but these appear to have limited usefulness. As mentioned above, relative diversity within *G. arboreum* is largely unclear, including diversity compared to tetraploid cultivars (see below). That is, while there exists some evidence that the amount of diversity in the diploid species is roughly equivalent to that found in the tetraploids (Wendel et al. 1989; Stanton et al. 1994), others suggest that the diploid cultivars retain more diversity than do the polyploid species (Jena et al. 2012), possibly due to more intense selection in the latter. Clearly, additional research on diversity within and among these two species is warranted, to facilitate an understanding of their modern gene pools, patterns of interspecific introgression (Wendel et al. 1989), and their domestication bottlenecks.

2.6.3 Domesticated Polyploids

Most famously, the cotton genus has also given us two domesticated polyploid species, *G. hirsutum* and *G. barbadense*, collectively responsible for the vast majority of the cotton trade. As noted above, both species are derived from a single allopolyploidization event arising from the chance dispersal of an A-genome ancestor to the Americas, one that fortuitously bore long fibers on its seeds. Interestingly, while the contribution of the African A-genome is apparent in both wild and domesticated fiber phenotypes of all seven wild allopolyploid species, the contribution of the short-fibered paternal parent to the allopolyploid fiber is more enigmatic, although potentially important (Reinisch et al. 1994; Jiang et al. 1998, 2000; Lacape et al. 2005; Han et al. 2006; Rong et al. 2007; Said et al. 2015). Although independently domesticated, the naturally occurring range of *G. hirsutum* and *G. barbadense* overlaps in parts of Central and South America (Brubaker et al. 1999a; Wendel and Cronn 2003; Chee et al. 2016), leading to natural introgression in addition to the intentional post-domestication introgression (Wang et al. 1995; Wayne Smith and Tom Cothren 1999; May 2001; Zhang et al. 2005, 2014; Reddy et al. 2017; Fang et al. 2017a). Introgression between the two cultivars is not straightforward, as there exist partial reproductive barriers (Stephens 1946), segregation distortion (Jiang et al. 2000), and eventual hybrid breakdown (Stephens 1946; Brown and Ware 1958; Zhang et al. 2014). Nevertheless, interest remains in combining the most desirable traits from each species (Zhang et al. 2014; Cao et al. 2015).

G. hirsutum or "upland cotton" is the most widely cultivated cotton species, accounting for 98–99% of the total cotton-growing area (Kranthi 2018). Native to Mexico and Central America, the initial domestication center of *G. hirsutum* is unclear, although northern Mesoamerica, specifically the Yucatán Peninsula, *circa* 5000 years ago, has been proposed (Wendel and Albert 1992; Wendel et al. 1992; Brubaker and Wendel 1994; Chee et al. 2016). The history of transformation from dooryard cultigen into a modern row crop is better documented, having occurred in the southern United States of America in what is known as the "Cotton Belt" (Ware 1951). Domestication of *G. hirsutum* involved the transformation of wild forms into those that bore commonly domesticated traits, e.g., compact plant architecture, day-length neutrality, morphological exaggeration of desirable phenotype (here, fiber), etc. Naturally, the intense selective pressure during domestication resulted in a reduction in genetic diversity; however, the reduction in diversity for domesticated cotton exceeds what is commonly observed for domesticated species (Wendel et al. 1992; Brubaker and Wendel 1994). The low levels of diversity in *G. hirsutum* have been characterized using various markers (Hutchinson 1951, 1959; Wendel 1989; Wendel et al. 1992; Brubaker and Wendel 1994, 2001; Iqbal et al. 1997, 2001; Brubaker et al. 1999a; Lubbers and Chee 2009; Ahmad et al. 2012; Fang et al. 2013; Tyagi et al. 2014; Zhao et al. 2015; Grover et al. 2017b), which all support a very narrow germplasm base.

While *G. barbadense* represents a significantly smaller share of cotton grown, it is frequently desired for its longer, stronger, and finer fiber (Liu et al. 2015b).

Colloquially known as "Egyptian" or "Pima" cotton, *G. barbadense* was domesticated in the dry coastal regions of northern Peru and southern Ecuador at least 7500 years ago (Percy and Wendel 1990; Westengen et al. 2005; Splitstoser et al. 2016). Following a primary domestication west of the Andes, the range of *G. barbadense* was expanded into northern South America through a trans-Andean dispersal and then into Central America, the Caribbean, and the Pacific (Percy and Wendel 1990; Rossen et al. 1996; Piperno and Pearsall 1998; Westengen et al. 2005). Modern cultivars of *G. barbadense* were first developed on the coastal plains and islands of the southeastern USA (as "sea island cotton") and were subsequently introduced to Egypt, where *G. barbadense* acquired the designation "Egyptian cotton" (Khadi et al. 2010). Subsequent reintroduction of *G. barbadense* (as "pima cotton") into the southwestern USA gave rise to the modern elite cultivars, also known as "extra-long-staple" cotton (Khadi et al. 2010). While the overall market share of *G. barbadense* is less than that of *G. hirsutum*, some countries (e.g., Egypt and Israel) favor the superior fiber quality, and plant this species mostly or entirely exclusively (Kranthi 2018). Like *G. hirsutum*, multiple accessions that span the wild to domesticate continuum are available, including early dooryard cultigens and landraces (Percy and Wendel 1990); however, truly wild forms of *G. barbadense* are rare. Diversity in this species has been characterized (Percy and Wendel 1990; Westengen et al. 2005; Lacape et al. 2007; Boopathi et al. 2008; Wang et al. 2011b; Abdellatif et al. 2012; Hinze et al. 2015, 2016; Grover et al. 2017b), although the number of modern datasets is low (Hinze et al. 2016). In general, it is accepted that *G. barbadense* also has low diversity that is either similarly low (Page et al. 2016; Grover et al. 2017b) or lower than (Abdalla et al. 2001; Lacape et al. 2007; Hinze et al. 2015, 2016) that found in *G. hirsutum* (however see Van Deynze et al. 2009).

2.7 Whole-Genome Resequencing Studies and Insights into the Genetic Structure of the Domesticated Species

Advances in next-generation sequencing (NGS) and third-generation sequencing technologies have greatly facilitated the sequencing of the domesticated cotton genomes (Wang et al. 2012, 2019; Paterson et al. 2012; Li et al. 2014, 2015; Zhang et al. 2015; Liu et al. 2015b; Yuan et al. 2015; Du et al. 2018). These studies collectively provide an important, enabling resource for a wealth of opportunities in plant breeding, genomic research, and gene discovery, among other applications. Following the initial completion of reference genome sequences (reviewed by Wang et al. 2015b), several whole-genome resequencing (WGR) studies have been reported (Page et al. 2013, 2016; Fang et al. 2017a, b; Wang et al. 2017; Du et al. 2018; Ma et al. 2018), which provide unprecedented insights into the evolutionary relationships between different gene pools, the structuring of genome variation, and revealing variable types of diversity (such as single-nucleotide polymorphism and InDels) within populations. Here, we summarize the recent progress in cotton WGR studies with a special emphasis on their applications to understand the genetic basis of phenotypic variation (Table 2.2).

The first studies of whole-genome resequencing in diploid and tetraploid cotton were reported by Page et al. (2013, 2016). A total of 10 accessions of diploid cotton and 34 accessions of tetraploid cotton were resequenced separately, with an average coverage of about 37× and 23× per accession, respectively. In 2017, WGR efforts in cotton accelerated, with three massive datasets being released. Fang et al. (2017a) sequenced 147 accessions of cotton (a total of 1.8 terabases) with an average of approximately 5× coverage per accession, including 33 accessions of *G. hirsutum* landraces, 52 *G. hirsutum*, 52 *G. barbadense* cultivars, and 10 other tetraploids (NCBI Project ID: PRJNA257154). Fang et al. (2017b) next sequenced 258 *G. hirsutum* cultivars (3.96 Tb total) at about 2.5× coverage each (PRJNA375965). Wang et al. (2017) sequenced 321 *G. hirsutum* cultivars (6.1 Tb total) at an average of 6.9-fold coverage (PRJNA336461). Until 2018, a total of 419 accessions of core upland cotton were sequenced which produced 6.55 Tb of data with 6.55× coverage (PRJNA399050). Complementing these efforts in tetraploid cotton, Du et al. (2018) resequenced 243 *G. arboreum* and *G. herbaceum* accessions at about a 6× coverage (PRJNA349094).

These WGR studies have permitted unprecedented insights into the patterns of variation in the cultivated cotton genomes, such as SNPs, InDels, copy number variations (CNV), and presence/absence variations (PAV), and the overall structuring of genetic diversity within and between different germplasm groups. For example, Fang et al. (2017a) found 16,377,749 SNPs in the interspecific populations of *G. hirsutum* and *G. barbadense*, among which 7,993,856 are common SNPs with allele frequency larger than 5% and a missing data rate less than 10%. The authors also discovered 144,662 InDels, which ranged in length between 1 bp and 8 kbp, and also reported 16,879 structural variations longer than 50 bp. Wang et al. (2017) detected 7,497,568 SNPs in their study of 321 *G. hirsutum* cultivars, as well as 351,013 small InDels (1–10 bps) and 93,783 structural variations (longer than 10 bps). Similar results were obtained in other WGR studies of *G. hirsutum*; for example, using 318 *G. hirsutum* cultivars, 8,621,073 SNPs were reported (Fang et al. 2017b), whereas 3,665,030 SNPs were detected for 419 core accessions of *G. hirsutum* cultivars (Ma et al. 2018). From two A-genome diploid cottons, Du et al. (2018) discovered 17,883,108 SNPs and 2,470,515 InDels.

The large sample size and sufficient sequence coverage of these WGR data allow a robust estimation of nucleotide diversity within and between populations. In the germplasm studied by Fang et al. (2017a), nucleotide diversity was reported as 0.00216 for *G. hirsutum* landraces and ~0.0007 for cultivars, indicating a strong genetic bottleneck during upland cotton domestication. Similar estimates of *G. hirsutum* cultivars were reported by the other two studies (Wang et al. 2017; Fang et al. 2017b), which consistently suggest that cultivated gene pools contain appreciable but low genetic diversity. By comparing genetic diversity and divergence between landraces and cultivars, over 200 selective sweeps—109 by Fang et al. (2017a), 93 by Wang et al. (2017), and 25 by Fang et al. (2017b)—were identified as the potential targets of human selection during crop domestication and later improvement.

Table 2.2 List of whole-genome resequencing projects

	Page et al. (2013, 2016) *G3*	Page et al. (2016) *PLOS Genetics*	Fang et al. (2017a) *Genome Biology*	Wang et al. (2017) *Nature Genetics*	Fang et al. (2017b) *Nature Genetics*	Du et al. (2018) *Nature Genetics*	Ma et al. (2018) *Nature Genetics*
Number of accessions	10	34	147	321[a]	258[b]	243	419
Species	*G. herbaceum*, *G. arboreum*, *G. raimondii*	Seven species of tetraploid cotton	*G. hirsutum* *G. barbadense*	*G. hirsutum*	*G. hirsutum*	*G. arboreum* *G. herbaceum*	*G. hirsutum*
Germplasm type	Wild, landrace, and cultivar	Wild, landrace, and cultivar	Landrace Cultivar	Cultivar	Cultivar		Cultivar
Total bases	~500 Gb	902 Gb	1.8 Tb	6.1 Tb	3.96 Tb	2.29 Tb	6.35 Tb
Sequencing depth	~37	23	~5	6.9	~2.5	~6.0	6.55
NCBI SRA ID	PRJNA202235, PRJNA202236, and PRJNA202239	PRJNA280597	PRJNA257154	PRJNA336461	PRJNA375965	PRJNA349094	PRJNA399050
Nucleotide variations (SNPs)	23,859,893 homoeo-SNPs	19.2 and 28.5 million homoeo-SNPs (35.6 million total unique loci)	16,377,749 (at least 2 accessions), 7,993,856 (MAF 0.05, missing rate 10%)	7,497,568 (depth at least 8, MAF 1%), 2,020,834 (MAF 5%)	8,621,073 2,167,186 (MAF 0.05)	17,883,108	3,665,030 (MAF 0.05, missing rate 20%), 1,980,539 (missing 10%)
InDels	–	–	144,662	351,013	–	2,470,515	–
Structural variation (SV)	–	–	16,879	93,783	–	–	–
Related traits and discovered loci	–	–	109 selective sweeps, 384 introgression events	93 domestication sweeps, 19 GWAS loci for fiber-quality-related traits, asymmetric subgenome domestication and cis-regulatory	25 improvement-selective sweeps, 119 GWAS loci (71 for yield-related traits, 45 for fiber qualities, and 3 for resistance to Verticillium wilt)	98 significant peak associations for 11 agronomically important traits	7383 GWAS SNPs for 13 fiber-related traits

[a]A total of 352 accessions, including 321 cultivated accessions and 31 wild accessions (public data)
[b]A total of 318 accessions, included 258 cultivars and other wild accessions (released previously)

An additional perspective of sequence evolution in allopolyploid cotton is provided by the comparisons of nucleotide diversity between the co-resident A and D subgenomes. In both *G. hirsutum* and *G. barbadense* cultivars, approximately equal levels of nucleotide diversity were reported (A-0.00075 vs D-0.00073 and A-0.00061 vs D-0.00051, respectively) by Fang et al. (2017a), which is consistent with a previous survey based on targeted sequence capture (Grover et al. 2012b). Interestingly, a significantly higher diversity of A than D subgenome (0.00072 vs 0.00056) in *G. hirsutum* was reported by Wang et al. (2017), which may reflect the sampling difference in the germplasms surveyed by different studies. Given that the external ecological and population-level influences should affect the co-resident genomes simultaneously, equal nucleotide diversity between subgenome is expected unless there has been selection favoring homoeologous genes from one subgenome over the other; this doesn't seem to be the case in cotton domestication.

An additional use of the resequencing data is to detect past episodes of interspecific gene flow between *G. hirsutum* and *G. barbadense*. It has long been known that these two species experienced massive historical introgression in colonial or precolonial times, particularly in the Caribbean and other areas where both species became intermingled (Wendel et al. 1992; Brubaker et al. 1993; Brubaker and Wendel 1994), as well as intentionally during crop improvement in the last couple of centuries (see below section: Natural and Artificial Introgression in the Genus). Some of this history is evident in the WGR data. Fang et al. (2017a), for example, reported 384 introgressed regions between *G. hirsutum* races and *G. barbadense* cultivars. They also found that the introgression events were significantly biased toward gene flow from *G. hirsutum* into *G. barbadense*, rather than the reverse. Whether these results will be confirmed by additional studies and using different germplasm datasets remains to be determined.

A central task of modern plant genetics and breeding is to understand the genetic basis of phenotypic variations. The abundant WGR resources listed above have allowed the construction of genetic variation maps and made possible genome-wide analysis studies (GWAS) as well as experiments designed to detect quantitative trait loci (QTLs) for agronomic traits. For example, Fang et al. (2017b) reported that 54.8% of the elite genome-wide association study (GWAS) alleles of upland cotton cultivars in China were transferred from three founders—Deltapine 15, Stoneville 2B, and Uganda Mian. A total of 119 GWAS loci—71 for yield-related traits, 45 for fiber qualities, and 3 for resistance to *Verticillium wilt*—were identified. Among those, the authors noted that more associated loci for lint yield and fiber quality were located in the A subgenome (70) than in the D subgenome (49), and there are more associated loci for lint yield than for fiber quality. In the study of Wang et al. (2017), a total of 19 association signals were detected for fiber-quality-related traits, including 8 and 11 in the A and D subgenomes, respectively. More recently, Ma et al. (2018) conducted a GWAS for 13 fiber-related traits, which identified 7383 unique SNPs that were significantly associated with these traits and located within or near 4820 genes; in contrast to Fang et al. (2017b), there are more associated loci identified for fiber quality than for yield, and more fiber genes detected in the D than the A subgenome. For diploid cotton, Du et al. (2018) reported 98 significant peak associations for 11 agronomically important traits.

Overall, these examples offer a glimpse into the types of insights that will emerge from focused WGR efforts in cotton (Table 2.2). Using increasingly precise and focused high coverage datasets, an entire spectrum of questions that once were inconceivable will now become experimentally feasible. This include the many details regarding the parallel domestication of the four cultivated cotton species, insights into the genetic and genomic basis of key ecological and agronomic traits, and paths forward for continued cotton improvement in an ever-changing agricultural landscape.

2.8 Wild Cotton Species as Sources of Desirable Breeding Traits

As noted in an earlier section, the process of domestication and crop improvement were accompanied by bottlenecks that have substantially reduced genetic variation in cultivated cottons, thus limiting their potential for developing novel varieties with improved traits. In principle, their wild progenitors and other wild relatives have far more genetic diversity and likely preserve many valuable genetic variants and associated phenotypes that are not present in the crop gene pool. In the context of climate change, crop wild relatives also serve a reservoir of genetic diversity adapted to a wide range of environmental conditions that plant breeders are increasingly likely to need to create new varieties able to cope with new and possibly exceptional abiotic conditions (Dempewolf et al. 2017; Zhang et al. 2017a; Mammadov et al. 2018). In *Gossypium*, although many wild species are narrow endemics with likely low levels of within-species genetic diversity, their worldwide distribution collectively encompasses a full range of geographic and ecological variation, suggesting that there are ample resources to be explored for use in cotton breeding. Cotton improvement programs have exploited many diploid species in association with specific morphological traits, disease resistance, cytoplasmic male sterility, and fertility restoration, whereas genes for stress tolerance, disease resistance, and nectariless and glandless cotton have been deliberately introduced from wild and feral tetraploids. These genetic enhancements are mainly obtained through conventional breeding approaches involving interspecific hybridization and backcrossing. We have collected and summarized these breeding efforts (reviewed in Mammadov et al. 2018) and potentially valuable ecological traits (Hutchinson 1959; Fryxell 1979; Wendel et al. 2010) in Table 2.1, along with a list of the relevant wild species used.

For example, *G. bickii* and other Australian diploid species in section *Sturtia* and *Hibiscoidea* have greatly lessened deposition of gossypol (toxic to human beings and nonruminant animals) in seeds but retain higher levels in aboveground plant tissues (Fryxell 1965). This "glandless-seed and glanded-plant" trait was introgressed into upland cotton cultivars for maximizing the utilization of cotton seed oil and proteins while retaining resistance to many diseases and phytophagous pests

(Zhu et al. 2005). The African species *G. anomalum* provides a unique source for developing cultivars with finer and stronger fibers (Mehetre 2010), as well as improved resistance to diseases like bacteria blight and cotton rust (Fryxell et al. 1984; Endrizzi et al. 1985).

Introgression of pest and disease resistance to cultivated cotton is also exemplified by using *G. longicalyx* as a source of reniform nematode resistance (Robinson et al. 2007), using *G. stocksii* for cotton leaf curl virus (CLCuV), using *G. australe* (Liu et al. 2015a), *G. bickii* (Wang et al. 2004), *G. sturtianum* (McFadden et al. 2004), and *G. thurberi* (Zhao et al. 2012) to transfer resistance to Fusarium wilt and Verticillium wilt diseases. With respect to abiotic stress tolerance, the majority of wild diploid species are more or less xerophytically adapted to the drier environments of desert areas, except *G. longicalyx* and several little-known species from the northwest Kimberley region of Australia (Fryxell 1979). Specifically, the American D-genome species have evolved different strategies for drought tolerance: *G. harknessii*, *G. armourianum*, and *G. turneri* have evolved reduced leaves with thick cuticles and a double layer of palisade cells for reducing water loss; *G. aridum*, *G. lobatum*, and *G. laxum* stay dormant during the dry season to circumvent the aridity of the habitat. Perhaps less obvious as an adaptation, the aggressive and deeply penetrating root system enables *G. gossypioides* and *G. thurberi* to grow on steep rocky slopes far from any watercourses. Sharing many similar physiological and molecular responses to drought stress, salt tolerance was previously noted for *G. aridum*, *G. davidsonii*, and *G. klotzschianum*, and many stress-responsive genes have been identified by transcriptomic analyses (Fan et al. 2015; Zhang et al. 2016; Wei et al. 2017).

Because of the differences in ploidy, genetic incompatibility, growth habit, and other agronomic traits, transferring beneficial traits from wild diploid species into cultivated tetraploid cottons is always challenging. The most effective strategy has been developing interspecific hybrids through bridge crosses. For example, to facilitate the introgression of reniform nematode resistance, Bell and Robinson (2004) developed synthetic tetraploid triple-species hybrids between *G. hirsutum*, *G. longicalyx*, and *G. armourianum* (or *G. herbaceum*) as bridges, from which successful introgression was accomplished by recurrent backcrosses to *G. hirsutum* (Robinson et al. 2007). Such conventional breeding programs are usually time-intensive and resource-intensive. With recent advances in CRISPR/Cas genome editing in cotton (Chen et al. 2017a; Gao et al. 2017; Zhang et al. 2018; Long et al. 2018; Li and Zhang 2019; Li et al. 2019a, b), transferring specific genes from wild cotton species with improved efficiency and precision might soon become universal. Understanding the evolutionary relationship from primitive species to crops and among wild relatives will continue to provide insights into the biology of cotton and agronomic improvement.

2.9 Natural and Artificial Introgression in the Genus

The extent of natural interspecific introgression in *Gossypium* is remarkable given the number of pre- and post-reproductive barriers that exist among species (Cronn and Wendel 2003). Although the natural range of cotton species collectively encompass much of the drier tropics and subtropics worldwide, most species exist in small, scattered populations that are geographically isolated from other regional species, making interspecific contact a rare occurrence (Cronn and Wendel 2003). For those species that are not geographically isolated, further reproductive barriers exist. Reduced pollen germination and abnormal pollen tube growth are present in some interspecific crosses (Ram et al. 2008), cytogenetic incompatibilities often lead to univalent formation, and there exist several sources of cytoplasmic male sterility (e.g., *G. harknessii* and *G. sturtianum*) that likewise inhibit interspecific crosses (Meyer 1975; Stewart 1992; Meshram 1994; Zhang and Stewart 1999, 2001; Liu et al. 2003; Suzuki et al. 2013; Chen et al. 2017d). Successful interspecific crosses, however, may also become a dead end; i.e., F1 hybrids between genome groups tend to be sterile (Endrizzi et al. 1985).

These natural barriers to introgression notwithstanding, evidence suggests that natural introgression is prevalent, although often cryptic (Cronn and Wendel 2003). Cytoplasmic introgression (i.e., introgression involving the chloroplast) appears to be most prominent. Both *G. cunninghamii* (Australian K-genome) and *G. bickii* (Australian G-genome) possess chloroplast genomes most similar to those species found in the Australian C-genome (e.g., *G. robinsonii*) (Wendel et al. 1991; Cronn and Wendel 2003), and the entire African B-genome has Australian-like chloroplasts, despite being nestled within a monophyletic African clade (i.e., A-, B-, E-, F-genomes) according to nuclear markers (Cronn et al. 2002; Cronn and Wendel 2003). In these examples, nuclear introgression has not been reported, possibly due to rapid elimination of one parental genome, which has been observed in interspecific cotton crosses (Stephens 1949, 1950).

Alternatively, nuclear introgression may be cryptically present. The first example of cryptic nuclear introgression in cotton was reported for *G. gossypioides*, involving two independent introgression events. The more recent introgression involved transfer of the *G. raimondii* cytoplasm into *G. gossypioides*, resulting in the placement of *G. gossypioides* at a terminal phylogenetic position within the D-genome chloroplast phylogeny (Small and Wendel 2000; Cronn et al. 2003; Cronn and Wendel 2003; Grover et al. 2019a). Nuclear ribosomal DNA sequences, however, strongly place *G. gossypioides* as a member of the African clade (A-, B-, E-, F-genome (Wendel et al. 1995a; Seelanan et al. 1999; Cronn et al. 2003)), which is supported by the presence of African-specific repetitive sequences in *G. gossypioides* (Zhao et al. 1998). Low-copy nuclear sequences, however, suggest that *G. gossypioides* is the first lineage to diverge within the D-genome (Small and Wendel 2000; Liu et al. 2001b; Cronn et al. 2003; Grover et al. 2019a). Together, these suggest a more ancient introgression of an African cotton into *G. gossypioides* that includes nuclear introgression, followed by a more recent cytoplasmic introgression of the *G. raimondii* chloroplast into *G. gossypioides*.

Similarly, populations of *G. aridum sect. Erioxylum* from the Mexican state of Colima have experienced cryptic introgression from section *Houzingenia*. This population of *G. aridum* is distinct from other populations (e.g., those from Jalisco, Mexico) in that the chloroplast is more similar to the sister species *G. klotzschianum* and *G. davidsonii* than it is to the remaining members of section *Erioxylum* (Alvarez and Wendel 2006; Grover et al. 2019a). Recently, genome-wide analyses of *G. aridum* with the remainder of the D-genome species also detected remnants of nuclear introgression (Grover et al. 2019a), which was not previously reported. Other instances of cryptic nuclear introgression in diploid *Gossypium* may become apparent as resequencing within the genus becomes commonplace. Candidates include *G. triphyllum*, one of the B-genome cottons with Australian cytoplasm that also exhibits some morphological similarities to the Australian cottons (Fryxell 1979; Cronn and Wendel 2003), and *G. bickii*, which as noted above appears to have a hybrid ancestry.

Naturally occurring introgression among polyploid species also has been reported, although this seems unlikely for more restricted island endemics like *G. tomentosum* (Hawaii) and *G. darwinii* (Galapagos Islands). Perhaps the best characterized is the introgression between *G. hirsutum* and *G. barbadense* in the Caribbean and northern South America (Stephens 1967; Percy and Wendel 1990; Wendel et al. 1992; Brubaker et al. 1993; Brubaker and Wendel 1994; Coppens d'Eeckenbrugge and Lacape 2014; Hinze et al. 2016). Despite the reproductive barriers mentioned above, bidirectional introgression has been detected for populations of both species in their overlapping ranges. The amount of introgression, however, is not equivalent between the two species; that is, *G. hirsutum* naturally carries more *G. barbadense* alleles than the converse (Chee et al. 2016). This may be due to the aforementioned phenomenon of segregation distortion (Jiang et al. 2000) and/or biased genome elimination (Stephens 1949, 1950). This ability of *G. hirsutum* to integrate *G. barbadense* alleles may have led to the creation of the botanical race Marie-Galante (Stephens 1967, 1974), and further introgression of the Brazilian endemic *G. mustelinum* into some populations of Marie-Galante may have subsequently resulted in the Brazilian "moco" cotton (Pinto de Menezes et al. 2010).

In addition to natural interspecific gene flow, humans have introduced intention introgression into breeding stocks for cultivar development (Saha et al. 2006; Wang et al. 2011a; Zhang et al. 2014; Chee et al. 2016). As mentioned above, some are from diploid sources. In particular, there is much interest in using introgression between *G. hirsutum* and *G. barbadense* to enhance fiber traits. To date, it appears that there has been more success introgressing *G. hirsutum* into *G. barbadense* than the converse (Chee et al. 2016), converse to the natural direction.

2.10 Cotton Genomics and Fiber

From an economic point of view, the most agronomically important traits of cotton are the quantity and quality of fiber produced. Cotton fiber is a single-celled, epidermal, ovular trichome with a well-established developmental program (Haigler et al. 2009, 2012) that is largely conserved, but whose variations produce the remarkable differences in fiber between cotton species and between wild and domesticated accessions (Applequist et al. 2001). *Gossypium barbadense* fiber, for example, has an extended elongation phase resulting in the longer, finer phenotype for which

Egyptian cotton is famous (Tu et al. 2007). Conversely, subtle changes in the developmental program for the diploid cultivars result in fiber that is shorter, coarser, and weaker, with less convolution or "spinnability," thereby reducing commercial value (Haigler et al. 2005).

Generally, cotton fiber develops over a 40-day period, during which the fiber first elongates (primary cell wall synthesis) and subsequently undergoes cell wall thickening (secondary cell wall synthesis) until maturity. The pace and duration of each phase of cotton fiber development are crucial for various aspects of fiber quality, such as length (elongation phase) and spinnability (cell wall thickening). Accordingly, research has been focused on characterize either the specific genes or specific developmental time points relevant to fiber development.

Decades of cell biology have characterized many aspects of fiber developmental biology. In particular, much is now known about fiber cell wall patterning, deposition, and composition in wild versus domesticated species, as well as many other aspects (reviewed by Fang et al. 2018; Fang 2018). A series of master transcriptional factors involved in regulation of cotton fiber development, such as MYBs (Shangguan et al. 2008; Walford et al. 2011; Wan et al. 2016; Tan et al. 2016; Wu et al. 2018), HOX3 (Shan et al. 2014; Rombolá-Caldentey et al. 2014), and PRE1 (Zhao et al. 2018), have been reported. Genes such as cellulose synthase (Tuttle et al. 2015; Nixon et al. 2016; Cho et al. 2017) and profilin (Wang et al. 2010; Bao et al. 2011; Argiriou et al. 2012; Gallagher et al. 2016) have been the subject of comparative analyses to determine their roles in fiber development. In other cases, while the genes and/or mechanisms of control are not yet identified, the accumulated results from molecular biological research suggest important roles for hormones (Chen et al. 1997; Dasani and Thaker 2006; Liao et al. 2010; Zhang et al. 2017b), such as ethylene (Shi et al. 2006) and auxin (Singh et al. 2009), and for mediators of sugar flux (Delmer and Haigler 2002; Ruan et al. 2003; Shi et al. 2006) within the developing fiber.

While fiber development has been studied extensively, most research on the *controls* of fiber development is focused on *G. hirsutum*. QTL analysis of a *G. hirsutum* F_2 cross showed that many fiber QTLs were located on the D-genome. While this is perhaps a surprising result, as D-genome *Gossypium* does not produce spinnable fiber (Grover et al. 2019b), it also suggests that factors contributing to the special properties of polyploid cotton may be found in D-genome species, pointing to a potential avenue of improvement via introgression of genes from these species (Jiang et al. 1998; Paterson et al. 2003; Lacape et al. 2005; Han et al. 2006; Rong et al. 2007; Qin et al. 2008; Said et al. 2015). Regulation of fiber-related genes has also been studied in the context of fiber coexpression networks, whose modules tend to exhibit D-genome bias although individual modules may experience the converse (Gallagher et al. 2020). Notably, the A- and D-homoeologous networks are generally similar in structure, despite the lack of spinnable fiber in the D-genome cotton species, suggesting that the foundation exists in those species. The most significant change, however, in the cotton fiber network is the increased co-regulation among genes and homoeologs that results in more tightly regulated gene networks in domesticated cotton (Gallagher et al. 2020). Network analysis could provide insights into nongenetic avenues of improvement for cotton, or means of altering entire networks within the fiber by editing transcription factors or other "hub" genes, or perhaps by altering their expression levels (Gallagher et al. 2016). It is an exciting prospect that the genomic and technological resources needed to affect these new insights are now becoming mature.

Historically, genetic modification and even breeding of cotton have been challenging due to its polyploid nature and general recalcitrance to regeneration after transformation. Despite this, strides have been made in these areas, with new technologies constantly developing and improving. Diploid cotton is a rich source of variation that could be used to improve any number of traits in commercial cotton, and in some cases this has occurred in the wild (Wendel and Cronn 2003). Because *G. hirsutum* and *G. barbadense* are allopolyploid, introgressing genetic material from diploid species is challenging, although there has been some success historically through hexaploid bridging (Becerra Lopez-Lavalle and Brubaker 2007). Producing knockout or knockdown lines also requires thoughtful consideration to insure both homoeologs are targeted, but here there has also been recent success using virus-induced gene silencing (VIGS) to silence genes in polyploid cotton (Idris et al. 2010; Pang et al. 2013). Perhaps the most challenging aspect of genetic modification in cotton are the hurdles of regeneration after transformation (Juturu et al. 2015). Traditionally, success has been limited to particular genotypes of *G. hirsutum* (Jin et al. 2006, 2012); however, many of the challenges and recent developments in cotton transformation have recently been outlined (Juturu et al. 2015). Some work has been done on ovule culture transformation and with floral dip transformation leading to single-gene transformants (Johar Campus 2005; Zhang and Chen 2012), and strides in large-scale genetic modification are being made, particularly with CRISPR technology (Chen et al. 2017a; Li et al. 2017; Janga et al. 2017; Wang et al. 2018b). New transformation technologies have also shown promise, as in the case of pollen magnetofection (Zhao et al. 2017). These recent advances involving genetic engineering in cotton provide exciting prospects for using the substantial molecular research in *Gossypium* in breeding programs and in generating genetically modified cultivars.

2.11 Polyploidy, Stress Tolerance, and Implications for Crop Improvement in a Changing Environment

As is the case for *G. hirsutum* and *G. barbadense*, many important crops are contemporary polyploids, including wheat, potato, oats, oilseed rape, strawberries, banana, kiwifruit, watermelon, and many others. In these cases, physiological and morphological traits are often distinct in polyploid plants relative to their diploid progenitors. At the cellular level, the increment in cell size with the increase of ploidy level is referred to as "gigas" effect (Müntzing 1936; Stebbins et al. 1971).

Correlated with larger cell size, polyploid individuals often are transgressive relative to their diploid conspecifics at the tissue or organ level, with larger structures of roots, leaves, flowers, and seeds (Stebbins 1950; Dudits et al. 2016). Many of these scaling and stoichiometric issues are poorly understood, as detailed in the lengthy and insightful review by Doyle and Coate (2019). In natural settings, polyploidy often is associated with ecologically marginal areas or niche expansion due to their adaptive capacity to adverse or extreme environments (Lewis 1979; Ehrendorfer 1980; Stebbins 1985; Otto and Whitton 2000; Ni et al. 2009; McIntyre 2012). One generalization that has emerged from these and other papers is that in some environments and at certain periods, polyploids have a fitness advantage relative to their diploid cousins (Leitch and Leitch 2008; Ainouche and Wendel 2014; Soltis et al. 2016; de Peer et al. 2017). This is often suggested to be due to the increased "buffering" capacity afforded by duplicated genes and enhanced vigor from the "fixed" heterozygosity (two homoeologs providing a minimum of two different alleles), together with a diverse suite of genetic and genomic interactions, as discussed above. This remarkable property of polyploids has long attracted the attention of plant breeders, who often have used artificial polyploids as a tool for crop improvement (reviewed in Sattler et al. 2016). As mentioned above, the extraordinary advances in genomic resources combined with gene editing technologies such as CRISPR/Cas9 promise an entirely new opportunity to harness the phenotypic and ecological potentiality of polyploid genomes for specific breeding objectives.

In the case of *Gossypium*, perhaps the most notable ecological consequence of allopolyploidy is that it led to an apparent invasion of new ecological niches. In contrast to diploid species that mostly are found inland, away from coastal margins, allopolyploid species in their native habitats typically occur in coastal and insular environments, which, as pointed out by Fryxell (1965, 1979), is associated with the oceanic dispersal of salt-water-tolerant seeds in the newly evolved allopolyploid lineage. This raises the prospect that allopolyploid cotton inherently has a higher tolerance to salt stress. Recently, a diverse panel of eight diploid (from A- and D-genome groups) and four allopolyploid cotton species were screened for their morphophysiological responses to different levels of salinity stress in hydroponics (Dong et al. 2020). Among these species, the Brazilian allopolyploid *G. mustelinum* exhibits the highest level of salt tolerance; however, no significant difference was noted in allopolyploids relative to either A- or D-genome diploids, due to a wide range of variations within each genome group. Although these observations do not support the speculation of higher salt tolerance in polyploid cotton, there is insufficient evidence to reject this hypothesis because salt stress affects all cotton developmental stages and only the seeding stage was surveyed in this study. Moreover, hydroponic evaluation for salinity tolerance cannot account for the temporal and spatial variations in soil physical and chemical properties under field conditions or in natural settings, which are linked to other types of environmental constraints such as drought and mineral nutrient deficiency. By this means, perhaps, the ecological innovations of allopolyploid cotton are beyond salt tolerance alone, perhaps entailing an increased level of phenotypic plasticity to cope with altered environmental conditions. Other polyploids have provided some empirical support for this notion, for example, in spotted knapweed (*Centaurea stoebe* L); as compared to the diploid *C. stoebe*, the tetraploids exhibit a generally broader ecological tolerance which may have contributed to their success as invasive species in novel environments (Hahn et al. 2012).

In the context of climate change, ever-increasing variability in weather patterns is threatening cotton production globally due to temperature extremes, irregular rainfalls, and drought, which collectively comprise novel abiotic and biotic stresses. Therefore, it is a pressing matter to develop climate-resilient and stress-tolerant crops to sustain world cotton production. Although some progress has been made in addressing the physiological and molecular aspects of abiotic stress tolerance in cotton (Khan et al. 2018; Zafar et al. 2018; Harkess 2018; Abdelraheem et al. 2019; Chen et al. 2019, 2020), to date this understanding has not led to comparable advances in breeding. To the best of our knowledge, no cotton cultivars with a markedly enhanced abiotic stress tolerance, with high yield and fiber quality, are commercially available (Abdelraheem et al. 2019). This likely reflects the complex genetic basis for these traits, limited resources from the core germplasm pool, and difficulties in incorporated complex physiological capacities into breeding programs. While powerful new transgenic and gene editing methods are beginning to revolutionize technical aspects developing new cotton varieties, it seems likely that the knowledge foundation for this important endeavor will continue to emerge from our understanding of the extraordinary natural diversity in *Gossypium* and the still mysterious consequences of allopolyploid formation.

References

Abdalla AM, Reddy OUK, El-Zik KM, Pepper AE (2001) Genetic diversity and relationships of diploid and tetraploid cottons revealed using AFLP. *Theor Appl Genet* 102:222–229

Abdellatif KF, Khidr YA, El-Mansy YM et al (2012) Molecular diversity of Egyptian cotton (*Gossypium barbadense* L.) and its relation to varietal development. *J Crop Sci Biotechnol* 15:93–99

Abdelraheem A, Esmaeili N et al (2019) Progress and perspective on drought and salt stress tolerance in cotton. *Ind Crop Prod* 130:118–129

Abdul Kadir ZB (1976) DNA evolution in the genus *Gossypium*. *Chromosoma* 56:85–94

Abraham P et al (1940) Cytological studies in Gossypium: chromosome behaviour in the interspecific hybrid *G. arboreum* x *G. stocksii*. *Indian J Agric Sci* 10:285–298

Adams KL, Wendel JF (2013) Dynamics of duplicated gene expression in polyploid cotton. In: Polyploid and hybrid genomics. pp 187–194

Adams KL, Cronn R, Percifield R, Wendel JF (2003) Genes duplicated by polyploidy show unequal contributions to the transcriptome and organ-specific reciprocal silencing. *Proc Natl Acad Sci U S A* 100:4649–4654

Adams KL, Percifield R, Wendel JF (2004) Organ-specific silencing of duplicated genes in a newly synthesized cotton allotetraploid. *Genetics* 168:2217–2226

Adams KL, Flagel L, Wendel JF (2009) Responses of the cotton genome to polyploidy. In: Paterson AH (ed) Plant Genetics and Genomics: Crops and Models. Springer, New York, NY, pp 419–4

Ågren JA, Huang H-R, Wright SI (2016) Transposable element evolution in the allotetraploid *Capsella bursa-pastoris*. *Am J Bot* 103:1197–1202

Ahmad MQ, Khan SH, Azhar FM (2012) Decreasing level of genetic diversity in germplasm and cultivars of upland cotton (*Gossypium hirsutum*) in Pakistan. *J Agric Soc Sci* 8:92–96
Ainouche ML, Wendel JF (2014) Polyploid speciation and genome evolution: lessons from recent allopolyploids. In: Pontarotti P (ed) Evolutionary biology: genome evolution, speciation, coevolution and origin of life. Springer International Publishing, Cham, pp 87–113
Alvarez I, Wendel JF (2006) Cryptic interspecific introgression and genetic differentiation within *Gossypium aridum* (Malvaceae) and its relatives. *Evolution* 60:505–517
An Z, Tang Z, Ma B et al (2014) Transposon variation by order during allopolyploidisation between *Brassica oleracea* and *Brassica rapa*. *Plant Biol* 16:825–835
Applequist WL, Cronn R, Wendel JF (2001) Comparative development of fiber in wild and cultivated cotton. *Evol Dev* 3:3–17
Argiriou A, Kalivas A, Michailidis G, Tsaftaris A (2012) Characterization of *PROFILIN* genes from allotetraploid (*Gossypium hirsutum*) cotton and its diploid progenitors and expression analysis in cotton genotypes differing in fiber characteristics. *Mol Biol Rep* 39:3523–3532
Bao Y, Hu G, Flagel LE et al (2011) Parallel up-regulation of the profilin gene family following independent domestication of diploid and allopolyploid cotton (*Gossypium*). *Proc Natl Acad Sci U S A* 108:21152–21157
Basu AK (1996) Current genetic research in cotton in India. *Genetica* 97:279–290
Beasley JO (1940) The origin of American tetraploid *Gossypium* species. *Am Nat* 74:285–286
Beasley JO (1942) Meiotic chromosome behavior in species, species hybrids, haploids, and induced polyploids of *Gossypium*. *Genetics* 27:25–54
Becerra Lopez-Lavalle LA, Brubaker CL (2007) Frequency and fidelity of alien chromosome transmission in *Gossypium* hexaploid bridging populations. *Genome* 50:479–491
Bell AA (1984) Morphology, chemistry, and genetics of *Gossypium* adaptations to pests. In: Timmermann BN, Steelink C, Loewus FA (eds) Phytochemical adaptations to stress. Springer US, Boston, pp 197–230
Bell A, Robinson AF (2004) Development and characteristics of triple species hybrids used to transfer reniform nematode resistance from *Gossypium longicalyx* to *Gossypium hirsutum*. In: Proceedings of the beltwide cotton conferences. pp 422–426
Bell AA, Forest Robinson A, Quintana J et al (2014) Registration of LONREN-1 and LONREN-2 germplasm lines of upland cotton resistant to Reniform Nematode. *J Plant Regist* 8:187
Bell AA, Robinson AF, Quintana J et al (2015) Registration of BARBREN-713 germplasm line of upland cotton resistant to Reniform and Root-Knot Nematodes. *J Plant Regist* 9:89
Ben-David S, Yaakov B, Kashkush K (2013) Genome-wide analysis of short interspersed nuclear elements SINES revealed high sequence conservation, gene association and retrotranspositional activity in wheat. *Plant J* 76:201–210
Boopathi NM, Gopikrishnan A, Selvam NJ et al (2008) Genetic diversity assessment of *G. barbadense* accessions to widen cotton (*Gossypium* spp.) gene pool for improved fibre quality. *J Cotton Res Dev* 22:135–138
Bowman DT, McCarty JC (1997) Thrips (Thysanoptera: Thripidae) Tolerance in cotton: sources and heritability. *J Entomol Sci* 32:460–471
Brown HB, Ware JO (1958) Cotton. McGraw-Hill, New York
Brubaker CL, Wendel JF (1994) Reevaluating the origin of domesticated cotton (*Gossypium hirsutum*; Malvaceae) using nuclear restriction fragment length polymorphisms (RFLPs). *Am J Bot* 81:1309–1326
Brubaker CL, Wendel JF (2001) RFLP diversity in cotton. In: Jenkins JN, Saha S (eds) Genetic improvement of cotton: emerging technologies. Science Publishers, Enfield, pp 81–102
Brubaker CL, Koontz JA, Wendel JF (1993) Bidirectional cytoplasmic and nuclear introgression in the New World cottons *Gossypium barbadense* and *G. hirsutum* (Malvaceae). *Am J Bot* 80:1203–1208
Brubaker CL, Benson CG, Miller C, Leach DN (1996) Occurrence of terpenoid aldehydes and lysigenous cavities in the "glandless" seeds of Australian *Gossypium* species. *Aust J Bot* 44:601

Brubaker CL, Borland FM, Wendel JF (1999a) The origin and domestication of cotton. In: Smith CW, Cothren JT (eds) Cotton: origin, history, technology and production. Wiley, New York, pp 3–31
Brubaker CL, Paterson AH, Wendel JF (1999b) Comparative genetic mapping of allotetraploid cotton and its diploid progenitors. *Genome* 42:184–203
Bruggmann R, Bharti AK, Gundlach H et al (2006) Uneven chromosome contraction and expansion in the maize genome. *Genome Res* 16:1241–1251
Cao Z, Zhu X, Chen H, Zhang T (2015) Fine mapping of clustered quantitative trait loci for fiber quality on chromosome 7 using a *Gossypium barbadense* introgressed line. *Mol Breed* 35:215
Chaudhary B, Flagel L, Stupar RM et al (2009) Reciprocal Silencing, Transcriptional Bias and Functional Divergence of Homeologs in Polyploid Cotton (*Gossypium*). *Genetics* 182:503–517
Chee PW, Paterson AH, Udall JA, Wendel JF (2016) Interspecific hybridization for upland cotton improvement. In: Mason AS (ed) Polyploidy and hybridization for crop improvement. pp 1–20
Chen ZJ (2007) Genetic and epigenetic mechanisms for gene expression and phenotypic variation in plant polyploids. *Annu Rev Plant Biol* 58:377–406
Chen J-G, Du X-M, Zhou X, Zhao H-Y (1997) Levels of cytokinins in the ovules of cotton mutants with altered fiber development. *J Plant Growth Regul* 16:181–185
Chen Z, Feng K, Grover CE et al (2016) Chloroplast DNA structural variation, phylogeny, and age of divergence among diploid cotton species. *PLoS One* 11:e0157183
Chen X, Lu X, Shu N et al (2017a) Targeted mutagenesis in cotton (*Gossypium hirsutum* L.) using the CRISPR/Cas9 system. Sci Rep 7:44304. https://doi.org/10.1038/srep44304
Chen Z, Grover CE, Li P et al (2017b) Molecular evolution of the plastid genome during diversification of the cotton genus. *Mol Phylogenet Evol* 112:268–276
Chen Z, Nie H, Grover CE et al (2017c) Entire nucleotide sequences of *Gossypium raimondii* and *G. arboreum* mitochondrial genomes revealed A-genome species as cytoplasmic donor of the allotetraploid species. *Plant Biology* 19:484–493
Chen Z, Zhao N, Li S et al (2017d) Plant mitochondrial genome evolution and cytoplasmic male sterility. *CRC Crit Rev Plant Sci* 36:55–69
Chen H, Li Y, Ma X et al (2019) Analysis of potential strategies for cadmium stress tolerance revealed by transcriptome analysis of upland cotton. *Sci Rep* 9:86
Chen ZJ, Sreedasyam A, Ando A, et al (2020) Genomic diversifications of five *Gossypium* allopolyploid species and their impact on cotton improvement. *Nat Genet* 52:525–533
Cheng F, Wu J, Fang L et al (2012) Biased gene fractionation and dominant gene expression among the subgenomes of *Brassica rapa*. *PLoS One* 7:e36442
Cho SH, Purushotham P, Fang C et al (2017) Synthesis and self-assembly of cellulose microfibrils from reconstituted cellulose synthase. *Plant Physiol* 175:146–156
Conover JL, Karimi N, Stenz N et al (2019) A Malvaceae mystery: a mallow maelstrom of genome multiplications and maybe misleading methods? *J Integr Plant Biol* 61:12–31
Coppens d'Eeckenbrugge G, Lacape J-M (2014) Distribution and differentiation of wild, feral, and cultivated populations of perennial upland cotton (*Gossypium hirsutum* L.) in Mesoamerica and the Caribbean. *PLoS One* 9:e107458
Craven LA, Stewart JMD, Brown AHD, Grace JP (1994) The Australian wild species of *Gossypium*. In: Challenging the future: proceedings of the world cotton research conference I. CSIRO. pp 278–281
Cronn R, Wendel JF (2003) Cryptic trysts, genomic mergers, and plant speciation: research review. *New Phytol* 161:133–142
Cronn RC, Zhao X, Paterson AH, Wendel JF (1996) Polymorphism and concerted evolution in a tandemly repeated gene family: 5S ribosomal DNA in diploid and allopolyploid cottons. *J Mol Evol* 42:685–705
Cronn RC, Small RL, Wendel JF (1999) Duplicated genes evolve independently after polyploid formation in cotton. *Proc Natl Acad Sci* 96:14406–14411
Cronn RC, Small RL, Haselkorn T, Wendel JF (2002) Rapid diversification of the cotton genus (*Gossypium*: Malvaceae) revealed by analysis of sixteen nuclear and chloroplast genes. *Am J Bot* 89:707–725

Cronn R, Small RL, Haselkorn T, Wendel JF (2003) Cryptic repeated genomic recombination during speciation in *Gossypium gossypioides*. *Evolution* 57:2475–2489
Dasani SH, Thaker VS (2006) Role of abscisic acid in cotton fiber development. *Russ J Plant Physiol* 53:62–67
Davie JH, Hugh Davie J (1933) Cytological studies in the Malvaceae and certain related families. *J Genet* 28:33–67
de Peer YV, Van de Peer Y, Mizrachi E, Marchal K (2017) The evolutionary significance of polyploidy. *Nat Rev Genet* 18:411–424
DeJoode DR, Wendel JF (1992) Genetic diversity and origin of the Hawaiian Islands cotton, *Gossypium tomentosum*. *Am J Bot* 79:1311–1319
Delmer DP, Haigler CH (2002) The regulation of metabolic flux to cellulose, a major sink for carbon in plants. *Metab Eng* 4:22–28
Dempewolf H, Baute G, Anderson J et al (2017) Past and future use of wild relatives in crop breeding. *Crop Sci* 57:1070–1082
Ding M, Chen ZJ (2018) Epigenetic perspectives on the evolution and domestication of polyploid plant and crops. *Curr Opin Plant Biol* 42:37–48
Dong Y, Hu G, Yu J, et al (2020) Salt-tolerance diversity in diploid and polyploid cotton (*Gossypium*) species. *Plant J* 101:1135–1151
Doyle JJ, Coate JE (2019) Polyploidy, the nucleotype, and novelty: the impact of genome doubling on the biology of the cell. *Int J Plant Sci* 180:1–52
Du X, Huang G, He S et al (2018) Resequencing of 243 diploid cotton accessions based on an updated A genome identifies the genetic basis of key agronomic traits. *Nat Genet* 50:796–802
Dudits D, Török K, Cseri A et al (2016) Response of organ structure and physiology to autotetraploidization in early development of energy willow *Salix viminalis* L. *Plant Physiol* 170(3):1504–1523. https://doi.org/10.1104/pp.15.01679
Ehrendorfer F (1980) Polyploidy and distribution. In: Polyploidy. pp 45–60
Endrizzi JE, Turcotte EL, Kohel RJ (1985) Genetics, cytology, and evolution of *Gossypium*. In: Caspari EW, Scandalios JG (eds) Advances in genetics. Academic, New York, pp 271–375
Endrizzi JE, Katterman FRH, Geever RF (Department of Plant Sciences, University of Arizona, Tucson, AZ 85721 (USA)) (1989) DNA hybridization and time of origin of three species of *Gossypium*. In: Evolutionary trends in plants (United Kingdom). p 3
Fan HH, Wei J, Li TC et al (2013) DNA methylation alterations of upland cotton (*Gossypium hirsutum*) in response to cold stress. *Acta Physiol Plant* 35:2445–2453
Fan X, Guo Q, Xu P et al (2015) Transcriptome-wide identification of salt-responsive members of the *WRKY* gene family in *Gossypium aridum*. *PLoS One* 10:e0126148
Fang DD (2018) Cotton fiber: physics, chemistry and biology. Springer, New York
Fang DD, Hinze LL, Percy RG et al (2013) A microsatellite-based genome-wide analysis of genetic diversity and linkage disequilibrium in Upland cotton (*Gossypium hirsutum* L.) cultivars from major cotton-growing countries. *Euphytica* 191:391–401
Fang L, Gong H, Hu Y et al (2017a) Genomic insights into divergence and dual domestication of cultivated allotetraploid cottons. *Genome Biol* 18:33
Fang L, Wang Q, Hu Y et al (2017b) Genomic analyses in cotton identify signatures of selection and loci associated with fiber quality and yield traits. *Nat Genet* 49:1089–1098
Fang DD, Naoumkina M, Kim HJ (2018) Unraveling cotton fiber development using fiber mutants in the post-genomic era. *Crop Sci* 58:2214–2228
Feldman M, Liu B, Segal G et al (1997) Rapid elimination of low-copy DNA sequences in polyploid wheat: a possible mechanism for differentiation of homoeologous chromosomes. *Genetics* 147:1381–1387
Feng C, Ulloa M, Perez-M C, Stewart JM (2011) Distribution and molecular diversity of arborescent *Gossypium* species. *Botany* 89:615–624
Flagel LE, Wendel JF (2010) Evolutionary rate variation, genomic dominance and duplicate gene expression evolution during allotetraploid cotton speciation. *New Phytol* 186:184–193

Flagel L, Udall J, Nettleton D, Wendel J (2008) Duplicate gene expression in allopolyploid *Gossypium* reveals two temporally distinct phases of expression evolution. *BMC Biol* 6:16

Flagel LE, Wendel JF, Udall JA (2012) Duplicate gene evolution, homoeologous recombination, and transcriptome characterization in allopolyploid cotton. *BMC Genomics* 13:302

Freeling M, Thomas BC (2006) Gene-balanced duplications, like tetraploidy, provide predictable drive to increase morphological complexity. *Genome Res* 16:805–814

Fryxell PA (1965) A revision of the Australian species of *Gossypium* with observations on the occurrence of *Thespesia* in Australia (Malvaceae). *Aust J Bot* 13:71–102

Fryxell PA (1968) A redefinition of the tribe Gossypieae. *Bot Gaz* 129:296–308

Fryxell PA (1979) The natural history of the cotton tribe (Malvaceae, tribe Gossypieae). TAMU Press, College Station

Fryxell PA (1986) Ecological adaptations of *Gossypium* species. In: Cotton Physiology. The Cotton Foundation, Memphis, TN, pp 1–7

Fryxell PA (1992) A revised taxonomic interpretation of *Gossypium* L (Malvaceae). Rheeda 2:108–165

Fryxell PA, Kohel RJ, Lewis CF (1984) Taxonomy and germplasm resources. In: Cotton. American Society of Agronomy, Crop Science Society of America, Soil Science Society of America, pp 27–57

Fryxell PA, Koch SD, Koch SD (1987) New or noteworthy species of flowering plants from the Sierra Madre de Sur of Guerrero and Michoacán, Mexico. *J Syst Evol Bot* 11:539–561

Galau GA, Wilkins TA (1989) Alloplasmic male sterility in AD allotetraploid *Gossypium hirsutum* upon replacement of its resident A cytoplasm with that of D species *G. harknessii*. *Theor Appl Genet* 78:23–30

Gallagher JP, Grover CE, Hu G, Wendel JF (2016) Insights into the ecology and evolution of polyploid plants through network analysis. *Mol Ecol* 25:2644–2660

Gallagher JP, Grover CE, Rex K et al (2017) A new species of cotton from wake atoll, *Gossypium stephensii* (Malvaceae). *Syst Bot* 42:115–123

Gallagher JP, Grover CE, Hu G, et al (2020) Conservation and divergence in duplicated fiber coexpression networks accompanying domestication of the polyploid *Gossypium hirsutum* L. *G3* 10:2879–2892

Gao W, Long L, Tian X et al (2017) Genome editing in cotton with the CRISPR/Cas9 system. *Front Plant Sci* 8:1364

Gerstel DU (1953) Chromosomal translocations in interspecific hybrids of the genus *Gossypium*. *Evolution* 7:234–244

Grover CE, Yu Y, Wing RA et al (2008) A phylogenetic analysis of indel dynamics in the cotton genus. *Mol Biol Evol* 25:1415–1428

Grover CE, Gallagher JP, Szadkowski EP et al (2012a) Homoeolog expression bias and expression level dominance in allopolyploids. *New Phytol* 196:966–971

Grover CE, Salmon A, Wendel JF (2012b) Targeted sequence capture as a powerful tool for evolutionary analysis1. *Am J Bot* 99:312–319

Grover CE, Gallagher JP, Jareczek JJ et al (2015a) Re-evaluating the phylogeny of allopolyploid *Gossypium* L. *Mol Phylogenet Evol* 92:45–52

Grover CE, Gallagher JP, Wendel JF (2015b) Candidate gene identification of flowering time genes in cotton. *Plant Genome* 8: plantgenome2014.12.0098

Grover CE, Zhu X, Grupp KK et al (2015c) Molecular confirmation of species status for the allopolyploid cotton species, *Gossypium ekmanianum* Wittmack. *Genet Resour Crop Evol* 62:103–114

Grover CE, Arick MA 2nd, Conover JL et al (2017a) Comparative genomics of an unusual biogeographic disjunction in the cotton tribe (*Gossypieae*) yields insights into genome downsizing. *Genome Biol Evol* 9:3328–3344

Grover CE, Gallagher JP, Szadkowski EP et al (2017b) Nucleotide diversity in the two co-resident genomes of allopolyploid cotton. *Plant Syst Evol* 303:1021–1042

Grover CE, Arick MA 2nd, Thrash A et al (2019a) Insights into the evolution of the New World diploid cottons (*Gossypium*, Subgenus Houzingenia) based on genome sequencing. *Genome Biol Evol* 11:53–71

Grover CE, Yoo M-J, Gore MA et al (2019b) Genetic analysis of the transition from wild to domesticated cotton (*G. hirsutum*). bioRxiv 616763

Guan X, Pang M, Nah G et al (2014) miR828 and miR858 regulate homoeologous *MYB2* gene functions in *Arabidopsis* trichome and cotton fibre development. *Nat Commun* 5:3050

Gulati AN, Turner AJ (1928) A note on the early history of cotton. Indian Central Cotton Committee, Technological Laboratory

Guo W-Z, Zhou B-L, Yang L-M et al (2006) Genetic diversity of landraces in *Gossypium arboreum* L. race sinense assessed with simple sequence repeat markers. *J Integr Plant Biol* 48:1008–1017

Guo H, Wang X, Gundlach H et al (2014) Extensive and biased intergenomic nonreciprocal DNA exchanges shaped a nascent polyploid genome, *Gossypium* (Cotton). *Genetics* 197:1153–1163

Guseinov VA, Kiryanov GI, Vanyushin BF (1975) Intragenome distribution of 5-methylcytosine in DNA of healthy and wilt-infected cotton plants (*Gossypium hirsutum* L.). *Mol Biol Rep* 2:59–63

Hahn MA, van Kleunen M, Müller-Schärer H (2012) Increased phenotypic plasticity to climate may have boosted the invasion success of polyploid Centaurea stoebe. *PLoS One* 7:e50284

Haigler CH, Zhang D, Wilkerson CG (2005) Biotechnological improvement of cotton fibre maturity. *Physiol Plant* 124:285–294

Haigler CH, Singh B, Wang G, Zhang D (2009) Genomics of cotton fiber secondary wall deposition and cellulose biogenesis. In: Paterson AH (ed) Genetics and genomics of cotton. Springer US, New York, pp 385–417

Haigler CH, Betancur L, Stiff MR, Tuttle JR (2012) Cotton fiber: a powerful single-cell model for cell wall and cellulose research. *Front Plant Sci* 3:104

Han Z, Wang C, Song X et al (2006) Characteristics, development and mapping of *Gossypium hirsutum* derived EST-SSRs in allotetraploid cotton. *Theor Appl Genet* 112:430–439

Hanson R, Zhao XP, Islam-Faridi MN et al (1998) Evolution of interspersed repetitive elements in *Gossypium* (Malvaceae). *Am J Bot* 85:1364–1368

Hanson RE, Islam-Faridi MN, Crane CF et al (2000) *Ty1-copia*-retrotransposon behavior in a polyploid cotton. *Chromosome Res* 8:73–76

Harkess A (2018) Handling the heat: methylome variation underlying heat tolerance in cotton. *Plant Cell* 30:1947–1948

Hawkins JS, Kim H, Nason JD et al (2006) Differential lineage-specific amplification of transposable elements is responsible for genome size variation in *Gossypium*. *Genome Res* 16:1252–1261

Hawkins JS, Hu G, Rapp RA et al (2008) Phylogenetic determination of the pace of transposable element proliferation in plants: *copia* and LINE-like elements in *Gossypium*. *Genome* 51:11–18

Hendrix B, Stewart JM (2005) Estimation of the nuclear DNA content of *Gossypium* species. *Ann Bot 95*:789–797

Hinze LL, Fang DD, Gore MA et al (2015) Molecular characterization of the *Gossypium* diversity reference set of the US National Cotton Germplasm Collection. *Theor Appl Genet* 128:313–327

Hinze LL, Gazave E, Gore MA et al (2016) Genetic diversity of the two commercial tetraploid cotton species in the *Gossypium* diversity deference set. *J Hered* 107:274–286

Hovav R, Udall JA, Chaudhary B et al (2008) Partitioned expression of duplicated genes during development and evolution of a single cell in a polyploid plant. *Proc Natl Acad Sci U S A* 105:6191–6195

Hovav R, Faigenboim-Doron A, Kadmon N et al (2015) A transcriptome profile for developing seed of polyploid cotton. *Plant Genome* 8

Hu G, Wendel JF (2019) Cis-trans controls and regulatory novelty accompanying allopolyploidization. *New Phytol* 221:1691–1700

Hu G, Hawkins JS, Grover CE, Wendel JF (2010) The history and disposition of transposable elements in polyploid *Gossypium*. *Genome* 53:599–607

Hu G, Houston NL, Pathak D et al (2011) Genomically biased accumulation of seed storage proteins in allopolyploid cotton. *Genetics* 189:1103–1115

Hu G, Koh J, Yoo M-J et al (2013) Proteomic profiling of developing cotton fibers from wild and domesticated *Gossypium barbadense*. *New Phytol* 200:570–582
Hu G, Koh J, Yoo M-J et al (2015) Gene-expression novelty in allopolyploid cotton: a proteomic perspective. *Genetics* 200:91–104
Hu G, Hovav R, Grover CE et al (2016) Evolutionary conservation and divergence of gene coexpression networks in *Gossypium* (cotton) seeds. *Genome Biol Evol* 8:3765–3783
Hu Y, Chen J, Fang L et al (2019) *Gossypium barbadense* and *Gossypium hirsutum* genomes provide insights into the origin and evolution of allotetraploid cotton. *Nat Genet* 51(4):739–748. https://doi.org/10.1038/s41588-019-0371-5
Hutchinson JB (1951) Intra-specific differentiation in *Gossypium hirsutum*. *Heredity* 5:161–193
Hutchinson JB (1954) New evidence on the origin of the old world cottons. *Heredity* 8:225–241
Hutchinson JB (1959) The application of genetics to cotton improvement, 1st edn. Cambridge University Press, Cambridge
Hutchinson JB, Silow RA, Stephens SG (1947) The evolution of *Gossypium* and the differentiation of the cultivated cottons. Geoffrey Cumberlege, Oxford University Press, London
Idris AM, Tuttle JR, Robertson D et al (2010) Differential Cotton leaf crumple virus-VIGS-mediated gene silencing and viral genome localization in different *Gossypium hirsutum* genetic backgrounds. *Physiol Mol Plant Pathol* 75:13–22
Iqbal MJ, Aziz N, Saeed NA et al (1997) Genetic diversity evaluation of some elite cotton varieties by RAPD analysis. *Theor Appl Genet* 94:139–144
Iqbal MJ, Reddy OUK, El-Zik KM, Pepper AE (2001) A genetic bottleneck in the 'evolution under domestication' of upland cotton *Gossypium hirsutum* L. examined using DNA fingerprinting. *Theor Appl Genet* 103:547–554
Janga MR, Campbell LM, Rathore KS (2017) CRISPR/Cas9-mediated targeted mutagenesis in upland cotton (*Gossypium hirsutum* L.). *Plant Mol Biol* 94:349–360
Jena SN, Srivastava A, Singh UM et al (2012) Analysis of genetic diversity, population structure and linkage disequilibrium in elite cotton (*Gossypium* L.) germplasm in India. *Crop Pasture Sci* 62:859–875
Jiang C, Wright RJ, El-Zik KM, Paterson AH (1998) Polyploid formation created unique avenues for response to selection in *Gossypium* (cotton). *Proc Natl Acad Sci U S A* 95:4419–4424
Jiang CX, Chee PW, Draye X et al (2000) Multilocus interactions restrict gene introgression in interspecific populations of polyploid *Gossypium* (cotton). *Evolution* 54:798–814
Jiao Y, Wickett NJ, Ayyampalayam S et al (2011) Ancestral polyploidy in seed plants and angiosperms. *Nature* 473:97–100
Jin S, Zhang X, Nie Y et al (2006) Identification of a novel elite genotype for in vitro culture and genetic transformation of cotton. *Biol Plant* 50:519–524
Jin S-X, Liu G-Z, Zhu H-G et al (2012) Transformation of upland cotton (*Gossypium hirsutum* L.) with *gfp* gene as a visual marker. *J Integr Agric* 11:910–919
Jin X, Pang Y, Jia F et al (2013) A potential role for *CHH* DNA methylation in cotton fiber growth patterns. *PLoS One* 8:e60547
Johar Campus H (2005) In-ovule embryo culture: a novel method of cotton transformation. *Pak J Biol Sci* 8:297–301
Johnson BL (1975) *Gossypium palmeri* and a polyphyletic origin of the New World cottons. *Bull Torrey Bot Club* 102:340–349
Juturu VN, Mekala GK, Kirti PB (2015) Current status of tissue culture and genetic transformation research in cotton (*Gossypium* spp.). *Plant Cell Tissue Organ Cult* 120:813–839
Kashkush K, Feldman M, Levy AA (2003) Transcriptional activation of retrotransposons alters the expression of adjacent genes in wheat. *Nat Genet* 33:102–106
Katterman FRH, Ergle DR (1970) A study of quantitative variations of nucleic acids in *Gossypium*. *Phytochemistry* 9:2007–2010
Kawakami T, Strakosh SC, Zhen Y, Ungerer MC (2010) Different scales of *Ty1/copia*-like retrotransposon proliferation in the genomes of three diploid hybrid sunflower species. *Heredity* 104:341–350
Keyte AL, Percifield R, Liu B, Wendel JF (2006) Infraspecific DNA methylation polymorphism in cotton (*Gossypium hirsutum* L.). *J Hered* 97:444–450

Khadi BM, Santhy V, Yadav MS (2010) Cotton: an introduction. In: Cotton: biotechnological advances. Springer, Berlin, Heidelberg, pp 1–14
Khan A, Pan X, Najeeb U et al (2018) Coping with drought: stress and adaptive mechanisms, and management through cultural and molecular alternatives in cotton as vital constituents for plant stress resilience and fitness. *Biol Res* 51:47
Konan NO, Baudoin J-P, D'Hont A, Mergeai G (2009) Bridging classical and molecular cytogenetics of *Gossypium*. In: Paterson AH (ed) Genetics and genomics of cotton. Springer US, New York, pp 257–281
Kranthi KR (2018) Cotton production practices: snippets from global data 2017. The ICAC Recorder XXXVI:4–14
Krapovickas A, Seijo G, Seijo JG (2008) *Gossypium ekmanianum* (Malvaceae), algodon silvestre de la Republica Dominicana. *Bonplandia* 55–63
Lacape J-M, Nguyen T-B, Courtois B et al (2005) QTL analysis of cotton fiber quality using multiple *Gossypium hirsutum Gossypium barbadense* backcross generations. *Crop Sci* 45:123–140
Lacape J-M, Dessauw D, Rajab M et al (2007) Microsatellite diversity in tetraploid *Gossypium* germplasm: assembling a highly informative genotyping set of cotton SSRs. *Mol Breed* 19:45–58
LaDuke JC, Doebley J (1995) A chloroplast DNA based phylogeny of the Malvaceae. *Syst Bot* 20:259
Lawrence WJC (1931) The secondary association of chromosomes. *Cytologia* 2:352–384
Leitch AR, Leitch IJ (2008) Genomic plasticity and the diversity of polyploid plants. *Science* 320:481–483
Lewis WH (1979) Polyploidy in species populations. *Basic Life Sci* 13:103–144
Li C, Zhang B (2019) Genome editing in cotton using CRISPR/Cas9 system. *Methods Mol Biol* 1902:95–104
Li F, Fan G, Wang K et al (2014) Genome sequence of the cultivated cotton *Gossypium arboreum*. *Nat Genet* 46:567–572
Li F, Fan G, Lu C et al (2015) Genome sequence of cultivated Upland cotton (*Gossypium hirsutum* TM-1) provides insights into genome evolution. *Nat Biotechnol* 33:524–530
Li C, Unver T, Zhang B (2017) A high-efficiency CRISPR/Cas9 system for targeted mutagenesis in Cotton (*Gossypium hirsutum* L.). *Sci Rep* 7:43902
Li B, Rui H, Li Y et al (2019a) Robust CRISPR/Cpf1(Cas12a) mediated genome editing in allotetraploid cotton (*G. hirsutum*). *Plant Biotechnol J* 17(10):1862–1864. https://doi.org/10.1111/pbi.13147
Li J, Manghwar H, Sun L et al (2019b) Whole genome sequencing reveals rare off-target mutations and considerable inherent genetic or/and somaclonal variations in CRISPR/Cas9-edited cotton plants. *Plant Biotechnol J* 17:858–868
Liao W, Zhang J, Xu N, Peng M (2010) The role of phytohormones in cotton fiber development. *Russ J Plant Physiol* 57:462–468
Lim KY, Kovarik A, Matyasek R et al (2007) Sequence of events leading to near-complete genome turnover in allopolyploid *Nicotiana* within five million years. *New Phytol* 175:756–763
Liu Z, Adams KL (2007) Expression partitioning between genes duplicated by polyploidy under abiotic stress and during organ development. *Curr Biol* 17:1669–1674
Liu B, Vega JM, Feldman M (1998a) Rapid genomic changes in newly synthesized amphiploids of *Triticum* and *Aegilops*. II. Changes in low-copy coding DNA sequences. *Genome* 41:535–542
Liu B, Vega JM, Segal G et al (1998b) Rapid genomic changes in newly synthesized amphiploids of *Triticum* and *Aegilops*. I. Changes in low-copy noncoding DNA sequences. *Genome* 41:272–277
Liu B, Brubaker CL, Mergeai G et al (2001a) Polyploid formation in cotton is not accompanied by rapid genomic changes. *Genome* 44:321–330
Liu Q, Brubaker CL, Green AG et al (2001b) Evolution of the FAD2-1 fatty acid desaturase 5' UTR intron and the molecular systematics of *Gossypium* (Malvaceae). *Am J Bot* 88:92–102

Liu L, Guo W, Zhu X, Zhang T (2003) Inheritance and fine mapping of fertility restoration for cytoplasmic male sterility in *Gossypium hirsutum* L. *Theor Appl Genet* 106:461–469
Liu Q, Chen Y, Wang Y et al (2015a) A new synthetic allotetraploid (A1A1G2G2) between *Gossypium herbaceum* and *G. australe*: bridging for simultaneously transferring favorable genes from these two diploid species into upland cotton. *PLoS One* 10:e0123209
Liu X, Zhao B, Zheng H-J et al (2015b) *Gossypium barbadense* genome sequence provides insight into the evolution of extra-long staple fiber and specialized metabolites. *Sci Rep* 5:14139
Long L, Guo D-D, Gao W et al (2018) Optimization of CRISPR/Cas9 genome editing in cotton by improved sgRNA expression. *Plant Methods* 14:85
Lopes FR, Jjingo D, da Silva CRM et al (2013) Transcriptional activity, chromosomal distribution and expression effects of transposable elements in *Coffea* genomes. *PLoS One* 8:e78931
Lubbers EL, Chee PW (2009) The worldwide gene pool of *G. hirsutum* and its improvement. In: Paterson AH (ed) Genetics and Genomics of Cotton. Springer US, New York, NY, pp 23–52
Lukens LN, Pires JC, Leon E et al (2006) Patterns of sequence loss and cytosine methylation within a population of newly resynthesized *Brassica napus* allopolyploids. *Plant Physiol* 140:336–348
Ma Z, He S, Wang X et al (2018) Resequencing a core collection of upland cotton identifies genomic variation and loci influencing fiber quality and yield. *Nat Genet* 50:803–813
Madlung A (2013) Polyploidy and its effect on evolutionary success: old questions revisited with new tools. *Heredity* 110:99–104
Madlung A, Tyagi AP, Watson B et al (2005) Genomic changes in synthetic *Arabidopsis* polyploids. *Plant J* 41:221–230
Mammadov J, Buyyarapu R, Guttikonda SK et al (2018) Wild relatives of maize, rice, cotton, and soybean: treasure troves for tolerance to biotic and abiotic stresses. *Front Plant Sci* 9:886
May OL (2001) Registration of PD 94045 germplasm line of Upland cotton. *Crop Sci* 41:279–280
McClintock B (1984) The significance of responses of the genome to challenge. *Science* 226:792–801
McFadden H, Beasley D, Brubaker CL (2004) Assessment of *Gossypium sturtianum* and *G. australe* as potential sources of Fusarium wilt resistance to cotton. *Euphytica* 138:61–72
McIntyre PJ (2012) Polyploidy associated with altered and broader ecological niches in the *Claytonia perfoliata* (Portulacaceae) species complex. *Am J Bot* 99:655–662
Mehetre SS (2010) Wild *Gossypium anomalum*: a unique source of fibre fineness and strength. *Curr Sci* 99:58–71
Menzel MY, Brown MS (1954) The significance of multivalent formation in three-species *Gossypium* hybrids. *Genetics* 39:546–557
Meredith WR (1991) Contributions of introductions to cotton improvement. In: Use of plant introductions in cultivar development part 1. Crop Science Society of America, Madison, pp 127–146
Meshram LD (1994) Development of male sterile system from various sources in cotton (*Gossypium* spp.). *PKV Res J* 18:83–86
Meyer VG (1975) Male sterility from *Gossypium harknessii*. *J Hered* 66:23–27
Meyer JR, Meyer VG (1961) Origin and inheritance of nectariless cotton. *Crop Sci* 1:167
Müntzing A (1936) The evolutionary significance of autopolyploidy. *Hereditas* 21:363–378
Muravenko OV, Fedotov AR, Punina EO et al (1998) Comparison of chromosome BrdU-Hoechst-Giemsa banding patterns of the A1 and (AD)2 genomes of cotton. *Genome* 41:616–625
Nazeer W, Ahmad S, Mahmood K et al (2014) Introgression of genes for cotton leaf curl virus resistance and increased fiber strength from *Gossypium stocksii* into upland cotton (*G. hirsutum*). *Genet Mol Res* 13:1133–1143
Ni Z, Kim E-D, Ha M et al (2009) Altered circadian rhythms regulate growth vigour in hybrids and allopolyploids. *Nature* 457:327–331
Nixon BT, Mansouri K, Singh A et al (2016) Comparative structural and computational analysis supports eighteen cellulose synthases in the plant cellulose synthesis complex. *Sci Rep* 6:28696

Oluoch G, Zheng J, Wang X et al (2016) QTL mapping for salt tolerance at seedling stage in the interspecific cross of *Gossypium tomentosum* with *Gossypium hirsutum*. *Euphytica* 209:223–235

Osabe K, Clement JD, Bedon F et al (2014) Genetic and DNA methylation changes in cotton (*Gossypium*) genotypes and tissues. *PLoS One* 9:e86049

Otto SP, Whitton J (2000) Polyploid incidence and evolution. *Annu Rev Genet* 34:401–437

Ozkan H, Levy AA, Feldman M (2001) Allopolyploidy-induced rapid genome evolution in the wheat (*Aegilops-Triticum*) group. *Plant Cell* 13:1735–1747

Page JT, Huynh MD, Liechty ZS et al (2013) Insights into the evolution of cotton diploids and polyploids from whole-genome re-sequencing. *G3* 3:1809–1818

Page JT, Liechty ZS, Alexander RH et al (2016) DNA sequence evolution and rare homoeologous conversion in tetraploid cotton. *PLoS Genet* 12:e1006012

Pang J, Zhu Y, Li Q et al (2013) Development of *Agrobacterium*-mediated virus-induced gene silencing and performance evaluation of four marker genes in *Gossypium barbadense*. *PLoS One* 8:e73211

Parisod C, Senerchia N (2012) Responses of transposable elements to polyploidy. In: Grandbastien M-A, Casacuberta JM (eds) Plant transposable elements: impact on genome structure and function. Springer, Berlin, Heidelberg, pp 147–168

Paterson AH, Bowers JE, Burow MD et al (2000) Comparative genomics of plant chromosomes. *Plant Cell* 12:1523

Paterson AH, Saranga Y, Menz M et al (2003) QTL analysis of genotype × environment interactions affecting cotton fiber quality. *Theor Appl Genet* 106:384–396

Paterson AH, Wendel JF, Gundlach H et al (2012) Repeated polyploidization of *Gossypium genomes* and the evolution of spinnable cotton fibres. *Nature* 492:423–427

Paun O, Fay MF, Soltis DE, Chase MW (2007) Genetic and epigenetic alterations after hybridization and genome doubling. *Taxon* 56:649–656

Percy RG, Wendel JF (1990) Allozyme evidence for the origin and diversification of *Gossypium barbadense* L. *Theor Appl Genet* 79:529–542

Phillips LL (1964) Segregation in new allopolyploids of *Gossypium*. V. multivalent formation in New World x Asiatic and New World x Wild American hexaploids. *Am J Bot* 51:324–329

Phillips LL (1966) The cytology and phylogenetics of the diploid species of *Gossypium*. *Am J Bot* 53:328–335

Phillips LL (1977) Interspecific incompatibility in *Gossypium*. Temperature-conditional lethality in hybrids of G. klotzschianum. *Am J Bot* 64:914–915

Phillips LL, Clement D (1967) Variation in the diploid *Gossypium* species of Baja, California. *Madrono* 19:137–147

Phillips LL, Strickland MA (1966) The cytology of a hybrid between *G. hirsutum* and *G. longicalyx*. *Can J Genet Cytol* 8:91–95

Phuphathanaphong L (2006) *Thepparatia* (Malvaceae), a new genus from. *Thailand* 1:195–200

Piednoël M, Carrete-Vega G, Renner SS (2013) Characterization of the LTR retrotransposon repertoire of a plant clade of six diploid and one tetraploid species. *Plant J* 75:699–709

Pinto de Menezes IP, Barroso PAV, Hoffmann LV et al (2010) Genetic diversity of mocó cotton (*Gossypium hirsutum* race marie-galante) from the northeast of Brazil: implications for conservation. *Botany* 88:765–773

Piperno DR, Pearsall DM (1998) Background of tropical agricultural origins. Academic, San Diego

Qin H, Guo W, Zhang Y-M, Zhang T (2008) QTL mapping of yield and fiber traits based on a four-way cross population in *Gossypium hirsutum* L. *Theor Appl Genet* 117:883–894

Rajendran TP, Venugopalan MV, Praharaj CS (2005) Cotton research towards sufficiency to Indian textile industry. *Indian J Agric Sci* 75:699–708

Ram SG, Thiruvengadam V, Ramakrishnan SH, Kannan Bapu JR (2008) Investigation on pre-zygotic barriers in the interspecific crosses involving *Gossypium barbadense* and four diploid wild species. *Euphytica* 159:241–248

Rambani A, Page JT, Udall JA (2014) Polyploidy and the petal transcriptome of *Gossypium*. *BMC Plant Biol* 14:3

Rapp RA, Udall JA, Wendel JF (2009) Genomic expression dominance in allopolyploids. *BMC Biol* 7:18

Reddy UK, Nimmakayala P, Abburi VL et al (2017) Genome-wide divergence, haplotype distribution and population demographic histories for *Gossypium hirsutum* and *Gossypium barbadense* as revealed by genome-anchored SNPs. *Sci Rep* 7:41285

Reinisch AJ, Dong JM, Brubaker CL et al (1994) A detailed RFLP map of cotton, *Gossypium hirsutum* x *Gossypium barbadense*: chromosome organization and evolution in a disomic polyploid genome. *Genetics* 138:829–847

Renny-Byfield S, Gong L, Gallagher JP, Wendel JF (2015) Persistence of subgenomes in paleopolyploid cotton after 60 my of evolution. *Mol Biol Evol* 32:1063–1071

Renny-Byfield S, Page JT, Udall JA et al (2016) Independent domestication of two Old World cotton species. *Genome Biol Evol* 8:1940–1947

Robinson AF, Bell AA, Dighe ND et al (2007) Introgression of resistance to nematode *Rotylenchulus reniformis* into upland cotton (*Gossypium hirsutum*) from *Gossypium longicalyx*. *Crop Sci* 47:1865–1877

Rombolá-Caldentey B, Rueda-Romero P, Iglesias-Fernández R et al (2014) *Arabidopsis* DELLA and two HD-ZIP transcription factors regulate GA signaling in the epidermis through the L1 box *cis*-element. *Plant Cell* 26:2905–2919

Rong J, Feltus FA, Waghmare VN et al (2007) Meta-analysis of polyploid cotton QTL shows unequal contributions of subgenomes to a complex network of genes and gene clusters implicated in lint fiber development. *Genetics* 176:2577–2588

Rossen J, Dillehay TD, Ugent D (1996) Ancient cultigens or modern intrusions?: evaluating plant remains in an Andean case study. *J Archaeol Sci* 23:391–407

Ruan Y-L, Llewellyn DJ, Furbank RT (2003) Suppression of sucrose synthase gene expression represses cotton fiber cell initiation, elongation, and seed development. *Plant Cell* 15:952–964

Saha S, Jenkins JN, Wu J et al (2006) Effects of chromosome-specific introgression in upland cotton on fiber and agronomic traits. *Genetics* 172:1927–1938

Said JI, Song M, Wang H et al (2015) A comparative meta-analysis of QTL between intraspecific *Gossypium hirsutum* and interspecific *G. hirsutum* × *G. barbadense* populations. *Mol Genet Genomics* 290:1003–1025

Salmon A, Flagel L, Ying B et al (2010) Homoeologous nonreciprocal recombination in polyploid cotton. *New Phytol* 186:123–134

Sarilar V, Marmagne A, Brabant P et al (2011) *BraSto*, a *Stowaway* MITE from *Brassica*: recently active copies preferentially accumulate in the gene space. *Plant Mol Biol* 77:59–75

Sarilar V, Palacios PM, Rousselet A et al (2013) Allopolyploidy has a moderate impact on restructuring at three contrasting transposable element insertion sites in resynthesized *Brassica napus* allotetraploids. *New Phytol* 198:593–604

Sattler MC, Carvalho CR, Clarindo WR (2016) The polyploidy and its key role in plant breeding. *Planta* 243:281–296

Saunders JH (1961) The wild species of *Gossypium* and their evolutionary history. Oxford University Press, New York

Schnable PS, Ware D, Fulton RS et al (2009) The B73 maize genome: complexity, diversity, and dynamics. *Science* 326:1112–1115

Schnable JC, Springer NM, Freeling M (2011) Differentiation of the maize subgenomes by genome dominance and both ancient and ongoing gene loss. Proc Natl Acad Sci U S A 108:4069–4074

Seelanan T, Brubaker CL, James McD. Stewart, et al (1999) Molecular systematics of Australian *Gossypium* section *Grandicalyx* (Malvaceae). *Syst Bot* 24:183–208

Senchina DS, Alvarez I, Cronn RC et al (2003) Rate variation among nuclear genes and the age of polyploidy in *Gossypium*. *Mol Biol Evol* 20:633–643

Senerchia N, Felber F, Parisod C (2014) Contrasting evolutionary trajectories of multiple retrotransposons following independent allopolyploidy in wild wheats. *New Phytol* 202:975–985

Shaked H, Kashkush K, Ozkan H et al (2001) Sequence elimination and cytosine methylation are rapid and reproducible responses of the genome to wide hybridization and allopolyploidy in wheat. *Plant Cell* 13:1749–1759

Shan C-M, Shangguan X-X, Zhao B et al (2014) Control of cotton fibre elongation by a homeodomain transcription factor *GhHOX3*. *Nat Commun* 5:5519

Shangguan X-X, Xu B, Yu Z-X et al (2008) Promoter of a cotton fibre MYB gene functional in trichomes of *Arabidopsis* and glandular trichomes of tobacco. *J Exp Bot* 59:3533–3542

Shepherd RL (1974) Breeding root-knot-resistant *Gossypium hirsutum* L. Using a resistant wild *G. barbadense* L.1. *Crop Sci* 14:687–691

Shi Y-H, Zhu S-W, Mao X-Z et al (2006) Transcriptome profiling, molecular biological, and physiological studies reveal a major role for ethylene in cotton fiber cell elongation. *Plant Cell* 18:651–664

Singh B, Cheek HD, Haigler CH (2009) A synthetic auxin (NAA) suppresses secondary wall cellulose synthesis and enhances elongation in cultured cotton fiber. *Plant Cell Rep* 28:1023–1032

Skovsted A (1933) Cytological studies in cotton I. The mitosis and the meiosis in diploid and triploid Asiatic cotton. *Ann Bot* os-47:227–251

Skovsted A (1937) Cytological studies in cotton. *J Genet* 34:97–134

Small RL, Wendel JF (1999) The mitochondrial genome of allotetraploid cotton (*Gossypium* L.). *J Hered* 90:251–253

Small RL, Wendel JF (2000) Phylogeny, duplication, and intraspecific variation of *Adh* sequences in New World diploid cottons (*Gossypium* l., malvaceae). *Mol Phylogenet Evol* 16:73–84

Soltis DE, Soltis PS (1999) Polyploidy: recurrent formation and genome evolution. Trends Ecol Evol 14:348–352

Soltis PS, Soltis DE (2016) Ancient WGD events as drivers of key innovations in angiosperms. *Curr Opin Plant Biol* 30:159–165

Soltis DE, Soltis PS, Rieseberg LH (1993) Molecular data and the dynamic nature of polyploidy. *CRC Crit Rev Plant Sci* 12:243–273

Soltis DE, Soltis PS, Tate JA (2004) Advances in the study of polyploidy since plant speciation. *New Phytol*

Soltis DE, Albert VA, Leebens-Mack J et al (2009) Polyploidy and angiosperm diversification. *Am J Bot* 96:336–348

Soltis DE, Visger CJ, Blaine Marchant D, Soltis PS (2016) Polyploidy: pitfalls and paths to a paradigm. *Am J Bot* 103:1146–1166

Song Q, Chen ZJ (2015) Epigenetic and developmental regulation in plant polyploids. *Curr Opin Plant Biol* 24:101–109

Song K, Lu P, Tang K, Osborn TC (1995) Rapid genome change in synthetic polyploids of *Brassica* and its implications for polyploid evolution. *Proc Natl Acad Sci U S A* 92:7719–7723

Song Q, Guan X, Chen ZJ (2015) Dynamic roles for small RNAs and DNA methylation during ovule and fiber development in allotetraploid cotton. *PLoS Genet* 11:e1005724

Song Q, Zhang T, Stelly DM, Chen ZJ (2017) Epigenomic and functional analyses reveal roles of epialleles in the loss of photoperiod sensitivity during domestication of allotetraploid cottons. *Genome Biol* 18:99

Splitstoser JC, Dillehay TD, Wouters J, Claro A (2016) Early pre-Hispanic use of indigo blue in Peru. *Sci Adv* 2:e1501623

Springer NM, Lisch D, Li Q (2016) Creating order from chaos: epigenome dynamics in plants with complex genomes. *Plant Cell* 28:314–325

Stanton MA, Stewart JM, Pervical AE, Wendel JF (1994) Morphological diversity and relationships in the A-Genome cottons, *Gossypium arboreum* and *G. herbaceum*. Crop Sci 34:519–527

Stebbins GL (1950) Variation and evolution in plants. New York, Columbia University Press. 643 pp.

Stebbins GL (1985) Polyploidy, hybridization, and the invasion of new habitats. *Ann Mo Bot Gard* 72:824–832

Stebbins GL et al (1971) Chromosomal evolution in higher plants. In: Chromosomal evolution in higher plants

Stephens SG (1944) The genetic organization of leaf-shape development in the genus *Gossypium*. *J Genet* 46:28–51

Stephens SG (1946) The genetics of corky; the New World alleles and their possible role as an interspecific isolating mechanism. *J Genet* 47:150–161

Stephens SG (1947) Cytogenetics of *Gossypium* and the problem of the origin of New World cottons. In: Demerec M (ed) Advances in genetics. Academic, New York, pp 431–442
Stephens SG (1949) The cytogenetics of speciation in *Gossypium*. I. Selective elimination of the donor parent genotype in interspecific backcrosses. *Genetics* 34:627–637
Stephens SG (1950) The internal mechanism of speciation in *Gossypium*. *Bot Rev* 16:115–149
Stephens SG (1958) Salt water tolerance of seeds of *Gossypium* species as a possible factor in seed dispersal. *Am Nat* 92:83–92
Stephens SG (1966) The potentiality for long range oceanic dispersal of cotton seeds. *Am Nat* 100:199–210
Stephens SG (1967) Evolution under domestication of the New World cotton (*Gossypium* spp.). *Cienciae Cultura* 19:118–134
Stephens SG (1974) Geographic and taxonomic distribution of anthocyanin genes in New World cottons. *J Genet* 61:128
Stewart JM (1992) A new cytoplasmic male sterile and restorer for cotton. In: Proceedings of the Beltwide cotton conference. National Cotton Council, Memphis. p 610
Stewart J (1995) Potential for crop improvement with exotic germplasm and genetic engineering. In: Proceeding of the world cotton research conference-I, Brisbane, Australia, February 14–17, Melbourne. pp 313–327
Sun Y, Wu Y, Yang C et al (2017) Segmental allotetraploidy generates extensive homoeologous expression rewiring and phenotypic diversity at the population level in rice. *Mol Ecol* 26:5451–5466
Suzuki H, Yu J, Ness SA et al (2013) RNA editing events in mitochondrial genes by ultra-deep sequencing methods: a comparison of cytoplasmic male sterile, fertile and restored genotypes in cotton. *Mol Gen Genomics* 288:445–457
Tan J, Walford S-A, Dennis ES, Llewellyn D (2016) Trichomes control flower bud shape by linking together young petals. *Nat Plants* 2:16093
Tang H, Woodhouse MR, Cheng F et al (2012) Altered patterns of fractionation and exon deletions in *Brassica rapa* support a two-step model of paleohexaploidy. *Genetics* 190:1563–1574
Tang M, Chen Z, Grover CE et al (2015) Rapid evolutionary divergence of *Gossypium barbadense* and *G. hirsutum* mitochondrial genomes. *BMC Genomics* 16:770
Tate JA, Ni Z, Scheen A-C et al (2006) Evolution and expression of homeologous loci in *Tragopogon miscellus* (Asteraceae), a recent and reciprocally formed allopolyploid. *Genetics* 173:1599–1611
Tu L-L, Zhang X-L, Liang S-G et al (2007) Genes expression analyses of sea-island cotton (*Gossypium barbadense* L.) during fiber development. *Plant Cell Rep* 26:1309–1320
Tuttle JR, Nah G, Duke MV et al (2015) Metabolomic and transcriptomic insights into how cotton fiber transitions to secondary wall synthesis, represses lignification, and prolongs elongation. *BMC Genomics* 16:477
Tyagi P, Gore MA, Bowman DT et al (2014) Genetic diversity and population structure in the US Upland cotton (*Gossypium hirsutum* L.). *Theor Appl Genet* 127:283–295
Udall JA, Long E, Ramaraj T, et al (2019) The genome sequence of *Gossypioides kirkii* illustrates a descending dysploidy in plants. *Frontiers in Plant Science* 10:1541
Van Deynze A, Stoffel K, Lee M et al (2009) Sampling nucleotide diversity in cotton. *BMC Plant Biol* 9:125
Vicient CM, Casacuberta JM (2017) Impact of transposable elements on polyploid plant genomes. *Ann Bot* 120:195–207
Vicient CM, Suoniemi A, Anamthawat-Jónsson K et al (1999) Retrotransposon *BARE-1* and its role in genome evolution in the genus *Hordeum*. *Plant Cell* 11:1769–1784
Vollesen K (1987) The native species of *Gossypium* (Malvaceae) in Africa, Arabia and Pakistan. *Kew Bull* 42:337–349
Walford S-A, Wu Y, Llewellyn DJ, Dennis ES (2011) GhMYB25-like: a key factor in early cotton fibre development. *Plant J* 65:785–797
Walker GP, Natwick ET (2006) Resistance to silverleaf whitefly, *Bemisia argentifolii* (Hem., Aleyrodidae), in *Gossypium thurberi*, a wild cotton species. *J Appl Entomol* 130:429–436
Wan Q, Guan X, Yang N et al (2016) Small interfering RNAs from bidirectional transcripts of *GhMML3_A12* regulate cotton fiber development. *New Phytol* 210:1298–1310

Wang GL, Dong JM, Paterson AH (1995) The distribution of *Gossypium hirsutum* chromatin in *G. barbadense* germ plasm: molecular analysis of introgressive plant breeding. *Theor Appl Genet* 91:1153–1161
Wang B, Brubaker CL, Burdon JJ (2004) Fusarium species and Fusarium wilt pathogens associated with native *Gossypium* populations in Australia. *Mycol Res* 108:35–44
Wang J, Wang H-Y, Zhao P-M et al (2010) Overexpression of a profilin (*GhPFN2*) promotes the progression of developmental phases in cotton fibers. *Plant Cell Physiol* 51:1276–1290
Wang F, Gong Y, Zhang C et al (2011a) Genetic effects of introgression genomic components from Sea Island cotton (*Gossypium barbadense* L.) on fiber related traits in upland cotton (*G. hirsutum* L.). *Euphytica* 181:41–53
Wang XQ, Feng CH, Lin ZX, Zhang XL (2011b) Genetic diversity of sea-island cotton (*Gossypium barbadense*) revealed by mapped SSRs. *Genet Mol Res* 10:3620–3631
Wang X, Wang H, Wang J et al (2011c) The genome of the mesopolyploid crop species *Brassica rapa*. *Nat Genet* 43:1035–1039
Wang K, Wang Z, Li F et al (2012) The draft genome of a diploid cotton *Gossypium raimondii*. *Nat Genet* 44:1098–1103
Wang M, Yuan D, Tu L et al (2015a) Long noncoding RNAs and their proposed functions in fibre development of cotton (*Gossypium* spp.). *New Phytol* 207:1181–1197
Wang M, Yuan D, Zhang X (2015b) Genome Sequencing. In: Fang DD, Percy RG (eds) Cotton. American Society of Agronomy, Inc.; Crop Science Society of America, Inc.; Soil Science Society of America, Inc., 5585 Guilford Road, Madison, WI 53711-1086 USA, pp 289–302
Wang B, Zhang M, Fu R et al (2016a) Epigenetic mechanisms of salt tolerance and heterosis in Upland cotton (*Gossypium hirsutum* L.) revealed by methylation-sensitive amplified polymorphism analysis. *Euphytica* 208:477–491
Wang X, Guo H, Wang J et al (2016b) Comparative genomic de-convolution of the cotton genome revealed a decaploid ancestor and widespread chromosomal fractionation. *New Phytol* 209:1252–1263
Wang X, Zhang H, Li Y et al (2016c) Transcriptome asymmetry in synthetic and natural allotetraploid wheats, revealed by RNA-sequencing. *New Phytol* 209:1264–1277
Wang M, Tu L, Lin M et al (2017) Asymmetric subgenome selection and cis-regulatory divergence during cotton domestication. *Nat Genet* 49:579–587
Wang K, Wendel JF, Jinping HUA (2018a) Designations for individual genomes and chromosomes in *Gossypium*. *J Cotton Res* 1:3
Wang P, Zhang J, Sun L et al (2018b) High efficient multisites genome editing in allotetraploid cotton (*Gossypium hirsutum*) using CRISPR/Cas9 system. *Plant Biotechnol J* 16:137–150
Wang M, Tu L, Yuan D et al (2019) Reference genome sequences of two cultivated allotetraploid cottons, *Gossypium hirsutum* and *Gossypium barbadense*. *Nat Genet* 51:224–229
Ware JO (1951) Origin, rise and development of American Upland cotton varieties and their status at present. University of Arkansas College of Agriculture, Agricultural Experiment Station, Fayetteville
Watt SG (1907) The wild and cultivated cotton plants of the world: a revision of the genus *Gossypium*, framed primarily with the object of aiding planters and investigators who may contemplate the systematic improvement of the cotton staple. Longmans, Green, and Co.
Wayne Smith C, Tom Cothren J (1999) Cotton: origin, history, technology, and production. Wiley, New York
Wei Y, Xu Y, Lu P et al (2017) Salt stress responsiveness of a wild cotton species (*Gossypium klotzschianum*) based on transcriptomic analysis. *PLoS One* 12:e0178313
Wendel JF (1989) New World tetraploid cottons contain Old World cytoplasm. *Proc Natl Acad Sci U S A* 86:4132–4136
Wendel JF (2015) The wondrous cycles of polyploidy in plants. *Am J Bot* 102:1753–1756
Wendel JF, Albert VA (1992) Phylogenetics of the cotton genus (*Gossypium*): character-state weighted parsimony analysis of chloroplast-DNA restriction site data and its systematic and biogeographic implications. *Syst Bot* 17:115–143
Wendel JF, Cronn RC (2003) Polyploidy and the evolutionary history of cotton. *Adv Agron* 78:139–186

Wendel JF, Grover CE (2015) Taxonomy and evolution of the cotton genus, *Gossypium*. In: Agronomy monograph

Wendel JF, Percival AE (1990) Molecular divergence in the Galapagos Islands—Baja, California species pair, *Gossypium klotzschianum* and *G. davidsonii* (Malvaceae). *Plant Syst Evol* 171:99–115

Wendel JF, Olson PD, Stewart JM (1989) Genetic diversity, introgression, and independent domestication of Old World cultivated cottons. *Am J Bot* 76:1795–1806

Wendel JF, Stewart JM, Rettig JH (1991) Molecular evidence for homoploid reticulate evolution among Australian species of *Gossypium*. *Evolution* 45:694–711

Wendel JF, Brubaker CL, Percival AE (1992) Genetic diversity in *Gossypium hirsutum* and the origin of Upland cotton. *Am J Bot* 79:1291–1310

Wendel JF, Rowley R, Stewart JM (1994) Genetic diversity in and phylogenetic relationships of the Brazilian endemic cotton, *Gossypium mustelinum* (Malvaceae). *Plant Syst Evol* 192:49–59

Wendel JF, Schnabel A, Seelanan T (1995a) An unusual ribosomal DNA sequence from *Gossypium gossypioides* reveals ancient, cryptic, intergenomic introgression. *Mol Phylogenet Evol* 4:298–313

Wendel JF, Schnabel A, Seelanan T (1995b) Bidirectional interlocus concerted evolution following allopolyploid speciation in cotton (*Gossypium*). *Proc Natl Acad Sci U S A* 92:280–284

Wendel JF, Cronn RC, Alvarez I et al (2002) Intron size and genome size in plants. *Mol Biol Evol* 19:2346–2352

Wendel JF, Brubaker C, Alvarez I et al (2009) Evolution and natural history of the cotton genus. In: Paterson AH (ed) Genetics and genomics of cotton. Springer US, New York, pp 3–22

Wendel JF, Brubaker CL, Seelanan T (2010) The origin and evolution of *Gossypium*. In: Physiology of cotton, pp 1–18

Wendel JF, Flagel LE, Adams KL (2012) Jeans, genes, and genomes: cotton as a model for studying polyploidy. In: Polyploidy and genome evolution

Westengen OT, Huamán Z, Heun M (2005) Genetic diversity and geographic pattern in early South American cotton domestication. *Theor Appl Genet* 110:392–402

Wittkopp PJ, Haerum BK, Clark AG (2004) Evolutionary changes in cis and trans gene regulation. *Nature* 430:85–88

Woodhouse MR, Schnable JC, Pedersen BS et al (2010) Following tetraploidy in maize, a short deletion mechanism removed genes preferentially from one of the two homeologs. *PLoS Biol* 8:e1000409

Woodhouse MR, Cheng F, Pires JC et al (2014) Origin, inheritance, and gene regulatory consequences of genome dominance in polyploids. *Proc Natl Acad Sci U S A* 111:5283–5288

Wu H, Tian Y, Wan Q et al (2018) Genetics and evolution of MIXTA genes regulating cotton lint fiber development. *New Phytol* 217:883–895

Xu Q, Xiong G, Li P et al (2012) Analysis of complete nucleotide sequences of 12 *Gossypium* chloroplast genomes: origin and evolution of allotetraploids. *PLoS One* 7:e37128

Xu C, Bai Y, Lin X et al (2014) Genome-wide disruption of gene expression in allopolyploids but not hybrids of rice subspecies. *Mol Biol Evol* 31:1066–1076

Yang SS, Samuel Yang S, Cheung F et al (2006) Accumulation of genome-specific transcripts, transcription factors and phytohormonal regulators during early stages of fiber cell development in allotetraploid cotton. *Plant J* 47:761–775

Yik CP, Birchfield W (1984) Resistant germplasm in *Gossypium* species and related plants to *Rotylenchulus reniformis*. *J Nematol* 16:146–153

Yoo M-J, Wendel JF (2014) Comparative evolutionary and developmental dynamics of the cotton (*Gossypium hirsutum*) fiber transcriptome. *PLoS Genetics* 10:e1004073

Yoo M-J, Szadkowski E, Wendel JF (2013) Homoeolog expression bias and expression level dominance in allopolyploid cotton. *Heredity* 110:171–180

Yoo M-J, Liu X, Pires JC et al (2014) Nonadditive gene expression in polyploids. *Annu Rev Genet* 48:485–517

Yuan D, Tang Z, Wang M et al (2015) The genome sequence of Sea-Island cotton (*Gossypium barbadense*) provides insights into the allopolyploidization and development of superior spinnable fibres. *Sci Rep* 5:17662

Yu-xiang W, Jin-hong C, Qiu-ling H, Shui-jin Z (2013) Parental origin and genomic evolution of tetraploid *Gossypium* species by molecular marker and GISH analyses. *Caryologia* 66:368–374
Zafar SA, Noor MA, Waqas MA et al (2018) Temperature extremes in cotton production and mitigation strategies. In: Mehboob-Ur-Rahman, Zafar Y (eds) Past, present and future trends in cotton breeding. InTech
Zhang T, Chen T (2012) Cotton pistil drip transformation method. *Methods Mol Biol* 847:237–243
Zhang T, Endrizzi JE (2015) Cytology and cytogenetics. In: Cotton. American Society of Agronomy, Inc., Crop Science Society of America, Inc., and Soil Science Society of America, Inc., Madison, pp 129–154
Zhang JF, Stewart JM (1999) Cytoplasmic male sterility based on *Gossypium sturtianum* cytoplasm (CMS-C1): Characterization and genetics of restoration. *Proc Cotton Res Meeting, Ark Agric Exp Stn Special Rep* 193:269–272
Zhang JF, Stewart JM (2001) Inheritance and genetic relationships of the D8 and D2-2 restorer genes for cotton cytoplasmic male sterility. *Crop Sci* 41:289–294
Zhang J, Lu Y, Adragna H, Hughs SE (2005) Genetic improvement of New Mexico Acala cotton germplasm and their genetic diversity. *Crop Sci* 45:2363–2373
Zhang J, Fang H, Zhou H et al (2013) Inheritance and transfer of thrips resistance from Pima cotton to Upland cotton. *J Cotton Sci* 17:163–169
Zhang J, Percy RG, McCarty JC (2014) Introgression genetics and breeding between Upland and Pima cotton: a review. *Euphytica* 198:1–12
Zhang T, Hu Y, Jiang W et al (2015) Sequencing of allotetraploid cotton (*Gossypium hirsutum* L. acc. TM-1) provides a resource for fiber improvement. *Nat Biotechnol* 33:531–537
Zhang F, Zhu G, Du L et al (2016) Genetic regulation of salt stress tolerance revealed by RNA-Seq in cotton diploid wild species, *Gossypium davidsonii*. *Sci Rep* 6:20582
Zhang H, Mittal N, Leamy LJ et al (2017a) Back into the wild-Apply untapped genetic diversity of wild relatives for crop improvement. *Evol Appl* 10:5–24
Zhang X, Hu D-P, Li Y et al (2017b) Developmental and hormonal regulation of fiber quality in two natural-colored cotton cultivars. *J Integr Agric* 16:1720–1729
Zhang Z, Ge X, Luo X et al (2018) Simultaneous editing of two copies of *Gh14-3-3d* confers enhanced transgene-clean plant defense against *Verticillium dahliae* in allotetraploid upland cotton. *Front Plant Sci* 9:842
Zhao X-P, Si Y, Hanson RE et al (1998) Dispersed repetitive DNA has spread to new genomes since polyploid formation in cotton. *Genome Res* 8:479–492
Zhao Y-L, Yu S-X, Ye W-W et al (2010) Study on DNA cytosine methylation of cotton (*Gossypium hirsutum* L.) genome and its implication for salt tolerance. *Agric Sci China* 9:783–791
Zhao F, Fang W, Xie D et al (2012) Proteomic identification of differentially expressed proteins in *Gossypium thurberi* inoculated with cotton *Verticillium dahliae*. *Plant Sci* 185-186:176–184
Zhao Y, Wang H, Chen W et al (2015) Genetic diversity and population structure of elite cotton (*Gossypium hirsutum* L.) germplasm revealed by SSR markers. *Plant Syst Evol* 301:327–336
Zhao X, Meng Z, Wang Y et al (2017) Pollen magnetofection for genetic modification with magnetic nanoparticles as gene carriers. *Nat Plants* 3:956–964
Zhao B, Cao J-F, Hu G-J et al (2018) Core cis-element variation confers subgenome-biased expression of a transcription factor that functions in cotton fiber elongation. *New Phytol* 218:1061–1075
Zheng D, Ye W, Song Q et al (2016) Histone modifications define expression bias of homoeologous genomes in allotetraploid cotton. *Plant Physiol* 172:1760–1771
Zhu SJ, Reddy N, Jiang YR (2005) Introgression of a gene for delayed pigment gland morphogenesis from *Gossypium bickii* into upland cotton. *Plant Breed* 124:590–594

Chapter 3
Chloroplast Genomics for Sustainable Cotton Production

Niaz Ahmad, Zhengyi Wei, Muhammad Sarwar Khan, and Brent L. Nielsen

Abbreviations

GM	Genetically modified
Indels	Insertions and deletions
IR	Inverted repeat
LSC	Large single copy
ptDNA	Plastid DNA
SNPs	Single nucleotide polymorphism
SSC	Small single copy
SSRs	Simple sequence repeats
TSP	Total soluble proteins

3.1 Introduction

Cotton is a leading fiber crop, known as "white gold" due to its economic importance, and is currently grown in 90 countries of the world. The economy of many developing countries is directly dependent upon cotton and its products. Out of 50 cotton species, only 4—*G. hirsutum* L., *G. barbadense* L., *G. herbaceum* L., and *G.*

N. Ahmad (✉)
Agricultural Biotechnology Division, National Institute for Biotechnology and Genetic Engineering (NIBGE), Faisalabad, Pakistan

Z. Wei
Institute of Agricultural Biotechnology, Jilin Academy of Agricultural Science, Changchun, Jilin Province, P.R. China

M. S. Khan
Centre of Agricultural Biochemistry and Biotechnology (CABB), University of Agriculture, Faisalabad, Pakistan

B. L. Nielsen
Department of Microbiology and Molecular Biology, Brigham Young University, Provo, UT, USA

Mehboob-ur-Rahman et al. (eds.), *Cotton Precision Breeding*,
https://doi.org/10.1007/978-3-030-64504-5_3

arboreum L.—are cultivated (Wendel and Cronn 2003). These species are highly susceptible to various biotic and abiotic stresses, which incur significant crop losses, particularly in developing counties. Several efforts have been made to develop high-yielding resistant varieties through breeding and genetic engineering tools (Chapman et al. 2001; Zhao et al. 2006; Shen et al. 2014; Chen et al. 2017a). The latter has been the most successful in introducing genes of different traits including insect resistance, even from evolutionarily distant taxa, which is otherwise impossible (Ahmad and Mukhtar 2017). For example, GM crops were planted on ~190 million hectares in the year 2018 (ISAAA 2018), an increase of 113-fold in area since their commercialization in 1995. Genetic engineering allows isolation, manipulation, and introduction of recombinant DNA molecules into plant hosts for a number of traits including herbicide tolerance, insect resistance, and drought/salt tolerance (Shen et al. 2014; Zhang et al. 2016, 2018; Chen et al. 2017a). However, these GM crops have been developed by manipulating nuclear genomes. Engineering of the nuclear genome faces several challenges, notably poor gene expression, silencing of transgenes due to interactions with nonallelic genes, and variable gene expression due to the Mendelian laws of inheritance. The most serious of these is linked with the natural property of plants in which they undergo gene transfer via the pollen to their relatives and vice versa (Stewart et al. 2003). Therefore, it is feared that transgenic plants may transfer their phenotype to non-transformed or weedy relatives. If this transfer involves fitness-enhancing traits such as herbicide tolerance, drought tolerance, or salt tolerance, then it may fuel the evolution of herbicide, drought, or salt-tolerant weeds. These so-called "superweeds" may then become difficult to control. Therefore, open-field cultivation of transgenic crops is viewed as deliberately arming the enemy, and consequently many countries have totally banned their cultivation (Ahmad and Mukhtar 2017).

Higher plants contain two additional genomes in mitochondria and plastids. With the advancement of molecular techniques, two of these genomes, nuclear and chloroplastic, can be engineered for different applications. Chloroplasts are green plastids, which house an important biochemical reaction, photosynthesis. In addition, chloroplasts are also involved in many other vital cellular reactions such as amino acid synthesis, sucrose metabolism, N assimilation, and the production of a variety of metabolites. Engineering the chloroplast genome has emerged as an alternative platform to develop transgenic plants to effectively address the challenges faced by the conventional transgenesis approaches (Maliga and Bock 2011; Bock 2015; Ahmad et al. 2016).

Due to maternal inheritance in most field crops, transplastomic plants (transgenic plants developed by transforming the plastid genome) always produce a progeny with a uniform gene expression. Chloroplasts as cellular organelles form a distinct compartment within the cell; protein expression is therefore confined to this compartment, which helps prevent interference with other biochemical reactions in the cell. For example, even small amounts of cholera toxin β-subunit (CTB; 0.3% TSP) resulted in stunted plant growth when expressed in the nucleus (Arakawa et al. 1997), whereas the same CTB toxin had no observable drastic effect on plant growth when produced in tobacco chloroplasts, despite 14-times higher expression (Daniell et al. 2001). The insertion of a transgene into the nuclear genome is a random process and cannot be predetermined. If insertion occurs within an open reading frame encoding an essential protein, its reading frame would be disrupted, and consequently its function would be lost. On the other hand, transgene integration in chloroplasts takes place via evolutionary-conserved homologous recombination, which allows precise transgene insertion at a chosen location in the chloroplast genome without disturbing the reading frames of the resident genes. In addition to these advantages, chloroplasts have the remarkable capacity to express transgenes at extraordinary levels. Several studies have reported transgene expression reaching as high as >70% TSP (Oey et al. 2009; Ruhlman et al. 2010). The high expression level is often beneficial in traits that need a high concentration of proteins such as insect resistance or drought tolerance. Therefore, the transformation of the chloroplast genome is viewed as a promising tool for developing transgenic cotton with improved agronomic traits of economic importance.

In addition to engineering agronomic traits and production of recombinant proteins, chloroplast transformation has been used to study the functions of different genes, biogenesis of plastids, plant homeostasis, and signaling mechanisms and to estimate evolutionary history of higher plants. The chloroplast genome exists in a high copy number (up to 10,000 copies per plant cell). This high copy number ensures the conserved organizational arrangement of the genome to be preserved, resulting in relatively low mutational rates compared to the nucleus. It has been estimated that the chloroplast genome evolves at an ~10× lower rate than the nuclear genome (Wolfe et al. 1987; Rousseau-Gueutin et al. 2018). The highly conserved order, low mutation recombination rate, and low level of nucleotide substitution rate make it an excellent tool for phylogenetic and comparative genome evolution studies (Twyford and Ness 2017). Plastid inheritance in most plant species is predominantly uniparental which makes it an ideal tool for studying gene flow and population differentiation (Li et al. 2014). In this chapter, we discuss the different features of the cotton chloroplast genome and highlight the potential of transforming the cotton chloroplast genome for sustaining cotton production.

3.2 The Chloroplast Genome

Like other plant species, the cotton chloroplast genome exhibits a quadripartite structure of 160,301 bp in length, which is divided into large single-copy (~88,816 bp) and small single-copy (~20,269 kb) regions by the presence of two inverted repeats ($IR_{a/b}$), each 25,608 bp long—the hallmark of higher plant chloroplast genomes (Lee et al. 2006). These IR regions are rich in ribosomal genes. It is worth mentioning that these inverted repeats have been lost in different lineages, particularly lower plants. Figure 3.1 shows a generalized map of the *Gossypium hirsutum* chloroplast genome. The genome contains 83 protein-coding genes, 9 of which are found in the repeats, 8 genes coding for rRNA, and 37 for tRNA molecules. There are 17 genes that contain introns. The gene order and the number of genes in the cotton chloroplast genome allocated for different cellular functions are highly

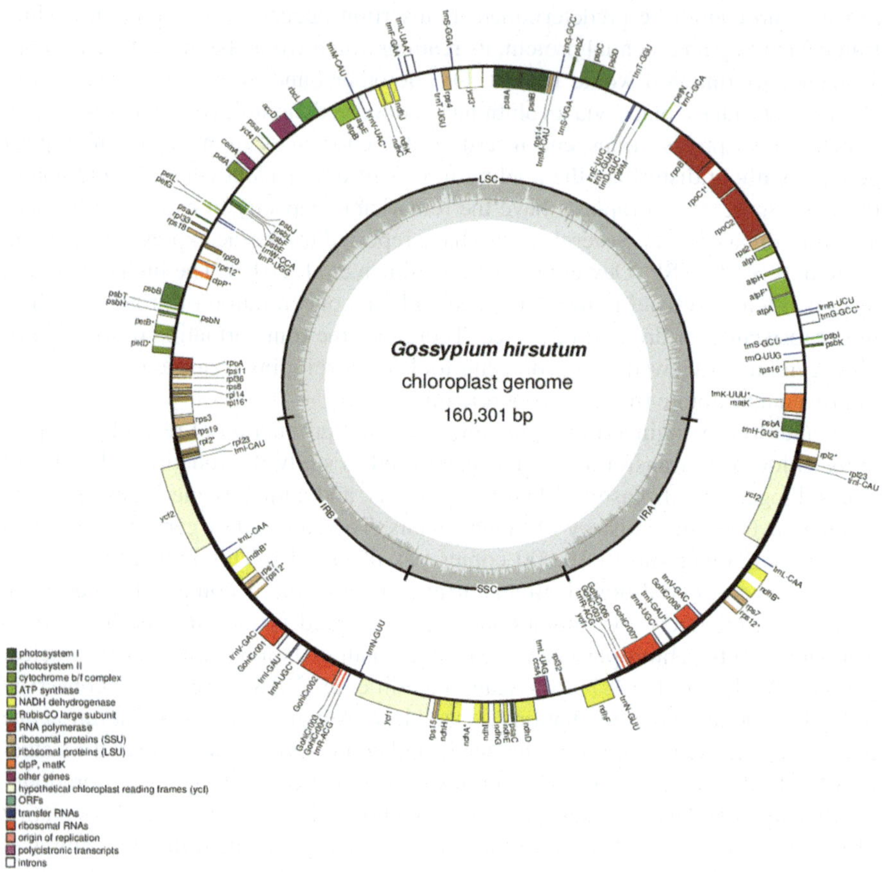

Fig. 3.1 Physical map of the *Gossypium hirsutum* chloroplast genome (NC_007944.1) showing the organization of different plastid genes. Genes on the inside of the circle are transcribed clockwise, whereas those on the outside are transcribed counterclockwise. The open reading frames of unknown function are shown by ycf plus a designation number. The inner circle shows the boundaries of LSC, SSC, and IRs, whereas the circular bar in gray shows the AT and GC content of the genome, while the darker vertical lines represent the GC content. The thin black circle line in GC graph marks the 50% level of GC content of the genome. Genes containing introns are marked by an asterisk (*). The map was constructed using a web-based tool, OGDRAW v1.3.1 (Greiner et al. 2019). Abbreviations: *LSC* large single copy, *SSC* small single copy, *IR* inverted repeat

similar to tobacco, but cotton lacks *rpl22* and *infA* genes (Lee et al. 2006) (Fig. 3.2). The "AT" content is 62.8%, and the coding region makes up 48.9% of the genome, with an average coding length of 944 bp/gene, and 2.1% of the genome is repetitive with an average repeat length of 50 bp. The non-protein-coding regions have been reported to have higher AT content (~68.5%) compared to protein-coding regions (Lee et al. 2006; Xu et al. 2012). The cotton chloroplast genome has been found to contain a number of dispersed direct and inverted repeats. Analyses of the chloroplast genome of various species show that the genome organization in *Gossypium* is highly conserved (Xu et al. 2012; Wu et al. 2018).

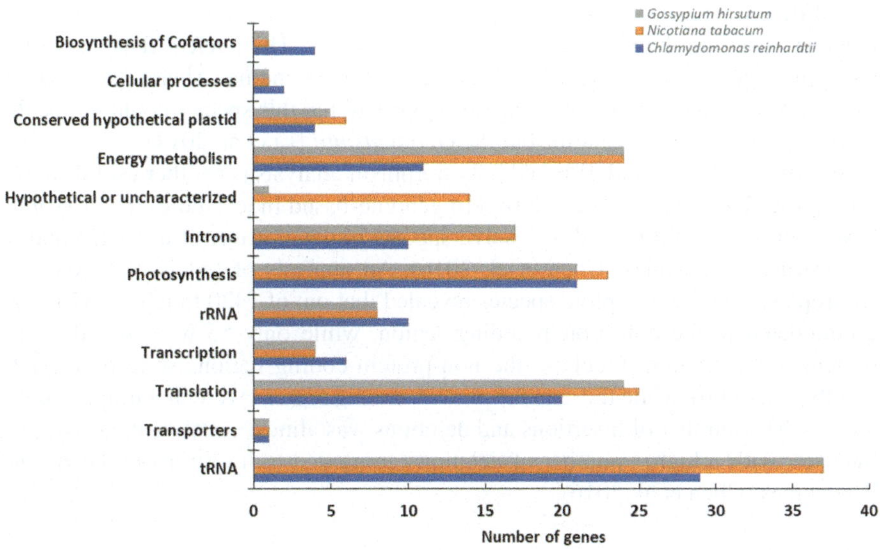

Fig. 3.2 Comparison of gene distribution of *Gossypium hirsutum, Chlamydomonas reinhardtii,* and *Nicotiana tabacum* chloroplast genomes for different cellular functions

It has been observed that the SC regions (LSC + SSC) of the cotton chloroplast genome harbor higher polymorphic content than the IR regions (Li et al. 2014). In one study, Li et al. (2014) found that all polymorphic markers were located in the noncoding regions, indicating that the intergenic regions and introns are more suitable for diversity studies using the cotton chloroplast genome. Another study showed that the diversity of the non-protein-coding region was 2–3× higher than that of the protein-coding region (Chen et al. 2016). In *Gossypium*, the non-protein-coding region contains around 112 intergenic regions and 17 introns. It is also worth mentioning that recombination of the chloroplast genome in *Gossypium* is restricted to the IR regions. Therefore, the four junctions (LSC-IR_a, LSC-IR_b, IR_b-SSC, SSC-IR_a) are fast evolving to the point that they are unsuitable for phylogenetic studies (Chen et al. 2016).

Chloroplast-based marker studies of the *Gossypium* genus have led to increased understanding of the genetic relatedness among different genome groups (Chen et al. 2017b). Simple sequence repeats (SSRs) are often used as molecular markers in genetics. SSRs usually have a higher rate of mutation compared to other neutral regions of DNA due to slipped strand mispairing (slippage) during DNA replication on a single DNA strand. A large number of SSR loci have been reported in the *Gossypium* chloroplast genome. The SSRs are in the non-protein-coding sequences of single-copy regions, which allow the reconstruction of phylogeny independent of the nuclear genome. Earlier studies suggest a rapid and early diversification of the cotton lineages, with many branch points unresolved. Both the entire chloroplast genome and individual genes have been a major source of cotton phylogenic inferences. For example, the analysis of 41 *Gossypium* accessions using SSR markers from the chloroplast genome predicted that *G. aridum* and *G. lobatum* were actually two different species (Li et al. 2014). These species previously were clustered together based on the nuclear ribosomal ITS regions. The nuclear data suggested *Gossypium gossypioides* as the basal species of the D-genome (Álvarez et al. 2005). The analysis of the chloroplast genome suggested that this species could not be the earlier species of the D-genome but the *G. davidsonii* (Li et al. 2014).

Insertion-deletion (Indel)-based polymorphism analysis is another useful source for phylogenetic studies. The chloroplast genome-based Indel studies in *Gossypium* have been successfully used to resolve species issues (Grover et al. 2015) and to reconstruct phylogenies (Chen et al. 2016). An analysis of Indels of *Gossypium* chloroplast genome of diploid species revealed that out of 1420 Indels, ~1370 were accumulated in the non-protein-coding region, while only 55 were found in the protein-coding region. Indels in the non-protein-coding regions were of variable length (1–272 bp), while those in the protein-coding regions were in multiples of the 3 bases. The number of insertions and deletions was almost equal in all *Gossypium* lineages, with a higher number of deletions compared to insertions in the B- and E-genomes (Chen et al. 2016).

3.3 Why Chloroplast Transformation in Cotton

The advent of recombinant DNA technology realized the idea to express insect toxins within the plants instead of spraying, which would protect the plant in a better and more consistent way and minimize the insecticidal sprays being released into the environment. This idea caught attention as the whole plant could be immunized from insect attack and it appeared environmentally friendly too. The idea first met its practical realization in the year 1987 when §-endotoxins of *Bacillus thuringiensis* (Bt) specifically lethal to Lepidopteran insects could be successfully produced in tobacco plants (Barton et al. 1987). While the initial attempts met with limited success, the first breakthrough was made by Monsanto by achieving transformation in a field crop, cotton (*G. hirsutum*) in 1987 (Umbeck et al. 1987). Since its commercialization in 1995, Bt cotton has been cultivated in many countries including Australia, China, and India, and the area under GE (genetically engineered) cotton is increasing every year. According to the International Service for the Acquisition of Agri-biotech Applications (ISAAA), GM cotton was grown in 15 countries on an area of ~325 million hectares in the year 2018 (ISAAA 2018). The adoption of Bt cotton is also benefiting the environment. For example, since its adoption (in 1995), the technology has saved ~470 million kg of active ingredient from being sprayed/released into the environment (Gilbert 2013).

Efforts to engineer insect resistance in plants by expressing insecticidal proteins in plants via the nucleus have met with several issues including poor and variable gene expression and evolution of resistance in insects to these toxins. For example, the first publication on the expression of *cry* genes, encoding insecticidal crystal proteins, reported 0.001% total soluble proteins (TSP) and could only be maximized to 0.2–1% TSP by using synthetic *cry* gene constructs (Jouanin et al. 1998). Variations in the concentration of Bt toxins within the plants as well as among different plant species have been observed (Knight et al. 2016; Khan et al. 2017;

Carrière et al. 2018), which are believed to accelerate evolution of resistance in insects, particularly those species which show low susceptibility (Carrière et al. 2018). Due to evolution of resistance in insects, researchers are trying to pyramid multiple genes in order to suppress the evolving resistance as well as engineering broad-spectrum resistance against insects. However, the durability and efficacy of such pyramids are reduced due to cross-contamination and cross-resistance (Ni et al. 2017). This surge in insect resistance is overshadowing the benefits of insect-resistant crops (Tabashnik and Carrière 2017). Despite the benefits, the use of GM crops has generated a controversy; consequently many countries have banned the open-field cultivation of GM crops (Baffes 2011). This controversy is the result of the use of conventional methods for engineering crops where foreign genes are inserted in the nuclear genome. Plants undergo introgression of their genetic material into their relatives and vice versa through the pollen. This is a natural phenomenon which provides organisms with genetic variations to adapt to their environments. If foreign genes are related to fitness-enhancing of the plants under certain environmental conditions, then the traits might be transferred to the non-transformed or weedy relatives. For example, transfer of genes conferring resistance to three different types of herbicides in weeds in Northern Alberta, Canada, was reported to be allegedly transferred from genetically modified canola (MacArthur 2000), making these weeds resistant to those herbicides. Containment of the transgenes has therefore become central to future applications of genetic engineering in agriculture.

Keeping in view the challenges that conventional genetic engineering approaches are facing, transgenic plants could be developed by introducing foreign genes into nonnuclear genomes such as chloroplasts, as their inheritance is predominantly maternal. Due to their maternal mode of inheritance, the ptDNA (plastid DNA) is not present in the pollen or is degraded before fertilization occurs, thereby providing an absolute natural gene containment system for the transgenes. Therefore, using this technology several issues like low-level gene expression, risk of transgene outflow to weedy relatives or non-transformed plants, and gene silencing could be avoided. As chloroplasts do not follow the Mendelian laws of inheritance, the progeny of a transplastomic plant line always shows a uniform gene expression level. In the case of nuclear transformation, the new gene may interact with the native nuclear genes, and as a result of this nonallelic interaction, the function of native genes could be masked or vice versa (Finnegan and McElroy 1994). For example, 50% of transgenic *Arabidopsis thaliana* plants carrying a hygromycin resistance transgene failed to exhibit the transgenic phenotype to their progeny (Scheid et al. 1991). Such potential challenges have not been observed when foreign genes were introduced into the chloroplast genome (Ahmad and Mukhtar 2017; Leister et al. 2017; Khan et al. 2019).

Studies have shown that transgenes inserted in the chloroplast genome often result in a high level of gene expression (Oey et al. 2009; Ruhlman et al. 2010; Michoux et al. 2011; Ahmad et al. 2012). For example, when an operon of bacterial origin consisting of *cry2Aa2* was expressed in tobacco chloroplasts, expression as high as 46.7% TSP was recorded (De Cosa et al. 2001). While the high AT content in Bt genes could be a potential cause of low expression when these genes are inserted into the nucleus, possibly due to the presence of splice sites, several fortuitous polyadenylation sites, ATTTA sequences, mRNA degradation signals, and the transcription termination sites (Perlak et al. 1991; Strizhov et al. 1996), expression of these genes directly inside the chloroplasts could be beneficial. For example, AT-rich *Clostridium tetani* sequences could not be expressed in *Saccharomyces cerevisiae* and resulted in the accumulation of several truncated mRNAs (Romanos et al. 1991). Expression was only possible when the codons were optimized to high GC content. On the other hand, both versions—with high "AT" or high "GC" content of the transgenes—could be successfully expressed in tobacco chloroplasts (Tregoning et al. 2003, 2005). The technology is increasingly being applied to field crops (Khan and Maliga 1999; Daniell et al. 2001; Ahmad et al. 2016; Ahmad and Mukhtar 2017).

Since their inclusion into a eukaryotic cell, chloroplasts have taken up a number of new roles while retaining many key prokaryotic features. One of these is the organization of genes in the form of operons and their transcription as polycistrons, allowing multiple transgenes to be expressed. Advances in genomics revealed that many agronomic traits are controlled by multiple genes (Khan 2015; Ahmad and Mukhtar 2017). Engineering such traits via the nucleus would require multiple transformation rounds, making it a lengthy and challenging task. For example, engineering the entire β-carotene biosynthetic pathway into rice to fortify rice endosperm with vitamin A took around 7 years (Ye et al. 2000). Transformation of chloroplasts has been employed to engineer different agronomic traits such as insect resistance, drought and salt tolerance, herbicide tolerance, and cytoplasmic male sterility in higher plants (Table 3.1). The recent demonstration of producing long intact RNA molecules in chloroplasts for controlling insect pests through RNA interference has opened up new ways of protecting crops from insect damage (Jin et al. 2015; Zhang et al. 2015). However, the potential of this technology has not been fully realized in field crops such as cotton (Kumar et al. 2004b). Transformation of the cotton chloroplast genome, therefore, holds a great potential for engineering agronomic traits of economic importance more effectively.

Table 3.1 List of agronomic traits engineered using chloroplast transformation technology

Trait	Gene of interest	Host plant	Expression level	Reference
Abiotic stress tolerance	*γ-TMT*	Tobacco	7.7% TLP	Jin and Daniell (2014)
Aphid/whitefly resistance	*Bgl-1*	Tobacco	ND	Jin et al. (2011)
Antiviral/antibacterial/phloem-feeding insects	*pta*	Tobacco	9.2% TSP	Jin et al. (2012)
Antiviral/antimicrobial	*RC101* *PG1*	Tobacco	32–38% TSP 17–26% TSP	Lee et al. (2011)
Cytoplasmic male sterility	*phaA*	Tobacco	ND	Ruiz and Daniell (2005)

Table 3.1 (continued)

Trait	Gene of interest	Host plant	Expression level	Reference
Bacterial/fungal resistance	*msi-99*	Tobacco	ND	DeGray et al. (2001)
	Chloroperoxidase	Tobacco	10–15 μg per mL of leaf extract	Ruhlman et al. (2014)
	msi-99	Tobacco	ND	Wang et al. (2015)
Drought tolerance	*tps1* (yeast)	Tobacco	ND	Lee et al. (2003)
	ArDH	Tobacco	ND	Khan et al. (2015)
Herbicide resistance	*aroA*	Tobacco	ND	Daniell et al. (1998)
	bar	Tobacco	ND	Iamtham and Day (2000)
	crtY	Tomato/tobacco	ND	Wurbs et al. (2007)
	dao	Tobacco	ND	Gisby et al. (2012)
Insect resistance	*cry1A(c)*	Tobacco	5% TSP	McBride et al. (1995)
	cry2Aa2	Tobacco	3% TSP	Kota (1999)
	cry2Aa2 operon	Tobacco	45.3% TSP	De Cosa et al. (2001)
	cry1Aa10	Oilseed rape	ND	Hou et al. (2003)
	cry1Ab	Soybean	ND	Dufourmantel et al. (2005)
	cry9Aa2	Tobacco	10% TSP	Chakrabarti et al. (2006)
	dsRNA of p450 monooxygenase, V-ATPase, and chitin synthase-coding genes	Tobacco (against *Helicoverpa armigera*)	ND	Jin et al. (2015)
	dsRNA of CPB, ACT, and *SHR* genes	Tobacco (against Colorado potato beetle (*Leptinotarsa decemlineata*)	ND	Zhang et al. (2015)
	cry1Ab	Cabbage	4.8 11.1% TSP	Liu et al. (2008)
Multiple biotic and abiotic stresses	Simultaneous expression of protease inhibitors and chitinase	Tobacco	ND	Chen et al. (2014)
Oxidative stress resistance	Trx m	Tobacco	ND	Rey et al. (2013)

(continued)

Table 3.1 (continued)

Trait	Gene of interest	Host plant	Expression level	Reference
Altered photosynthesis	Cyanobacterial Rubisco along with assembly factors RbcX, CcM35 replaced endogenous Rubisco Large subunit in tobacco chloroplasts	Tobacco	12–18% Rubisco of Wt level	Lin et al. (2014)
	Cyanobacterial Rubisco without RbcX or CcM35 replaced tobacco large subunit	Tobacco	10× lower Rubisco than Wt	Occhialini et al. (2016)
Phytoremediation	*merA*/*merB*	Tobacco	ND	Ruiz et al. (2003); Hussein et al. (2007)
	mt1	Tobacco	ND	Ruiz et al. (2011)
Salt tolerance	*badh*	Carrot	ND	Kumar et al. (2004a)
	ArDH	Tobacco	ND	Khan et al. (2015)

TLP total leaf protein, *TSP* total soluble proteins, *ND* not determined, *Wt* wild type

3.4 Sequencing of the Chloroplast Genome and Its Implications for Crop Improvement

As discussed earlier, DNA sequences of the chloroplast genome of cotton have provided many useful insights into phylogenetic relationships among different genome groups in the genus *Gossypium* (Cronn et al. 2002). The highly conserved gene order and reduced recombination make the genome of this organelle highly popular for designing universal primers to amplify homologous regions from evolutionary divergent species. Further, high copy number per cell allows easy access to the genome compared to single-copy nuclear loci. With advances in sequencing technologies, the chloroplast genomics in cotton is now moving away from genes to whole genomes for the reconstruction of phylogenies at lower taxonomic level, divergence different groups, as well as structural analysis of the chloroplast genomes (Xu et al. 2012; Chen et al. 2016, 2017b; Wu et al. 2018). Complete sequences of the chloroplast genome also facilitate studies on mechanisms of gene loss, gene transfer to the nucleus, and evolution in different genome lineages (Twyford and Ness 2017). Cotton species present an excellent model system for studying polyploidization. The usefulness of the chloroplast genome for evolutionary studies coupled with improved sequencing techniques has led to a surge in whole-genome sequencing. The first chloroplast genome of cotton (*G. hirsutum*) was sequenced by the Daniell Laboratory in the year 2006 followed by *G. barbadense* in the same year

(Ibrahim et al. 2006). After that, several studies have reported the complete chloroplast sequences of different species of genus *Gossypium*. To date, the NCBI organelle genome database contains 36 chloroplast genomes belonging to the *Gossypium* genus (Table 3.2). Advances in obtaining complete nucleotide sequences of

Table 3.2 List of cotton chloroplast genomes of genus *Gossypium* that have been fully sequenced

Species	Size (bp)	GC content (%)	Accession number	CDS	Genes	Released
Gossypium lobatum	160205	37.2741	NC_039569.1/MG891802.1	83	128	2018
Gossypium nandewarense	160205	37.1312	NC_039568.1/MG779276.1	86	131	2018
Gossypium schwendimanii	160205	37.2668	NC_039570.1/MG891803.1	85	130	2018
Gossypium aridum	160205	37.2645	NC_033396.1/KP170502.1	84	129	2017
Gossypium armourianum	160205	37.2904	NC_033400.1/KP221926.1	84	129	2017
Gossypium australe	160205	37.1561	NC_033401.1/KP221928.1	84	129	2017
Gossypium davidsonii	160205	37.3101	NC_033395.1/KP170501.1	84	129	2017
Gossypium harknessii	160205	37.2968	NC_033333.1/KP221927.1	84	129	2017
Gossypium klotzschianum	160205	37.3036	NC_033394.1/KP170500.1	84	129	2017
Gossypium nelsonii	160205	36.8011	NC_033399.1/KP221925.1	84	129	2017
Gossypium populifolium	160205	37.1961	NC_033398.1/KP221924.1	84	129	2017
Gossypium trilobum	160205	37.2902	NC_033397.1/KP170503.1	84	129	2017
Gossypium turneri	160205	37.2614	NC_026835.1/JQ742090.1	83	128	2015
Gossypium anomalum	160205	37.3256	NC_023213.1/JF317351.1	86	133	2014
Gossypium bickii	160205	37.1969	NC_023214.1/JF317352.1	87	132	2014
Gossypium herbaceum	160205	37.308	NC_023215.1/JF317353.1	86	132	2014
Gossypium longicalyx	160205	37.242	NC_023216.1/JF317354.1	86	133	2014
Gossypium stocksii	160205	37.4128	NC_023217.1/JF317355.1	86	132	2014
Gossypium sturtianum	160205	37.1334	NC_023218.1/JF317356.1	86	131	2014
Gossypium areysianum	160205	37.3687	NC_018112.1/JN019795.1	83	128	2012
Gossypium capitis-viridis	160205	37.3206	NC_018111.1/JN019794.1	83	128	2012
Gossypium incanum	160205	37.392	NC_018109.1/JN019792.1	83	128	2012
Gossypium robinsonii	160205	37.1601	NC_018113.1/JN019791.1	83	128	2012
Gossypium somalense	160205	37.3652	NC_018110.1/JN019793.1	83	128	2012
Gossypium gossypioides	160205	37.3121	NC_017894.1/HQ901195.1	83	128	2012
Gossypium raimondii	160205	37.3075	NC_016668.1/	83	128	2012
Gossypium arboreum	160205	37.2296	Pltd:NC_016712.1/	83	128	2012
Gossypium darwinii	160205	37.2339	NC_016670.1/HQ325741.1	83	128	2012
Gossypium mustelinum	160205	37.2209	NC_016711.1/HQ325743.1	83	128	2012
Gossypium tomentosum	160205	37.2043	NC_016690.1/HQ325745.1	83	128	2012
Gossypium herbaceum subsp. *africanum*	160205	37.218	NC_016692.1/HQ325742.1	83	128	2012
Gossypium raimondii	160205	37.3075	HQ325744.1	83	128	2012
Gossypium thurberi	160205	37.218	NC_015204.1/GU907100.1	85	134	2011
Gossypium barbadense	160205	37.2294	NC_008641.1/AP009123.1	84	129	2006
Gossypium hirsutum	160205	37.2449	Pltd:NC_007944.1/	83	129	2006

chloroplast genomes hold great implications for crop improvement using chloroplast transformation technology. In the next section, we will discuss how chloroplast genome sequencing could be useful for extending chloroplast transformation technology to cotton.

A large number of recombinant proteins have been produced using chloroplast transformation, the majority of which use tobacco chloroplasts (Ahmad et al. 2016). Although new crops such as tomato, petunia, potato, soybean, lettuce, and cabbage have been added, there is a need to include more species to fully exploit the potential of this technology. Transgene integration in the chloroplast genome takes place through homologous recombination—a hallmark feature of the technology for transgene integration at predetermined locations in the genome. The amplification of flanking regions from the wild-type chloroplast genome requires prior knowledge of the insertion region for the synthesis of primers. Expression vectors for chloroplast transformation are therefore constructed with two flanking regions from the plastome (chloroplast genome) in order to make a crossover event successful. The expression vectors made for a particular species cannot be used for the transformation of the chloroplast genome in other species, necessitating the construction of a separate set of expression vectors for every plant species. This is indeed a bottleneck in the expansion of chloroplast transformation technology to those crops whose plastome has not been sequenced yet. Several attempts were made to use universal transformation vectors to maximize the potential of chloroplast transformation; the transformation efficiency was, however, compromised (Ruhlman et al. 2010).

Keeping in view the economic importance of cotton as a crop, chloroplast transformation was attempted by the Daniell Laboratory in the year 2004, at the time when the chloroplast sequence for any species of cotton was not available. The chloroplast transformation vectors used in this report were constructed using sequence information of the tobacco chloroplast genome including gene expression regulatory elements such as promoters and terminators and untranslated regions. The construct carried a selectable marker gene, *aphA-6*, to detoxify kanamycin. The transformation efficiency was quite low. To increase the transformation efficiency, the authors added another selectable marker gene, *npt-II*, coding for the neomycin phosphotransferase, which also detoxifies kanamycin. The double gene/single selection (DGSS) strategy sufficiently increased transformation efficiency. However, this report did not engineer any agronomic trait but was rather limited to proof of concept and patenting the methodology for obtaining transplastomic plants from nongreen plastids using somatic embryogenesis. Although successful the study has not been reproduced by any laboratory or by the same laboratory for engineering agronomic traits in cotton for unknown reasons.

With the availability of entire chloroplast genome sequences, it should now be possible to use endogenous regulatory elements from the *Gossypium* chloroplast genome in a bid to increase transformation efficiency. We have started using chloroplast transformation technology to engineer insect resistance in cotton using the indigenous chloroplast gene expression regulatory elements as well as native cotton chloroplast DNA sequences for flanking regions (unpublished work).

3.5 Conclusions and Future Perspectives

Developing transgenic plants by transforming the chloroplast genome holds a great potential to address problems such as poor gene expression, variable gene expression, random integration of transgenes, and transgene outcrossing into untransformed plants via the pollen. However, this technology has not been yet extended to field crops. One of the major reasons is that the technology, due to its applied nature, is heavily patented. Apart from legal barriers, there are technical challenges that hinder the application of this technology to field crops. For example, the lengthy regeneration protocol to recover transplastomic cotton plant lines from nongreen plastids via somatic embryogenesis takes around 2 years. Resulting plants require an additional round of selection and regeneration to achieve homoplasmy, the state in which all chloroplasts and their genome copies contain the transgene. The absence of complete chloroplast genome sequences also hampered the progress of this technology in field crops. However, the pace with which sequencing technologies are advancing is likely to revolutionize chloroplast transformation technology. The current regeneration protocols for cotton are limited to nongreen plastids; therefore, there is a need to develop robust regeneration protocols for obtaining transgene cotton plants from green plastids via organogenesis.

Acknowledgments The authors thank the Higher Education Commission (HEC) and Punjab Agricultural Research Board (PARB) for funding their work.

References

Ahmad N, Mukhtar Z (2017) Genetic manipulations in crops: challenges and opportunities. *Genomics* 109:494–505

Ahmad N, Michoux F, McCarthy J, Nixon PJ (2012) Expression of the affinity tags, glutathione-*S*-transferase and maltose-binding protein, in tobacco chloroplasts. *Planta* 235:863–871

Ahmad N, Michoux F, Lossl AG, Nixon PJ (2016) Challenges and perspectives in commercializing plastid transformation technology. *J Exp Bot* 67:5945–5960

Álvarez I, Cronn R, Wendel JF (2005) Phylogeny of the New World diploid cottons (*Gossypium* L., *Malvaceae*) based on sequences of three low-copy nuclear genes. *Plant Syst Evol* 252:199–214

Arakawa T, Chong DKX, Lawrence Merritt J, Langridge WHR (1997) Expression of cholera toxin B subunit oligomers in transgenic potato plants. *Transgenic Res* 6:403–413

Baffes J (2011) Cotton, biotechnology, and economic development. Policy Research Working Paper. 5896. The World Bank, Washington, DC. http://wwwwds.worldbank.org/servlet/WDSContentServer/WDSP/IB/2011/12/05/000158349_20111205112253/Rendered/PDF/WPS5896.pdf

Barton KA, Whiteley HR, Yang NS (1987) *Bacillus thuringiensis* section sign-endotoxin expressed in transgenic *Nicotiana tabacum* provides resistance to Lepidopteran insects. *Plant Physiol* 85:1103–1109

Bock R (2015) Engineering plastid genomes: methods, tools, and applications in basic research and biotechnology. *Annu Rev Plant Biol* 66:211–241

Carrière Y, Degain BA, Unnithan GC, Harpold VS, Heuberger S, Li X, Tabashnik BE (2018) Effects of seasonal changes in cotton plants on the evolution of resistance to pyramided cotton producing the Bt toxins Cry1Ac and Cry1F in *Helicoverpa zea*. *Pest Manag Sci* 74:627–637

Chakrabarti SK, Lutz KA, Lertwiriyawong B, Svab Z, Maliga P (2006) Expression of the *cry9Aa2 Bt* gene in tobacco chloroplasts confers resistance to potato tuber moth. *Transgenic Res* 15:481–488

Chapman KD, Austin-Brown S, Sparace SA, Kinney AJ, Ripp KG, Pirtle IL, Pirtle RM (2001) Transgenic cotton plants with increased seed oleic acid content. *J Am Oil Chem Soc* 78:941–947

Chen PJ, Senthilkumar R, Jane WN, He Y, Tian Z, Yeh KW (2014) Transplastomic *Nicotiana benthamiana* plants expressing multiple defence genes encoding protease inhibitors and chitinase display broad-spectrum resistance against insects, pathogens and abiotic stresses. *Plant Biotechnol J* 12:503–515

Chen Z, Feng K, Grover CE, Li P, Liu F, Wang Y, Xu Q, Shang M, Zhou Z, Cai X, Wang X, Wendel JF, Wang K, Hua J (2016) Chloroplast DNA structural variation, phylogeny, and age of divergence among diploid cotton species. *PLoS One* 11:e0157183–e0157183

Chen W-b, Lu G-q, Cheng H-m, Liu C-x, Xiao Y-t, Xu C, Shen Z-c, Soberón M, Bravo A, Wu K-m (2017a) Transgenic cotton co-expressing chimeric Vip3AcAa and Cry1Ac confers effective protection against Cry1Ac-resistant cotton bollworm. *Transgenic Res* 26:763–774

Chen Z, Grover CE, Li P, Wang Y, Nie H, Zhao Y, Wang M, Liu F, Zhou Z, Wang X, Cai X, Wang K, Wendel JF, Hua J (2017b) Molecular evolution of the plastid genome during diversification of the cotton genus. *Mol Phylogenet Evol* 112:268–276

Cronn RC, Small RL, Haselkorn T, Wendel JF (2002) Rapid diversification of the cotton genus (*Gossypium*: Malvaceae) revealed by analysis of sixteen nuclear and chloroplast genes. *Am J Bot* 89:707–725

Daniell H, Datta R, Varma S, Gray S, Lee SB (1998) Containment of herbicide resistance through genetic engineering of the chloroplast genome. *Nat Biotechnol* 16:345–348

Daniell H, Lee SB, Panchal T, Wiebe PO (2001) Expression of the native cholera toxin B subunit gene and assembly as functional oligomers in transgenic tobacco chloroplasts. *J Mol Biol* 311:1001–1009

De Cosa B, Moar W, Lee SB, Miller M, Daniell H (2001) Overexpression of the *Btcry2Aa2* operon in chloroplasts leads to formation of insecticidal crystals. *Nat Biotechnol* 19:71–74

DeGray G, Rajasekaran K, Smith F, Sanford J, Daniell H (2001) Expression of an antimicrobial peptide via the chloroplast genome to control phytopathogenic bacteria and fungi. *Plant Physiol* 127:852–862

Dufourmantel N, Tissot G, Goutorbe F, Garcon F, Muhr C, Jansens S, Pelissier B, Peltier G, Dubald M (2005) Generation and analysis of soybean plastid transformants expressing *Bacillus thuringiensis* Cry1Ab protoxin. *Plant Mol Biol* 58:659–668

Finnegan J, McElroy D (1994) Transgene inactivation: plants fight back! *Nat. Biotechnol* 12:883

Gilbert N (2013) A hard look at GM crops. *Nature* 497:24–26

Gisby MF, Mudd EA, Day A (2012) Growth of transplastomic cells expressing D-amino acid oxidase in chloroplasts is tolerant to D-alanine and inhibited by D-valine. *Plant Physiol* 160:2219–2226

Greiner S, Lehwark P, Bock R (2019) OrganellarGenomeDRAW (OGDRAW) version 1.3.1: expanded toolkit for the graphical visualization of organellar genomes. *Nucleic Acids Res* 47(W1):W59–W64. https://doi.org/10.1093/nar/gkz1238

Grover CE, Zhu X, Grupp KK, Jareczek JJ, Gallagher JP, Szadkowski E, Seijo JG, Wendel JF (2015) Molecular confirmation of species status for the allopolyploid cotton species, *Gossypium ekmanianum* Wittmack. *Genet Resour Crop Evol* 62:103–114

Hou BK, Zhou YH, Wan LH, Zhang ZL, Shen GF, Chen ZH, Hu ZM (2003) Chloroplast transformation in oilseed rape. *Transgenic Res* 12:111–114

Hussein S, Ruiz ON, Terry N, Daniell H (2007) Phytoremediation of mercury and organomercurials in chloroplast transgenic plants: enhanced root uptake, translocation to shoots, and volatilization. *Environ Sci Technol* 41:8439–8446

Iamtham S, Day A (2000) Removal of antibiotic resistance genes from transgenic tobacco plastids. *Nat Biotechnol* 18:1172–1176

Ibrahim RI, Azuma J, Sakamoto M (2006) Complete nucleotide sequence of the cotton (*Gossypium barbadense* L.) chloroplast genome with a comparative analysis of sequences among 9 dicot plants. *Genes Genet Syst* 81:311–321

ISAAA (2018) Brief 54: global status of commercialized Biotech/GM crops: 2018. In: James C, editor. International Service for the Acquisition of Agri-biotech Applications (ISAAA)
Jin S, Daniell H (2014) Expression of gamma-tocopherol methyltransferase in chloroplasts results in massive proliferation of the inner envelope membrane and decreases susceptibility to salt and metal-induced oxidative stresses by reducing reactive oxygen species. *Plant Biotechnol J* 12:1274–1285
Jin S, Kanagaraj A, Verma D, Lange T, Daniell H (2011) Release of hormones from conjugates: chloroplast expression of β-glucosidase results in elevated phytohormone levels associated with significant increase in biomass and protection from aphids or whiteflies conferred by sucrose esters. *Plant Physiol* 155:222–235
Jin S, Zhang X, Daniell H (2012) *Pinellia ternata* agglutinin expression in chloroplasts confers broad spectrum resistance against aphid, whitefly, Lepidopteran insects, bacterial and viral pathogens. *Plant Biotechnol J* 10:313–327
Jin S, Singh ND, Li L, Zhang X, Daniell H (2015) Engineered chloroplast dsRNA silences cytochrome p450 monooxygenase, V-ATPase and chitin synthase genes in the insect gut and disrupts Helicoverpa armigera larval development and pupation. *Plant Biotechnol J* 13:435–446
Jouanin L, Bonadé-Bottino M, Girard C, Morrot G, Giband M (1998) Transgenic plants for insect resistance. *Plant Sci* 131:1–11
Khan MS (2015) Transplastomics: a convergence of genomics and biotechnology. In: Barh D, Khan MS, Davies E (eds) Plant omics: the omics of plant science. Springer, pp 559–571
Khan MS, Maliga P (1999) Fluorescent antibiotic resistance marker to track plastid transformation in higher plants. *Nat Biotechnol* 17:910–915
Khan MS, Kanwal B, Nazir S (2015) Metabolic engineering of the chloroplast genome reveals that the yeast ArDH gene confers enhanced tolerance to salinity and drought in plants. *Front Plant Sci* 6:725. https://doi.org/10.3389/fpls.2015.00725
Khan MS, Nazir S, Amjad H, Jamil S, Khan IA (2017) Genotype-dependent regulation of δ-endotoxin reveals that insect-resistant cotton contains toxin predominantly in seeds. *Pak J Agric Sci* 54:395–406
Khan MS, Mustafa G, Joyia FA (2019) Technical advances in chloroplast biotechnology. technical advances in chloroplast biotechnology. In: Khan MS, Malik KA (eds) Transgenic crops: emerging trends and future perspectives. IntechOpen. https://doi.org/10.5772/intechopen.81240
Knight K, Head G, Rogers J (2016) Relationships between Cry1Ac and Cry2Ab protein expression in field-grown Bollgard II® cotton and efficacy against *Helicoverpa armigera* and *Helicoverpa punctigera* (Lepidoptera: Noctuidae). *Crop Prot* 79:150–158
Kota M (1999) Overexpression of the *Bacillus thuringiensis* (*Bt*) Cry2Aa2 protein in chloroplasts confers resistance to plants against susceptible and *Bt*-resistant insects. *Proc Natl Acad Sci U S A* 96:1840–1845
Kumar S, Dhingra A, Daniell H (2004a) Plastid-expressed betaine aldehyde dehydrogenase gene in carrot cultured cells, roots, and leaves confers enhanced salt tolerance. *Plant Physiol* 136:2843–2854
Kumar S, Dhingra A, Daniell H (2004b) Stable transformation of the cotton plastid genome and maternal inheritance of transgenes. *Plant Mol Biol* 56:203–216
Lee SB, Kwon HB, Kwon SJ, Park SC, Jeong MJ, Han SE, Byun MO, Daniell H (2003) Accumulation of trehalose within transgenic chloroplasts confers drought tolerance. *Mol Breeding* 11:1–13
Lee S-B, Kaittanis C, Jansen RK, Hostetler JB, Tallon LJ, Town CD, Daniell H (2006) The complete chloroplast genome sequence of *Gossypium hirsutum*: organization and phylogenetic relationships to other angiosperms. *BMC Genomics* 7:61–61
Lee S-B, Li B, Jin S, Daniell H (2011) Expression and characterization of antimicrobial peptides Retrocyclin-101 and Protegrin-1 in chloroplasts to control viral and bacterial infections. *Plant Biotechnol J* 9:100–115
Leister D, Wang L, Kleine T (2017) Organellar gene expression and acclimation of plants to environmental stress. *Front Plant Sci* 8:387
Li P, Li Z, Liu H, Hua J (2014) Cytoplasmic diversity of the cotton genus as revealed by chloroplast microsatellite markers. *Genet Resour Crop Evol* 61:107–119

Lin MT, Occhialini A, Andralojc PJ, Parry MA, Hanson MR (2014) A faster Rubisco with potential to increase photosynthesis in crops. *Nature* 513:547–550

Liu CW, Lin CC, Yiu JC, Chen JJ, Tseng MJ (2008) Expression of a *Bacillus thuringiensis* toxin (Cry_1A_b) gene in cabbage (*Brassica oleracea* L. var. capitata L.) chloroplasts confers high insecticidal efficacy against *Plutella xylostella*. *Theor Appl Genet* 117:75–88

MacArthur M (2000) Triple-resistant canola weeds found in Alta. https://www.producer.com/news/tripleresistant-canola-weeds-found-in-alta/

Maliga P, Bock R (2011) Plastid biotechnology: food, fuel, and medicine for the 21st century. *Plant Physiol* 155:1501–1510

McBride KE, Svab Z, Schaaf DJ, Hogan PS, Stalker DM, Maliga P (1995) Amplification of a chimeric *Bacillus* gene in chloroplasts leads to an extraordinary level of an insecticidal protein in tobacco. *Nat Biotechnol* 13:362–365

Michoux F, Ahmad N, McCarthy J, Nixon PJ (2011) Contained and high-level production of recombinant proteins in plant chloroplasts using a temporary immersion bioreactor. *Plant Biotechnol J* 9:575–584

Ni M, Ma W, Wang X, Gao M, Dai Y, Wei X, Zhang L, Peng Y, Chen S, Ding L (2017) Next-generation transgenic cotton: pyramiding RNAi and *Bt* counters insect resistance. *Plant Biotechnol J* 15:1204–1213

Occhialini A, Lin MT, Andralojc PJ, Hanson MR, Parry MAJ (2016) Transgenic tobacco plants with improved cyanobacterial Rubisco expression but no extra assembly factors grow at near wild-type rates if provided with elevated CO_2. *Plant J* 85:148–160

Oey M, Lohse M, Kreikemeyer B, Bock R (2009) Exhaustion of the chloroplast protein synthesis capacity by massive expression of a highly stable protein antibiotic. *Plant J* 57:436–445

Perlak FJ, Fuchs RL, Dean DA, McPherson SL, Fischhoff DA (1991) Modification of the coding sequence enhances plant expression of insect control protein genes. *Proc Natl Acad Sci* 88:3324–3328

Rey P, Sanz-Barrio R, Innocenti G, Ksas B, Courteille A, Rumeau D, Issakidis-Bourguet E, Farran I (2013) Overexpression of plastidial thioredoxins f and m differentially alters photosynthetic activity and response to oxidative stress in tobacco plants. *Front Plant Sci* 4:390

Romanos MA, Makoff AJ, Fairweather NF, Beesley KM, Slater DE, Rayment FB, Payne MM, Clare JJ (1991) Expression of tetanus toxin fragment C in yeast: gene synthesis is required to eliminate fortuitous polyadenylation sites in AT-rich DNA. *Nucleic Acids Res* 19:1461–1467

Rousseau-Gueutin M, Keller J, De Carvalho JF, Aïnouche A, Martin G (2018) The intertwined chloroplast and nuclear genome coevolution in plants. In Plant growth and regulation-alterations to sustain unfavorable conditions. IntechOpen

Ruhlman T, Verma D, Samson N, Daniell H (2010) The role of heterologous chloroplast sequence elements in transgene integration and expression. *Plant Physiol* 152:2088–2104

Ruhlman TA, Rajasekaran K, Cary JW (2014) Expression of chloroperoxidase from *Pseudomonas pyrrocinia* in tobacco plastids for fungal resistance. *Plant Sci* 228:98–106

Ruiz ON, Daniell H (2005) Engineering cytoplasmic male sterility via the chloroplast genome by expression of β-ketothiolase. *Plant Physiol* 138:1232–1246

Ruiz ON, Hussein HS, Terry N, Daniell H (2003) Phytoremediation of organomercurial compounds via chloroplast genetic engineering. *Plant Physiol* 132:1344–1352

Ruiz ON, Alvarez D, Torres C, Roman L, Daniell H (2011) Metallothionein expression in chloroplasts enhances mercury accumulation and phytoremediation capability. *Plant Biotechnol J* 9:609–617

Scheid OM, Paszkowski J, Potrykus I (1991) Reversible inactivation of a transgene in *Arabidopsis thaliana*. *Mol Gen Genet* 228:104–112

Shen G, Wei J, Qiu X, Hu R, Kuppu S, Auld D, Blumwald E, Gaxiola R, Payton P, Zhang H (2014) Co-overexpression of AVP1 and *At*NHX1 in cotton further improves drought and salt tolerance in transgenic cotton plants. *Plant Mol Biol Rep* 33:167–177

Stewart CN, Halfhill MD, Warwick SI (2003) Transgene introgression from genetically modified crops to their wild relatives. *Nat Rev Genet* 4:806–817

Strizhov N, Keller M, Mathur J, Koncz-Kálmán Z, Bosch D, Prudovsky E, Schell J, Sneh B, Koncz C, Zilberstein A (1996) A synthetic *cryIC* gene, encoding a *Bacillus thuringiensis* δ-endotoxin, confers Spodoptera resistance in alfalfa and tobacco. *Proc Natl Acad Sci* 93:15012–15017

Tabashnik BE, Carrière Y (2017) Surge in insect resistance to transgenic crops and prospects for sustainability. *Nat Biotechnol* 35:926

Tregoning JS, Nixon P, Kuroda H, Svab Z, Clare S, Bowe F, Fairweather N, Ytterberg J, van Wijk KJ, Dougan G, Maliga P (2003) Expression of tetanus toxin fragment C in tobacco chloroplasts. *Nucl Acids Res* 31:1174–1179

Tregoning JS, Clare S, Bowe F, Edwards L, Fairweather N, Qazi O, Nixon PJ, Maliga P, Dougan G, Hussell T (2005) Protection against tetanus toxin using a plant-based vaccine. *Eur J Immunol* 35:1320–1326

Twyford AD, Ness RW (2017) Strategies for complete plastid genome sequencing. *Mol Ecol Resour* 17:858–868

Umbeck P, Johnson G, Barton K, Swain W (1987) Genetically transformed cotton (Gossypium hirsutum L.) plants. *Biotechnology (NY)* 5:263–266

Wang YP, Wei ZY, Zhang YY, Lin CJ, Zhong XF, Wang YL, Ma JY, Ma J, Xing SC (2015) Chloroplast-expressed MSI-99 in tobacco improves disease resistance and displays inhibitory effect against rice blast fungus. *Int J Mol Sci* 16:4628–4641

Wendel JF, Cronn RC (2003) Polyploidy and the evolutionary history of cotton. *Adv Agron* 78:139–186

Wolfe KH, Li WH, Sharp PM (1987) Rates of nucleotide substitution vary greatly among plant mitochondrial, chloroplast, and nuclear DNAs. *Proc Natl Acad Sci U S A* 84:9054–9058

Wu Y, Liu F, Yang D-G, Li W, Zhou X-J, Pei X-Y, Liu Y-G, He K-L, Zhang W-S, Ren Z-Y, Zhou K-H, Ma X-F, Li Z-H (2018) Comparative chloroplast genomics of gossypium species: insights into repeat sequence variations and phylogeny. *Front Plant Sci* 9:376–376

Wurbs D, Ruf S, Bock R (2007) Contained metabolic engineering in tomatoes by expression of carotenoid biosynthesis genes from the plastid genome. *Plant J* 49:276–288

Xu Q, Xiong G, Li P, He F, Huang Y, Wang K, Li Z, Hua J (2012) Analysis of complete nucleotide sequences of 12 *Gossypium* chloroplast genomes: origin and evolution of allotetraploids. *PLoS One* 7:e37128–e37128

Ye X, Al-Babili S, Klöti A, Zhang J, Lucca P, Beyer P, Potrykus I (2000) Engineering the provitamin A (*β*-carotene) biosynthetic pathway into (carotenoid-free) rice endosperm. *Science* 287:303–305

Zhang J, Khan SA, Heckel DG, Bock R (2015) Full crop protection from an insect pest by expression of long double-stranded RNAs in plastids. *Science* 347(6225):991–994

Zhang K, Song J, Chen X, Yin T, Liu C, Li K, Zhang J (2016) Expression of the *Thellungiella halophila* vacuolar H^+-pyrophosphatase gene (TsVP) in cotton improves salinity tolerance and increases seed cotton yield in a saline field. *Euphytica* 211:231–244

Zhang Z, Ge X, Luo X, Wang P, Fan Q, Hu G, Xiao J, Li F, Wu J (2018) Simultaneous editing of two copies of Gh14-3-3d confers enhanced transgene-clean plant defense against *Verticillium dahliae* in Allotetraploid Upland cotton. *Front Plant Sci* 9:842

Zhao FY, Li YF, Xu P (2006) Agrobacterium-mediated transformation of cotton (*Gossypium hirsutum* L. cv. Zhongmian 35) using glyphosate as a selectable marker. *Biotechnol Lett* 28:1199–1207

Chapter 4
Cotton Genomes Are Sequenced Due to Their Economic and Agriculture Importance

Yuxin Pan, Fanbo Meng, and Xiyin Wang

Cotton is the world's economic crop for its natural textile fiber, averaging about 25% of total world fiber use. And also, cottonseeds are rich in oil and proteins and used for cottonseed oil production, as a feed supplement for cattle and sheep and producing industrial products such as soaps and cosmetics.

More than 80 countries produce cotton, distributed in arid and semiarid regions of the tropics and subtropics. The top three cotton producers are India, China, and the United States. World consumption of cotton fiber and seed oil is approximately 119.4 million bales and 43.67 million metric tons until 2018, respectively (US Department of Agriculture, https://usda.library.cornell.edu/).

Cotton fiber is an outstanding model system for the plant cell elongation, cell wall, cellulose biosynthesis, and polyploidization research (Kim and Triplett 2001). Cotton (*Gossypium* spp.) is from Malvaceae family and includes four subgenera, *G.* subg. *Gossypium*, *G*. subg. *Houzingenia*, *G*. subg. *Karpas*, and *G*. subg. *Sturtia*. *Gossypium* includes approximately 45 diploid ($2n = 2\times = 26$) and 7 tetraploid ($2n = 4\times = 52$) species (Hawkins et al. 2006; Salmon et al. 2010; Grover et al. 2015;Ditta et al., 2018). About 12.5 million years ago (mya), *Gossypium* diverged from its closer relatives *Kokia* and *Gossypioides* during the Miocene (Cronn et al. 2002; Amer. J. Bot. 89, 707–725). Eight diploid cytogenetic genomes, designated as A to G and K, have been found in the world (Wang et al. 2012). The A, B, E, and F genomes (Cronn et al. 2003) occur naturally in Africa and Asia. The D genome occurs in the Americas. The C, G, and K genomes are found in Australia (Chen et al.

Y. Pan · X. Wang (✉)
School of Life Sciences, North China University of Science and Technology, Tangshan, Hebei, China

Center for Genomics and Computational Biology, North China University of Science and Technology, Tangshan, Hebei, China

F. Meng
School of Life Sciences, North China University of Science and Technology, Tangshan, Hebei, China

Mehboob-ur-Rahman et al. (eds.), *Cotton Precision Breeding*,
https://doi.org/10.1007/978-3-030-64504-5_4

2007). The haploid genome size diverged from 2500 Mb in the K genome to less than 800 Mb in the D genome (Hawkins et al. 2006; Hendrix and Stewart 2005).

Five widely recognized tetraploid species include *G. hirsutum* (AD1), or "upland cotton," *G. barbadense* (AD2), "Pima" or "Egyptian" cotton, and three other exclusively wild polyploid species endemic to coastal and island habitats: *G. tomentosum* (AD3), *G. mustelinum* (AD4), and *G. darwinii* (AD5) (Salmon et al. 2010). *G. hirsutum* and *G. barbadense*, widely cultivated tetraploid cotton species, arose in the New World. Four cultivated cotton species, *G. hirsutum* (AD1) in Mexico, *G. barbadense* (AD2) in Peru, *G. herbaceum* (A1) in Pakistan, and *G. arboreum* (A2) in Sudan, had different origins, which illustrated that parallel and convergent domestication occurred.

In view of cotton's central importance in human life, three diploid cottons and two cultivated tetraploid cottons have been sequenced, including *G. arboreum* (Li et al. 2014; Du et al., 2018; Huang et al., 2020), *G. raimondii* (Lin et al. 2010; Wang et al. 2012; Paterson et al., 2012; Udall et al., 2019), *G. herbaceum* (Huang et al., 2020), *G. hirsutum* (Li et al. 2015; Zhang et al. 2015; Hu et al.,2019; Wang M. et al., 2019; Chen et al., 2020; Huang et al., 2020)), and *G. barbadense* (Yuan et al. 2015; Liu et al. 2015; Hu et al., 2019; Wang M. et al., 2019; Chen et al., 2020).). Two prominent databases include these information, CottonGen (https://www.cottongen.org/) (Yu et al. 2014) and CottonFGD (https://cottonfgd.org/). In *G. raimondii*, the putative D genome ancestral species of tetraploid cottons, its first whole-genome physical map was sequenced, anchoring 1585 contigs to a cotton consensus genetic map (Lin et al. 2010). *G. arboreum* has a genome size that is almost twice than that of *G. raimondii*, which had been sequenced by some different institutions, using the whole-genome shotgun (WGS), Illumina, PacBio, HiC technology. A total of 1,694 Mb length was assembled, and 41,330 protein-coding genes were identified in the *G. arboreum* genome (Li et al. 2014). An improved genome sequence of *G. arboreum* was assembled to 1,637Mb (with 92% oriented) (Huanget al., 2020). The detailed information is listed in Table 4.1.

Based on the cotton consensus map marker, the CAAS also detected the *G. raimondii* genome information through Illumina HiSeq 2000 platform. A 775.2-Mb *G. raimondii* genome was assembled (Wang et al. 2012). Detailed *G.hirsutum* and *G.barbadense* genome characteristics are given in Table 4.2. *G. hirsutum*, upland or American cotton, known for high lint production than any other cultivated cotton species, accounts for more than 90% of commercial cotton production worldwide (Cronn et al. 2003). Acc. Texas Marker-1 (TM-1) is a genetic standard *for G. hirsutum*. Two academic institutes revealed its genome mystery in 2015. A total of 2173-Mb genome sequence was assembled, and 88.5% was anchored and oriented to 26 pseudochromosomes by the CAAS. The anchored At subgenome is 1170 Mb with 35,056 genes, and the Dt subgenome is 753 Mb with 37,086 genes. In 2020, by using PacBio, Illumina HiSeq and Hi-C,a much improved *G. hirsutum* genome sequence was published (Huang et al., 2020). Finally, 43,952 genes were annotated, including 32,032 in the A subgenome and 34,402 in the D subgenome. *G. barbadense* (Pima cotton), island cotton, is famous for its extra-long strength and fineness fiber. *G. barbadense*, acc. 3–79, was sequenced by HZAU, including 80,876

Table 4.1 Characteristics of *G. arboreum*, *G. herbaceum* and *G. raimondii* genome sequences

Species	*G. arboretum* (Li et al. 2014)	*G. arboretum* (Du et al. 2018)	*G. arboretum* (Huang et al. 2020)	*G. herbaceum* (Huang et al. 2020)	*G. raimondii* (Wang et al. 2012)	*G. raimondii* (Paterson et al. 2012)	*G. raimondii* (Udall et al. 2019)
Total contigs	40,381	8223	2432	1781	41,307	19,735	187
Contig N50 (kb)	72	1100	1832	1915	44.9	135.6	6291
Total contig length (Mb)	1561	1710	1637	1556	744.4	748.1	734.9
Anchored contigs (Mb)	NA	1573	1509	1489	NA	NA	NA
Total scaffolds	7914	4516	1269	732	4715	1084	NA
Scaffold N50 (kb)	665.8	NA	NA	NA	2284	18,800	58,819
Total scaffold length (Mb)	1694	NA	NA	NA	775.2	761.4	734.9
Anchored and oriented scaffolds (Mb)	1532	NA	NA	NA	406.3	748.7	NA
Total genes	41,330	40,960	43,278	43,952	40,976	NA	41,030

NA, not available

Table 4.2 Characteristics of *G. barbadense* and *G. hirsutum* genome sequences

Species	*G. hirsutum* (TM-1,V1.0) (Li et al.2015)	*G. hirsutum* (TM-1,V1.1) (Zhang et al. 2015)	*G. hirsutum* (TM-1) (Wang et al. 2019b)	*G. hirsutum* (TM-1,V2.1) (Hu et al. 2019)	*G. hirsutum* (TM-1,updated V1) (Huang et al. 2020)	*G. hirsutum* (TM-1) (Chen et al. 2020)	*G. barbadense* (Xinhai21) (Liu et al. 2015)	*G. barbadense* (3-79) (Yuan et al. 2015)	*G. barbadense* (3-79) (Wang et al. 2019b)	*G. barbadense* (Hai7124,V1.1) (Hu et al. 2019)	*G. barbadense* (3-79,V1.1) (Chen et al. 2020)
Scaffold number	8591	40,407	2190	48	342	1025	NA	29,751	3032	11,701	4748
Scaffold length (Mb)	2173	2432.7	2347	2295.3	2290	2305.2	NA	2573.2	2266.7	2224.98	2195.8
Scaffold N50 (Mb)	0.764	1.6	97.8	15.5	NA	108.1	0.503	0.26	92.9	23.44	93.8
Contig number	44,816	265,279	4746	NA	1235	6733	NA	NA	4930	75,898	4767
Contig length (Mb)	2090.4	2068.1	2281.9	2267.9	NA	2302.3	NA	NA	2222.5	2192.5	2193.9
Contig N50 (kb)	80.38	34.0	1891.9	113.3	5020	783.9	72	NA	2151	77.66	1769.6
Total gene	76,943	70,478	70,199	72,761	74,350	75,376	77,526	80,876	71,297	75,071	74,561
Transposable element (Mb)	1445.06	1339	1640	1460.1	1467.55	NA	1391.48	1778.62	1582.8	1374.61	NA

NA, not available

protein-coding genes. It was shown that the A subgenome (or At) (1.50 Gb) has a double size than that of the D subgenome (or Dt) (853 Mb). And also, researchers sequenced *G. barbadense* cv. Xinhai21 through deploying three next-generation sequencing platforms, Roche 454, Illumina HiSeq 2000, and PacBio SMRT (Liu et al. 2015). This genome covered 1.395 Gigabases (Gb) of the A subgenome (A_t) and 0.776 Gb of the D subgenome (D_t).

The draft genome sequences of allotetraploid cotton species are still highly fragmented and incomplete. So, HZAU sequenced and annotated for *G. hirsutum* accession TM-1 and *G. barbadense* accession 3–79 by integrating single-molecule real-time sequencing, BioNano optical mapping, and high-throughput chromosome conformation capture techniques. In the new genome, 70,199 genes in *G. hirsutum* and 71,297 genes in *G. barbadense* were predicted (Wang et al. 2019a). In the same year, NAU assembled *G. hirsutum* accession TM-1 and *G. barbadense* accession Hai7124 by integrating non-PCR-based short-read sequencing, long-read-based gap closure, scaffolding, and orientation based on 3D proximity information derived from chromosome conformation capture (Hi-C) data and from optical and genetic maps. 72,761 and 75,071 proteins were identified in TM-1 and Hai7124, separately. Contiguity and completeness for regions with high content of repeats were improved in the two researches, especially, centromeric regions for each chromosome (Hu et al. 2019). Based on single-molecule real-time, Illumina andHi-C, another five allotetraploid cotton genomes were sequenced.The assembled genomes range in size from 2.2 to 2.3 Gb (Chenet al., 2020).

4.1 Phylogeny, Paleohexaploidization, and Whole-Genome Duplication

Polyploidy is a significant evolutionary process in higher organisms (Bowers et al. 2003; Tang et al. 2008a). Cotton is ideal for investigating polyploidy for its typical polyploidy and large-scale duplication events. *Gossypium* had the ancient hexaploidization event shared among the eudicots about 115–146 mya (Tang et al. 2008b). Molecular phylogenetic analyses suggested that the common ancestor of *G. arboreum* and *G. raimondii* diverged from *T. cacao* around 18–58 million years ago (Wikstrom et al. 2001).

By examining 745 single-copy gene families from nine sequenced plant genomes, *G. raimondii* and *T. cacao* probably diverged approximately 33.7 million years ago from a common ancestor (Argout et al. 2011; Wang et al. 2012). Nonreciprocal homoeologous exchanges have occurred throughout the polyploid divergence and speciation in cotton (Salmon et al. 2010). Shortly after divergence from cacao, the *Gossypium* lineage experienced a five- to sixfold ploidy increase (Paterson et al. 2012). Further, through detecting gene colinearity, five-times of duplicated regions in cotton to those in cacao and grape suggested a paleo-decaploidy, or penta-plication of the ancestral genome, nature of the event (Wang et al. 2016). Besides, the event was not shared with durian (*Durio zibethinus*), a plant from the Helicteroideae subfamily in the Malvaceae family, which was affected by an independent paleohexaploidization (Wang et al. 2019b).

All diploid cotton species retain 13 chromosomes and largely collinear gene order. *G. arboreum* (A2) and *G. raimondii* (D5), the receptive descendants to create all cultivated allopolyploid cotton species, exhibit a twofold difference in genome size. A-genome diploids native to Africa and Mexican D-genome diploids shared a common ancestor about 5–10 million years ago (Wendel and Albert 1992). The level of divergence between homologous nucleotide sequences is widely used as a molecular clock to estimate the relative age of their separation. Synonymous substitutions (Ks) between two homologous sequences are generally used to assume the time of species divergence and evolutionary distances among different plant species. With the development of next-generation sequencing technology, two diploid *Gossypium* evolution time was revised. *G. arboreum* and *G. raimondii* gene families are observed the similar divergence time (Argout et al. 2011; Wang et al. 2012). Using Ks values obtained from 3195 paralogous gene pairs in the *G. raimondii* and *T. cacao* genomes, two Ks value peaks were observed at 0.40–0.60 and 1.5–1.90. So, a more recent polyploidization event was proposed at approximately 16.6 (13.3–20.0) million years ago in the *Gossypium* lineage. The second peak appeared at approximately 130.8 (115.4–146.1) million years ago, corresponding to the paleohexaploidization event shared among the eudicots (Tang et al. 2008b; Van de Peer et al. 2009). In *G. arboreum*, two coincident whole-genome duplication (WGD) events occurred in *G. raimondii* which were observed using Ks of 1917 paralogous gene pairs of similar age. One peaked around 0.17 synonymous transversions per site, and a second peak at about 0.54 was observed. It also suggested the two polyploidization events occurred at 13–20 and 115–146 million years ago (Li et al. 2014).

A mount of collinearity are conserved between *G. hirsutum* and *G. barbadense* genome, suggesting that both were originated from a common allotetraploid ancestor. Polyploidization was estimated to have occurred 1–2 million years ago. A genome, African species much like modern *G. herbaceum* (A1) and *G. arboreum* (A2), and D genome like American diploid species, *G. raimondii* (D5), were reunited by transoceanic dispersal and chromosome doubling to give rise to allotetraploid cotton species (Paterson et al. 2012; Paterson and Wendel 2003).

G. hirsutum and *G. arboreum* yielded coincident Ks peaks at 0.4–0.45 (Fawcett et al. 2009). Sequenced *G. hirsutum* genome had proved this divergence time. *G. hirsutum* had two typical duplication events, one is the ancient hexaploidization event shared among the eudicots and the other is the progenitor of *G. arboreum* and *G. raimondii*. And also, *G. hirsutum* had an additional event about 1.5 mya, which corresponds to the predicted hybridization, and subsequent polyploidization event was observed only in *G. hirsutum*. But by comparing the sequenced *G. barbadense*, *G. arboreum*, and *G. raimondii* orthologous genes, the divergence time between the A- and D-progenitor genomes was estimated to be 6.0–6.3 mya (*Ks* peaks at 0.031 and 0.033, respectively) (Liu et al. 2015). The allotetraploid formed around 1.7–1.9 mya, and the divergence of *G. barbadense* and *G. hirsutum* was 0.4–0.6 mya (Ks peaks at 0.002 and 0.003, respectively) (Hu et al. 2019). There are some unique structural variants generated after the polyploidization in the two tetraploids. A genome of diploid cotton was reorganized following allopolyploidization, leading to large chromosome inversions in different tetraploids (Wang et al. 2019a). A faster evolution rate was observed in allotetraploid cottons, compared with their diploid progenitors (Hu et al. 2019).

4.2 Asymmetric Evolution of the A and D Genomes

The A genome species produce spinnable fiber, whereas the D genome species do not (Applequist et al. 2001). Two allotetraploid cottons, *G. hirsutum* and *G. barbadense*, had a similar evolution character with asymmetric evolution of the A and D.

In *G. hirsutum* chromosomes, the D or Dt genome has higher mutation rates than the A or At genome. There were 100 possible HEs, of which 54 were Dt to At and 46 were At to Dt. Also, the single-nucleotide variation rate for Dt versus D was greater than that for At versus A in intergenic collinear regions. Further analysis revealed that A and At genomes are undergoing a greater positive selection than the D and Dt genomes. For example, the average Ks values for collinearity-supported gene pairs in At and Dt subgenomes were substantially lower than those of the A and D diploid genomes, respectively. In addition, Dt versus *T. cacao* or D versus *T. cacao* had lower dN/dS ratios than At versus *T. cacao* or AA versus *T. cacao*. These data suggest that the genetic redundancy generated by allotetraploidy may have allowed relaxed purifying selection in both At and Dt subgenomes (Li et al. 2015). There was similar rearrangement between chromosomes and the number of rearrangements between the A and D subgenomes (19 versus 18). However, the length of total rearrangements was larger in the A subgenome (372.6 Mb) than in the D subgenome (82.6 Mb). The SNP frequency between *G. hirsutum* and *G. barbadense* is slightly larger in the A subgenome (At) with 8,131,276 (5.95 per kb) than that in the D subgenome (Dt), with 4,685,422 (5.81 per kb) (Wang et al. 2019a). All the above information revealed that A subgenome has experienced relaxed selection pressure (Liu et al. 2015; Yuan et al. 2015). In the same way, allele-SNPs within subgenomes, between accessions of six cotton tetraploid species, were also higher in the At than Dt genomes. In both AD1 and AD2, the number of At-genome allele-SNPs was about 1.5× the number of DT-genome allele-SNPs (Page et al. 2016).

A systematic characterization of presence/absence variations (PAVs) between the two tetraploid accessions was also detected. In the PAVs, a lot of genes unique to *G. barbadense* are highly expressed during fiber development. Besides that, there were also inversions between *G. hirsutum* and *G. barbadense*, including 120.4 Mb of At subgenome and 49.8 Mb of Dt. Through Hi-C technique, large inversions were shown including paracentric and pericentric inversions in the A06 and D12 (Wang et al. 2019a). Except that, introgressions were detected in AD1 and AD2 cultivars' genome owing to the attempts of breeders for transfer of genes for disease resistance, fiber quality, and other traits between AD1 and AD2 (Page et al. 2016).

Some different duplication phenomenon was discovered through comparing accessions among *G. hirsutum and G. barbadense*. The study showed that

AT-genome duplications were more (2×) conserved than Dt-genome duplications in *G. hirsutum* cultivars, although not in *G. barbadense*. Also, At-genome deletions were more conserved than Dt-genome deletions in *G. barbadense* but not in *G. hirsutum*. Both these findings illuminate the existence of independent domestication events for these two species (Page et al. 2016).

Asymmetric in A- and D-genome is also reflected in biased A- or D-homoeolog expression (Yoo and Wendel 2014). For example, more transcription factor genes (such as *MYB* family members) were expressed in the A homoeologs, suggesting important roles in fiber development. This may lead to subfunctionalization (Zhang et al. 2015).

4.3 Expansion of Transposable Elements

Transposable elements (TEs) have important roles in driving genome evolution (Galindo-Gonzalez et al. 2017). There are 57% (441 Mb) transposable elements of the *G. raimondii* genome, mainly the *gypsy* and *copia*-like LTRs, which are partially accountable for the expansion of the *G. raimondii* genome. Phylogenetic analysis shows a larger expansion of specific LTR retrotransposon clades had occurred in *G. raimondii than T. cacao and Arabidopsis* (Lin et al. 2010). Data analysis indicated that the *G. arboreum* genome tended to harbor more LTRs inserted than G. *raimondii* genome during the last 0.5 million years. *G. arboreum* genome has the greatest amount of repeat-containing sequences among sequenced cottons (Cheng et al. 2019). Long terminal repeat (LTR) retrotransposons accounted for 95.12% of all repeat sequences. In comparison with the *G. raimondii* genome, the *G. arboreum* genome had noticeable proliferation of *Gorge* elements. LTR retrotransposons in *G. arboreum* appeared to cluster near the centromere (Li et al. 2014). Thus, LTR activities substantially contributed to the twofold increase in the size of the *G. arboreum* genome.

G. hirsutum is the main cultured cotton crop. Zhang and Li analyzed the *G. hirsutum* genome in the same year (Li et al. 2015; Zhang et al. 2015). They used different methods and get similar TE proportion of *G. hirsutum* genome (1339 Mb versus 1445 Mb, 64.8% versus 66%). There were more TEs in the A subgenome (at least 843.5 Mb) than in the D subgenome (at least 433 Mb). Among them, the number of Gypsy retroelements was threefold higher in the A subgenome than in the D subgenome. The TE types and relative proportions of A and D subgenome were similar with their corresponding genomes of *G. arboreum* and *G. raimondii*, whereas the retrotransposon frequencies were different (52.29% versus 62.81%) between the two *G. hirsutum* genomes. The TE divergence time is older than the 1.5 mya, which suggests that most TEs expanded before the formation of allopolyploid cotton. It is likely that the TE expansion occurred in the progenitor genomes and was retained after allopolyploid formation. *Copia* elements were remarkably more active than *gypsy*; at the same time, *copia* was located near coding genes with higher proportions than *gypsy* type. This indicates that the TEs of the Dt subgenome tend to be more active than that of the At subgenome after the tetraploidization.

In 2015, two *G. barbadense* genomes were sequenced. One observed 83.5% and 82.2% LTR in At and Dt (Liu et al. 2015); the other detected 1778.6 Mb of TE (69.1% of the tetraploid cotton) were identified, including 1098.0 Mb of TE sequences in At (representing 73.5% of the subgenome) and 541.6 Mb of TE sequences in Dt (representing 63.5% of the subgenome) (Yuan et al. 2015). LTR retrons insertion rates in the A genome appear consistently higher than those in the D genome. Liu et al. consider a large number (9.15%) of LTR retrons burst at 5 mya and decreased thereafter in At, whereas a substantially lower and flat peak appeared 3–5 mya in Dt (Liu et al. 2015). Whereas the timing of insertion for LTR retrotransposons peaked within 1.9 mya in Dt, whereas, approximately 3.1 Myr ago in At (Yuan et al. 2015). Although the results had different timing estimation, these suggest that most expansions of extant LTR retrotransposons independently occurred after the lineage separation but before allotetraploidization (Wu et al. 2017).

4.3.1 Genomic Insight for Cotton Fiber Gene Expression and Development

Cotton fibers, which are highly elongated epidermal cells, undergo four overlapping development, initiation, elongation (primary cell wall (PCW)), secondary cell wall (SCW) synthesis, and maturation (Basra and Malik 1984). The elongation and SCW synthesis stages are major for fiber length and strength. At the SCW stage, cross-linking of cellulose microfibrils and non-cellulosic matrices presumably "fixes" the structure of the PCW, resulting in the first significant increase in fiber strength (Wilkins and Arpat 2005).

The different development in *Gossypium* genus may lead to their typical fiber traits. *G. barbadense*, Hai7124, had an extended time than *G. hirsutum*, TM-1, in fiber elongation period. The stage lasts about 15 days from 5 to 25DPA in TM-1, whereas this lasts from 5 to 30 DPA in Hai7124. During the PCW-SCW transition period, expression of most genes was earlier in TM-1 than in Hai7124 (Zhu et al. 2011).

There are a large number of genes in fiber development. Based on genomic sequence information, more and more genes with structure and qualitative transcript differences in fiber development were found. Fiber elongates more than 2000-fold after initiation, which is regulated by cell turgor. The plasmodesmata on/off switch, together with sucrose and potassium ion (K+) transporter, is crucial for fiber cells (Ruan et al. 2004). Several genes associated with membrane transport, transcription, and glycan biosynthesis and carbon metabolism were significantly expressed in Hai7124 as compared to TM-1 fibers. Furthermore, some gene expression of Hai7124 fibers, the sucrose transporter (*GbTST1*), Na+/H+ antiporter (*GbNHX1*), aluminum-activated malate transporter (*GbALMT16*), the vacuole-localized vacuolar (GbVIN1), and plasmodesmata (PD) opening had a longer period of expression than the corresponding genes in TM-1 (Hu et al. 2019).

The fiber SCW consisted of >94% cellulose (Wilkins and Arpat 2005). Sucrose synthase gene (sus) expressed in the later stage of fiber development plays a role in cell wall cellulose (suppression of sucrose synthase gene expression represses cotton fiber cell initiation, elongation, and seed development). Between *G. raimondii* and *G. hirsutum* cotton ovules at 3 days post-anthesis (DPA), three Sus genes (*SusB*, *Sus1*, and *SusD*) were expressed at substantially higher levels in *G. hirsutum* than in *G. raimondii* (Wang et al. 2012). *Cellulose synthase A* (*CesA*), synthesis cellulose, acts a key role in the formation of the secondary cell wall thickening. Through genome-wide analysis, 32 CesA and 64 cellulose synthase-like (CSL) genes were detected in *G. hirsutum, and* 37 *CesA* genes were identified in *G. barbadense.* The CesA genes can be classified into two major groups with six branches. One group was expressed during primary cell wall development, whereas the other group was expressed during secondary cell wall biosynthesis. CesA, UGD, UGP, and UER, important for cotton fiber growth, which originated from the At subgenome, were expressed highly during either the primary or the secondary cell wall biosynthesis stages (Li et al. 2015; Yuan et al. 2015; Zhang et al. 2015).

Transcription factors are important for cotton fiber development. MYB are one of the most abundant transcription factors in cotton, and these play diverse roles during cotton growth and evolution (Salih et al. 2016). A total of 219 and 524 MYB genes were identified in the *G. raimondii* and *G. hirsutum* genome (Salih et al. 2016; Wang et al. 2012). In *G. hirsutum*, two groups of myb genes are expressed. One exhibited low or undetectable expression levels, and the other expressed higher during −1 to 10 DPA. A large number of MYB genes were expressed more predominantly in *G. hirsutum* ovules than *G. raimondii.* Some of the MYB MIXTA-like (GhMML) homoeologs were also highly expressed during fiber initiation in *G. hirsutum*, but downregulated in fiberless mutants. All the above indicated that some of the MYB genes may be required for early fiber development. WRKY, another transcription factor, participates in fiber development. In total, 112 and 109 WRKY genes were identified in *G. raimondii* and *G. arboreum*, respectively. Many SNPs are distributed unequally in exon and intron regions in these genes (Ding et al. 2015).

Ethylene is a key signaling modulator of cotton fiber cell growth (Pang et al. 2010; Shi et al. 2006). Among these, 1-aminocyclopropane-1-carboxylic acid oxidase (ACO) is the last enzyme in the ethylene synthesis. High amounts of transcripts encoding ACO activities were recovered from *G. raimondii* at the 3-DPA stage. ACO3 and ACO1 are 500- and 1000-fold higher than *G. arboreum*, respectively. CaACO loss of MYB-binding sites may lead to lower transcripts in *G. arboreum*. Compared with *G. raimondii*, ACO expression and ethylene were both lower in *G. hirsutum*. And also, the ACO expression peak and ethylene burst occurred in the later stage of *G. hirsutum*. Very high levels of ACO transcripts give rise to ethylene burst which might force an early fiber senescence phenotype, whereas the inactivation of ACO in *G. arboreum* ovules might be responsible for the short-fiber phenotype in this species. A compromise ACO expression might be the power for fiber traits in *G. hirsutum*. So, ACO is suggestive of a major role for the plant hormone ethylene during early fiber cell development (Li et al. 2014; Wang et al. 2012).

A total of 591 PSG (72.8%) genes with the codons subjected to positive selection were expressed during the fiber development. Many of the A-homoeologous PSGs were enriched in the synthesis of ethylene and very-long-chain fatty acids, sucrose metabolism, and beta-d-glucan biosynthetic pathway to produce UDP-glucose, whereas the D-homoeologous PSGs were enriched in carbohydrate transport, response to superoxide, and other abiotic stresses. These results suggest that allotetraploid cotton domestication is associated with intensive human selection for fiber yield and quality on the A homoeologs from fiber-producing species and for wider adaptation on the D homoeologs from nonpoor species. These also suggest that A and D were asymmetric and A subgenome contribute to fiber improvement (Zhang et al. 2015).

4.4 Conclusions

The availability of multiple cotton genomes shed precious light on their genome structures and evolution, especially genome instability after recursive rounds of paleopolyploidization, and paves the way for efforts to clone key functional genes and cultivate cotton plants to produce high-yield, high-quality cotton fiber and, at the meantime, those being highly resistant to notorious pathogens and harsh environmental climates.

References

Applequist WL, Cronn R, Wendel JF (2001) Comparative development of fiber in wild and cultivated cotton. *Evol Dev* 3:3–17

Argout X, Salse J, Aury JM, Guiltinan MJ, Droc G, Gouzy J, Allegre M, Chaparro C, Legavre T, Maximova SN, Abrouk M, Murat F, Fouet O, Poulain J, Ruiz M, Roguet Y, Rodier-Goud M, Barbosa-Neto JF, Sabot F, Kudrna D, Ammiraju JS, Schuster SC, Carlson JE, Sallet E, Schiex T, Dievart A, Kramer M, Gelley L, Shi Z, Berard A, Viot C, Boccara M, Risterucci AM, Guignon V, Sabau X, Axtell MJ, Ma Z, Zhang Y, Brown S, Bourge M, Golser W, Song X, Clement D, Rivallan R, Tahi M, Akaza JM, Pitollat B, Gramacho K, D'Hont A, Brunel D, Infante D, Kebe I, Costet P, Wing R, McCombie WR, Guiderdoni E, Quetier F, Panaud O, Wincker P, Bocs S, Lanaud C (2011) The genome of Theobroma cacao. *Nat Genet* 43:101–108

Basra AS, Malik CP (1984) Development of the cotton fiber. *Int Rev Cytol* 89:65–113

Bowers JE, Chapman BA, Rong J, Paterson AH (2003) Unravelling angiosperm genome evolution by phylogenetic analysis of chromosomal duplication events. *Nature* 422:433–438

Chen ZJ, Scheffler BE, Dennis E, Triplett BA, Zhang T, Guo W, Chen X, Stelly DM, Rabinowicz PD, Town CD, Arioli T, Brubaker C, Cantrell RG, Lacape JM, Ulloa M, Chee P, Gingle AR, Haigler CH, Percy R, Saha S, Wilkins T, Wright RJ, Van Deynze A, Zhu Y, Yu S, Abdurakhmonov I, Katageri I, Kumar PA, Mehboob Ur R, Zafar Y, Yu JZ, Kohel RJ, Wendel JF, Paterson AH (2007) Toward sequencing cotton (Gossypium) genomes. *Plant Physiol* 145:1303–1310

Chen ZJ, Sreedasyam A, Ando A, Song Q, De Santiago, L M, Hulse-Kemp, AM, Ding M, Ye W, Kirkbride RC, Jenkins J, Plott C, Lovell J, Lin Y M, Vaughn R, Liu B, Simpson S, Scheffler BE, Wen L , Saski CA, Grover C, Hu G, Conover JL, Carlson JW, Shu SS, Boston LB, Williams M, Peterson DG, McGee K, Jones DC, Wendel JF, Stelly DM, Grimwood J, Schmutz J (2020) Genomic diversifications of five Gossypium allopolyploid species and their impact on cotton improvement. *Nat Genet* 52:525–533

Cheng H, Sun G, He S, Gong W, Peng Z, Wang R, Lin Z, Du X (2019) Comparative effect of allopolyploidy on transposable element composition and gene expression between Gossypium hirsutum and its two diploid progenitors. *J Integr Plant Biol* 61:45–59

Cronn RC, Small RL, Haselkorn T, Wendel JF (2002) Rapid diversification of the cotton genus (Gossypium: Malvaceae) revealed by analysis of sixteen nuclear and chloroplast genes. *Am J Bot* 89:707–725

Cronn R, Small RL, Haselkorn T, Wendel JF (2003) Cryptic repeated genomic recombination during speciation in Gossypium gossypioides. *Evolution* 57:2475–2489

Ding M, Chen J, Jiang Y, Lin L, Cao Y, Wang M, Zhang Y, Rong J, Ye W (2015) Genome-wide investigation and transcriptome analysis of the WRKY gene family in Gossypium. *Mol Gen Genomics* 290:151–171

Ditta A, Zhou Z, Cai X, Wang X, Okubazghi KW, Shehzad M, Xu Y, Hou Y, Iqbal MS, Khan M K, Wang K, Liu F (2018) Assessment of genetic diversity, population structure, and evolutionary relationship of uncharacterized genes in a novel germplasm collection of diploid and allotetraploid Gossypium accessions using EST and genomic SSR markers. *Int J Mol Sci* 19:2401

Du X, Huang G, He S, Yang Z. Sun G, Ma X, li N, Zhang X, Sun J, Liu M, Jia Y, Pan Z, Gong W, Liu Z, Zhu H, Ma L, Liu F, Yang D, Wang F, Fan W, Gong Q, Peng Z, Wang L, Wang X, Xu S, Shang H, Lu C, Zheng H, Huang S, Lin T, Zhu Y, Li F (2018) Resequencing of 243 diploid cotton accessions based on an updated A genome identifies the genetic basis of key agronomic traits. *Nat Genet* 50:796–802

Fawcett JA, Maere S, Van de Peer Y (2009) Plants with double genomes might have had a better chance to survive the Cretaceous-Tertiary extinction event. *Proc Natl Acad Sci U S A* 106:5737–5742

Galindo-Gonzalez L, Mhiri C, Deyholos MK, Grandbastien MA (2017) LTR-retrotransposons in plants: engines of evolution. *Gene* 626:14–25

Grover CE, Zhu X, Grupp KK, Jareczek JJ, Gallagher JP, Szadkowski E, Seijo JG, Wendel JF (2015) Molecular confirmation of species status for the allopolyploid cotton species, Gossypium ekmanianum Wittmack. *Genet Resour Crop Evol* 62:103–114

Hawkins JS, Kim H, Nason JD, Wing RA, Wendel JF (2006) Differential lineage-specific amplification of transposable elements is responsible for genome size variation in Gossypium. *Genome Res* 16:1252–1261

Hendrix B, Stewart JM (2005) Estimation of the nuclear DNA content of gossypium species. *Ann Bot* 95:789–797

Hu Y, Chen J, Fang L, Zhang Z, Ma W, Niu Y, Ju L, Deng J, Zhao T, Lian J, Baruch K, Fang D, Liu X, Ruan YL, Rahman MU, Han J, Wang K, Wang Q, Wu H, Mei G, Zang Y, Han Z, Xu C, Shen W, Yang D, Si Z, Dai F, Zou L, Huang F, Bai Y, Zhang Y, Brodt A, Ben-Hamo H, Zhu X, Zhou B, Guan X, Zhu S, Chen X, Zhang T (2019) Gossypium barbadense and Gossypium hirsutum genomes provide insights into the origin and evolution of allotetraploid cotton. *Nat Genet* 51:739–748

Huang G, Wu Z, Percy RG, Bai M, Li Y, Frelichowski JE, Hu J, Wang K, Yu J Z, Zhu Y (2020) Genome sequence of Gossypium herbaceum and genome updates of Gossypium arboreum and Gossypium hirsutum provide insights into cotton A-genome evolution. *Nat Genet* 52:516–524

Kim HJ, Triplett BA (2001) Cotton fiber growth in planta and in vitro. Models for plant cell elongation and cell wall biogenesis. *Plant Physiol* 127:1361–1366

Li F, Fan G, Wang K, Sun F, Yuan Y, Song G, Li Q, Ma Z, Lu C, Zou C, Chen W, Liang X, Shang H, Liu W, Shi C, Xiao G, Gou C, Ye W, Xu X, Zhang X, Wei H, Li Z, Zhang G, Wang J, Liu K, Kohel RJ, Percy RG, Yu JZ, Zhu YX, Wang J, Yu S (2014) Genome sequence of the cultivated cotton Gossypium arboreum. *Nat Genet* 46:567–572

Li F, Fan G, Lu C, Xiao G, Zou C, Kohel RJ, Ma Z, Shang H, Ma X, Wu J, Liang X, Huang G, Percy RG, Liu K, Yang W, Chen W, Du X, Shi C, Yuan Y, Ye W, Liu X, Zhang X, Liu W, Wei H, Wei S, Huang G, Zhang X, Zhu S, Zhang H, Sun F, Wang X, Liang J, Wang J, He Q, Huang L, Wang J, Cui J, Song G, Wang K, Xu X, Yu JZ, Zhu Y, Yu S (2015) Genome sequence of cultivated Upland cotton (Gossypium hirsutum TM-1) provides insights into genome evolution. *Nat Biotechnol* 33:524–530

Lin L, Pierce GJ, Bowers JE, Estill JC, Compton RO, Rainville LK, Kim C, Lemke C, Rong J, Tang H, Wang X, Braidotti M, Chen AH, Chicola K, Collura K, Epps E, Golser W, Grover C, Ingles J, Karunakaran S, Kudrna D, Olive J, Tabassum N, Um E, Wissotski M, Yu Y, Zuccolo A, ur Rahman M, Peterson DG, Wing RA, Wendel JF, Paterson AH (2010) A draft physical map of a D-genome cotton species (Gossypium raimondii). *BMC Genomics* 11:395

Liu X, Zhao B, Zheng HJ, Hu Y, Lu G, Yang CQ, Chen JD, Chen JJ, Chen DY, Zhang L, Zhou Y, Wang LJ, Guo WZ, Bai YL, Ruan JX, Shangguan XX, Mao YB, Shan CM, Jiang JP, Zhu YQ, Jin L, Kang H, Chen ST, He XL, Wang R, Wang YZ, Chen J, Wang LJ, Yu ST, Wang BY, Wei J, Song SC, Lu XY, Gao ZC, Gu WY, Deng X, Ma D, Wang S, Liang WH, Fang L, Cai CP, Zhu XF, Zhou BL, Jeffrey Chen Z, Xu SH, Zhang YG, Wang SY, Zhang TZ, Zhao GP, Chen XY (2015) Gossypium barbadense genome sequence provides insight into the evolution of extra-long staple fiber and specialized metabolites. *Sci Rep* 5:14139

Page JT, Liechty ZS, Alexander RH, Clemons K, Hulse-Kemp AM, Ashrafi H, Van Deynze A, Stelly DM, Udall JA (2016) DNA sequence evolution and rare homoeologous conversion in tetraploid cotton. *PLoS Genet* 12:e1006012

Pang CY, Wang H, Pang Y, Xu C, Jiao Y, Qin YM, Western TL, Yu SX, Zhu YX (2010) Comparative proteomics indicates that biosynthesis of pectic precursors is important for cotton fibre and Arabidopsis root hair elongation. *Mol Cell Proteomics* 9:2019–2033

Paterson AH, Wendel JF (2003) Polyploidy and the evolutionary history of cotton. *Adv Agron* 78:139–186

Paterson AH, Wendel JF, Gundlach H, Guo H, Jenkins J, Jin D, Llewellyn D, Showmaker KC, Shu S, Udall J, Yoo MJ, Byers R, Chen W, Doron-Faigenboim A, Duke MV, Gong L, Grimwood J, Grover C, Grupp K, Hu G, Lee TH, Li J, Lin L, Liu T, Marler BS, Page JT, Roberts AW, Romanel E, Sanders WS, Szadkowski E, Tan X, Tang H, Xu C, Wang J, Wang Z, Zhang D, Zhang L, Ashrafi H, Bedon F, Bowers JE, Brubaker CL, Chee PW, Das S, Gingle AR, Haigler CH, Harker D, Hoffmann LV, Hovav R, Jones DC, Lemke C, Mansoor S, ur Rahman M, Rainville LN, Rambani A, Reddy UK, Rong JK, Saranga Y, Scheffler BE, Scheffler JA, Stelly DM, Triplett BA, Van Deynze A, Vaslin MF, Waghmare VN, Walford SA, Wright RJ, Zaki EA, Zhang T, Dennis ES, Mayer KF, Peterson DG, Rokhsar DS, Wang X, Schmutz J (2012) Repeated polyploidization of Gossypium genomes and the evolution of spinnable cotton fibres. *Nature* 492:423–427

Ruan YL, Xu SM, White R, Furbank RT (2004) Genotypic and developmental evidence for the role of plasmodesmatal regulation in cotton fiber elongation mediated by callose turnover. *Plant Physiol* 136:4104–4113

Salih H, Gong W, He S, Sun G, Sun J, Du X (2016) Genome-wide characterization and expression analysis of MYB transcription factors in Gossypium hirsutum. *BMC Genet* 17:129

Salmon A, Flagel L, Ying B, Udall JA, Wendel JF (2010) Homoeologous nonreciprocal recombination in polyploid cotton. *New Phytol* 186:123–134

Shi YH, Zhu SW, Mao XZ, Feng JX, Qin YM, Zhang L, Cheng J, Wei LP, Wang ZY, Zhu YX (2006) Transcriptome profiling, molecular biological, and physiological studies reveal a major role for ethylene in cotton fiber cell elongation. *Plant Cell* 18:651–664

Tang H, Bowers JE, Wang X, Ming R, Alam M, Paterson AH (2008a) Synteny and collinearity in plant genomes. *Science* 320:486–488

Tang H, Wang X, Bowers JE, Ming R, Alam M, Paterson AH (2008b) Unraveling ancient hexaploidy through multiply-aligned angiosperm gene maps. *Genome Res* 18:1944–1954

Udall JA, Long E, Hanson C, Daojun Y, Thiruvarangan R, Justin LC, Lei G, Mark AA, Corrinne EG, Daniel GP, Jonathan FW (2019) De Novo genome sequence assemblies of Gossypium raimondii and Gossypium turneri. *G3* 9:3079–3085

Van de Peer Y, Fawcett JA, Proost S, Sterck L, Vandepoele K (2009) The flowering world: a tale of duplications. *Trends Plant Sci* 14:680–688

Wang K, Wang Z, Li F, Ye W, Wang J, Song G, Yue Z, Cong L, Shang H, Zhu S, Zou C, Li Q, Yuan Y, Lu C, Wei H, Gou C, Zheng Z, Yin Y, Zhang X, Liu K, Wang B, Song C, Shi N, Kohel RJ, Percy RG, Yu JZ, Zhu YX, Wang J, Yu S (2012) The draft genome of a diploid cotton Gossypium raimondii. *Nat Genet* 44:1098–1103

Wang X, Guo H, Wang J, Lei T, Liu T, Wang Z, Li Y, Lee TH, Li J, Tang H, Jin D, Paterson AH (2016) Comparative genomic de-convolution of the cotton genome revealed a decaploid ancestor and widespread chromosomal fractionation. *New Phytol* 209:1252–1263

Wang M, Tu L, Yuan D, Zhu D, Shen C, Li J, Liu F, Pei L, Wang P, Zhao G, Ye Z, Huang H, Yan F, Ma Y, Zhang L, Liu M, You J, Yang Y, Liu Z, Huang F, Li B, Qiu P, Zhang Q, Zhu L, Jin S, Yang X, Min L, Li G, Chen LL, Zheng H, Lindsey K, Lin Z, Udall JA, Zhang X (2019a) Reference genome sequences of two cultivated allotetraploid cottons, Gossypium hirsutum and Gossypium barbadense. *Nat Genet* 51:224–229

Wang J, Yuan J, Yu J, Meng F, Sun P, Li Y, Yang N, Wang Z, Pan Y, Ge W, Wang L, Li J, Liu C, Zhao Y, Luo S, Ge D, Cui X, Feng G, Wang Z, Ji L, Qin J, Li X, Wang X, Xi Z (2019b) Recursivepaleohexaploidization shaped the durian genome. *Plant Physiol* 179:209–219.

Wendel JF, Albert VA (1992) Phylogenetics of the Cotton genus (Gossypium): character-state weighted parsimony analysis of chloroplast-DNA restriction site data and its systematic and biogeographic implications. *Syst Bot* 17:115–143

Wikstrom N, Savolainen V, Chase MW (2001) Evolution of the angiosperms: calibrating the family tree. *Proc Biol Sci* 268:2211–2220

Wilkins TA, Arpat AB (2005) The cotton fiber transcriptome. *Physiol Plant* 124:295–300

Wu Z, Yang Y, Huang G, Lin J, Xia Y, Zhu Y (2017) Cotton functional genomics reveals global insight into genome evolution and fiber development. *J Genet Genomics* 44:511–518

Yoo MJ, Wendel JF (2014) Comparative evolutionary and developmental dynamics of the cotton (Gossypium hirsutum) fiber transcriptome. *PLoS Genet* 10:e1004073

Yu J, Jung S, Cheng CH, Ficklin SP, Lee T, Zheng P, Jones D, Percy RG, Main D (2014) CottonGen: a genomics, genetics and breeding database for cotton research. *Nucleic Acids Res* 42:D1229–D1236

Yuan D, Tang Z, Wang M, Gao W, Tu L, Jin X, Chen L, He Y, Zhang L, Zhu L, Li Y, Liang Q, Lin Z, Yang X, Liu N, Jin S, Lei Y, Ding Y, Li G, Ruan X, Ruan Y, Zhang X (2015) The genome sequence of Sea-Island cotton (Gossypium barbadense) provides insights into the allopolyploidization and development of superior spinnable fibres. *Sci Rep* 5:17662

Zhang T, Hu Y, Jiang W, Fang L, Guan X, Chen J, Zhang J, Saski CA, Scheffler BE, Stelly DM, Hulse-Kemp AM, Wan Q, Liu B, Liu C, Wang S, Pan M, Wang Y, Wang D, Ye W, Chang L, Zhang W, Song Q, Kirkbride RC, Chen X, Dennis E, Llewellyn DJ, Peterson DG, Thaxton P, Jones DC, Wang Q, Xu X, Zhang H, Wu H, Zhou L, Mei G, Chen S, Tian Y, Xiang D, Li X, Ding J, Zuo Q, Tao L, Liu Y, Li J, Lin Y, Hui Y, Cao Z, Cai C, Zhu X, Jiang Z, Zhou B, Guo W, Li R, Chen ZJ (2015) Sequencing of allotetraploid cotton (Gossypium hirsutum L. acc. TM-1) provides a resource for fiber improvement. *Nat Biotechnol* 33:531–537

Zhu H, Han X, Lv J, Zhao L, Xu X, Zhang T, Guo W (2011) Structure, expression differentiation and evolution of duplicated fiber developmental genes in Gossypium barbadense and G. hirsutum. *BMC Plant Biol* 11:40

Part II
Genomic-Assisted Breeding

Chapter 5
Genomics-Assisted Breeding for Biotic Stress Syndrome Resistance in Cotton

Ammad Abbas and Quddoos H. Muqaddasi

5.1 Introduction

Cotton is highly sensitive to both biotic and abiotic stresses. Biotic stresses are major limiting factors for cotton production globally. Under favorable climatic conditions, the spread of pathogen/s accelerates and drastically reduces the yield and lint quality of cotton. More than 50 bacteria, fungi, viruses, and nematodes have been implicated in causing multiple cotton diseases (Table 5.1). This makes cotton biotic stress (CBS) an important target of cotton breeding programs. Traditional breeding approaches rely on exploiting the natural genetic variations to improve resistance to CBS. Despite achieving success, CBS resistance remains a major threat mainly due to the qualitative and quantitative genetic nature of resistance, and narrow genetic diversity—both of which have imposed a bottleneck in the evolutionary history of cotton (Iqbal et al. 2001; Rahman et al. 2008). To tackle these challenges, genomic approaches combined with traditional phenotypic selection are at the forefront of modern cotton breeding. Identification of molecular markers linked with loci controlling the CBS resistance traits, together with the exploitation of genome-wide selection, provides an opportunity to speed up gain from the selection. Particularly, marker-assisted selection (MAS) or marker-assisted backcrossing (MABC) are now regularly deployed against CBS. However, the improvement of resistance to CBS—like other quantitative traits—remains challenging, primarily

A. Abbas
Institute for Molecular Plant Science, School of Biological Sciences, Daniel Rutherford Building, Max Born Crescent, Kings Buildings, University of Edinburgh, Edinburgh, UK

Q. H. Muqaddasi (✉)
Leibniz Institute of Plant Genetics and Crop Plant Research (IPK), Stadt Seeland OT Gatersleben, Germany

European Wheat Breeding Center, BASF Agricultural Solutions GmbH, Stadt Seeland OT Gatersleben, Germany
e-mail: muqaddasi@ipk-gatersleben.de

Mehboob-ur-Rahman et al. (eds.), *Cotton Precision Breeding*,
https://doi.org/10.1007/978-3-030-64504-5_5

Table 5.1 List of major cotton biotic stress syndrome components and their causal parasites

Syndrome components	Parasite	Parasite name
Cotton bacterial blight	Bacterium	*Xanthomonas campestris* pv. *malvacearum*
Crown gall	Bacterium	*Agrobacterium tumefaciens*
Xanthomonas boll rot	Bacterium	*Xanthomonas campestris* pv. *malvacearum*
Alternaria leaf spot	Fungus	*Alternaria alternata*
Alternaria leaf spot	Fungus	*Alternaria gossypina*
Alternaria leaf spot	Fungus	*Alternaria macrospora*
Ascochyta blight	Fungus	*Ascochyta gossypii*
Internal boll infection	Fungus	*Ashbya gossypii*
Aflatoxin	Fungus	*Aspergillus flavus*
Aspergillus boll rot	Fungus	*Aspergillus flavus*
Aflatoxin	Fungus	*Aspergillus nomius*
Aflatoxin	Fungus	*Aspergillus parasiticus*
Cercospora leaf spot	Fungus	*Cercospora gossypina*
Anthracnose boll rot	Fungus	*Colletotrichum* spp.
Fusarium wilt	Fungus	*Fusarium oxysporum* f.sp. *vasinfectum*
Seedling disease	Fungus	*Fusarium* spp.
Fusarium boll rot	Fungus	*Fusarium* spp.
Anthracnose	Fungus	*Glomerella gossypii*
Lasiodiplodia (Diplopoda) boll rot	Fungus	*Lasiodiplodia theobromae*
Powdery mildew	Fungus	*Leveillula taurica*
Charcoal rot	Fungus	*Macrophomina phaseolina*
Myrothecium leaf spot	Fungus	*Myrothecium roridum*
Internal boll infection	Fungus	*Nematospora coryli*
Nigrospora lint rot	Fungus	*Nigrospora oryzae*
Pantoea boll rot	Fungus	*Pantoea agglomerans*
Tropical cotton rust	Fungus	*Phakopsora gossypii*
Phomopsis leaf spot	Fungus	*Phomopsis* spp.
Phymatotrichum root rot	Fungus	*Phymatotrichopsis omnivora*
Phytophthora boll rot	Fungus	*Phytophthora* spp.
Southwestern cotton rust	Fungus	*Puccinia cacabata*
Tumblegrass rust	Fungus	*Puccinia schedonnardi*
Seedling disease	Fungus	*Pythium* spp.
Areolate or false mildew	Fungus	*Ramularia gossypii*
Seedling disease	Fungus	*Rhizoctonia solani*
Rhizoctonia leaf spot	Fungus	*Rhizoctonia solani*
Powdery mildew	Fungus	*Salmonia malachrae*
Southern blight	Fungus	*Sclerotium rolfsii*
Stemphylium leaf spot	Fungus	*Stemphylium solani*
Black root rot	Fungus	*Thielaviopsis basicola*
Verticillium wilt	Fungus	*Verticillium dahliae*

(continued)

Table 5.1 (continued)

Syndrome components	Parasite	Parasite name
Sting nematodes	Nematode	*Belonolaimus* spp.
Lance nematodes	Nematode	*Hoplolaimus columbus*
Root-knot nematodes	Nematode	*Meloidogyne incognita*
Reniform nematodes	Nematode	*Rotylenchulus reniformis*
African cotton mosaic	Virus	African cotton mosaic virus
Cotton anthocyanosis	Virus	Cotton anthocyanosis virus
Cotton blue disease	Virus	Cotton leafroll dwarf virus
Cotton leaf curl disease	Virus	Cotton leaf curl virus
Cotton yellow vein disease	Virus	Cotton yellow vein virus
Flavescence	Phytoplasma	Phytoplasma
Flower virescence	Phytoplasma	Phytoplasma
Psyllosis	Phytoplasma	Phytoplasma
Small leaf	Phytoplasma	Phytoplasma
Bronze wilt	Unknown	Unknown

due to the difficulty in detecting large-effect quantitative trait loci (QTLs) with fairly stable effects across genetic backgrounds and environments. Besides, in some cases, the effects of most QTL controlling CBS resistance traits are too small to be detected either through genome-wide linkage mapping or association mapping. Genomic selection (GS) appears to circumvent this issue by utilizing an index for the selection of unmapped QTLs with minor individual effects, but, when selected cumulatively, with an otherwise substantial effect at the whole plant level. The GS has become a major part of crop improvement programs in recent years. However, its potential still remains largely untapped in cotton. With the availability of genome sequence information of cotton, development of multi-omics platforms, and advancements in molecular breeding approaches, it has become possible to overcome the challenges that previously seem daunting. A successful strategy for CBS resistance will ultimately rely on effective exploitation of marker-assisted breeding by close integration between conventional breeding and other disciplines.

5.2 Genome Architecture of Genus *Gossypium* in the Face of Biotic Stresses

Understanding the cotton genome organization provides useful information about its evolutionary relationship and functional significance among different species (Chen et al. 2007). Most textile fiber is derived from four species, including two Old-World Asian-African diploids *G. arboreum* L. and *G. herbaceum* L. (A-genome) and two New-World tetraploid species, *G. hirsutum* L. and *G. barbadense* L. (AD-genome) species. The primitive diploid species evolved into multiple genomes designated A–G and K, based on their chromosomal pairing affinity during meiosis (Wendel and Cronn 2003). The New-World tetraploid species evolved ~1–2 million years ago resulting from hybridization of A-genome diploids *G. herbaceum*

Table 5.2 List of traits introgressed from wild relatives to cultivated cottons for biotic stresses

Trait	Wild relative	Reported cause of resistance
Disease tolerance		
Cotton bacterial blight	*G. arboreum*	Bacterial blight resistance gene B6
Cotton leaf curl disease resistance	*G. stocksii*	*Unknown*
	G. herbaceum	*Unknown*
	G. arboreum	*Unknown*
Fusarium wilt resistance	*G. austral*	*Unknown*
	G. sturtianum	*Unknown*
	G. darwinii	*Unknown*
Rust resistance	*G. anomalum*	*Unknown*
Verticillium wilt resistance	*G. austral*	*Unknown*
	G. thurberi	*Unknown*
	G. darwinii	*Unknown*
Nematode tolerance		
Reniform nematode resistance	*G. longicalyx*	Resistance gene on Chr-11
	G. somalense	*Unknown*
	G. stocksii	*Unknown*
	G. arboreum	*Unknown*
	G. barbadense L. GB713	Resistance gene in Chr-21
Root-knot nematode resistance	*G. hirsutum* L. M315 RNR	Resistance genes in Chr-11 and Chr-14

and *G. arboreum* ($2n = 2\times = 26$), with D-genome diploids *G. raimondii* Ulbrich and *G. gossipioides* L. ($2n = 2\times = 26$) (Wendel 1989; Wendel and Cronn 2003).

The availability of diversity in the form of wild species and cultivated cotton is a valuable resource for plant breeders to accelerate the selection of agronomically valuable traits, e.g., biotic and abiotic stress tolerance, yield, plant height, etc. Although the transfer of these traits from wild species into cultivated tetraploid species is a daunting task, many successes have been achieved through the development of interspecific hybrids involving donor and recipient genotypes. (Rahmat et al. 2019). The interspecific hybrids have been generated through bridge crosses or other means, including the use of cytogenetic tools to find the genetic recombinants between the chromosomes of donor and recipient genotypes (Konan et al. 2007; Robinson et al. 2004). A few examples of biotic stress tolerance traits transferred from wild species into cultivated cotton species are summarized in Table 5.2.

5.3 Cotton Biotic Stress Syndrome

The parts of cotton plant harvested per unit area that translate into the economic yield are mainly cotton fiber, seed, and wood to some extent. Cotton pathogen and pests—that include bacteria, fungi, nematodes, insects, viruses, and weeds—lead to

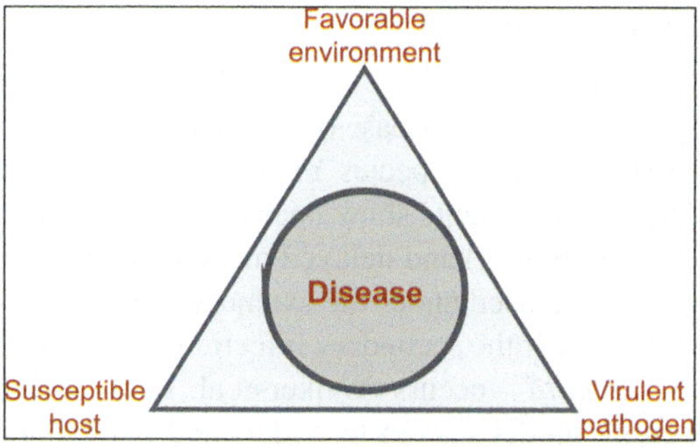

Fig. 5.1 Disease triangle shows the prerequisites for a successful disease infestation. The impact of disease can be reduced by tailoring the essentials, i.e., by decreasing the host susceptibility, altering the environment that is favorable for successful pathogen infestation, or by agrochemical practices to mitigate pathogen virulence

pre- and post-harvest damages by reducing plant vigor and often to plant death. This difficulty is further compounded by the fact that plants generally lack an adaptive immune system. Although a host of pathogen pests infect cotton, some major pathogens cause severe devastation. Furthermore, cotton plant often encounters a combination of pathogen-pest infections simultaneously during the entire life cycle, and therefore, we call this simultaneous infestation by a host of pathogen pests as cotton biotic stress (CBS) syndrome. For successful pathogen infestation, favorable environment coupled with a susceptible host is needed. In the absence of any one of these three components, disease progression does not take place, as shown in Fig. 5.1. In the following, a summary of some major pathogens and the economic losses associated with them is given.

5.3.1 Cotton Bacterial Blight

Bacterial blight (BB) of cotton is caused by a bacterial pathogen *Xanthomonas campestris* pv. *malvacearum* (*Xcm*) that occurs in all cotton-growing regions around the world (Pkania et al. 2014). It is among the most devastating diseases of cotton. Depending upon the severity of infection, the lint yield losses can range from 5 to 35%. The bacterium is mainly a seed-borne pathogen, but it can also enter through the leaf stomatal opening or open wounds caused by crop debris from a previous cropping season (Hillocks 1992). Early symptoms of the disease start as water-soaked lesions on leaves, which then progress into angular lesions as the leaf veins restrict bacterial movement. As the disease progresses, the leaf veins darken and produce a "blighting" appearance, and eventually, defoliation takes place. The same lesions appear on cotton bolls causing rot. Bacterial blight can rapidly spread in the other parts of the field through irrigation or wind-driven rain. Currently, no effective chemical treatments exist to control the diseases once it spreads in the field, nor does the economic value of crop allow for season-long preemptive applications of bacteriocides.

5.3.2 Cotton Black Root Rot

Black root rot (BRR) is a seedling disease of cotton caused by a soilborne pathogen *Thielaviopsis basicola*, a fungal species distributed across the globe (Raj and Kendrick 2006). The infected plants show necrosis at the tap and lateral roots that lead to reduced growth (stunting) and delayed maturity (Minton and Garber 1983). The scale of damage is increased under the extended period of cooler weather conditions (Rothrock 1992), or if the secondary infection—caused by a root-knot nematode *Meloidogyne incognita*— occurs (Walker et al. 1998). Black root rot has been identified as an exciting model system to understand the evolutionary differences resulting in *R*-gene formation and divergence in cotton. Cultivated cotton varieties do not harbor resistance against BRR primarily due to the lack of resistance in cultivated tetraploid cotton (*G. hirsutum and G. barbadense*). However, the genetic resistance was found in *G. herbaceum,* one of the nearest possible parent species of tetraploids (Wheeler et al. 1999), and was transferred to cultivated species.

5.3.3 **Fusarium** *Wilt of Cotton*

Fusarium oxysporum f.sp. *vasinfectum* (Fov) is a soil fungal pathogen that causes wilt in all cotton-producing regions. Typical disease symptoms include brown discoloration of the vascular system, growth stunting, wilting, necrosis, and eventually plant death. The symptoms can appear at any developmental stage and are further aggravated by pathogen virulence, host defense, soil type and fertility, climatic conditions, and nematode populations (Davis et al. 2006). More than eight races have been identified based on their pathogenicity in cotton and other plant species, including soybean, okra, alfalfa, and tobacco (Sanogo and Zhang 2016; Ulloa et al. 2013). Among these, race 1 (Fov1) was primarily detected with root-knot nematodes leading to severe disease symptoms (Hillocks 1983, 1992). However, the virulent Fov4 strain alone is sufficient to cause disease in the absence of nematodes (Davis et al. 1996). The Fov is challenging to control in cotton as fungal hyphae penetrate in the woody vascular tissues and make chlamydospores which may allow prolonged survival in soil (Cianchetta and Davis 2015; Sanogo and Zhang 2016). Control strategies mainly rely on the breeding of resistant germplasm against *Fusarium* wilt. Besides crop rotation with nonhost species and control of nematode population, there are no other control measures available for virulent Fov races.

5.3.4 **Verticillium** *Wilt of Cotton*

Verticillium wilt is one of the most important diseases prevalent in all major cotton-growing regions. The disease is caused by the soilborne fungus *Verticillium dahliae Kleb*, detectable in most soils, and, therefore, has been the focus of intensive research initiatives in recent years. The primary disease symptoms include the presence of necrotic patches on leaves, leaf yellowing, wilting, and discoloration of vascular tissues. Under humid conditions, infected plants produced white spores of *V. dahlia*. The appearance of a mosaic pattern with interveinal yellowing is the characteristic symptom of the affected plants before they become necrotic and eventually quickly die (Fradin and Thomma 2006).

The disease has also been found to damage bolls, leading to hyphal penetration into seeds. It significantly influences fiber quality by reducing micronaire and span length (Zhang et al. 2012). Disease symptoms are influenced by several environmental factors such as temperature, light intensity, photoperiod, level of soil moisture, irrigation, and cultural practices (Karademir et al. 2012; Robb 2007). It progresses well at 22–25 °C and infects *G. hirsutum* more severally than *G. barbadense,* which is believed to carry resistance genes. So, it is crucial for researchers to identify the responsible gene/s in *G. barbadense* and ultimately harness them for developing wilt-resistant germplasm resources (Cun et al. 1997; Shaban et al. 2018).

5.3.5 Cotton Nematodes

Nematodes were recognized as a threat to cotton in the late nineteenth century. Earlier work established the pathogenicity of root-knot nematode to cotton and its parasitic role in wilt disease of cotton (Atkinson 1892). Many species of plant-parasitic nematodes have been implicated in cotton. However, reniform nematode (*Rotylenchulus reniformis*) and southern root-knot nematode (*Meloidogyne incognita*) cause most damage in the United States and also other parts of the world (Lawrence et al. 2017). Cotton plants infected with reniform nematode show stunted growth, delayed flowering and fruiting, reduction in size and number of fruits, and decrease in lint quality (Robinson 2007). On the other hand, cotton plants affected with root-knot nematodes produce galls on the root system. Galls result from the swelling of root tissues associated with the feeding of parasitic nematodes inside the root. Aboveground symptoms include chlorosis and growth stunting. The foliar symptoms may also be linked with a secondary infection caused by bacteria or fungi after the root damage caused by the nematodes (Starr et al. 2007). The losses due to plant-parasitic nematodes in cotton have been attributed to numerous factors, viz., (1) the lack of resistant varieties, (2) inadequate use of crop rotation practices, (3) lesser awareness of pathogenic nematodes as a threat to cotton production, and (4) the limited availability of highly effective, low-cost, fumigant nematicides.

5.3.6 Cotton Leaf Curl Disease Virus

The cotton leaf curl disease (CLCuD) is a major and increasingly growing threat to worldwide cotton. The disease was first reported in Nigeria in 1912 (Kirkpatrick 1931), where it infected local cultivated cotton species (*G. peruvianum* and

G. vitiforum). Later in 1924, it was first time reported to infect *G. hirsutum* in Nigeria. The same year, the disease also affected cotton fields in Sudan that was followed by an outbreak of the disease in Northern Africa (Hussain et al. 1991). The CLCuD was first reported in Pakistan in 1967 near a village in Multan. Despite its sporadic presence in the cotton belt of Pakistan, CLCuD remained neglected until it appeared in an epidemic form in the early 1990s (Briddon and Markham 2000). The first phase of epidemic cost 5 billion US$ losses to the economy of Pakistan between 1992 and 1997 (Idris 1990). The disease was also reported in the cotton fields of India (Reddy et al. 2005) which were at the periphery of the cotton belt of Pakistani Punjab and later spread to the northern cotton-growing regions of India. Recently, the disease has also been reported in southeastern parts of China (Cai et al. 2010) and the Philippines—spread is blamed to the cross-border trade of infected plant samples (e.g., cuttings of hibiscus).

The characteristic symptoms of the disease are vein thickening and upward or downward leaf curling. Under severe infection, leaves develop enation and, eventually, growth retardation of the cotton plant—a state that reduces the yield by 15–70% (Brown 2002; Idris 1990). The CLCuD is caused by a complex of geminiviruses that are transmitted through whitefly (Brown 1992). The new classification suggests that five different species of begomovirus are responsible for causing the diseases in different parts of the world, i.e., cotton leaf curl Kokharan virus (CLCuKoV), cotton leaf curl Allahabad virus (CLCuAlV), cotton leaf curl Multan virus (CLCuMuV), cotton leaf curl Gezira virus (CLCuGeV), and cotton leaf curl Bangalore virus (CLCuBaV; Muhire et al. 2014). The complex comprises of a monopartite begomovirus DNA A, betasatellite, and alphasatellite. Added to this complexity is the higher rate of genetic recombination of the virus complex, which makes it challenging to develop a durable resistance against the disease. Currently, Burewala viral strain, also known as cotton leaf curl Burewala virus (CLCuBuV), infected all previously known resistant varieties in Pakistan (Rahman et al. 2017). The CLCuBuV has also been reported in India, replacing the old strains of CLCuRV and CLCuKoV (Rajagopalan et al. 2012). India, China, and Pakistan produce more than 60% of the world's cotton. Therefore, cotton is at risk in these countries and highlights the need for developing control measures to counter the disease spread to other cotton-growing regions (Rahman et al. 2017).

5.4 CBS Syndrome Variation

Agrochemical applications and crop management practices, viz., crop rotation, and crop mixtures are generally the methods of choice to mitigate the effects of biotic stresses in cotton. However, these methods bring along substantial crop production penalties. Moreover, the rigorous chemical application coupled with intensive use of fertilizers and artificial irrigation is not deemed sustainable. Genetic improvement of cotton is, therefore, a suitable alternative that could reduce the environmental impact and improve sustainability by harnessing the natural allelic diversity and minimizing the vulnerability to pathogen and pests.

Genetic (allelic) diversity is defined as the difference in genotypic (allelic) composition among the individuals of a given population (Lynch and Walsh 1998). For genetic improvement of cotton, similar to any other crop, a large and significant genetic variance in breeding germplasm is essential. However, a large and significant environmental variation and genotype-by-environmental interaction also exist for CBS syndrome traits which are typical because (1) disease pressure is not always consistent across the environments and (2) environmental effects are not predictable. For the release of a brilliant variety, genotype-by-environment interaction has to be minimum in a target population of environments. Breeders usually target diverse environments to develop a variety; and, therefore, of the abovementioned three components of variation, only genetic variation is mostly exploited in breeding programs.

Diverse crosses can create genetic variations (Rahman et al. 2002) and, for that purpose, different heterotic groups or patterns may be created (Melchinger and Gumber 1998). However, as described by Moll et al. (1965), an increase in genetic divergence beyond specific optimal results in the decrease in heterosis. In breeding programs, therefore, not all genetic variations are useful. For example, a cross between two highly genetically diverse lines results in trade-offs for important agronomic traits. Gene banks across the globe host a rich source of allelic variation. However, most of this variation cannot be utilized in current breeding programs because of adverse effects on the adaptation of plants to agricultural environments (Varshney et al. 2005).

5.5 Classical and Marker-Assisted Breeding for CBS Syndrome Resistance: Potential and Limits

5.5.1 Potentials and Limits of Classical Breeding in Cotton

Artificial selection forms the basis of every plant breeding program (Acquaah 2007). It involves the selection, preservation, and propagation of plants possessing the most desirable characters from a diverse population. The success of selection of desirable characters in the subsequent generations depends principally on their heritable genetic nature. Since its independent domestication in the Old and the New Worlds, cotton breeding has relied on the mass and pure-line selection methodologies in populations of genetically distinct landraces. Mass selection involves the removal of highly heterogeneous plants from a population. The remaining homogeneous plants are propagated in the following generations that eventually increases the uniformity of the crop. Pure-line selection method typically involves the recognition and selection of the best-performing plant and its propagation in the subsequent generation—a process that helps develop homogeneity and homozygosity in a short period. However, being homozygous and homogeneous, pure lines become dead-end products for further improvement.

Plant breeding has benefited from the understanding and utilization of principles of hybridity and inheritance at large (Bernardo 2010). Selection, by nature, works effectively only on heritable differences and does not create a new genetic variation—a key for a successful breeding program (Baenziger and Depauw 2009). Genetic variation, however, can be created through hybridization or crossing purelines—a procedure that has led to the development of improved breeding methods, e.g., pedigree selection, bulk selection, single seed descent, doubled-haploid production, backcrossing, synthetic development, and hybrid breeding. Creating genetic variation and understanding the heritability of desired traits are vital for successful breeding.

In principle, cotton breeding involves the genetic improvement based on an index of traits with the final goal to improve the economic yield that includes traits associated with fiber and seed. Nevertheless, these traits are the ultimate targets of biotic stresses. Mitigating the biotic stresses in cotton via genetic improvement is the most sustainable option which has been achieved via traditional breeding by crossing two parents, one having genetic factors responsible for CBS syndrome trait resistance (the donor variety) and the other with improved economic yield (elite variety). The donor variety is hybridized with the elite, and several cycles of backcrossing with the elite variety result in a variety having the genetic background of elite plus the resistant loci from the donor variety. This process is time-consuming and may result in the genetic erosion mainly because many loci act in a gene-for-gene relationship, i.e., host-parasite interaction is governed by pairs of matching genes, namely, resistant (*R*) avirulence (*Avr*) genes harbored by the host plant and the parasite, respectively (Flor 1955). The risk of genetic erosion via elite-by-elite crosses, however, can be avoided by improving the genetic base of a breeding program which can be achieved by establishing pre-breeding programs.

5.5.2 Potentials and Limits of Marker-Assisted Breeding in Cotton

In the past, cotton breeding mainly relied upon the phenotypic selection, e.g., mass selection or progeny selection. Molecular markers are identifiable sequences of DNA located ubiquitously in the genomes of organisms. There are many types of molecular markers, such as restriction fragment length polymorphisms (RFLP), simple sequence repeat (SSR), amplified fragment length polymorphisms (AFLP), randomly amplified polymorphic DNA (RAPD), and single nucleotide polymorphisms (SNP), each having strengths and weaknesses in identifying genetic polymorphisms in cotton (Rahman et al. 2008). The availability of abundant genetic markers has, nevertheless, helped the geneticists and breeders to exploit the variation among organisms at a genetic scale.

It is known that most agronomically important traits harbor quantitative genetic nature. Variation in quantitative traits is wide and caused by multiple loci (quantitative trait loci or QTL) segregating in the population with small effects and imparting minor to modest phenotypic variation to the trait. Also, fluctuating prevailing environments may alter the expression of the quantitative traits. Due to the varying degree of phenotype, detection of QTL conferring small effects on the trait is not possible by studying the segregation ratios in crosses or pedigrees. Mapping QTL relies on co-inheritance of the causative genes and the nearby genetic markers. It is used to identify the genetic markers by statistically estimating the mean and variance associated with a specific locus. A general prerequisite to identify QTL, therefore, is to find marker loci in tight linkage to the QTL. The marker loci in close vicinity to the QTL show larger effects and, by showing Mendelian segregation, are amenable to selection. These large-effect markers help breeders to speed up the process of breeding as they contribute to selection gain per unit time and cost. The benefits of MAS are, therefore, only realized when the markers are tightly linked to the genetic value of the trait, are of large effect, and impart large phenotypic variance. Depending upon the genetic nature of populations (e.g., biparental, multiparental, and diverse genetic populations), various types of QTL mapping procedures can be employed to detect the loci underlying the trait (Shaheen et al. 2012).

Traditionally there are two types of QTL mapping approaches, (1) genome-wide linkage mapping and (2) genome-wide association mapping. Genome-wide linkage mapping is performed to map the trait-linked loci in the artificially developed genetic population derived mainly from F_1 hybrid that results from a cross between two inbred parents. It is, therefore, performed on F_2 or its derived populations, e.g., F_3, F_4, etc., backcrossed lines, doubled-haploid lines, recombinant inbred lines, immortalized F_2 lines, and nearly isogenic lines. There are different methods of linkage mapping, e.g., simple interval mapping, composite interval mapping, multiple interval mapping, etc., that can be performed on these populations to identify the QTL underlying the quantitative traits. Genome-wide association mapping studies (GWAS), on the other hand, is a procedure to identify QTL in diverse populations consisting of individuals having different genetic background and geographic history. The QTL identification through GWAS is robust because it does not involve the time to develop the artificial populations via designed mapping experiments. The GWAS, therefore, detects QTL using existing phenotypic and genotypic datasets. Comparison between the two approaches is given in Fig. 5.2. The reader is suggested to read Bernardo (2010) and Lynch and Walsh (1998) to study the mapping methods in detail. In the following, we describe the use of linkage mapping and GWAS to identify the QTL for use in MAS to improve CBS syndrome resistance.

5.5.3 Breeding for Resistance to Cotton Bacterial Blight

Agronomic practices, e.g., seed cleaning schemes, may mitigate CBB infection to some degree, but genetic improvement serves as a durable option. Genetic resistance to CBB varies considerably within the genus *Gossypium*: diploid species grown in Indian sub-continent, viz., *G. arboreum* and *G. herbaceum* are immune or

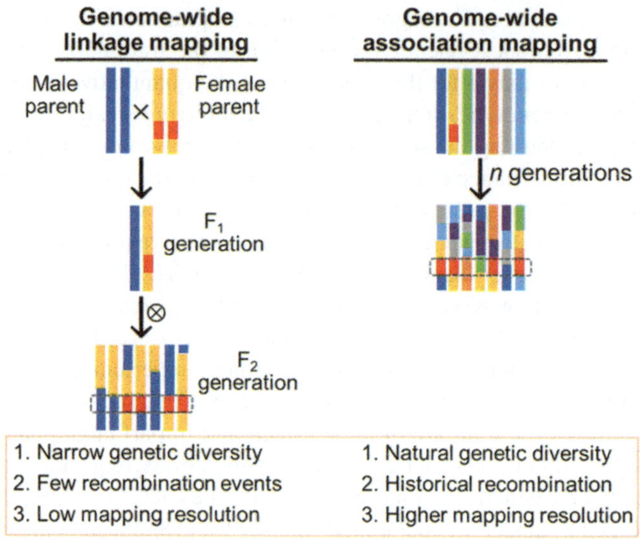

Fig. 5.2 A simplified comparison of genome-wide linkage mapping and genome-wide association mapping approaches to identify loci underlying the trait being investigated. Three main genetic features of both populations are given in the box. Genome-wide linkage mapping is performed on populations mainly derived from crosses between two parents harboring contrasting phenotypes. Genome-wide association mapping is performed on populations containing allelic diversity

highly resistant to CBB, tetraploid *G. hirsutum* exhibits a wide range of genetic variations for CBB resistance, and tetraploid *G. barbadense* shows very marginal or no resistance to the disease (Wallace and El-Zik 1990). Efforts for establishing the genetic of resistance (synonymous with blackarm) started in the 1930s (Knight and Clouston 1939). So far, 18 major resistance genes (also called *B*-genes) with fully or partially dominant genetic nature have been identified to be associated with CBB (Delannoy et al. 2005; Jalloul et al. 2015; Knight and Hutchinson 1950; Wright et al. 1998). These genes work as either in a race-specific manner or in concerted action to provide resistance; however, no single gene provides complete immunity or durable resistance because of continuous evolutionary shifts in pathogen (Brinkerhoff 1970; Brinkerhoff et al. 1984). This was further supported by the data from nearly isogenic lines of upland cotton harboring resistance genes for CBB, viz., B_2, B_4, B_{ln}, and b_7, that were developed to study the genes' action (Essenberg et al. 2002). The race-specific genes, viz., B_4, B_{ln}, and b_7, were pyramided in all possible combinations to observe whether the individual genes or their combinations provide high resistance compared to an exceptional line "lm2016" that was developed by L.A. Brinkerhoff (Essenberg et al. 2014). However, it was concluded that the pyramiding of *B*-genes achieved a broader but not always higher resistance to CBB. Nevertheless, tetraploid varieties, since they occupy almost all cotton-growing areas worldwide, have been developed via crossing with other *Gossypium* species containing CBB resistance genes: most of the *B*-genes or *B*-gene complexes were used via classical breeding approaches to improve resistance in elite lines (Bird 1982; El-Zik and Bird 1970).

With the advent of abundant molecular markers and development of genetic linkage maps, it was possible to locate the *B* resistance genes' chromosomal locations as well as identify new QTL underlying CBB resistance (Wright et al. 1998). Linkage maps have also provided the idea of the sub-genomic distribution of the resistance loci. For example, almost all resistance genes are located on D-subgenome of cotton. It was postulated that, among others, the possible reasons of D-subgenome being highly populated with the resistance genes are (1) biogeographic considerations, (2) differences in the evolutionary history of A- and D-subgenomes, and (3) low power of linkage mapping due to which some of the loci underlying CBB might have missed (Wright et al. 1998). Since then, multiple reports regarding the identification of CBB causal loci in various populations, including biparental and diverse mapping populations, have appeared (Shi et al. 2011). A comprehensive view of the genetic architecture of CBB is that the disease is controlled by large-effect loci that impart large phenotypic variance to trait—a situation that suits the use of MAS to efficiently complement the phenotypic selection. Because it is hard to phenotype disease traits, MAS can manifest better advantage than phenotypic selection to increase the genetic gains per unit time and cost.

5.5.4 Breeding for Resistance to Cotton Black Root Rot

Despite global initiatives toward developing black root rot resistance in cultivated tetraploid cotton, *G. hirsutum* and *G. barbadense*, no resistant cultivar is available for commercial use owing to the lack of resistance in tetraploid cotton germplasm. However, resistance has been traced in tetraploid progenitor species, *G. herbaceum* (AA genome) (Wheeler et al. 1999). Employing genetic mapping, QTLs linked with resistance to black root rot were identified in an F_2 population of *G. herbaceum* x G. *arboreum*. Genetic component that confers resistance could be explained by three QTLs, BRR5.1, BRR9.1, and BRR13.1, collectively explaining 32% of phenotypic variance (Niu et al. 2008). Since AA genome is the progenitor of cultivated tetraploid species AADD, it is likely that AA genome may harbor *R*-genes for black rot resistance. However, work on isolating *R*-genes in AA genome, particularly for *Thielaviopsis basicola,* is still evolving (Wright et al. 2009).

Comparative mapping is a powerful tool for quickly identifying genomic regions associated with the trait of interest by establishing the syntenic relationship between the well-studied genome and the genome under study. Thus the information generated on the well-studied genome can be translated to the genome under exploration. In multiple reports, several genes, as well as DNA makers, were identified using *Arabidopsis* (a well-studied genome) or any other closely related species within the corresponding family (Paterson 2010). Comparative analysis revealed several regions of correspondence between black root rot QTLs and *Arabidopsis* pathogen defense genes, suggesting that the position of 56 *Arabidopsis* genes could be placed within black root rot QTL region (Rong et al. 2007). Even though the position of all *Arabidopsis* genes could not be inferred in cotton, some stress and pathogen defense response genes could be potentially located within black root rot QTL intervals. The synteny-based comparison holds tremendous potential to identify underlying defense response genes in cotton and as well as establishing the evolutionary link with a common ancestor.

5.5.5 Breeding for Resistance to Cotton Leaf Curl Disease

Cotton leaf curl disease (CLCuD) is an emerging global threat to cotton production; hence, efforts are underway to find durable resistance against the virus. The critical prerequisites of designing breeding strategies for disease resistance include comprehensive information on the genetic basis of resistance and its inheritance in the subsequent generation (Hutchinson et al. 1950; Rahman et al. 2005a).

Knowledge on the molecular basis of resistance against CLCuD is limited. An early study has suggested that the resistance is conferred by a single major gene (Knight 1948). The inheritance of disease resistance and disease susceptible reactions were studied in an F_2 cross between susceptible *G. barbadense* L. (Giza-45) and resistant *G. hirsutum* L. (Reba P-288) parents. Individual plant reactions to CLCuD were bimodal (223 resistant: 62 susceptible) and did not deviate significantly from a single-gene model ($P < 0.2$). These observations were consistent with prior research suggesting that a single dominant gene in *G. hirsutum* confers resistance to the disease (Ali 1997). A region on chromosome 4 near the DNA markers A1215, A1826, and pGH318 accounted for resistant and susceptible phenotypic reactions to CLCuV. Two DNA marker loci, detected by probes A1215 and A1826, essentially co-segregate with the locus (0.0 and 0.1 cM). A third marker locus (pGH318) is 11.6 cM from the resistance locus. Since the publication of this research, two separate research groups have published findings that suggest the inheritance is controlled by three loci (Rahman et al. 2005a). Tolerance to CLCuD has been acknowledged as a complex trait with an incomplete expression.

Breeding of virus-resistant cotton varieties with adequate genetic diversity has been recommended as the most effective and durable strategy for controlling the disease (Rahman et al. 2002, 2005a). Several disease-resistant varieties have been developed using recombination breeding approaches. These approaches primarily relied on the transfer of resistance from two famous sources, "LRA-5166" and "CP-15/2," and therefore resulted in a narrow genetic base that led to the breakdown of resistance within five years by the evolution of new Burewala strain of virus (Arshad et al. 2009; Rahmat et al. 2014). To date, none of the commercially available cotton varieties is asymptomatic. Nevertheless, tolerant cotton genotypes, viz., "FH-142," "NN-3" (Rahman and Zafar 2012), IR-NIBGE-11 (derived resistance from Mac07), and "NIBGE-115" (Rahman and Zafar 2007), have been identified which can control the disease.

Diploid cotton species, *G. arboreum* and *G. herbaceum*, are resistant or immune to CLCuD (Rahman et al. 2005a). *G. robinsonii* has also been recognized as a new resistant species (Azhar et al. 2011). Efforts were made to transfer the resistance from diploid into *G. hirsutum*; however, the rate of success was limited because of extensive linkage drag. Alternatively, introgression from *G. hirsutum* traits into *G. arboreum* was carried out, but no commercial variety was developed using this technique (Nazeer et al. 2014).

The use of DNA markers in developing resistant lines against CLCuD has been handicapped due to the limited genetic diversity in available genetic resources. For instance, genetic similarity among cotton cultivars pre and post the first epidemic was 81.5–93.41% and 81.45–94.90%, respectively (Iqbal et al. 1997; Rahman et al. 2002, 2008). Nevertheless, DNA marker studies have been conducted by several researchers to map genomic regions linked to the CLCuD resistance. In an early study, RFLP markers were used to detect three loci linked to resistance in an interspecific F_2 population (Aslam et al. 2000). In a different study, three RAPD markers were found associated with the CLCuD resistance (Rahman et al. 2005b). A similar using simple sequence repeat (SSR) markers identified two SSR markers, PR-91 and CM-43, significantly associated with resistance to CLCuD and its viral causal agents (Abbas et al. 2015). QTL mapping in an interspecific population developed by crossing a highly tolerant genotype var. 2472-3 of *G. hirsutum* with the highly sensitive genotype var. PGMB-66 of *G. barbadense* resulted in the identification of two QTLs, i.e., QCLCuD25 and QCLCuD26, associated with CLCuD resistance (Rahman et al. 2014a, b). A detailed genetic linkage map, however, is not yet available and requires further genotyping of F_2 population.

5.6 Genome-Wide Selection for CBS Syndrome Resistance

The introgression of minor-effect loci is not beneficial and possible in marker-assisted breeding programs primarily due to low genetic gains per unit time and substantial operational costs. MAS is profitable when a few genes with large effects control much of a quantitative phenotypic variation of a trait. QTLs exerting significant effects on the trait of interest are identified by linkage mapping or GWAS and are used for introgression, F_2 enrichment, or marker-assisted recurrent selection by introducing or pyramiding them into elite germplasm to develop improved cultivars (Anderson et al. 2007; Bernardo 2008). However, MAS is unrealistic in cases where the trait is highly quantitative because of the challenge of stacking of many genes in a single cultivar. The significance test used to identify the trait-associated markers implies that only a subset of markers will be used in the MAS, i.e., markers crossing the significance threshold. As a result, MAS completely neglects the markers which do not cross the significance criteria however close their effects may appear to the markers crossing the significance threshold.

To circumvent the limitation of significance tests in mapping studies, genome-wide selection or genomic selection (GS) can be exploited for the selection of highly quantitative traits. In genomic selection, instead of using a subset of molecular markers with large effects, both large and small effects of all the markers on the trait are computed to predict the total genetic value (Meuwissen et al. 2001). The genetic value is estimated as the sum of the individual's genetic values across all the markers. Subsequent selection is performed based on these genomic-estimated genetic values (GEGVs) on the individuals that are not phenotyped (Fig. 5.3). The GEGVs, therefore, become particularly useful for traits which are highly complex and laborious to phenotype. Earlier studies indicated that the GS leads to high correlations between GEGVs and observed genetic values for quantitative traits (Meuwissen et al. 2001; Schaeffer 2006). In dairy cattle, for example, the application of genomic-estimated breeding values reduced the cost by 92% and increased genetic gain per year by twofold compared with the traditional progeny testing (Schaeffer 2006). Recent extensive studies in plant species like wheat, maize, barley, etc. also showed that the genetic values for quantitative traits could be predicted with high accuracy (Bernardo and Yu 2007; Heffner et al. 2010; Jain et al. 2017; Lorenzana and Bernardo 2009). Popular approaches for genomic selection are based on genomic (GBLUP) or ridge regression best linear unbiased predictions (RRBLUP) and Bayesian models (Meuwissen et al. 2001). Overall, in plant breeding, BLUP remains to be the method of choice because of its simplicity and swiftness (Lorenzana and Bernardo 2009).

In cotton, studies exploring the prospects of GS are limited. Recently, a study examined the accuracies of genome-wide prediction for cotton fiber quality phenotypes (fiber length and strength) in historical datasets and found the support for GS implementation in cotton breeding (Gapare et al. 2018). To our knowledge, the potential of GS has so far not been studied against CBS syndrome traits. Nevertheless, based on results from other major crops, it can be assumed that implementing GS via predicting the genetic values in cotton will improve the genetic gains per unit time and cost.

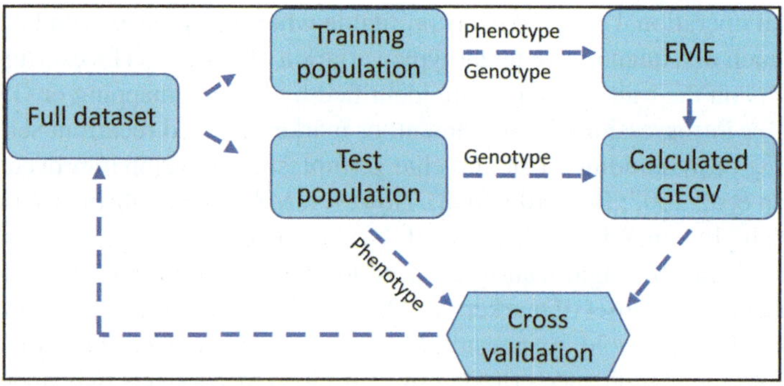

Fig. 5.3 A simplified overview of genomic selection coupled with cross-validation by employing a training population to estimate marker effects (EME) in order to get a genomic-estimated genetic value (GEGV) of the lines in the test population

Most of the CBS syndrome traits depict quantitative genetic variations that make them good candidates to improve via GS. There are, however, limitations to implementing GS, especially in small breeding programs. For example, (1) the new lines have to be genotyped to predict the performance of unobserved lines, and (2) functional pipelines for the whole-genome genotyping, genome-wide prediction analyses, and selection have to be established in order to make quick decisions.

5.7 Omics Approaches for CBS Syndrome Resistance

Recent years have seen unprecedented advances in next-generation sequencing-driven mass production of genomic data that, together with multi-omics approaches, have significantly contributed toward broadening our understanding of molecular systems in living organisms. These developments have not only reduced the cost of sequencing but also made it more accessible and practical to be routinely deployed in crop improvement programs. These advances have also led to the generation of reference-grade genome sequencing of tetraploid cotton, *G. hirsutum* and *G. barbadense* (Hu et al. 2019; Wang et al. 2019b), and diploid cotton, *G. arboreum*, *G. raimondii*, and *G. turneri* (Du et al. 2018). The lack of reliable genomic information in the past has been a key limiting factor in cotton breeding. However, with the access to genome-scale information, cotton functional genomics research and breeding have been revolutionized.

Multi-omics approaches have shown tremendous potential in identifying genes for breeding targets in crops (Fig. 5.4). However, their potential in cotton breeding is not fully realized yet. A successful example of deployment of such approaches is the identification of *GhHOX3* gene, which regulates cotton fiber elongation. Comparative analysis with *Arabidopsis* genome identified cotton orthologs of *Arabidopsis GLABRA2*, which is linked with trichome development in *Arabidopsis*. Transcriptome analysis revealed AtGL2 orthologs, *GhHOX1/2/3*, as highly expressed in cotton fiber. QTL mapping linked *GhHOX123* with fiber length. Subsequently, the overexpression, RNAi, and transcriptome analysis of transgenic line identified *GhHOX3* as a key regulator of fiber elongation (Shan et al. 2014). A similar approach resulted in the identification of *GaGSTF9* gene, which confers resistance against *Fusarium* wilt of cotton (Du et al. 2018). In another study, comparative proteomics identified Gh14-3-3d, a negative regulator of resistance to *Verticillium dahliae*. Subsequent analysis of CRISPR-Cas9-based homozygous loss of function mutant of Gh14-3-3d exhibited higher resistance to *V. dahliae* infestation (Gao et al. 2013).

A high-resolution transcriptional landscape of *G. arboreum* is now available that provides a valuable resource for functional genomic studies in cotton (Wang et al. 2019a). In addition, several transcriptome studies, coupled with functional analysis, provide new insights into host-pathogen interaction, particularly host response to infection. An RNA-Seq analysis of *G. arboreum* infected by graft inoculated using CLCuD-infected scion of *G. hirsutum* followed by sampling of asymptomatic and

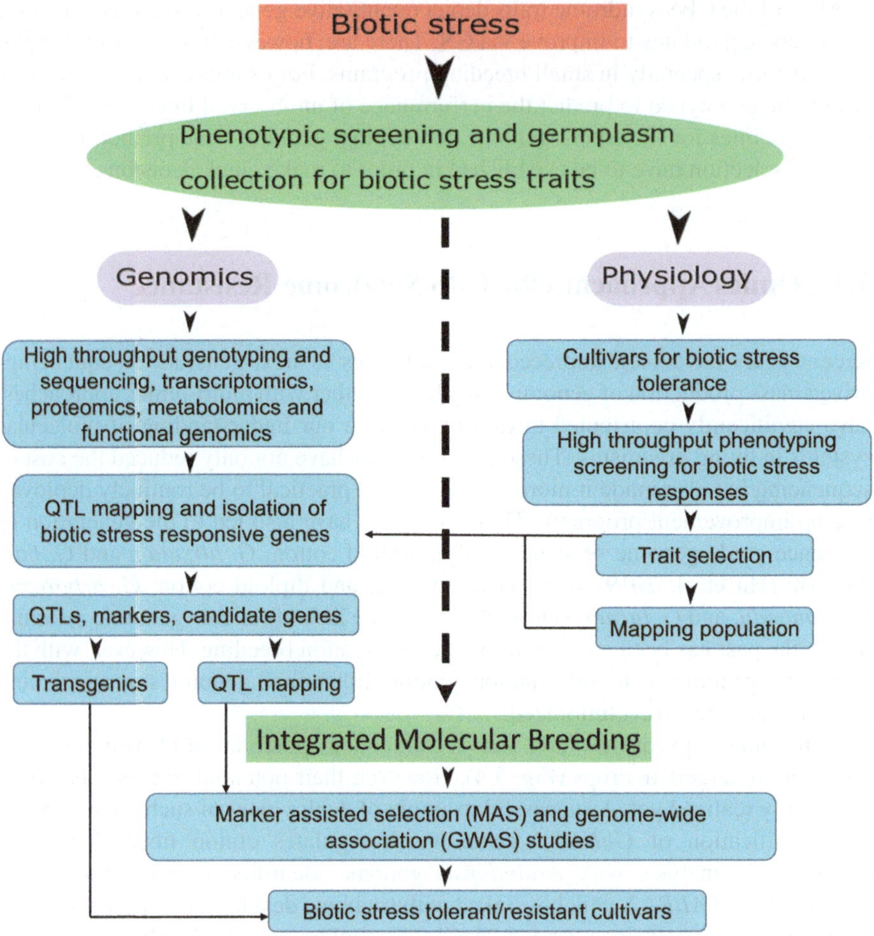

Fig. 5.4 Multi-omics approaches to increase plant breeding efficiency against biotic stresses

symptomatic plants revealed 1,062 differentially expressed genes (DEGS) in *G. arboreum*. Further analysis showed the existence of a co-expression network consisting of 50 hub genes (Naqvi et al. 2017). While most of the genes appeared to be regulating transport processes, it remains elusive to establish their role in host defense against the virus. Nonetheless, this fundamental work fuels the understanding of virus-host interaction. In another study, evolutionary analysis coupled with transcriptome analysis identified a cluster of nine glutathione S-transferase (GST) genes that participate in *Verticillium* wilt resistance. Overexpression of major genes of this cluster resulted in enhanced resistance, whereas suppression led to significantly increased susceptibility suggesting that this gene cluster plays a key role in resistance to *Verticillium* wilt (Li et al. 2019).

Metabolomics has emerged as a powerful complementary tool in the post-genomics era. It can be used to reveal physiological and biochemical responses of host plant under biotic and abiotic stresses (Fiehn and Weckwerth 2003). In particular, non-targeted metabolomics has been widely used to understand the biochemical mechanisms of quantitative resistance in the crop against many pathogens (Kushalappa and Gunnaiah 2013). However, the applications of metabolomics approaches in cotton, specifically in relation to CBS, remain limited. A comparative study of the metabolomic profile of cotton showed that different species of genus *Gossypium* have different levels of glycosides; for example, rhamnoglucosides are abundant in *G. hirsutum* but are found in trace amounts in *G. barbadense,* whereas quercetin-7-glycosides and kaempferol-3-glucoside are present in abundant amount in *G. barbadense*, as compared to *G. hirsutum* (Bolton 2009). In another study, ultrahigh-performance liquid chromatography-mass spectrometry (UPLC-MS)-based metabolomic profiling revealed an accumulation of key metabolites including terpenoids, flavonoids, phenylpropanoids, carbohydrates, and fatty acids in response to leaf spot of cotton. The analysis further showed an altered primary metabolism resulting in the upregulation of pyruvate and malate pathways that is coupled with elevated levels of carbohydrates like cellobiose and inulobiose suggesting that the general metabolism is more active in resistant cultivar, in comparison with sensitive cultivar. These differences in the metabolomics profiles of both cultivars underlie the molecular basis of respective tolerance and susceptibility to leaf spot disease (Khizar et al. 2020).

Omics data is being generated at an unprecedented rate. However, there is a dire need to integrate and translate this data for crop improvement. Therefore, integrated multi-omics approaches can play a pivotal role in order to bridge the information gap and accelerate precision breeding.

5.8 Conclusion and Future Outlook

Resistance to CBS is often complicated and influenced by genetic background and environmental factors. Conventional approaches are still practiced to breed cotton varieties against CBS. Nevertheless, these approaches often struggle to be fruitful due to limited genetic variability in *G. hirsutum* germplasm and difficulty in finding QTLs with large effect, hence posing a serious concern for a continued genetic gain. Recent advancements in cotton genomics, however, open new opportunities to address CBS resistance and associated challenges. With the availability of reference-grade genome sequence, precise base editing and targeted mutagenesis can be successfully achieved for CBS traits in *G. hirsutum*. Therefore, one of the significant bottlenecks associated with traditional breeding approaches can be addressed. At the same time, it opens the possibilities for transferring resistant alleles from diploid progenitors to the cultivated cotton species. The GS, on the other hand, has emerged as a powerful tool for the improvement of quantitative traits in crops, yet it remains virtually unexploited in cotton. Compared to the recent burst of cotton genomic sequence data, the availability of phenotypic data is still limited. The development of high-throughput, accurate, and cost-effective phenomics platforms are needed for cotton precision phenotyping. Genotyping coupled with lab and field-based phenotyping would help to minimize phenotype-to-genotype gap under multiple ecosystems, a key factor for breeding durable resistance against CBS. The availability of multi-omics approaches presents a considerable benefit in the current and future landscape of cotton genomics. To take full advantage of the available genomic information of cotton for CBS resistance breeding, it will become inevitable to use an integrated molecular breeding approach.

References

Abbas A, Iqbal MA, Rahman M-u, Paterson AH (2015) Estimating genetic diversity among selected cotton genotypes and the identification of DNA markers associated with resistance to cotton leaf curl disease. *Turk J Bot* 39:1033–1041

Acquaah G (2007) Principles of plant genetics and breeding. Blackwell Publishing Ltd

Ali M (1997) Breeding of cotton varieties for resistance to cotton leaf curl virus. *Pak J Phytopathol* 9:1–7

Anderson JA, Chao S, Liu S (2007) Molecular breeding using a major QTL for Fusarium head blight resistance in wheat. *Crop Sci* 47:S-112–S-119

Arshad M, Khan MI, Ali CR, Afzal M, ur-Rahman M (2009) Registration of 'CIM-496' cotton. *J Plant Regist* 3:231–235

Aslam M, Jiang C, Wright R, Paterson A (2000) Identification of molecular markers linked to leaf curl virus disease resistance in cotton. *Journal of Sciences Islamic Public of Iran* 11: 277–280

Atkinson G (1892) Some diseases of cotton. 3. Frenching. In: Some diseases of cotton 3 Frenching. pp 19–29

Azhar M, Amin I, Anjum Z, Mansoor S (2011) *Gossypium robinsonii*, an Australian wild cotton species is an asymptomatic host of the cotton leaf curl disease pathogen complex. *Aust Plant Dis Notes* 6:7–10

Baenziger PS, Depauw RM (2009) Wheat breeding: procedures and strategies. In: Wheat science and trade. pp 273–308

Bernardo R (2008) Molecular markers and selection for complex traits in plants: learning from the last 20 years. *Crop Sci* 48:1649–1664

Bernardo R (2010) Breeding for quantitative traits in plants, 2nd edn. Stemma Press, Woodbury

Bernardo R, Yu J (2007) Prospects for genomewide selection for quantitative traits in maize. *Crop Sci* 47:1082–1090

Bird L (1982) The MAR (multi-adversity resistance) system for genetic improvement of cotton. *Plant Dis* 66(2):172–176

Bolton MD (2009) Primary metabolism and plant defense—fuel for the fire. *Mol Plant-Microbe Interact* 22:487–497

Briddon RW, Markham PG (2000) Cotton leaf curl virus disease. *Virus Res* 71:151–159

Brinkerhoff LA (1970) Variation in *Xanthomonas malvacearum* and its relation to control. *Annu Rev Phytopathol* 8:85–110

Brinkerhoff LA, Verhalen LA, Johnson WM, Essenberg M, Richardson PE (1984) Development of immunity to bacterial blight of cotton and its implications for other diseases. *Plant Dis* 68:169

Brown JK (1992) Virus diseases of cotton. In: Hillocks RJ (ed) Cotton diseases. Commonwealth Agricultural Bureaux International, Wallingford, pp 275–330

Brown J (2002) Plant virus and mycoplasma diseases of cotton. Cotton Compendium Amer Phytopath Soc, St Paul

Cai JH, Xie K, Lin L, Qin B, Chen B, Meng J, Liu Y (2010) Cotton leaf curl Multan virus newly reported to be associated with cotton leaf curl disease in China. *Plant Pathol* 59:794–795

Chen ZJ, Scheffler BE, Dennis E, Triplett BA, Zhang T, Guo W, Chen X, Stelly DM, Rabinowicz PD, Town CD (2007) Toward sequencing cotton (Gossypium) genomes. *Plant Physiol* 145:1303–1310

Cianchetta AN, Davis R (2015) Fusarium wilt of cotton: management strategies. *Crop Prot* 73:40–44

Cun M, Guiliang J, Wenji S (1997) Current status, problem and countermeasure on resistance breeding to verticillium wilt of cotton in China. *Zhongguo Nong ye ke xue = Zhongguo Nongye Kexue* 30:58–64

Davis R, Moore N, Kochman J (1996) Characterisation of a population of *Fusarium oxysporum* f. sp. vasinfectum causing wilt of cotton in Australia. *Aust J Agric Res* 47:1143–1156

Davis R, Colyer P, Rothrock C, Kochman J (2006) Fusarium wilt of cotton: population diversity and implications for management. *Plant Dis* 90:692–703

Delannoy E, Lyon BR, Marmey P, Jalloul, A, Daniel JF, Montillet JL, Essenberg M, Nicole M (2005) Resistance of Cotton Towards *Xanthomonas campestris* pv. Malavacearum. *Annu Rev Phytopathol* 43: 63–82

Du X, Huang G, He S, Yang Z, Sun G, Ma X, Li N, Zhang X, Sun J, Liu M, Jia Y, Pan Z, Gong W, Liu Z, Zhu H, Ma L, Liu F, Yang D, Wang F, Fan W, Gong Q, Peng Z, Wang L, Wang X, Xu S, Shang H, Lu C, Zheng H, Huang S, Lin T, Zhu Y, Li F (2018) Resequencing of 243 diploid cotton accessions based on an updated A genome identifies the genetic basis of key agronomic traits. *Nat Genet* 50:796–802

El-Zik K, Bird L (1970) Effectiveness of specific genes and gene combinations in conferring resistance to races of *Xanthomonas malvacearum* in Upland cotton. *Phytopathology* 60:441–447

Essenberg M, Bayles MB, Samad RA, Hall JA, Brinkerhoff L, Verhalen LM (2002) Four near-isogenic lines of cotton with different genes for bacterial blight resistance. *Phytopathology* 92:1323–1328

Essenberg M, Bayles MB, Pierce ML, Verhalen LM (2014) Pyramiding *B* genes in cotton achieves broader but not always higher resistance to bacterial blight. *Phytopathology* 104:1088–1097

Fiehn O, Weckwerth W (2003) Deciphering metabolic networks. *Eur J Biochem* 270:579–588

Flor H (1955) Host-parasite interactions in flax rust-its genetics and other implications. *Phytopathology* 45:680–685

Fradin EF, Thomma BP (2006) Physiology and molecular aspects of Verticillium wilt diseases caused by *V. dahliae* and *V. albo-atrum*. *Mol Plant Pathol* 7:71–86

Gao W, Long L, Zhu L-F, Xu L, Gao W-H, Sun L-Q, Liu L-L, Zhang X-L (2013) Proteomic and virus-induced gene silencing (VIGS) analyses reveal that gossypol, brassinosteroids, and jasmonic acid contribute to the resistance of cotton to *Verticillium dahliae*. *Mol Cell Proteomics* 12:3690

Gapare W, Liu S, Conaty W, Zhu Q-H, Gillespie V, Llewellyn D, Stiller W, Wilson I (2018) Historical datasets support genomic selection models for the prediction of cotton fiber quality phenotypes across multiple environments. G3: Genes, Genomes, Genetics 8:1721–1732

Heffner EL, Lorenz AJ, Jannink JL, Sorrells ME (2010) Plant breeding with genomic selection: gain per unit time and cost. *Crop Sci* 50:1681–1690

Hillocks R (1983) Infection of cotton seed by Fusarium oxysporum f. sp. vasinfectum in cotton varieties resistant or susceptible to Fusarium wilt. *Trop Agric* 60:141–143

Hillocks R (1992) Bacterial blight. In: Cotton diseases. pp 39–85

Hu Y, Chen J, Fang L, Zhang Z, Ma W, Niu Y, Ju L, Deng J, Zhao T, Lian J, Baruch K, Fang D, Liu X, Ruan Y-L, Rahman M-u, Han J, Wang K, Wang Q, Wu H, Mei G, Zang Y, Han Z, Xu C, Shen W, Yang D, Si Z, Dai F, Zou L, Huang F, Bai Y, Zhang Y, Brodt A, Ben-Hamo H, Zhu X, Zhou B, Guan X, Zhu S, Chen X, Zhang T (2019) *Gossypium barbadense* and *Gossypium hirsutum* genomes provide insights into the origin and evolution of allotetraploid cotton. *Nat Genet* 51:739–748

Hussain T, Tahir M, Mehmood T (1991) Cotton leaf curl virus. *Pak J Plant Pathol* 3:57–61

Hutchinson J, Knight RL, Pearson E (1950) Response of cotton to leaf-curl disease. *J Genet* 50:100
Idris A (1990) Cotton leaf curl virus disease in the Sudan. Mededeling van de Faculteit Landbouwwetenschappen, Rijksuniversiteit Gent, vol 55. pp 263–267
Iqbal MJ, Aziz N, Saeed N, Zafar Y, Malik K (1997) Genetic diversity evaluation of some elite cotton varieties by RAPD analysis. *Theor Appl Genet* 94:139–144
Iqbal MJ, Reddy OUK, El-Zik KM, Pepper AE (2001) A genetic bottleneck in the 'evolution under domestication' of upland cotton *Gossypium hirsutum* L. examined using DNA fingerprinting. *Theor Appl Genet* 103:547–554
Jain A, Roorkiwal M, Pandey MK, Varshney RK (2017) Current status and prospects of genomic selection in legumes. Genomic selection for crop improvement. Springer, pp 131–147
Jalloul A, Sayegh M, Champion A, Nicole M (2015) Bacterial blight of cotton. *Phytopathologia Mediterranea* 54(1): 3–20
Karademir E, Karademir C, Ekinci R, Baran B, Sagir A (2012) Effect of *Verticillium dahliae* Kleb. on cotton yield and fiber technological properties. *International Journal of Plant Production* 6:387–407
Khizar M, Shi J, Saleem S, Liaquat F, Ashraf M, Latif S, Haroon U, Hassan SW, Rehman S, Chaudhary HJ (2020) Resistance associated metabolite profiling of Aspergillus leaf spot in cotton through non-targeted metabolomics. *PLoS One* 15:e0228675
Kirkpatrick TW (1931) Further studies on leaf-curl of cotton in the Sudan. *Bull Entomol Res* 22: 323–363
Knight R (1948) The role of major genes in the evolution of economic characters. *J Genet* 48:370
Knight RL, Clouston T (1939) The genetics of blackarm resistance-I. Factors B1 and B2. *J Genet* 38:133–159
Knight RL, Hutchinson JB (1950) The evolution of blackarm resistance in cotton. *J Genet* 50: 36–58
Konan ON, d'Hont A, Baudoin JP, Mergeai G (2007) Cytogenetics of a new trispecies hybrid in cotton:[(*Gossypium hirsutum* L.× *G. thurberi* Tod.) 2× G. longicalyx Hutch. & Lee]. *Plant Breed* 126:176–181
Kushalappa AC, Gunnaiah R (2013) Metabolo-proteomics to discover plant biotic stress resistance genes. *Trends Plant Sci* 18:522–531
Lawrence KS, Hagen A, Norton R, Faske T, Hutmacher R, Muller J, Mehl H (2017) Cotton disease loss estimate committee report, 2016. In: Proc. 2017 beltwide cotton conference, Dallas, vol 1. pp 150–152
Li Z-K, Chen B, Li X-X, Wang J-P, Zhang Y, Wang X-F, Yan Y-Y, Ke H-F, Yang J, Wu J-H, Wang G-N, Zhang G-Y, Wu L-Q, Wang X-Y, Ma Z-Y (2019) A newly identified cluster of glutathione S-transferase genes provides Verticillium wilt resistance in cotton. *Plant J* 98:213–227
Lorenzana RE, Bernardo R (2009) Accuracy of genotypic value predictions for marker-based selection in biparental plant populations. *Theor Appl Genet* 120:151–161
Lynch M, Walsh B (1998) Genetics and analysis of quantitative traits. Sinauer Associates, Inc., Sunderland
Melchinger AE, Gumber RK (1998) Overview of heterosis and heterotic groups in agronomic crops. Concepts and breeding of heterosis in crop plants. Crop Science Society of America, Madison, pp 29–44
Meuwissen T, Hayes B, Goddard M (2001) Prediction of total genetic value using genome-wide dense marker maps. *Genetics* 157:1819–1829
Minton EB, Garber RH (1983) Controlling the seedling complex of cotton. *Plant Dis* 67:115
Moll RH, Lonnquist JH, Fortuno JV, Johnson EC (1965) The relationship of heterosis and genetic divergence in maize. *Genetics* 52:139–144
Muhire BM, Varsani A, Martin DP (2014) SDT: a virus classification tool based on pairwise sequence alignment and identity calculation. *PLoS One* 9:e108277
Naqvi RZ, Zaidi SS-e-A, Akhtar KP, Strickler S, Woldemariam M, Mishra B, Mukhtar MS, Scheffler BE, Scheffler JA, Jander G, Mueller LA, Asif M, Mansoor S (2017) Transcriptomics reveals multiple resistance mechanisms against cotton leaf curl disease in a naturally immune cotton species, Gossypium arboreum. *Sci Rep* 7:15880

Nazeer W, Tipu A L, Ahmad S, Mahmood K, Mahmood A, Zhou B (2014) Evaluation of cotton leaf curl virus resistance in BC1, BC2 and BC3 progenies from an interspecific cross between *Gossypium arboreum* and *Gossypium hirsutum*. 9(11): e111861

Niu C, Lister H, Nguyen B, Wheeler T, Wright R (2008) Quantitative trait loci controlling resistance to Thielaviopsis basicola in diploid cotton. *Theor Appl Genet* 117:1313–1323

Paterson AH (2010) Comparative genomics in crop plants. Molecular techniques in crop improvement. Springer, pp 23–61

Pkania KC, Venneman J, Audenaert K, Kiplagat O, Gheysen G, Haesaert G (2014) Present status of bacterial blight in cotton genotypes evaluated at Busia and Siaya counties of Western Kenya. *Eur J Plant Pathol* 139:863–874

Rahman M-u, Zafar Y (2007) Registration of NIBGE-115 cotton. *J Plant Regist* 1:51–52

Rahman M, Hussain D, Zafar Y (2002) Estimation of genetic divergence among elite cotton cultivars–genotypes by DNA fingerprinting technology. *Crop Sci* 42:2137–2144

Rahman M, Hussain D, Malik T, Zafar Y (2005a) Genetics of resistance to cotton leaf curl disease in Gossypium hirsutum. *Plant Pathol* 54:764–772

Rahman M, Asif M, Ullah I, Malik KA, Zafar Y (2005b) Overview of cotton genomic studies in Pakistan. In: Plant & animal genome conference XIII. San Diego. p 81

Rahman M, Yasmin T, Tabbasam N, Ullah I, Asif M, Zafar Y (2008) Studying the extent of genetic diversity among *Gossypium arboreum* L. genotypes/cultivars using DNA fingerprinting. *Genet Resour Crop Evol* 55:331–339

Rahman M, Rahmat Z, Mahmood A, Abdullah K, Zafar Y (2014a) Cotton germplasm of Pakistan. In: Abdurakhmonov I (ed) World cotton germplasm resources. In-Tech. ISBN: 978-953-51-1622-6. https://doi.org/10.5772/58620

Rahman M, Ahmad A, Khan AQ, Abbas A, Rahmat Z, Sarfraz Z et al (2014b) Use of genetic and genomic approaches for controlling cotton leaf curl disease complex in Pakistan. In: Proceedings of the international cotton genome initiative conference, Wuhan, pp 26–28

Rahman M-u, Khan AQ, Rahmat Z, Iqbal MA, Zafar Y (2017) Genetics and genomics of cotton leaf curl disease, its viral causal agents and whitefly vector: a way forward to sustain cotton fiber security. *Front Plant Sci* 8:1157

Rahman M, Zafar Y (2012). Registration of NN-3 cotton. *J Plant Regist* 6, 342–347

Rahmat Z, Mahmood A, Abdullah K, Zafar Y (2014) Cotton germplasm of Pakistan. In World Cotton Germplasm Resources. IntechOpen

Rahmat Z, Ahmad S, Zia Z, Rahman M (2019) Genetic monitoring of introgressed alleles from *Gossypium arboreum* L. into *G. hirsutum* L. using SSR markers: a potential approach for bringing new alleles under cultivation. *Pak J Bot* 51:479–486

Raj TN, Kendrick B (2006) A monograph of Chalara and allied genera. Wilfrid Laurier Univ Press

Rajagopalan PA, Naik A, Katturi P, Kurulekar M, Kankanallu RS, Anandalakshmi R (2012) Dominance of resistance-breaking cotton leaf curl Burewala virus (CLCuBuV) in northwestern India. *Arch Virol* 157:855–868

Reddy RC, Muniyappa V, Colvin J, Seal S (2005) A new begomovirus isolated from *Gossypium barbadense* in Southern India. *Plant Pathol* 54:570

Robb J (2007) Verticillium tolerance: resistance, susceptibility, or mutualism? Botany 85:903–910

Robinson AF (2007) Reniform in US cotton: when, where, why, and some remedies. *Annu Rev Phytopathol* 45:263–288

Robinson A, Bell A, Dinghe N, Stelly D (2004) Status report on introgression of reniform nematode resistance from Gossypium longicalyx. In: National cotton council beltwide cotton conference, San Antonio

Rong J, Feltus FA, Waghmare VN, Pierce GJ, Chee PW, Draye X, Saranga Y, Wright RJ, Wilkins TA, May OL (2007) Meta-analysis of polyploid cotton QTL shows unequal contributions of subgenomes to a complex network of genes and gene clusters implicated in lint fiber development. *Genetics* 176:2577–2588

Rothrock C (1992) Influence of soil temperature, water, and texture on *Thielaviopsis basicola* and black root rot of cotton. *Phytopathology* 82:1202–1206

Sanogo S, Zhang J (2016) Resistance sources, resistance screening techniques and disease management for Fusarium wilt in cotton. *Euphytica* 207:255–271

Schaeffer L (2006) Strategy for applying genome-wide selection in dairy cattle. *J Anim Breed Genet* 123:218–223

Shaban M, Miao Y, Ullah A, Khan AQ, Menghwar H, Khan AH, Ahmed MM, Tabassum MA, Zhu L (2018) Physiological and molecular mechanism of defense in cotton against *Verticillium dahliae*. *Plant Physiol Biochem* 125:193–204

Shaheen T, Tabbasam N, Iqbal MA, Ashraf M, Zafar Y, Paterson AH (2012) Cotton genetic resources. A review. *Agron Sustain Dev* 32:419–432

Shan C-M, Shangguan X-X, Zhao B, Zhang X-F, Chao L-M, Yang C-Q, Wang L-J, Zhu H-Y, Zeng Y-D, Guo W-Z, Zhou B-L, Hu G-J, Guan X-Y, Chen ZJ, Wendel JF, Zhang T-Z, Chen X-Y (2014) Control of cotton fibre elongation by a homeodomain transcription factor GhHOX3. *Nat Commun* 5:5519

Shi C, Navabi A, Yu K (2011) Association mapping of common bacterial blight resistance QTL in Ontario bean breeding populations. *BMC Plant Biol* 11:52

Starr J, Koenning S, Kirkpatrick T, Robinson A, Roberts P, Nichols R (2007) The future of nematode management in cotton. *J Nematol* 39:283

Ulloa M, Hutmacher RB, Roberts PA, Wright SD, Nichols RL, Davis RM (2013) Inheritance and QTL mapping of Fusarium wilt race 4 resistance in cotton. *Theor Appl Genet* 126:1405–1418

Varshney RK, Graner A, Sorrells ME (2005) Genomics-assisted breeding for crop improvement. *Trends Plant Sci* 10:621–630

Walker NR, Kirkpatrick T, Rothrock C (1998) Interaction between Meloidogyne incognita and *Thielaviopsis basicola* on cotton (*Gossypium hirsutum*). *J Nematol* 30:415

Wallace T, El-Zik K (1990) Quantitative analysis of resistance in cotton to three new isolates of the bacterial blight pathogen. *Theor Appl Genet* 79:443–448

Wang K, Wang D, Zheng X, Qin A, Zhou J, Guo B, Chen Y, Wen X, Ye W, Zhou Y (2019a) Multi-strategic RNA-seq analysis reveals a high-resolution transcriptional landscape in cotton. *Nat Commun* 10:1–15

Wang M, Tu L, Yuan D, Zhu D, Shen C, Li J, Liu F, Pei L, Wang P, Zhao G, Ye Z, Huang H, Yan F, Ma Y, Zhang L, Liu M, You J, Yang Y, Liu Z, Huang F, Li B, Qiu P, Zhang Q, Zhu L, Jin S, Yang X, Min L, Li G, Chen L-L, Zheng H, Lindsey K, Lin Z, Udall JA, Zhang X (2019b) Reference genome sequences of two cultivated allotetraploid cottons, *Gossypium hirsutum* and *Gossypium barbadense*. *Nat Genet* 51:224–229

Wendel JF (1989) New World tetraploid cottons contain Old World cytoplasm. *Proc Natl Acad Sci* 86:4132–4136

Wendel JF, Cronn RC (2003) Polyploidy and the evolutionary history of cotton. *Adv Agron* 78:139

Wheeler T, Gannaway J, Keating K (1999) Identification of resistance to *Thielaviopsis basicola* in diploid cotton. *Plant Dis* 83:831–833

Wright RJ, Thaxton PM, El-Zik KM, Paterson AH (1998) D-subgenome bias of Xcm resistance genes in tetraploid Gossypium (cotton) suggests that polyploid formation has created novel avenues for evolution. *Genetics* 149:1987–1996

Wright RJ, Niu C, Nguyen B (2009) Bridging classical and molecular genetics of cotton disease resistance. In: Genetics and genomics of cotton. Springer, pp 313–336

Zhang J, Sanogo S, Flynn R, Baral JB, Bajaj S, Hughs S, Percy RG (2012) Germplasm evaluation and transfer of Verticillium wilt resistance from Pima (*Gossypium barbadense*) to Upland cotton (*G. hirsutum*). *Euphytica* 187:147–160

Chapter 6
Genomic-Assisted Breeding for Abiotic Stress Tolerance

Mehboob-ur-Rahman, Aqsa Majeed, Sana Zulfiqar, Shumila Ishfaq, Muhammad Mohsan, and Niaz Ahmad

6.1 Introduction

Sustainable cotton production is a prime objective of all cotton improvement programs worldwide. Various biotic and abiotic stresses impact negatively cotton yield as well as fiber quality (Rahman and Zafar 2018). Escalated temperature, prolonged drought periods, excessive salts in soils, etc. are major abiotic stresses that together reduce cotton productivity up to 50% (Bita and Gerats 2013). With the changing climate, these stresses are feared to be intensified in future, ultimately ending up with huge yield losses.

The cultivated cotton species particularly *G. hirsutum* (upland cotton) is sensitive to different stresses at all growth stages. For example, high salinity and decrease in optimum temperature may retard cotton growth at early stages including germination and seedling (Ashraf and Ahmad 2000; Bolek et al. 2014). It is estimated that about 20% of the irrigated land is affected by salts (Negrao et al. 2016). In contrary to this, escalation in heat and excessive salts (salinity) are serious challenges for a few cotton growing countries; however, their impact will be devastating in regions

Mehboob-ur-Rahman (✉)
Agricultural Biotechnology Division, National Institute for Biotechnology & Genetic Engineering (NIBGE)-College, Pakistan Institute for Engineering and Applied Science (PIEAS), Faisalabad, Pakistan
e-mail: mehboob@nibge.org

A. Majeed · S. Zulfiqar
Plant Genomics and Mol. Breeding Lab, National Institute for Biotechnology and Genetic Engineering (NIBGE), Faisalabad, Pakistan

Department of Biotechnology, Pakistan Institute of Engineering and Applied Sciences (PIEAS), Nilore, Islamabad, Pakistan

S. Ishfaq · M. Mohsan · N. Ahmad
Plant Genomics and Mol. Breeding Lab, National Institute for Biotechnology and Genetic Engineering (NIBGE), Faisalabad, Pakistan

Mehboob-ur-Rahman et al. (eds.), *Cotton Precision Breeding*,
https://doi.org/10.1007/978-3-030-64504-5_6

experiencing the issue of climate change, having less efficient irrigation system and using excessive fertilization (Rahimian and Poormohammadi 2012). Cotton is a tropical crop. Thus it is also vulnerable to low temperatures. It has been demonstrated that low temperature reduces germination percentage and also leaf expansion that results in wilting and ultimately necrosis (Mahajan and Tuteja 2005).

To mitigate the devastating impact of these stresses, it is important to develop cotton varieties with improved resilience to abiotic stresses. Narrow genetic base is a major obstacle for not improving tolerance to stresses. Cotton, a crop that grows in tropical and subtropical region, is dominantly a heat-loving plant; however, it is exposed to several other related stresses including drought and excessive heat. Salinity is another emerging threat which may impact the cotton production in the future (Zhang et al. 2014a, b; Rahman Talukder et al. 2015). Also, the continued salinization of the cultivated lands together with the limited availability of fresh water will further aggravate the situation (Peng et al. 2014). In this chapter, focus has been made to discuss abiotic stresses, their impact on cotton production, and strategies to address these stresses using precision breeding approaches including DNA markers.

6.2 Drought and Its Impacts on Cotton

Drought—non-availability of adequate amount of water supply, which affects plant growth—reduces cotton yield and deteriorates fiber quality (Ullah et al. 2006, 2008; Rahman et al. 2008; Dahab et al. 2012). Non-availability of water in cells triggers stomatal closure to avoid further water losses, which results in reduced CO_2 intake, lower CO_2 assimilation rates, escalated synthesis of reactive oxygen species (ROS), and deterioration of cell components (Oosterhuis and Snider 2011; Loka and Oosterhuis 2012; Deeba et al. 2012; Sekmen et al. 2014).

Drought stress induces substantive variations in cotton plant morphology and physiology and also brings changes in cell biochemistry that badly impair cotton growth and productivity (Ullah et al. 2008, 2017; Sarwar et al. 2012). Drought stress has been reported to affect a number of traits such as root development, plant height, biomass, fiber quality, and yield (Loka and Oosterhuis 2014; Hejnak et al. 2015).

6.3 Heat Stress and Its Impacts

Heat stress is the most common cause of cotton yield reduction (Talukder et al. 2014; Saleem et al. 2018b). This factor is one of the major devastating factors for not achieving high yield potential in the subcontinent (Sarwar et al. 2017). It has been reported that each 1-degree rise in daily maximum temperature results in the loss of 110 kg/hectare of cotton yield (Singh et al. 2007). All vegetative and reproductive stages of cotton plant are vulnerable to high temperature. At early stages of plant development, rise in temperature affects the photosynthesis rate and biomass production, while at maturity it causes the abortion of small fruiting bodies (floral buds) and flowers (Young et al. 2004). It was also reported that rise in temperature causes overproduction of ROS, which ultimately damages the cell membranes, and results in peroxidation of membrane lipids and degradation of long-chain proteins, leading to premature leaf as well as boll shedding (Hemantaranjan et al. 2014; Ahmed et al. 2017). Since the mean global temperature is increasing (Dinar et al. 2012), vulnerability of cotton to temperature stress particularly at reproductive stage is expected to result in drastic reduction in cotton yield in the future (Gür et al. 2010).

6.4 Impact of Soil Salinity on Cotton

Salinity is another major threat that reduces crop productivity as well as quality (Deinlein et al. 2014). It was reported that over 800 million hectares are salt-affected lands which is ~6% of the total world land area (Munns 2005). Development of salt-resistant crops will help in fostering saline agriculture in the salt-affected areas, reclaiming these lands, and ultimately improving the livelihoods of the community living in those regions (Saeed et al. 2012; Deinlein et al. 2014; Peng et al. 2014). Before developing salt-resistant cotton cultivars, it is important to understand the mechanism evolved in tolerance to salt stress in halophytes, which can be transferred to salt-sensitive field crops. Cotton (*Gossypium* spp.) shows moderate tolerance to salt stress (~7.7 dS m^{-1}; Zhang et al. 2013) and therefore can be grown on saline-alkali lands. However, the yield decreases drastically as the accumulation of salts increases in the soil (Guo et al. 2015). Like other crops, excessive salts in soil severely affect germination and seedling stages; both are the most sensitive stages in the life of a plant (Nematzadeh 2018; Frouin et al. 2018; Ahmad et al. 2020). It has been reported that germination stage of a plant is much more sensitive to salt-induced toxicity as compared to the seedling stage (Wang et al. 2011; Ahmad et al. 2020). Salinity not only reduces germination but it also delays the emergence stage. For example, at a salt concentration over 15 dSm^{-1}, emergence phase was delayed up to 5 days. Loss in germination reduces the plant population per unit area, which then results in reduced crop yields (Saqib et al. 2002). Salt-induced toxicity disturbs ionic as well as osmotic homeostasis, reduces photosynthetic activities, and results in perturbed redox status, ultimately ending up with an abnormal plant growth (Zhang et al. 2016). At maturity, salt-induced toxicity also reduces the number of mature bolls (Anagholi et al. 2005). Majority of the sucrose in the developing bolls is supplied from subtending leaf of cotton boll (LSCB). Although, the synthesis of sucrose remains unaffected under salt stress, its transport from LSCB to developing bolls is severely affected, resulting in the reduction of number of bolls and their weight (Peng et al. 2016a). It has been reported that salinity also affects fiber quality. For example, overall cellulose content and the conversion of sucrose to cellulose were severely reduced in salt-sensitive cultivars under salt stress, resulting in the deterioration of fiber quality (Peng et al. 2016b). It was confirmed that the activities of several enzymes involved in cellulose synthesis including acidic invertase, sucrose phosphatase, and alkaline invertases were significantly reduced by salt-induced toxicity (Peng et al. 2016b). Several studies have reported reduction in photosynthesis due to reduction in chlorophyll *a* and chlorophyll *b* content as well as alteration in chloroplast ultra-structures (Lee et al. 2013; Zhang et al. 2014a, b). Cotton has shown moderate level of salt tolerance and thus can be used as a good candidate for the development of augmented level of salt tolerance to effectively utilize the saline-alkaline soils (Zhang et al. 2011).

6.5 Mitigating the Abiotic Stresses

6.5.1 Other Than Genetic Solution

Various strategies are used for ensuring a sustainable crop production when a plant undergoes different stresses. Exogenous application of nutrients that is called a "shotgun approach" has been carried out to ameliorate the undesirable changes of abiotic stresses (Upadhyaya et al. 2012). It is well established that ethylene production at maturity confers leaf and fruit abscission. Restriction of ethylene production under stressful conditions therefore could be a way to increase cotton yield. Various chemicals including aminoethoxyvinyl glycine (AVG) (Najeeb et al. 2015a, b), aminoethoxycetic acid (AOA), and 1-methylcyclopropene (1-MCP) have been reported to minimize the yield losses in cotton under stress (Yang and Hoffman 1984).

The growth of cotton plant and yield is positively correlated with the uptake of nutrients (Bange et al. 2004). Therefore, increasing nutrient supply under stress may help plants to recover from the stress impact (Percival and Keary 2008; Zhang et al. 2012). In contrary to this, if plants are grown under normal conditions, the excessive use of nutrients particularly nitrogen might have adverse effect on plants (Iqbal et al. 2015). Application of N fertilizers has been shown to confer tolerance to waterlogging by enhancing root growth and plant vigor and improving photosynthetic rate (Guo et al. 2010). Likewise, application of potassium fertilizer has shown positive effects on cotton plant growth and chlorophyll content when grown under waterlogged environments. Potassium also has a role in stomatal regulation and renders the cotton plant to scavenge more macro- and micronutrients (Ashraf et al. 2011).

The major challenge for profitable cotton cultivation under salt stress is the establishment of optimum plant population. The germination is severely affected by osmotic stress imposed by Na^+ and Cl^- (Sattar et al. 2010). Therefore, approaches which can protect germination under stress were quite popular for obtaining optimum crop yield under salt stress. For example, seed priming assay may help in overcoming the seed germination under saline environment (Bradford 1986). Foliar application of KCl and NH_4NO_3 mixture, osmoprotectants, and plant hormones (ABA, GA3, SA) helps in boosting vegetative and reproductive growth. Hence, exogenous application of such chemicals and hormones alleviates the adverse effects of soil salinity by the overexpression of antioxidant enzymes (Jabeen and Ahmad 2009).

6.5.2 Genomic for Tolerance to Abiotic Stresses

A number of genes conferring salt tolerance such as MKK (Lu et al. 2013), ZFP (Guo et al. 2009), NAC (Meng et al. 2009), ERF (Johnson et al. 2003), DREB (Guo et al. 2009), *GhMT3a* (Xue et al. 2008), MPK (Zhang et al. 2011), and tonoplast Na^+/H^+ antiporter (Wu et al. 2004) have been identified in cotton. Two cotton species, *G. barbadense* and *G. davidsonii* (wild species), are relatively more tolerant to salt (Ahmad et al. 2002; Zhang et al. 2016). A number of genes conferring salinity tolerance have been identified in *G. hirsutum* on its D-subgenome (Li et al. 2014). In another study, 109 WRKY genes (*GarWRKYs*) have been identified in a wild species *G. aridum* (Fan et al. 2015). Overexpression of enzymes involved in the detoxification of ROS like *GhMT3a*, *GhSOD1*, and *GhCAT1* has been shown to increase salt tolerance in cotton (Luo et al. 2013). It has been demonstrated that the transgenic cotton plant co-expressing the *AtNHX1* and *TsVP* genes exhibited higher seed cotton yield than untransformed controls when grown under saline conditions (Chen et al. 2017a, b; Cheng et al. 2018). Suppressing the expression of sucrose non-fermenting-1-related protein kinase 2 (GhSnRK2) revealed a positive correlation of this gene with stress tolerance particularly drought and low temperature (Giambattista et al. 2014). Downregulation of *PHYA1*-coding genes in cotton using RNAi increased photosynthesis and improved drought, salt, and heat tolerance (Abdurakhmonov et al. 2014).

Another study reported that annexin-encoding gene (*GhAnn1*) enhanced chlorophyll content in cotton and also increased the peroxidase synthesis while decreasing the lipid peroxidation levels under stressful conditions. Overexpression of this gene can be useful in conferring tolerance to salt and drought in cotton plant (Zhang et al. 2015). In a few earlier reports, it was shown that the genes responsible for the synthesis of CBL-interacting protein kinase (*GhCIPK6*) and sucrose nonfermenting 1-related protein kinase 2 (*SnRK2*) conferred tolerance to abiotic stresses in cotton (He et al. 2013; Giambattista et al. 2014). In another study, a total of 1528 genes expressed in leaf and 1128 root-related genes are involved in several (28) biological pathways triggered with drought stress (Ranjan and Sawant 2015), suggesting that mechanisms of stress tolerance operating in leaves are distinct from those found in roots.

Comparative analysis of genome-wide expression has shown that stress tolerance is a complex phenomenon, which involves the co-expression of several genes, transcription factors, and physiological processes (Ranjan et al. 2012). In addition to genes conferring salt tolerance, transcription factors also represent a good choice to engineer stress tolerance in cotton using genetic transformation approaches (Guo et al. 2015). Most notable among the transcription factors is the *WRKY* transcription factor family (one of the largest transcription factor families) which can downregulate or upregulate a number of different genes (Dou et al. 2014).

Likewise, several NAC genes including *GhNAC1–GhNAC6* (Meng et al. 2009) and *GhNAC7–GhNAC13* (Huang et al. 2013) have been found distinctively regulating in roots and leaves of cotton plants when grown under stress conditions. Two families—the basic region leucine zipper (*bZIP*) and ethylene response factors (*ERF*)—are highly diverse which respond to different stresses in many plant species. However, very little studies have been conducted in cotton (Abid et al. 2017; Ma et al. 2017).

6.5.3 Molecular Breeding

Earlier conventional breeding efforts were deployed for improving tolerance to abiotic stresses with major emphasis on drought tolerance. It involves the crossing of diverse parent genotypes followed by making early-generation selections using molecular tools. Improving cotton lint yield is still a major criterion used in conventional cotton breeding programs. Tolerance to drought is a very complex phenomenon (Singh 2004); therefore, improving fiber quality particularly under water-limited environments is quite challenging (Shakoor et al. 2010). Thus breeders are trying to adopt new breeding procedures for selecting the true-to-type drought-tolerant cotton plants in early generations. In the coming sections, we have discussed molecular tools used in cotton breeding to improve its tolerance against abiotic stresses.

6.5.3.1 Quantitative Trait Loci (QTLs)

Advent of DNA markers in the toolkit of a breeder has been quite instrumental in selecting desirable plants in early generations which is a prerequisite for developing high-yielding varieties. These markers have also been used in recognizing QTLs involved in conferring several traits including stress tolerance (Dahab et al. 2016). DNA marker studies have also helped in deciphering the genetic basis of complex physiological traits particularly drought and salt tolerance (YongSheng et al. 2009).

Among the different types of DNA fingerprinting assays, restriction fragment length polymorphism (RFLP), random amplified polymorphic DNA (RAPD), amplified fragment length polymorphism (AFLP), simple sequence repeats (SSRs), and single nucleotide polymorphism (SNPs) have been used for the identification of DNA markers, estimating the extent of genetic divergence, cultivar identification, etc. (Reinisch et al. 1994; Multani and Lyon 1995; Tatineni et al. 1996; Iqbal et al. 1997; Abdalla et al. 2001; Lu and Myers 2002; Rong et al. 2004; Rahman et al. 2002, 2009; Ahmad et al. 2007; Shaheen et al. 2009, 2010).

The first PCR-based markers used in cotton were RAPDs which were deployed on *G. hirsutum* L. var. NIAB-78, Deir-Ezzor-22 (DE-22), Deltapine-50 (DP-50), and Aleppo-118 (A-118). Polymorphism in RAPD profiles was measured as genomic template stability (GTS %). Two varieties, DP-50 and A-118, showed highest GTS values, 79.1% and 58.2%, respectively, while DE-22 showed the lowest value (36.7%). The RAPDs markers were found to be highly irreproducible and present throughout the genome (Tabbasam et al. 2014; Rahman et al. 2014). Compared to RAPD markers, SSR markers were used extensively in studying genetic diversity and improving the fiber quality and tolerance to stresses (Nguyen et al. 2004; Park et al. 2005; Tabbasam et al.2014; Han et al. 2006). Earlier studies have depicted that majority of SSRs have been mapped on D subgenome (63%) than that of A subgenome. A high degree of polymorphism was reported on D subgenome for various traits including plant structure (Jiang et al. 2000), drought tolerance (Paterson et al. 2000; Saranga et al. 2001), disease tolerance (Wright et al. 1998), and fiber quality (Jiang et al. 1998). To date, 78,340 SSRs primer pairs are available for using in cotton breeding program (https://www.cottongen.org/data/markers).

Quantitative trait loci (QTL) analyses have been deployed in cotton to explore the genetic basis of complex trait (Said et al. 2015; www.cottonqtldb.org). Evaluation of QTLs is carried out based on their chromosomal positions and their contribution to phenotypic variations of underlying a particular trait (Miles and Wayne 2008). Identification of stable QTLs in diverse environments, i.e., well-watered and water-limited environments could facilitate development of cotton cultivars containing genes for producing high lint yield as well as fine lint quality (Li et al. 2014; Lu et al. 2015; Parekh et al. 2016).

In cotton, only few QTL maps pertaining to drought tolerance have been constructed. The first drought linkage map was generated using RFLPs of F_2 population developed by attempting a cross between upland cotton and pima cotton (Reinisch et al. 1994). Since then several linkage maps have been developed (Saeed et al. 2011); however, majority of the linkage maps were focused on discovering the yield related QTLs.

The RFLPs were used to identify QTLs related to leaf morphology in a F_2 population produced by making an interspecific cross between *G. hirsutum* x *G. barbadense*. Out of 40 QTLs discovered, majority of QTLs were found on chromosome 6 (Jiang et al. 2000). In another study, a phenotypic correlation was observed between the physiological and yield traits by using a segregating populations derived by making crosses between *G. hirsutum* and *G. barbadense*. These parent genotypes and population were surveyed with RFLPs. A total of 161 QTLs were identified for 16 traits. Out of these, 13 QTLs showed significant impact on performance of productivity, physiology, and quality traits under well-watered conditions (Saranga et al. 2004).

Attempts have been made to search for QTLs associated with lint quality and high yield under stressful environments using SSRs (Wang et al. 2007, 2015; Zhang et al. 2013; Qin et al. 2015; Jamshed et al. 2016). Earlier, a segregating population was developed by crossing *G. hirsutum* with *G. barbadense* for the detection of QTLs associated with water stress tolerance. A total of 33 QTLs associated with tolerance to water-limited conditions were reported (Levi et al. 2009). Recently, around 30 QTLs in *G. hirsutum* for yield and drought tolerance were identified (Baytar et al. 2018). Among these, 15 QTLs were involved under well-irrigated conditions, while 23 of these were involved in water-limited conditions. Similarly, two QTLs conferring drought tolerance were identified using SSRs and EST-SSRs (Saleem et al. 2015). In another study, 1295 SSR markers were surveyed and 11 QTLs identified, which were distributed among 8 chromosomes (Oluoch et al. 2016).

Recently, a meta-analysis has been performed using markers to identify QTLs linked with a particular trait. For instance, a total of 1223 QTLs for lint yield and quality, and resistance to disease and tolerance to drought were compiled (Said et al. 2013). Similarly, 23 out of the 661 stress-related QTLs were discovered for water-limited environments (Zheng et al. 2016). These newly discovered 23 drought-tolerant QTLs were located on 15 different chromosomes. Next year, a meta-analysis was carried out on 661 QTLs for identifying QTL clusters and hotspots. It was demonstrated that there was non-uniform distribution of stress tolerance-related QTLs in cotton genome (Abdelraheem et al. 2017).

Recently, in total, 524 EST-SSRs and SSRs primer pairs were surveyed to study polymorphisms between drought-resistant and drought-sensitive cotton cultivars and their F_2 population. The polymorphic EST-SSRs and SSRs were used to construct a genetic linkage map. Subsequent analysis showed that two QTLs related to relative water content were located on chr-23 and chr-12, whereas one QTL for excised leaf water loss was located on chr-23. These QTLs can be transferred into drought stress cultivars using marker-assisted approaches to develop drought-tolerant cotton varieties (Saleem et al. 2018a).

To identify new QTLs, it is always desirable to run more markers as well as search for SNPs in candidate genes. Identification of genome-wide SNPs through genotyping by sequencing (GBS) is relatively a low-cost alternative approach that can be used in the identification of genotypic variations (Elshire et al. 2011). A total of 110 drought-responsive genes involved in drought tolerance were identified using microarray analysis (Rodriguez-Uribe et al. 2014).

Like drought, tolerance to salt stress is also a very complex quantitative trait, which means that the number of genes and genomic regions have a small additive effect toward the trait. QTLs associated with salt tolerance in multiple crop species have been reported. In this complex genetic background, selection of salt-tolerant plants through deployment of DNA markers is a promising approach (Sharif et al. 2019). However, studies for the identification of QTLs associated with tolerance to salinity are scanty in cotton (Tiwari et al. 2013; Oluoch et al. 2016). Major issue is the complex genetics of tolerance for the trait and it contains many genes which are scattered throughout the cotton genome. Also, the genetic diversity is quite low between the parent genotype used in developing populations. Moreover, backcrossing for developing backcross populations may take months or even years (Weigel 2012). A total of 274 SSR markers were screened on salt-tolerant cotton accessions and 10 markers were identified linked with salt tolerance (Zhang et al. 2010). In another study, 132 pairs of EST-SSR primers were designed which were associated with salt tolerance in cotton (Wang et al. 2014). Recently, in total, 66 QTLs linked with salt tolerance are identified using single nucleotide polymorphic (SNP) markers (Diouf et al. 2017, Table 6.1). Association analysis was also conducted using 145 SSRs; out of these 95 SSRs exhibited significant associations with salt-tolerant characters. Out of these, 41 were associated with seedling stage physiological index, 37 with biochemical index at four seedling stages, and 17 with germinative index (Du et al. 2016).

Table 6.1 Characterization of significant SNPs/QTLs for abiotic stresses in cotton

Species	Genetic resource	Markers	Traits	Chr location	# of QTLS/ SNPs	References
G. hirsutum	8 varieties	SNPs	Salt tolerance	A04, D01	1282 SNPs	Wang et al. (2016)
G. hirsutum	277 $F_{2:3}$ population	SNPs	Salt tolerance	A02, D02, A01,	5178 SNPs, 66 QTLs	Diouf et al. (2017)
G. hirsutum	97 $F_{5:9}$ RILs	SNPs	Drought and salt tolerance	NA	165 QTLs	Abdelraheem et al. (2018)
G. hirsutum × *G. tomentosum*	BC_2F_2 population	SNPs	Drought tolerance	NA	10,888 SNPs	Magwanga et al. (2018)
G. arboreum	2 genotypes,	SNPs	MT-sHSP gene	NA	21 SNPs	Shaheen et al. (2009)

NA not available

6.5.3.2 Genome-Wide Association Studies: SNPs Arrays

Association-mapping, based on linkage disequilibrium, has emerged as a new assay, which examines several hundreds and thousands of polymorphisms for quantifying QTL effect. Association mapping does not require a large population and hence it is considered more effective as compared to the linkage analysis. Other advantages of association mapping include a high resolution for mapping, no need to develop segregating populations, and capacity to study a large number of alleles in the shortest possible time. Therefore, it is a powerful technique for the identification of genomic regions conferring a particular phenotype under a certain set of environmental or growth condition(s) (Saeed et al. 2014). However, the major issue for GWAS is the low coverage of cotton genome by conventional DNA fingerprinting assays (RFLP, RAPDs, SSRs, etc.). It seriously hinders the progress toward identifying stable QTLs and associated DNA markers. For example, several hundred (>300) accessions of upland cotton were surveyed with 106 SSRs for the identification of QTLs associated with drought and salt tolerance through GWAS. Only 15 markers showed association with tolerance to drought condition, while three markers showed linkage with salt-tolerant trait (Jia et al. 2014b).

New DNA fingerprinting assay SNP has been used for discovering new polymorphisms for use in enriching the genetic maps as well as identifying new QTLs (Abdelraheem et al. 2019). The GWAS is capable of studying millions of SNPs and thousands of genotypes accessions (Huang et al. 2010; Chen et al. 2014). Other than cotton, GWAS has also been successfully deployed in different crops including *Arabidopsis*, rice, wheat, maize, barley, etc. and contributed toward discovering SNPs linked with different agronomic and biochemical traits (Atwell et al. 2010; Xue et al. 2013; Chen et al. 2016; McCouch et al. 2016). For example, GWAS was deployed for identifying QTLs and DNA markers associated with boron toxicity (de Abreu et al. 2017), drought tolerance in wheat (Gupta et al. 2017; Bhatta et al. 2018), thermal tolerance in rice (Lafarge et al. 2017; Chen et al. 2017a, b), salt tolerance (Shi et al. 2017; Wan et al. 2017), yield, and other agronomic traits (Garcia et al. 2019) and also for improvement of provitamin A in maize (Xiao et al. 2017).

About 376 upland cotton accessions were surveyed using CottonSNP63 K array through GWAS for identifying QTLs conferring abiotic stress tolerance in cotton (Abdelraheem 2017). Later on, >470,000 SNPs were surveyed. A total of 16 and 27 QTLs were identified for dry shoot weight and plant height, respectively, under both water-limited and saline environments, while 11 QTLs were found commonly linked with tolerance to drought and saline environments (Abdelraheem et al. 2018, Table 6.1).

For high-throughput genotyping, whole-genome (2Gb) SNPs have been developed for exploring allotetraploid cotton. Cotton genome mainly comprises of copy number variations (CNVs) instead of SNPs, which could be helpful in exploring genetic variations for complex traits (Ashraf et al. 2018). Recently ~990 genes related to gene expression mechanisms and structural components of the cell have been reported containing CNVs (Wang et al. 2017).

Several studies pertaining to association analysis for various agronomic traits including tolerance to biotic and abiotic stresses, lint yield and lint quality traits have been performed (Wang et al. 2013; Jia et al. 2014a; Zhao et al. 2014; Sethi et al. 2016; Ademe et al. 2017; Iqbal and Rahman 2017; Li et al. 2017; Ma et al. 2017). For example, a total of 107 associated loci for fiber length, fiber strength, and micronaire were identified by exploring 97 *G. hirsutum* accessions when grown under three environments using SSR-coupled association analysis (Cai et al. 2014). In another study, 109 *G. hirsutum* accessions were grown under salinity stress and 16 different markers were found linked with salinity tolerance (Saeed et al. 2014). Likewise, 278 polymorphic loci in cotton were found associated with drought and salt tolerance by surveying on 323 *G. hirsutum* genotypes (Cai et al. 2017). Also, 15 and 23 SSR markers linked with well-watered and water-stress conditions, respectively, were identified (Baytar et al. 2018).

The genotyping by sequencing was applied on multiple crop species including cotton for identifying genome-wide SNPs in several accessions representing diverse populations (Elshire et al. 2011; Poland and Rife 2012; Gore et al. 2014; Abdelraheem and Zhang 2016). A total of 11 QTLs linked with salt tolerance were identified by screening a population of 188 $F_{2:3}$ which were derived by crossing upland cotton with *G. tomentosum* Nutt. ex Seem (Oluoch et al. 2016). The same population was used in another study for the identification of QTLs conferring drought tolerance (Zheng et al. 2016). This population was also used for investigating QTLs associated with drought tolerance under field environments. In total 67 QTLs for drought tolerance were identified and mapped on chr-5, chr-8, chr-9, and chr-16 (Zhang et al. 2016).

More recently, SNPs in genes conferring relative survival rate (RSR) and salt tolerance level (STL) in *G. hirsutum* L. were identified using the Illumina Infinium CottonSNP63K array. About 23 SNPs originating from seven chromosomes, A01, A10, D02, D08, D09, D10, and D11, were identified. Of these, SNPs i46598Gh and i47388Gh located on D09 were found associated with both the RSR and STL (Sun et al. 2018). In another study, 17 stable SNPs were mapped in or near close

Table 6.2 Important SNPs and QTLs linked with salt tolerance in *G. hirsutum*

Species	Markers	Traits	Chr location	QTLs	References
G. hirsutum	TM3300	Salinity tolerance	A01	qGhST-c1	Yuan et al. (2019)
G. hirsutum	TM4974	_	A02	qGhST-c2	
G. hirsutum	TM8361	_	A03	qGhST-c3	
G. hirsutum	TM18816	_	A07	qGhST-c7–1	
G. hirsutum	TM19026	_	A07	qGhST-c7–2	
G. hirsutum	TM41814	_	A12	qGhST-c12–2	
G. hirsutum	TM63245	_	D07	qGhST-c16–1	
G. hirsutum	TM63387	_	D07	qGhST-c16–2	
G. hirsutum	TM73567	_	D10	qGhST-c20–1	
G. hirsutum	TM73573	_	D10	_	
G. hirsutum	TM73579	_	D10	qGhST-c20–2	
G. hirsutum	TM67763	_	D08	qGhST-c24–1	
G. hirsutum	TM68258	_	D08	qGhST-c24–2	

proximity of 13 QTLs, and this region also contained 98 candidate genes conferring salt tolerance (Table 6.2). Through conducting RNA-seq analysis, a total of 13 differentially expressed genes (out of the 98) were identified. Out of these, 12 genes were validated by conducting qRT-PCR.

More recently, targeted genome editing with the CRISPR/Cas9 assay was found to be the most powerful tool for modifying genes precisely, and the resultant mutated alleles are heritable (Liang et al. 2018; Ahmad et al. 2020). Till now reports on editing cotton genome are few, and these are aimed at exploring the genes related to stress tolerance. This section has been discussed in detail in another chapter (Chap. 14) related to CRISPR of this book.

6.6 Perspective

For a long period of time, DNA markers particularly SSRs have been used as "marker of choice" to find DNA polymorphisms among various cotton genotypes. These markers are still being used in several labs particularly in developing countries owing to their high informativeness, reproducibility, and user-friendly nature. However, SSRs are not as abundant as SNPs in genomes. The SNPs markers are particularly useful when the genetic diversity among the germplasm is low as witnessed in cotton owing to the long history of domestication and selection. Thus, these markers are much more suitable for high-throughput genotyping; however, utility of these markers for crops containing high ploidy level, for example, allotetraploid cotton, is really a challenging task. Hence, efforts are being devoted for the identification and validations of SNPs impacting a particular phenotype. Discovery of new SNPs by resequencing cotton accessions is extremely important for fabricating a high-density SNP array. It is also important to develop permanent mapping populations of cotton, which could be used to identify reliable and stable QTLs associated with drought and salt tolerance for the development of climate-smart cotton cultivars. This can be achieved by conducting replicated experiments using the same genetic populations in various environments or even in diverse geographical locations. Such studies will help in overcoming the future challenges of changing environments by initiating targeted breeding approaches.

Acknowledgments This activity is undertaken under the project "PSF/NSFC-AGR/P-NIBGE (12) entitled "Mining of elite genes for high yield, super fiber qualities and heat tolerance and their usages in improving Pakistan and China cotton cultivars." This project is funded by Pakistan Science Foundation (PSF).

References

Abdalla AM, Reddy OUK, El-Zik KM, Pepper E (2001) Genetic diversity and relationships of diploid and tetraploid cottons revealed using AFLP. *Theor Appl Genet* 102:222–229

Abdelraheem A (2017) Joint genetic linkage mapping and genome-wide association study of drought and salinity tolerance and *Verticillium Wilt* and Thrips resistance in cotton. Ph.D. Thesis, New Mexico State University, Las Cruces, p 121

Abdelraheem A, Zhang JF (2016) Quantitative trait locus analysis of drought and salt tolerance in an introgressed recombinant inbred line population of upland cotton. In: Western Soc Crop Sci Ann Meet, 12–13 July

Abdelraheem A, Liu F, Song M, Zhang JF (2017) A meta-analysis of quantitative trait loci for abiotic and biotic stress resistance in tetraploid cotton. *Mol Genet Genomics* 292:1221–1235

Abdelraheem A, Thyssen G, Jenkins J, McCarty J, Fang DD, Zhang JF (2018) A genome-wide association study of tolerance to biotic and abiotic stresses in a MAGIC population of upland cotton. In Proceedings of the 2018 beltwild cotton conferences, San Antonio, 3–5, p 565

Abdelraheem A, Esmaeili N, O'Connell M, Zhang J (2019) Progress and perspective on drought and salt stress tolerance in cotton. *Ind Crop Prod* 130:118–129

Abdurakhmonov IY, Buriev ZT, Saha S, Jenkins JN, Abdukarimov A, Pepper AE (2014) Phytochrome RNAi enhances major fibre quality and agronomic traits of the cotton *Gossypium hirsutum* L. *Nat Commun* 5:3062

Ademe MS, He S, Pan Z, Sun J, Wang Q, Qin H, Liu J, Liu H, Yang J, Xu D, Yang J, Ma Z, Zhang J, Li Z, Cai Z, Zhang X, Zhang X, Huang A, Yi X, Zhou G, Li L, Zhu H, Pang B, Wang L, Jia Y, Du X (2017) Association mapping analysis of fiber yield and quality traits in Upland cotton (*Gossypium hirsutum* L.). Mol Genet Genomics 292:1267–1280. https://doi.org/10.1007/s00438-017-1346-9

Abid MA, Liang C, Malik W, Meng Z, Tao Z, Ashraf J, Guo, S. et al. (2017) Cascades of ionic and molecular networks involved in expression of genes underpin salinity tolerance in cotton. *J. Plant Growth Regul.* 37(2):668–679

Ahmad S, Khan N, Iqbal MZ, Hussain A, Hassan M (2002) Salt tolerance of cotton (*Gossypium hirsutum* L.). *Asian J Plant Sci* 1:715–719

Ahmad S, Zhang T, Islam N, Shaheen T, Rahman M (2007) Identifying genetic variation in *Gossypium* based on single nucleotide polymorphism. *Pak J Bot* 39(4):1245–1250

Ahmad N, Khan MO, Islam E, Wei ZY, McAusland L, Lawson T, Johnson GN, Nixon PJ (2020) Contrasting responses to stress displayed by tobacco overexpressing an algal plastid terminal oxidase in the chloroplast. *Front Plant Sci* 11:501. https://doi.org/10.3389/fpls.2020.00501

Ahmed K, Shahid S, Chung ES, Ismail T, Wang XJ (2017) Spatial distribution of secular trends in annual and seasonal precipitation over Pakistan. *Clim Res* 74:95–107

Anagholi A, Esmaeili S, Soltani V, Khaffarian H (2005) Effects of salt stress on the growth and yield of cotton at different stages of development. http://agris.fao.org/agris-search/search.do?recordID=IR2012039023

Ashraf M, Ahmad S (2000) Influence of sodium chloride on ion accumulation, yield components and fibre characteristics in salt-tolerant and salt-sensitive lines of cotton (*Gossypium hirsutum* L.). *Field Crop Res* 66:115–127

Ashraf M, Ahmad A, Qurainy F, Ashraf MY (2011) Alleviation of waterlogging stress in upland cotton (*Gossypium hirsutum* L.) by exogenous application of potassium in soil and as a foliar spray. *Crop Pasture Sci* 62:25–38

Ashraf J, Zuo D, Wang Q, Malik W, Zhang Y, Abid MA, Cheng H, Yang Q, Song G (2018) Recent insights into cotton functional genomics: progress and future perspectives. *Plant Biotechnol J* 16:699–713

Atwell S, Huang YS, Vilhjalmsson BJ, Willems G, Horton M, Li Y (2010) Genome-wide association study of 107 phenotypes in *Arabidopsis thaliana* inbred lines. *Nature* 465:627–631. https://doi.org/10.1038/nature08800

Bange MP, Milroy SP, Thongbai P (2004) Growth and yield of cotton in response to waterlogging. *Field Crops Res* 88:129–142

Baytar AA, Peynircioğlu C, Sezener V, Basal H, Frary A, Frary A, Doganlar S (2018) Genome-wide association mapping of yield components and drought tolerance-related traits in cotton. *Mol Breed* 38:74

Bhatta M, Morgounov A, Belamkar V, Baenziger PS (2018) Genome-wide association study reveals novel genomic regions for grain yield and yield-related traits in drought-stressed synthetic hexaploid wheat. *Int J Mol Sci* 19(10):3011

Bita C, Gerats T (2013) Plant tolerance to high temperature in a changing environment: scientific fundamentals and production of heat stress-tolerant crops. *Front Plant Sci* 4:273

Bolek Y, Hayat K, Bardak A, Azhar MT (2014) Molecular breeding of cotton. Intech Open, London

Bradford KJ (1986) Manipulation of seed water relations via osmotic priming to improve germination under stress conditions. *Hort Sci (USA)* 21:1105–1112

Cai C, Ye W, Zhang T, Guo W (2014) Association analysis of fiber quality traits and exploration of elite alleles in upland cotton cultivars/accessions (*Gossypium hirsutum* L.). *J Integr Plant Biol* 56(1):51–62

Cai C, Wu S, Niu E, Cheng C, Guo W (2017) Identification of genes related to salt stress tolerance using intron-length polymorphic markers, association mapping and virus-induced gene silencing in cotton. *Sci Rep* 7:528. https://doi.org/10.1038/s41598-017-00617-7

Chen W, Gao Y, Xie W, Gong L, Lu K, Wang W, Li Y, Liu X, Zhang H, Dong X, Zhang W, Zhang L, Yu S, Wang G, Lian X, Luo J (2014) Genome wide association analyses provide genetic and biochemical insights into natural variation in rice metabolism. *Nat Genet* 46:714–721. https://doi.org/10.1038/ng.3007

Chen G, Zhang H, Deng Z, Wu R, Li D, Wang M, Tian J (2016) Genome-wide association study for kernel weight-related traits using SNPs in a Chinese winter wheat population. *Euphytica* 212:173–185. https://doi.org/10.1007/s10681-016-1750-y

Chen X, Lu X, Shu N, Wang D, Wang S, Wang J, Guo L, Guo X, Fan W, Lin Z, Ye W (2017a) *GhSOS1*, a plasma membrane Na^+/H^+ antiporter gene from upland cotton, enhances salt tolerance in transgenic *Arabidopsis thaliana*. *PLoS One* 12:e0181450

Chen J, Chopra R, Hayes C, Morris G, Marla S, Burke J, Xin Z, Burow G (2017b) Genome-wide association study of developing leaves' heat tolerance during vegetative growth stages in a Sorghum association panel. *Plant Genome* 10(2):1–15

Cheng C, Zhang Y, Chen X, Song J, Guo Z, Li K, Zhang K (2018) Co-expression of *AtNHX1* and *TsVP* improves the salt tolerance of transgenic cotton and increases seed cotton yield in a saline field. *Mol Breed* 38:19

Dahab AA, Mohamed BB, Husnain T, Saeed M (2012) Variability for drought tolerance in cotton (*Gossypium hirsutum* L.) for growth and productivity traits using selection index. *Afr J Agric Res* 7:4934–4942

Dahab AA, Saeed M, Hamza NB, Mohamed BB, Husnain T (2016) Linkage disequilibrium and association mapping of drought tolerance in cotton (*Gossypium hirsutum* L.) germplasm population from diverse regions of Pakistan. *Afr J Biotechnol* 15(46): 2603–2612.

de Abreu N, Hurtado-Perez JB, Wimmer MC, Frei MA (2017) Genetic factors underlying boron toxicity tolerance in rice: genome-wide association study and transcriptomic analysis. *J Exp Bot* 68(3):687–700

Deeba F, Pandey AK, Ranjan S, Mishra A, Singh R, Sharma YK, Shirke PA, Pandey V (2012) Physiological and proteomic responses of cotton (*Gossypium herbaceum L.*) to drought stress. *Plant Physiol Biochem* 53:6–18

Deinlein U, Stephan AB, Horie T, Luo W, Xu G, Schroeder JI (2014) Plant salt tolerance mechanisms. *Trends Plant Sci* 19(6):371–379

Dinar A, Hassan R, Mendelsohn R, Benhin J (2012) Climate change and agriculture in Africa: impact assessment and adaptation strategies. https://doi.org/10.4324/9781849770767

Diouf L, Pan Z, He SH, Gong WF, Jia YH, Magwanga RO, Romy KR, Rashid HO, Kirungu JN, Du X (2017) High-density linkage map construction and mapping of salt-tolerant QTLs at seedling stage in upland cotton using genotyping by sequencing (GBS). *Int J Mol Sci* 18:2622. https://doi.org/10.3390/ijms18122622

Dou LL, Zhang XH, Pang CY, Song MZ, Wei HL, Fan SL, Yu SX (2014) Genome-wide analysis of the WRKY gene family in cotton. *Mol Genet Genomics.* 289: 1103–1121

Du L, Cai C, Wu S, Zhang F, Hou S, Guo W (2016) Evaluation and exploration of favorable QTL alleles for salt stress related traits in cotton cultivars (*G. hirsutum L.*). *PLoS One* 11(3):e0151076

Elshire RJ, Glaubtiz JC, Sun Q, Poland JA, Kawamoto K, Bukler ES, Mitchell SE (2011) A robust, simple genotyping-by-sequencing (GBS) approach for high diversity species. *PLoS One* 6:e19379

Fan X, Guo Q, Xu P, Gong YY, Shu H, Yang Y, Ni W, Zhang X, Shen X (2015) Transcriptome-wide identification of salt-responsive members of the WRKY gene family in *Gossypium aridum. PLoS One* 10:e0126148

Frouin J, Languillaume A, Mas J, Mieulet D, Boisnard A, Labeyrie A, Bettembourg M, Bureau C, Lorenzini E, Portefaix M, Turquay P, Vernet A, Périn C, Ahmadi N, Courtois B (2018) Tolerance to mild salinity stress in japonica rice: a genome-wide association mapping study highlights calcium signaling and metabolism genes. *PLoS One* 13(1):e0190964

Garcia M, Eckermann P, Haefele S, Satija S, Sznajder B, Timmins A, Baumann U, Wolters P, Mather DE, Fleury D (2019) Genome-wide association mapping of grain yield in a diverse collection of spring wheat *(Triticum aestivum* L.) evaluated in southern Australia. *PLoS One* 14(2):e0211730

Giambattista B, Causse R, Lovrenc L, Dulčić J (2014) A proposed best practice approach to overcome unverified and unverifiable "first records" in ichthyology. *Cybium* 38:9–14

Gore MA, Fang DD, Poland JA, Zhang JF, Percy PG, Cantrell RG, Thyssen G, Lipka AE (2014) Linkage map construction and quantitative trait locus analysis of agronomic and fiber quality traits in cotton. *Plant Genome* 7:1–10

Guo YH, Yu YP, Wang D, Wu CA, Yang GD, Huang JG, Zheng CC (2009) *GhZFP1*, a novel CCCH-type zinc finger protein from cotton, enhances salt stress tolerance and fungal disease resistance in transgenic tobacco by interacting with *GZIRD21A* and *GZIPR5. New Phytol* 183:62–75

Guo WQ, Chen BL, Liu RX, Zhou ZG (2010) Effects of nitrogen application rate on cotton leaf antioxidant enzyme activities and endogenous hormone contents under short-term waterlogging at flowering and boll-forming stage. *Chin J Appl Ecol* 21:53–60

Guo J, Shi G, Guo X, Zhang L, Xu W, Wang Y, Su Z, Hua J (2015) Transcriptome analysis reveals that distinct metabolic pathways operate in salt-tolerant and salt-sensitive upland cotton varieties subjected to salinity stress. *Plant Sci* 238:33–45

Gupta PK, Balyan HS, Gahlaut V (2017) QTL analysis for drought tolerance in wheat: present status and future possibilities. *Agronomy* 7:5

Gür A, Demirel U, Özden M, Kahraman A, Çopur O (2010) Diurnal gradual heat stress affects antioxidant enzymes, proline accumulation and some physiological components in cotton (*Gossypium hirsutum* L.). *Afr J Biotechnol* 9:1008–1015

Han Z, Wang C, Song Z (2006) Characteristics, development and mapping of *Gossypium hirsutum* derived EST-SSRs in allotetraploid cotton. *Theor Appl Genet* 112(3):430–439

He L, Yang X, Wang L, Zhu L, Zhou T, Deng J (2013) Molecular cloning and functional characterization of a novel cotton CBL-interacting protein kinase gene (GhCIPK6) reveals its involvement in multiple abiotic stress tolerance in transgenic plants. *Biochem Biophysical Rese Commun.* 435: 209–215

Hejnak V, Tatar €O, Atasoy GD, Martinkova J, Elen C, Hnilicka F, Skalicky M (2015) Growth and photosynthesis of upland and pima cotton: response to drought and heat stress. *Plant Soil Environ* 61:507–514

Hemantaranjan A, Bhanu A, Singh M, Yadav D, Patel P, Singh R, Katiyar D (2014) Heat Stress responses and thermo tolerance. *Adv Plant Agric Res* 1:10

Huang X, Wei X, Sang T, Zhao Q, Feng Q, Zhao Y (2010) Genome wide association studies of 14 agronomic traits in rice landraces. *Nat Genet* 42:961–967. https://doi.org/10.1038/ng.695

Huang J, Levine A, Wang Z (2013) Plant abiotic stress. *Sci World J* 2013:432836. https://doi.org/10.1155/2013/432836

Iqbal MA, Rahman M (2017) Identification of marker-trait associations for lint traits in cotton. *Front Plant Sci* 8:86. https://doi.org/10.3389/fpls.2017.00086

Iqbal MA, Aziz N, Saeed NA, Zafar Y, Malik KA (1997) Genetic diversity evaluation of some elite cotton varieties by RAPD analysis. *Theor Appl Genet* 94:139–144

Iqbal N, Umar S, Khan NA (2015) Nitrogen availability regulates proline and ethylene production and alleviates salinity stress in mustard (*Brassica juncea*). *J Plant Physiol* 178:84–91

Jabeen R, Ahmad R (2009) Alleviation of the adverse effects of salt stress by foliar application of sodium antagonistic essential minerals on cotton (*Gossypium hirsutum*). *Pak J Bot* 41:2199–2208

Jamshed M, Jia F, Gong J, Palanga KK, Shi Y, Li J, Shang H, Liu A, Chen T, Zhang Z, Cai J, Ge Q, Liu Z, Lu Q, Deng X, Tan Y, Or Rashid H, Sarfra Z, Hassan M, Gong W, Yuan Y (2016) Identification of stable quantitative trait loci (QTLs) for fiber quality traits across multiple environments in *Gossypium hirsutum* recombinant inbred line population. *BMC Genomics* 17:197

Jia Y, Sun X, Sun J, Pan Z, Wang X, He S, Xiao S, Shi W, Zhou Z, Pang B, Wang L, Liu J, Ma J, Du X, Zhu J (2014a) Association mapping for epistasis and environmental interaction of yield traits in 323 cotton cultivars under 9 different environments. *PLoS One* 9:e95882

Jia Y, Sun J, Wang X, Zhou Z, Pan Z, He S, Pang B, Wang L, Du X (2014b) Molecular diversity and association analysis of drought and salt tolerance in *Gossypium hirsutum* L. germplasm. *J Integr Agric* 13(9):1845–1853

Jiang C, Wright RJ, El-Zik KM, Paterson AH (1998) Polyploid formation created unique avenues for response to selection in *Gossypium* (cotton). *Proc Natl Acad Sci U S A* 95(8):4419–4424

Jiang C, Wright R, Woo SS, Delmonte TA, Paterson AH (2000) QTL analysis of leaf morphology in tetraploid *Gossypium* (cotton). *Theor Appl Genet* 100:409–418

Johnson KL, Jones BJ, Bacic A, Schultz CJ (2003) The fasciclin-like arabinogalactan proteins of Arabidopsis. A multigene family of putative cell adhesion molecules. *Plant Physiol* 133:1911–1925

Lafarge T, Bueno C, Frouin J, Jacquin L, Courtois B, Ahmadi N (2017) Genome-wide association analysis for heat tolerance at flowering detected a large set of genes involved in adaptation to thermal and other stresses. *PLoS One* 12(2):e0171254

Lee MH, Cho EJ, Wi SG, Bae H, Kim JE, Cho J-Y, Lee S, Kim J-H, Chung BY (2013) Divergences in morphological changes and antioxidant responses in salt-tolerant and salt-sensitive rice seedlings after salt stress. *Plant Physiol Biochem* 70:325–335

Levi A, Paterson AH, Barak V, Yakir D, Wang B, Chee PW, Saranga Y (2009) Field evaluation of cotton near-isogenic lines introgressed with QTLs for productivity and drought related traits. *Mol Breed* 23:179–195

Li X, Gao W, Guo H, Zhang X, Fang DD, Lin Z (2014) Development of EST-based SNP and InDel markers and their utilization in tetraploid cotton genetic mapping. *BMC Genomics* 15(1):1046

Li CQ, Dong N, Fu YZ, Sun RR, Wang QL (2017) Marker detection and elite allele mining for yield traits in upland cotton (*Gossypium hirsutum* L.) by association mapping. *J Agric Sci* 155(4):613–628

Liang Z, Chen K, Yan Y, Zhang Y, Gao C (2018) Genotyping genome-edited mutations in plants using CRISPR ribonucleoprotein complexes. *J Plant Biotechnol* 16(12):2053–2062

Loka DA, Oosterhuis DM (2012) Water stress and reproductive development in cotton. In: Flowering and fruiting in cotton. pp 51–58

Loka DA, Oosterhuis DM (2014) Water-deficit stress effects on pistil biochemistry and leaf physiology in cotton (*Gossypium hirsutum* L.). *S Afr J Bot* 93:131–136

Lu H, Myers G (2002) Genetic relationships and discrimination of ten influential upland cotton varieties using RAPD markers. *Theor Appl Genet* 105(2–3):325–331

Lu W, Chu X, Li Y, Wang C, Guo X (2013) Cotton *GhMKK1* induces the tolerance of salt and drought stress, and mediates defense responses to pathogen infection in transgenic *Nicotiana benthamiana*. *PLoS One* 8:e68503

Lu C, Zou C, Zhang Y, Yu D, Cheng H, Jiang P, Song G (2015) Development of chromosome-specific markers with high polymorphism for allotetraploid cotton based on genome-wide characterization of simple sequence repeats in diploid cottons (*Gossypium arboretum* L. and *Gossypium raimondii Ulbrich*). *BMC Genomics* 16(1):55

Luo X, Wu J, Li Y, Nan Z, Guo X, Wang Y, Zhang A, Wang Z, Xia G, Tian Y (2013) Synergistic effects of *GhSOD1* and *GhCAT1* overexpression in cotton chloroplasts on enhancing tolerance to methyl viologen and salt stresses. *PLoS One* 8:e54002

Ma Q, Zhao J, Lin H, Ning X, Liu P, Deng F, Si A, Li J (2017) Association between SSR markers and fibre traits in sea island cotton (*Gossypium barbadense*) germplasm resources. *J Genet* 96:55–63

Magwanga RO, Lu P, Kirungu JN, Diouf L, Dong Q, Hu Y, Cai X, Xu Y, Hou Y, Zhou Z, Wang X, Wang K, Liu F (2018) GBS mapping and analysis of genes conserved between *Gossypium tomentosum* and *Gossypium hirsutum* cotton cultivars that respond to drought stress at the seedling stage of the BC_2F_2 Generation. *Int J Mol Sci* 19:1614

Mahajan S, Tuteja N (2005) Cold, salinity and drought stresses: an overview. *Arch Biochem Biophys* 444:139–158

McCouch SR, Wright MH, Tung CW, Maron LG, Mcnally KL, Fitzgerald M, Singh N, DeClerck G, Agosto-Perez F, Korniliev P, Greenberg AJ, Naredo MEB, Mercado SMQ, Harrington SE, Shi Y, Branchini DA, Kuser-Falcão PR, Leung H, Ebana K, Yano M, Eizenga G, McClung A, Mezey J (2016) Open access resources for genome-wide association mapping in rice. *Nat Commun* 7:10532

Meng C, Cai C, Zhang T, Guo W (2009) Characterization of six novel NAC genes and their responses to abiotic stresses in *Gossypium hirsutum* L. *Plant Sci* 176:352–359

Miles C, Wayne M (2008) Quantitative trait locus (QTL) analysis. *Nat Educ* 1:1–3

Multani DS, Lyon BR (1995) Genetic fingerprinting of Australian cotton cultivars with RAPD markers. *Genome* 38(5):1005–1008

Munns R (2005) Genes and salt tolerance: bringing them together. *New Phytol* 167(3):645–663

Najeeb U, Atwell B, Bange M, Tan DKY (2015a) Aminoethoxyvinylglycine (AVG) ameliorates waterlogging-induced damage in cotton by inhibiting ethylene synthesis and sustaining photosynthetic capacity. *Plant Growth Regul* 76:83–98

Najeeb U, Tan DKY, Bange MP (2015b) Inducing waterlogging tolerance in cotton via an anti-ethylene agent amino ethoxy vinyl glycine application. *Arch Agron Soil Sci* 62:1136–1146

Negrao S, Schmockel SM, Tester M (2016) Evaluating physiological responses of plants to salinity stress. *Ann Bot* 119:1–11

Nematzadeh GA (2018) Salt-related genes expression in salt-tolerant and salt-sensitive cultivars of cotton (*Gossypium* sp. L.) under NaCl Stress. *J Plant Mol Breed* 6(1):1–15

Nguyen TB, Giband M, Brottier P, Risterucci AM, Lacape JM (2004) Wide coverage of the tetraploid cotton genome using newly developed microsatellite markers. *Theor Appl Genet* 109:167–175

Oluoch G, Zheng J, Wang X, Khan MKR, Zhou Z, Cai X, Wang C, Wang Y, Li X, Wang H, Liu F, Wang K (2016) QTL mapping for salt tolerance at seedling stage in the interspecific cross of *Gossypium tomentosum* with *Gossypium hirsutum*. *Euphytica* 209:223–235

Oosterhuis DM, Snider JL (2011) High temperature stress on floral development and yield of cotton. In: Oosterhuis DM (ed) Stress physiology in cotton. Cotton Foundation, Memphis

Parekh MJ, Kumar S, Zala HN, Fougat RS, Patel CB, Bosamia TC, Kulkarni KS, Parihar A (2016) Development and validation of novel fiber relevant dbEST–SSR markers and their utility in revealing genetic diversity in diploid cotton (*Gossypium herbaceum* and *G. arboreum*). *Ind J Crop Prod Process* 83:620–629

Park YH, Alabady MS, Ulloa M, Sickler B, Wilkins TA, Yu J, Stelly DM, Kohel RJ, El-Shiny OM, Cantrell RG (2005) Genetic mapping of new cotton fiber loci using EST-derived microsatellites in an interspecific recombinant inbred (RIL) cotton population. *Mol Gen Genomics* 274:428–441

Paterson AH, Bowers JE, Burow MD, Draye X, Elsik GC, Jiang C, Katsar CS, Lan TH, Lin YR, Ming R, Wright RJ (2000) Comparative genomics of plant chromosomes. *Plant Cell* 12(9):1523–1540

Peng Z, He S, Gong W, Sun J, Pan Z, Xu F, Lu Y, Du X (2014) Comprehensive analysis of differentially expressed genes and transcriptional regulation induced by salt stress in two contrasting cotton genotypes. *BMC Genomics* 15(1):760

Peng J, Liu J, Zhang L, Luo J, Dong H, Ma Y, Zhao X, Chen B, Sui N, Zhou Z, Meng Y (2016a) Effects of soil salinity on sucrose metabolism in cotton leaves. *PLoS One* 11:e0156241

Peng J, Liu J, Zhang L, Luo J, Dong H, Ma Y, Zhao X, Sui N, Zhou Z, Meng Y (2016b) Effects of soil salinity on sucrose metabolism in cotton fiber. *PLoS One* 11:e0156398

Percival GC, Keary IP (2008) The influence of nitrogen fertilization on waterlogging stresses in *Fagus sylvatica L.* and *Quercus robur L. Arboric Urban For* 34:29

Poland JA, Rife TW (2012) Genotyping-by-sequencing for plant breeding and genetics. *Plant Genome* 5:92–102

Qin H, Chen M, Yi X, Bie S, Zhang C, Zhang Y, Lan J, Meng Y, Yuan Y, Jiao C (2015) Identification of associated SSR markers for yield component and fiber quality traits based on frame map and upland cotton collections. *PLoS One* 10:e0118073

Rahimian MH, Poormohammadi S (2012) Assessing the impact of climate change on evapotranspiration and soil salinization. In: Filho WL (ed) Climate change and the sustainable use of water resources. Climate change management. Springer, Berlin, Heidelberg, pp 69–76

Rahman Talukder MR, Rutherford S, Chu C (2015) Salinization of drinking water in the context of climate change and sea level rise: a public health priority for coastal Bangladesh. *Int J Clim Change* 8:21–32

Rahman M, Zafar Y (eds) (2018) Past, present and future trends in cotton breeding. Intech Open Limited, London

Rahman M, Hussain D, Zafar Y (2002) Estimation of genetic divergence among elite cotton (*Gossypium hirsutum L.*) cultivars/genotypes by DNA fingerprinting technology. *Crop Sci* 42:2137–2144

Rahman M, Ullah I, Ashraf M, Zafar Y (2008) Genotyping variation for drought tolerance in cotton. *Agron Sustain Dev* 28:439–447

Rahman M, Zafar Y, Paterson AH (2009) *Gossypium* DNA markers types, number and uses. In: Paterson AH (ed) Genomics of cotton. Springer, Dordrecht

Rahman M, Iqbal MA, Shaheen N, Zafar Y (2014) Microsatellites: methods & protocols. In: Miransari M (ed) Stress and plant biotechnology. Stadium Press LLC

Ranjan A, Pandey N, Lakhwani D, Dubey NK, Pathre UV, Sawant SV (2012) Comparative transcriptomic analysis of roots of contrasting gossypium herbaceum genotypes revealing adaptation to drought. *BMC Genom.* 13:680

Ranjan A, Sawant S (2015). Genome-wide transcriptomic comparison of cotton (Gossypium herbaceum) leaf and root under drought stress. *3 Biotech.* 5(4):585–596

Reinisch AJ, Dong J, Brubaker CL, Stelly DM, Wendel JF, Paterson AH (1994) A detailed RFLP map of cotton, *Gossypium hirsutum* × *Gossypium barbadense* chromosome organization and evolution in a disomic polyploidy genome. *Genetics* 138:829–847

Rodriguez-Uribe L, Abdelraheem A, Tiwari R, Sengupta-Gopalan C, Hughs S, Zhang J (2014) Identification of drought-responsive genes in a drought-tolerant cotton *(Gossypium hirsutum L.)* cultivar under reduced irrigation field conditions and development of candidate gene markers for drought tolerance. *Mol Breed* 34:1777–1796

Rong J, Abbey C, Bowers JE, Brubaker CL, Chang C, Chee PW, Delmonte TA, Ding X, Garza JJ, Marler BS, Park CH, Pierce GJ, Rainey KM, Rastogi VK, Schulze SR, Trolinder NL, Wendel JF, Wilkins TA, Williams-Coplin TD, Wing RA, Wright RJ, Zhao X, Zhu L, Paterson AH (2004) A 3347-locus genetic recombination map of sequence-tagged sites reveals features of genome organization, transmission and evolution of cotton (Gossypium). *Genet.* 166(1):389-417

Saeed M, Guo W, Ullah I, Tabbasam N, Zafar Y, Zhang T (2011) QTL mapping for physiology, yield and plant architecture traits in cotton (*Gossypium hirsutum* L.) grown under well-watered versus water stress conditions. *Electron J Biotechnol* 14:3–3

Saeed M, Dahab AHA, Guo WZ, Zhang TZ (2012) A cascade of recently discovered molecular mechanisms involved in abiotic stress tolerance of plants. *OMICS* 16:188–199

Saeed M, Guo W, Zhang T (2014) Association mapping for salinity tolerance in cotton (*Gossypium hirsutum* L.) germplasm from US and diverse regions of China. *Aust J Crop Sci* 8:338–346

Said JI, Lin ZX, Zhang XL, Song MZ, Zhang JF (2013) A comprehensive meta QTL analysis for fiber quality, yield, yield related and morphological traits, drought tolerance, and disease resistance in tetraploid cotton. *BMC Genomics* 14:776

Said JI, Song M, Wang H, Lin Z, Zhang X, Fang DD, Zhang J (2015) A comparative meta-analysis of QTL between intraspecific *Gossypium hirsutum* and interspecific *G. hirsutum* 9 G. *barbadense* populations. *Mol Gen Genomics* 290:1003–1025

Saleem M, Malik T, Shakeel A, Ashraf M (2015) QTL mapping for some important drought tolerant traits in upland cotton. *J Anim Plant Sci* 25:502–509

Saleem M, Amjid MW, Ahmad MQ, Noor E, Qayyum A, Awan MI, Asif M, Nauman M (2018a) Marker assisted selection for relative water content, excised leaf water loss and cell membrane stability in cotton. *Adv Life Sci* 5(2):56–60

Saleem M, Muhammad K, Shakeel A, Muhammad S, Muhammad R, Muhammad A (2018b) Improving the performance of Bt-cotton under heat stress by foliar application of selenium. *J Plant Nutr* 41:1711–1723

Saqib M, Akhtar J, Pervaiz S, Qureshi R, Aslam M (2002) Comparative growth performance of five cotton (*Gossypium hirsutum* L.) genotypes against different levels of salinity. *Pak J Agric Sci* 39:69–75

Saranga Y, Menz M, Jiang CX, Wright RJ, Yakir D, Paterson AH (2001) Genomic dissection of genotype × environment interactions conferring adaptation of cotton to arid conditions. *Genome Res* 11:1988–1995

Saranga Y, Jiang CX, Wright RJ, Yakir D, Paterson AH (2004) Genetic dissection of cotton physiological responses to arid conditions and their inter-relationships with productivity. *Plant Cell Environ* 27(3):263–277

Sarwar M, Ashraf MY, Rahman M, Zafar Y (2012) Genetic variability in different biochemical traits and their relationship with yield and yield parameters of cotton cultivars grown under water stress conditions. *Pak J Bot* 44(2):515–520

Sarwar M, Saleem MF, Ullah N, Shakeel A, Ali S, Bilal MF (2017) Hydrogen peroxide reduces heat-induced yield losses in cotton (*Gossypium hirsutum* L.) by protecting cellular membrane damage. *J Agron Crop Sci* 203:429–441

Sattar S, Hussnain T, Javaid A (2010) Effect of NaCl salinity on cotton (*Gossypium arboreum* L.) grown on MS medium and in hydroponic cultures. *J Anim Plant Sci* 20:87–89

Sekmen AH, Ozgur R, Uzilday B, Turkan I (2014) Reactive oxygen species scavenging capacities of cotton (*Gossypium hirsutum*) cultivars under combined drought and heat induced oxidative stress. *Environ Exp Bot* 99:141–149

Sethi K, Siwachi P, Verma SK (2016) Linkage disequilibrium and association mapping of fibre quality traits in elite Asiatic cotton (*Gossypium arboreum*) germplasm populations. *Czech J Genet Plant Breed* 53:159–167

Shaheen T, Asif M, Zafar Y, Rahman M (2009) Single nucleotide polymorphism analysis of MT-SHSP gene of *Gossypium arboreum* and its relationship with other diploid cotton genomes, *G. hirsutum* and *Arabidopsis thaliana*. *Pak J Bot* 41(1):117–183

Shaheen T, Zafar Y, Rahman M (2010) Detection of single nucleotide polymorphisms in the conserved ESTs regions of *Gossypium arboreum*. *Electron J Biotechnol*. https://doi.org/10.2225/vol13-issue5-fulltext-3

Shakoor MS, Malik TA, Azhar FM, Saleem MF (2010) Genetics of agronomic and fiber traits in upland cotton under drought stress. *Int J Agric Biol* 12:495–500

Sharif I, Aleem S, Farooq J, Rizwan M, Younas A, Sarwar G, Chohan SM (2019) Salinity stress in cotton: effects, mechanism of tolerance and its management strategies. *Physiol Mol Biol Plant* 25(4):807–820

Shi Y, Gao L, Wu Z, Zhang X, Wang M, Zhang C, Zhang F, Zhou Y, Li Z (2017) Genome wide association study of salt tolerance at the seed germination stage in rice. *BMC Plant Biol* 17(1):92

Singh P (2004) Cotton breeding. Kalyani Publishers, Ludhiana New Delhi Noida (U.P.), Hyderabad, Chennai, Kolkata, Cuttack, India, p 295

Singh R, Vara Prasad PV, Sunita K, Giri SN, Reddy K (2007) Influence of high temperature and breeding for heat tolerance in cotton: a review. *Adv Agron* 93:313–385

Sun Z, Li H, Zhang Y, Li Z, Ke H, Wu L, Zhang G, Wang X, Ma Z (2018) Identification of SNPs and candidate genes associated with salt tolerance at the seedling stage in cotton (*Gossypium hirsutum* L.). *Front Plant Sci* 9:1011

Tabbasam N, Zafar Y, Rahman M (2014) Pros and cons of using genomic SSRs and EST-SSRs for resolving phylogeny of the genus *Gossypium*. *Plant Syst Evol* 300(3):559–575

Talukder S, Babar M, Vijayalakshmi K, Poland J, Vara Prasad PV, Robert B, Allan F (2014) Mapping QTL for the traits associated with heat tolerance in wheat (*Triticum aestivum L.*). *BMC Genet* 15:97

Tatineni V, Cantrell RG, Davis DD (1996) Genetic diversity in elite cotton germplasm determined by morphological characteristics and RAPDs. *Crop Sci* 36(1):186–192

Tiwari RS, Picchioni GA, Steiner RL, Jones DC, Hughs SE, Zhang J (2013) Genetic variation in salt tolerance at the seedling stage in an interspecific backcross inbred line population of cultivated tetraploid cotton. *Euphytica* 194:1–11

Ullah I, Rahman M, Zafar Y (2006) Genotypic variation for drought tolerance in cotton (*Gossypium hirsutum* L.): seed cotton yield responses. *Pak J Bot* 38:1679–1687

Ullah I, Rahman M, Ashraf M, Zafar Y (2008) Genotypic variation for drought tolerance in cotton (*Gossypium hirsutum* L.): leaf gas exchange and productivity. *Flora J* 203:105–115

Ullah A, Sun H, Yang XY, Zhang XL (2017) Drought coping strategies in cotton: increased crop per drop. *Plant Biotechnol J* 15:271–284

Upadhyaya H, Dutta L, Sahooa Panda SK (2012) Comparative effect of Ca, K, Mn and B on post-drought stress recovery in tea *Camellia sinensis* (L.). *J Plant Sci* 3:443–460

Wan H, Chen L, Guo J, Li Q, Wen J, Yi B, Ma C, Tu J, Fu T, Shen J (2017) Genome-wide association study reveals the genetic architecture underlying salt tolerance-related traits in rapeseed (*Brassica napus L.*). *Front Plant Sci* 8:593

Wang CY, Isoda A, Li MS, Wang DL (2007) Growth and ecophysiological performance of cotton under water stress conditions. *Agric Sci China* 6(8):949–955

Wang J, Wang D, Fan W, Song G, Wang S, Ye W (2011) The characters of salt-tolerance at different growth stages in cotton. *Shengtai Xuebao Acta Ecol Sin* 31:3720–3727

Wang XQ, Yu Y, Li W, Guo HL, Lin ZX, Zhang XL (2013) Association analysis of yield and fiber quality traits in *Gossypium barbadense* with SSRs and SRAPs. *Genet Mol Res* 12(3):3353–3362

Wang BH, Zhu P, Yuan YL, Wang CB, Yu CM, Zhang HH, Zhu XY, Wang W, Yao CB, Zhuang ZM, Li P (2014) Development of EST-SSR markers related to salt tolerance and their application in genetic diversity and evolution analysis in *Gossypium*. *Genet Mol Res* 13(2):3732–3746

Wang H, Huang C, Guo H, Li X, Zhao W, Dai B (2015) QTL mapping for fiber and yield traits in upland cotton under multiple environments. *PLoS One* 10(6):e0130742

Wang X, Lu X, Wang J, Wang D, Yin Z, Fan W, Wang S, Ye W (2016) Mining and analysis of SNP in response to salinity stress in upland cotton (*Gossypium hirsutum* L.). *PLoS One* 11(6):e0158142

Wang M, Tu L, Lin M, Lin Z, Wang P, Yang Q, Ye Z, Shen C, Li J, Zhang L, Zhou X, Nie X, Li Z, Guo K, Ma Y, Huang C, Jin S, Zhu L, Yang X, Min L, Yuan D, Zhang Q, Lindsey K, Zhang X (2017) Asymmetric subgenome selection and cis-regulatory divergence during cotton domestication. *Nat Genet* 49:579–587

Weigel D (2012) Natural variation in Arabidopsis: from molecular genetics to ecological genomics. *Plant Physiol* 158:2–22

Wright RJ, Thaxton PM, El-Zik KM, Paterson AH (1998) D-subgenome bias of Xcm resistance genes in tetraploid *Gossypium* (cotton) suggests that polyploid formation has created novel avenues for evolution. *Genetics* 149(4):1987–1996

Wu CA, Yang GD, Meng QW, Zheng CC (2004) The cotton *GhNHX1* gene encoding a novel putative tonoplast Na^+/H^+ antiporter plays an important role in salt stress. *Plant Cell Physiol* 45:600–607

Xiao Y, Liu H, Wu L, Warburton M, Yan B (2017) Genome-wide association studies in maize: praise and stargaze. *Mol Plant* 10(3):359–374

Xue T, Li X, Zhu W, Wu C, Yang G, Zheng C (2008) Cotton metallothionein *GhMT3a*, a reactive oxygen species scavenger, increased tolerance against abiotic stress in transgenic tobacco and yeast. *J Exp Bot* 60:339–349

Xue Y, Warburton ML, Sawkins M, Zhang X, Setter T, Xu Y, Grudloyma P, Gethi J, Ribaut JM, Li W, Zhang X, Zheng Y, Yan J (2013) Genome-wide association analysis for nine agronomic traits in maize under well-watered and water-stressed conditions. *Theor Appl Genet* 126:2587–2596

Yang SF, Hoffman NE (1984) Ethylene biosynthesis and its regulation in higher plants. *Plant Physiol* 35:155–189

YongSheng Q, RenZhong L, HongXian M, TianZhen Z, WangZhen G (2009) QTL mapping for yield traits in upland cotton (*Gossypium hirsutum* L.). *Acta Agron Sin* 35:1812–1821

Young LW, Wilen RW, Bonham-Smith PC (2004) High temperature stress of *Brassica napus* during flowering reduces micro- and megagametophyte fertility, induces fruit abortion, and disrupts seed production. *J Exp Bot* 55:485–495

Yuan Y, Xing H, Zeng W, Xu J, Mao L, Wang L, Feng W, Tao J, Wang H, Zhang H, Wang Q, Zhang G, Song X, Sun XZ (2019) Genome-wide association and differential expression analysis of salt tolerance in *Gossypium hirsutum* L at the germination stage. *BMC Plant Biol* 19:394

Zhang LN, Ye WW, Wang JJ, Fan BX (2010) Studies of salinity-tolerance with SSR markers on *G. hirsutum* L. *J Cotton Sci* 22:175–180

Zhang Y, Wang XF, Li ZK, Zhang GY, Ma ZY (2011) Assessing genetic diversity of cotton cultivars using genomic and newly developed expressed sequence tag-derived microsatellite markers. *Genet Mol Res* 10:1462–1470

Zhang D, Li W, Xin C, Tang W, Eneji AE, Dong H (2012) Lint yield and nitrogen use efficiency of field-grown cotton vary with soil salinity and nitrogen application rate. *Field Crops Res* 138:63–70

Zhang L, Zhang G, Wang Y, Zhou Z, Meng Y, Chen B (2013) Effect of soil salinity on physiological characteristics of functional leaves of cotton plants. *J Plant Res* 126:293–304

Zhang L, Ma H, Chen T, Pen J, Yu S, Zhao X (2014a) Morphological and physiological responses of cotton (*Gossypium hirsutum* L.) plants to salinity. *PLoS One* 9:e112807

Zhang J, Zou D, Li Y, Sun X, Wang N-N, Gong S-Y, Zheng Y, Li X-B (2014b) *GhMPK17*, a cotton mitogen-activated protein kinase, is involved in plant response to high salinity and osmotic stresses and ABA signaling. *PLoS One* 9(4):e112807

Zhang F, Li S, Yang S, Wang L, Guo W (2015) Overexpression of a cotton annexin gene, *GhAnn1*, enhances drought and salt stress tolerance in transgenic cotton. *Plant Mol Biol* 87(1–2):47–67. https://doi.org/10.1007/s11103-014-0260-3

Zhang F, Zhu G, Du L, Shang X, Cheng C, Yang B, Hu Y, Cai C, Guo W (2016) Genetic regulation of salt stress tolerance revealed by RNA-Seq in cotton diploid wild species *Gossypium davidsonii*. *Sci Rep* 6:20582

Zhao Y, Wang H, Chen W, Li Y (2014) Genetic structure, linkage disequilibrium and association mapping of *Verticillium* wilt resistance in elite cotton (*Gossypium hirsutum* L.) germplasm population. *PLoS One* 9(1):e86308

Zheng JY, Oluoch G, Riaz KM, Wang XX, Cai XY, Zhou ZL, Wang CY, Wang YH, Li XY, Wang K (2016) Mapping QTLs for drought tolerance in an F2:3 population from an inter-specific cross between *Gossypium tomentosum* and *Gossypium hirsutum*. *Genet Mol Res* 15. https://doi.org/10.4238/gmr.15038477

Chapter 7
Genomics-Assisted Breeding for Fiber Quality Traits in Cotton

Muhammad Saeed, Xianliang Song, Muhammad Atif Iqbal, and Xuezhen Sun

7.1 Introduction

Upland cotton (*G. hirsutum* L.) accounts for more than 90% of the world cotton production (Xu et al. 2008). The need for improving cotton fiber quality is increasing rapidly with acceleration of spinning speeds. Therefore, one of the goals of modern upland cotton breeding is to improve fiber quality by breeding new cultivars with both high yield and elite fiber qualities. Cotton fibers are single-celled trichomes derived from epidermal cells of the ovule. Metabolism of sucrose and import is considered a major factor determining sink strength in tissues such as developing cotton seed and fibers; and, thus, influences cotton fiber quality (Ruan et al. 2003; Pugh et al. 2010). It is important to elucidate the molecular genetics of fiber quality traits in upland cotton. In view of uncertainty in environmental conditions as a result of climate change, fiber quality of cotton is adversely affected. In recent years, the occurrence of high temperatures, long dry spells, and sudden severe intense rains are frequent. These conditions have negative effects on cotton fiber quality. Conventional breeding efforts are not fruitful in producing elite cotton cultivars with suitable fiber quality traits under changing environmental conditions. Incorporation of genomics tools in breeding has increased efficiency and reliability of crop breeding programs (Cobb et al.

M. Saeed (✉)
Department of Botany, Government College University, Faisalabad, Pakistan
e-mail: saeed_pbg@gcuf.edu.pk

X. Song · X. Sun
Agronomy College/National Key Lab of Crop Science, Shandong Agricultural University, Shandong, Taian, China

M. A. Iqbal
National Institute for Biotechnology and Genetic Engineering (NIBGE), Faisalabad, Pakistan

Department of Biotechnology, Pakistan Institute of Engineering and Applied Sciences (PIEAS), Nilore, Islamabad, Pakistan

Mehboob-ur-Rahman et al. (eds.), *Cotton Precision Breeding*,
https://doi.org/10.1007/978-3-030-64504-5_7

2019). Recent availability of whole genome sequences of cultivated diploid and tetraploid species of cotton will greatly accelerate molecular breeding efforts to improve fiber quality in cotton (Zaidi et al. 2018). Such information would enable the subsequent improvement of cotton cultivars by pyramiding multiple breeding target traits (Rong et al. 2004) through genomics-assisted breeding (GAB). The GAB is a robust technique which has accelerated modern molecular plant breeding (MAB) (Xu et al. 2012; Cooper et al. 2014). It has brought precision and more accuracy in cultivar development programs. As a result, breeding practices have become more economical by saving time and cost. It is a promising tool to accelerate the generation of new crop varieties and promote the development of modern agriculture. It can help the breeders to select suitable parents for different crossing programs so as to have novel combinations leading to elite breeding lines. In the past, inadequate genomic resources, narrow genetic base in cultivated gene pool, and high costs associated with genomics research were the limiting factors to applying GAB for crop improvement (Varshney et al. 2013, 2016). However, in recent years, partly due to availability of low-cost high-throughput genotyping (HTPG), NGS technologies, and robust phenotyping platforms, GAB has become popular for crop improvement (Kole et al. 2015). Several successful examples of GAB are now available in a number of crop species (Varshney et al. 2012, 2014).

The main objective of this chapter is to summarize the recent progress of GAB for fiber quality in cotton by encompassing information about mapping/sequencing techniques, achievement through GAB, and its future prospects.

7.2 Genomics-Assisted Breeding Tools

For a successful GAB program, identification of genomic regions associated with traits of interest is the prerequisite (Shi et al. 2019). For this purpose, quantitative trait locus (QTL) mapping techniques are used.

7.2.1 QTL Mapping Techniques

World cotton breeding programs are mainly concerned with the improvement of yield under diverse environmental conditions and consideration of fiber quality at the same time. Key fiber quality traits are staple length, fiber fineness, and fiber strength. These traits are controlled by QTLs, which complicate conventional breeding for fiber improvement. Two mapping techniques are used to tag important chromosomal regions harboring loci for traits of interest. These techniques are (a) linkage mapping and (b) genome-wide association mapping.

Conventional linkage mapping depends upon the use of segregating populations. A number of linkage mapping studies for fiber quality in cotton were carried out (Table 7.1). A total of 79 QTLs for fiber quality (6 fiber length, 7 length uniformity,

Table 7.1 Salient QTLs for fiber quality identified through linkage mapping in cotton

Number of QTLs	Marker system used	Population used	R^2	Reference
2 QTLs	SNPs	F_2; $F_{2:3}$	8.52–22.44%	Liu et al. (2019)
153 QTLs	SSRs	BC_1F_1; BC_1S_1; BC_2F_1; BC_3F_0	4.98–19.80%	Shi et al. (2019)
13 QTLs	SNPs	F_2, $F_{2:3}$	1.71–22.03%	Zhang et al. (2019a, b)
28 QTLs	SSRs	Sealand lines; introgression lines	–	Kumar et al. (2019)
63 QTLs	SSRs, SNPs	196 RILs	4.04–13.55%	Zhang et al. (2017)
71 QTLs	SNPs	RILs	4.70–32.28%	Li et al. (2016)
12 QTLs	SSRs, SNPs, InDels	F_2, $F_{2:3}$	7.59–37.09%	Wang et al. (2015a)
13QTLs	SNPs	RILs	1.52–14.71%	Wang et al. (2015b)
4 QTLs	429 SSRs; 412 SNPs	95 $F_{5:7}$ RILs	14.33–22.30%	Gore et al. (2014)
2 QTLs	SSRs and RAPDS	$F_{2:3}$	14.9%	Shaheen et al. (2013)
79 QTLs	253 RFLPs	900 interspecific F_2 cotton plants; 214 F_3 families	2.1–30.3%	Saranga et al. (2001); Paterson et al. (2003)

9 fiber elongation, 21 fiber strength, 25 fiber fineness, 11 fiber color) were identified under well-watered versus water-limited growth conditions (Saranga et al. 2001; Paterson et al. 2003). Four QTLs for fiber quality traits (1 fiber strength, 1 2.5% span length, 2 length uniformity) were identified using 95 $F_{5:7}$ RILs population (Gore et al. 2014). Twelve QTLs were identified for fiber quality traits with phenotypic variance explained (R^2) values ranging from 7.59–37.09% (Wang et al. 2015a). Thirteen QTLs for fiber strength were identified by using an RIL population comprising 179 individuals. Fiber strength data was recorded from six environments, and RILs were genotyped with SNP markers, generated through restriction site-associated DNA (RAD) sequencing (Wang et al. 2015b). Seventy-one QTLs for fiber quality and yield traits were identified. These QTLs showed 4.70–32.28% phenotypic variance. Sixteen out of 71 QTLs were stably expressed across two environments (Li et al. 2016). Sixty-three QTLs for fiber strength by single-environment model and 16 QTLs by combined multiple-environment model were identified (Zhang et al. 2017). Two QTLs for fiber length were identified on A08 and D03 chromosomes (Liu et al. 2019). Further analyses of these QTLs and mapping population revealed two genes, *cytochrome b5* (*CB5*, *Gh_A08G1729*) and *microtubule end-binding 1C* (*EB1C*, *Gh_D03G0232*), regulating fiber length during fiber elongation stage. One hundred and fifty-three QTLs for fiber quality and yield traits were identified (Shi et al. 2019). These QTLs explained from 4.98–19.80% of the

observed phenotypic variations and 17 of these QTLs were novel. Twenty-eight QTLs for fiber quality (six for elongation, four for fineness, five for short fiber content, five for strength, three for length, and five for fiber uniformity) were identified (Kumar et al. 2019). Out of the identified 28 QTLs, 2 QTLs (*qFL-Chr25* and *qMIC-Chr24*) were valuable to improve fiber length and fiber fineness, respectively.

In recent years, genome-wide association study (GWAS) has been widely used in different crops, including cotton (Fig. 7.1; Table 7.2). GWAS is based on linkage disequilibrium (non-random association of alleles at two or more loci in a given population (Saeed et al. 2014; Yuan et al. 2019)). In earlier studies, the genome-wide LD mapping of fiber quality traits was done using a core set of 95 microsatellite markers in a total of 285 exotic *G. hirsutum* accessions, comprising 208 landrace stocks and 77 variety accessions of Mexican and African origin (Abdurakhmonov et al. 2008). In the subsequent year, Abdurakhmonov et al. (2009) reported genetic diversity, population characteristics, the extent of linkage disequilibrium (LD), and association mapping of fiber quality traits using 202 microsatellite marker primer pairs in 335 *G. hirsutum* germplasm accessions. To mine QTLs related to fiber quality in *G. arboreum*, 56 *G. arboreum* accessions introduced from nine regions of Africa, Asia, and Europe were evaluated with 98 simple sequence repeat (SSR) markers (Kantartzi and Stewart 2008). In another study, a total of 260 lines derived from multiple crosses among tetraploid species in *Gossypium* and 86 SSR markers were used for association mapping to analyze lint percent, boll weight, and fiber quality across three environments (Zeng et al. 2009). Association mapping of agronomic and fiber quality traits was studied in seven diverse environments using 81 *G. hirsutum* cultivars with 121 SSR markers (Zhang et al. 2013a). In recent years, a

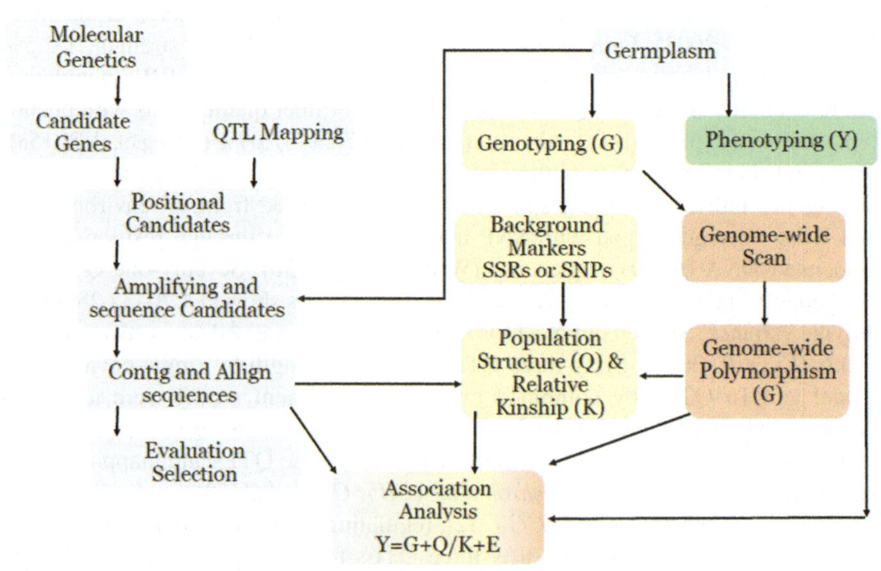

Fig. 7.1 Overall workflow of genome-wide association study (GWAS)

Table 7.2 Salient QTLs identified through genome-wide association mapping in cotton

Number of QTLs	Marker system used	Germplasm used	Reference
25 QTLs	57,413 SNPs	316 *G. hirsutum* accessions	Zhu et al. (2020)
17,264 SNPs	6.35 TB SNPs	419 *G. hirsutum* accessions	Yasir et al. (2019)
342 quantitative trait nucleotides (QTNs)	53,848 SNPs	169 diverse accessions	Li et al. (2018)
7383 SNPs	3.66 million SNPs	419 accessions	Ma et al. (2018)
119 QTLs	1.87 million SNPs	258 accessions	Fang et al. (2017)
75 QTLs	382 SSRs	546 exotic and locally bred cotton genotypes	Iqbal and Rahman (2017)
160 QTLs	63,058 SNPs	503 *G. hirsutum* accessions	Huang et al. (2017)
46 QTLs	10,511 SNPs	719 *G. hirsutum* accessions	Sun et al. (2017)
–	121 SSRs	81 *G. hirsutum* cultivars	Zhang et al. (2013a)
120 QTLs	202 SSRs	335 *G. hirsutum* germplasm accessions	Abdurakhmonov et al. (2009)
–	86 SSRs	260 lines derived from multiple crosses	Zeng et al. (2009)
103 QTLs	95 SSRs	285 exotic *G. hirsutum* accessions	Abdurakhmonov et al. (2008)

total of 342 quantitative trait nucleotides (QTNs) were identified by using 169 diverse *G. hirsutum* L. accessions genotyped with 53,848 high-quality SNPs and phenotyped in four environments. Nine out of 342 QTNs had R^2 values >10%. Four promising candidate genes were identified within the linkage disequilibrium regions of these nine QTNs (Li et al. 2018). A core collection consisting of 419 accessions was genotyped with 3.66 million SNPs and phenotyped across 12 environments. 7383 SNPs were found significantly associated with fiber traits. These associated SNPs were found located within or near 4820 genes. This study showed that D subgenome was most populated with fiber genes compared to A subgenome (Ma et al. 2018).

7.2.1.1 DNA Markers

Different DNA markers had been used for construction of molecular linkage maps and subsequent QTL mapping. Various DNA markers, used extensively for linkage maps construction in cotton, were restriction fragment length polymorphism (RFLP), amplified fragment length polymorphism (AFLP), random amplified polymorphic DNA (RAPD), and microsatellites or simple sequence repeats (SSRs). Recently, single nucleotide polymorphisms (SNPs) are employed extensively in molecular mapping studies in cotton. Molecular linkage maps based on DNA markers serve as the backbone for genetic analyses and are widely recognized as an essential tool for genetic research in many species (Tanksley et al. 1992; Reinisch et al. 1994; Ulloa and

Meredith 2000; Ulloa et al. 2002; Rong et al. 2004; Saeed et al. 2011). Well-saturated genetic maps are helpful in precise identification of DNA markers associated with traits of interest. The identified DNA markers linked to important traits are valuable tools for GAB and increase the efficiency of breeding and significantly decrease cost, time, and risk of subjective phenotypic assays. Molecular markers offer efficient tools for dissecting QTLs affecting traits with complex genetic inheritance and facilitate marker-assisted selection (MAS), GAB, and map-based cloning (Ulloa and Meredith 2000; Paterson et al. 2003; Ulloa et al. 2005; Saeed et al. 2011; Varshney et al. 2014, 2016). A high-density molecular map of fiber genes can assist breeding procedures by directly tagging the genes conferring traits of interest. Genetic maps have been used to partition traits into quantitative trait loci (QTLs) (Ulloa et al. 2005; Shi et al. 2015). Genome sequencing, physical alignment of genomic sequences into the chromosomal maps, and anchoring of the genetic maps are all steps that will improve the accuracy of the genetic maps and enable discovery of genes underlying QTLs and their functions in important biological processes in cotton. Precise marker location, order, and duplication of loci are obstacles to map-based cloning of sequences of interest (Ulloa et al. 2005). In the past, there were two major limiting factors in the use of molecular markers for both QTL analysis and GAB programs in cotton: (1) the limited number of suitable markers available in the public sector and (2) the lack of knowledge of how these markers are associated with economically important QTLs. Now, a wealth of information about DNA markers associated with important traits of cotton is available at CottonGen (https://www.cottongen.org/; Yu et al. 2014).

To date, several genetic maps of cotton genomes have been constructed using diverse molecular markers and mapping populations (Table 7.3). A genetic linkage map consisting of 705 RFLP loci and covering 4675cM was constructed by Reinisch et al. (1994). A linkage map harboring 253 RFLPs markers was constructed using 900 interspecific F_2 population individuals (Saranga et al. 2001; Paterson et al. 2003). A comprehensive reference map (CRM) of cotton was produced by combining information from 28 available public genetic maps of cotton. This CRM contained 7424 markers and carried over 93% combined mapping information of 28 individual genetic maps (Yu et al. 2010). SSRs have been extensively used in genetic maps of cotton. A genetic map consisting of 2763 SSR markers was constructed from an F_2 population of an interspecific *Gossypium hirsutum* and *G. darwinii* (Chen et al. 2015). A genetic map consisting of 2292 SSR was constructed. It spanned 5115.16 cM (centimorgan) of cotton AD genome (Shi et al. 2015). A genetic map consisting of 3093 SSR loci, distributed across 26 cotton chromosomes, was constructed. It spanned 4365.3 cM (Khan et al. 2016). Recently, a number of linkage maps in cotton were constructed by incorporating SNPs. A genetic map consisting of a total of 2072 loci, 1825 SSRs and 247 SNPs, was constructed (Yu et al. 2012). Another genetic map consisting of 429 SSRs and 412 SNPs was constructed. It covered ~2061 cM (Gore et al. 2014). A genetic linkage map containing 2393 loci was constructed. This map consisted of 77 SSRs and 2316 SNPs markers and covered 2865.73 cM distance. The average distance between consecutive markers was 1.20 cM (Zhang et al. 2017). A high-density molecular map composed of loci for fiber genes is still lacking in cotton, primarily due

Table 7.3 Information about linkage maps constructed

Number of markers spanning linkage map	Genome coverage	Population used	Reference
2393 (77 SSRs; 2316 SNPs) loci	2865.73 cM	196 $F_{6:8}$ RILs	Zhang et al. (2017)
3093 SSRs	4365.3 cM		Khan et al. (2016)
2618 SNP markers	63K SNP assay	188 F_8 RILs	Li et al. (2016)
2763 SSR markers		F_2 population of an interspecific cross of *Gossypium hirsutum* and *G. darwinii*	Chen et al. (2015)
2292 SSRs	5115.16 cM		Shi et al. (2015)
7000 upland intraspecific and 19,000 interspecific SNP markers	CottonSNP63K Illumina Infinium array	1156 samples	Hulse-kemp et al. (2015)
429 SSRs; 412 SNPs	~2061 cM	95 $F_{5:7}$ RILs	Gore et al. (2014)
1825 SSRs and 247 SNPs			Yu et al. (2012)
CRM (comprehensive reference map); 7424 markers			Yu et al. (2010)
253 RFLPs	5844.3cM	900 interspecific F_2 population	Saranga et al. (2001); Paterson et al. (2003)
705 RFLPs	4675cM		Reinisch et al. (1994)

to the lack of fiber gene sequences and the limited number of simple PCR-based DNA markers, such as microsatellites, available in the public domain.

Microsatellites or concatemeric repeats of short DNA sequences, resulting from mutational effects of replication slippage, are found in abundance and scattered throughout the genomes of eukaryotes (Tautz and Schlötterer 1994). Generally, microsatellites are classified into two types: simple sequence repeats (SSRs) and complex sequence repeats (CSRs). A SSR is composed of one type of repeating unit, while a CSR contains stretches of two or more different repeat motifs, which can also vary in the number of repeating units, and therefore leads to an additional source of polymorphism. CSRs represent the second largest group of microsatellites in eukaryotic genomes, accounting for about 10% of microsatellites in the human genome (Weber 1990). Amplification of SSRs or CSRs by PCR reveals length polymorphisms of microsatellites resulting from differences in the number of repeat units. This simple PCR-based marker system is very valuable because of the co-dominant, highly variable, and multi-allelic nature of the markers it provides (Gupta et al. 1996). The occurrence of microsatellite loci in EST collections has been observed in *Arabidopsis* and several major crops (Temnykh et al. 2000; Thiel et al. 2003), including tetraploid cotton (Nguyen et al. 2004; Qureshi et al. 2004; Han et al. 2004, 2006). ESTs containing microsatellites offer a relatively simple and efficient means for the development of PCR-based microsatellite markers with functional gene information (Thiel et al. 2003).

The genetic map of RFLP markers and QTLs in defined chromosomal regions was very helpful in transferring useful orthologous QTL loci among the cotton germplasm. Cytogenetic analyses indicated that cultivated cotton (*G. hirsutum* L.) is of polyploid origin and that, by virtue of the complementation between its two compensating homoeologous genomes, cotton generally tolerates deletions and deficiencies at the single or partial chromosome level. Currently, a large number of interspecific aneuploid chromosomal substitution stocks are available in cotton. The comparative analysis of the interspecific chromosomal substitution lines of tetraploid cotton provides an opportunity in developing a chromosome-based consensus map of important traits and molecular markers within a very short period. A chromosome-specific molecular mapping strategy will also serve as an anchor for integrating large genome segments originating from megabase technologies like the BAC (Tomkins et al. 2001). This approach will also be a key tool for map-based chromosome walking and cloning of desirable genes. In addition, the identification of chromosomal regions with economically important traits will be especially important in germplasm improvement and introgression programs using interspecific chromosome substitution lines.

The initial SNP marker development in cotton was slow and costly, and few SNP markers were made available in the past decade. In addition, initial efforts to develop SNP markers were hindered by the co-identification of SNP interlocus variants between the two subgenomes in the tetraploids or homeo-SNPs (Udall et al. 2006; Van Deynze et al. 2009).

With the availability of next-generation sequencing (NGS) technology, sequencing has become faster and cheaper, recently helping to identify larger number of SNP markers (Hulse-Kemp et al. 2015; Islam et al. 2015). Considerable progress has been made toward the development of new cotton genomic resources. The larger collection of SNPs (up to 90,000) was assembled from gene transcripts and genomic DNA of multiple cultivars, genotypes, and species (Cotton SNP Chip, Illumina BeadArray, Illumina Inc., Mira Loma, CA, USA, and public institutions). Recently, a CottonSNP63K Illumina Infinium array (Illumina Inc.) was validated with 1156 samples, providing more than 7000 upland intraspecific and 19,000 interspecific SNP markers that were amenable to mapping in 2 F_2 populations (Hulse-kemp et al. 2015). The development of linkage maps or genetic mapping in the last decade was primarily performed with SSR markers (Yu et al. 2012). However, distribution and numbers of SSRs are limited in a genome and have been primarily limited to inclusion of a couple hundred single nucleotide polymorphism (SNP) markers with SSRs (Yu et al. 2012). The cotton 63K single nucleotide polymorphism (SNP) assay was used to genotype 188 F_8 RILs, and a genetic map containing 2618 polymorphic SNP markers was constructed (Li et al. 2016).

Mapping multiple populations and developing consensus maps will help to reduce large gaps due to the lack of polymorphism in certain complex genomic regions, to increase the number of mapped loci, to validate marker order, and to increase marker genome coverage (Ulloa et al. 2002, 2005).

7.2.1.2 Genotyping by Sequencing (GBS)

Numerous genome complexity techniques are available, e.g., genotyping by sequencing (GBS) (Soto et al. 2015; Diouf et al. 2018; Fan et al. 2018; Majeed et al. 2019), restriction site-associated DNA (RAD) sequencing (Pegadaraju et al. 2013; Wang et al. 2015a, b), and specific locus amplified fragment sequencing (SLAF-seq) (Zhang et al. 2013b, 2016; Li et al. 2014; Xu et al. 2015; Wang et al. 2019; Yu et al. 2020). Among these techniques, GBS is a robust technique which has accelerated genotyping of plant materials for QTL mapping purposes. GBS has intensified SNP discoveries in different plant species (Verma et al. 2015; TorelloMarinoni et al. 2018; Luo et al. 2019; Yadav et al. 2019; Ballesta et al. 2020). Recently, GBS is also implemented in cotton genotyping procedures. A genetic map containing 9182 SNPs, derived by GBS, was constructed. It spanned 3462.8cM (Liu et al. 2019). There are other reports of the use of GBS in cotton genotyping such as that of Magwanga et al. (2018) who genotyped BC_2F_2 generations through GBS and developed a fine genetic map; Reddy et al. (2017) used GBS and generated 10,129 single nucleotide polymorphisms (SNPs) for 440 *G. hirsutum* and 219 *G. barbadense* cultivated and wild pools; and Qi et al. (2017) examined F_2 population using GBS with 5571 single nucleotide polymorphism (SNP) markers to construct a genetic linkage map which consisted of 3187 polymorphic markers. In recent years, a total of 143 recombinant inbred lines (RILs derived from a cross between Chinese *G. barbadense* cultivar 5917 and American Pima S-7), along with their parental genotypes, were evaluated in three replicated field tests for fiber yield traits and then genotyped by deploying GBS to develop polymorphic SNP markers (Fan et al. 2018). In another study, a total of 94 cotton accessions comprising of seven primitive races of upland cotton along with nine upland cotton and 12 Egyptian cotton were evaluated through GBS, and 146,558 SNPs were reported (Zhang et al. 2019b).

With the availability of reference genome sequences in cotton, resequencing of candidate genes or other chromosomal regions of interest has become a routine practice in cotton genotyping. The genome-wide resequencing of 318 landraces and improved cultivars or lines was carried out for a comprehensive genomic assessment and genome-wide association study. There were more associated loci for lint yield than for fiber quality (Fang et al. 2017). In cotton, breeders also investigated the genetic basis of the hybrid vigor by resequencing the 9053 and sGK9708 having 62.13× coverage depth—the parents of the elite hybrid cotton CCRI63 that had hybrid vigor in lint percentage and boll weight (Song et al. 2018).

7.2.2 Mapping Populations Used

Different mapping populations were used during mapping studies in cotton. In linkage mapping, four backcross generation populations were used to identify QTLs for fiber quality and yield (Shi et al. 2019). These populations included BC_1F_1, BC_1S_1, BC_2F_1, and BC_3F_0 and were developed from crossing CCRI36 (*Gossypium hirsutum*

L.) and Hai1 (*G. barbadense* L.). A total of 465 F_2 individuals were derived from a cross between *Gossypium hirsutum* var. *yucatanense* accession TX2094 as the maternal parent and the modern elite cultivar *G. hirsutum* cv. Acala Maxxa as the paternal parent (Grover et al. 2020). In a number of studies, F_2 populations were used for genetic maps construction and subsequent QTL mapping (Saeed et al. 2011; Chen et al. 2015; Khan et al. 2016). A high-density linkage map of tetraploid cotton was constructed by using F_2 population derived from interspecific cross of *G. hirsutum* and *G. darwinii* (Chen et al. 2015). An F_2 population derived from an interspecific cross *G. hirsutum* × *G. tomentosum* was used (Khan et al. 2016). Backcross populations were also used to construct genetic maps. A BC_1F_1 population was used to construct a genetic map. This BC_1F_1 population was derived from an interspecific cross *Gossypium hirsutum* × *Gossypium barbadense* and used to identify QTLs for lint percentage (Yu et al. 2012; Li et al. 2016). A high-density genetic map was constructed by genotyping 186 RILs. These RILs were derived from an interspecific cross of *Gossypium hirsutum* L. (TM-1) × *G. barbadense* L. (3–79) (Yu et al. 2012). QTLs for fiber quality and yield traits were mapped using 188 F_8 RILs, derived from intraspecific cross of HS46 and MARCABUCAG8US-1-88 (Li et al. 2016). In another study, a total of 185 (139 exotic, 20 Bt cotton and 26 *G. arboreum*) cotton genotypes were selected to study marker-trait association after evaluating 546 exotic and local bred cotton genotypes for fiber traits in 2009 (Iqbal and Rahman 2017). A genetic map comprising of 5178 SNPs was generated using 277 $F_{2:3}$ population from an intraspecific cross of two *G. hirsutum* accessions—CCR135 and Nan Dan Ba Di Da Hua (Diouf et al. 2018).

7.2.3 Methods and Softwares

In linkage mapping studies, for construction of genetic maps, different softwares were used. MapMaker 1.0 (Lander et al. 1987) was initially used to construct genetic maps. JoinMap 3.0 (Van Ooijen and Voorrips 2001) was extensively used for calculation of genetic linkage maps. It is based on version 2.0 developed by Van Ooijen (1995). PowerMarker has also a greater application in genetic map construction. In genetic mapping, QTL Cartographer is used to identify QTLs by incorporating genotypic and phenotypic data. In GWAS, STRUCTURE software is used to find sub-populations in the used plant material, and TASSEL software is used to find out marker-trait associations. In TASSEL software, two models—Mixed Linear Model (MLM) and General Linear Models (GLM)—are commonly used.

7.3 Achievements and Future Prospects

Due to recent progress in genomics approaches and success to sequence large crop plant genomes, GAB has been successfully incorporated in routine breeding programs. In cotton, complete genome sequences of diploid (*G. raimondii*, *G. arbo-*

reum) as well as tetraploid (*G. hirsutum, G. barbadense*) are available. In addition, for upland cotton, there is availability of a large collection of fine-quality EST datasets related to fiber quality traits. This has added a significant number of reliable SNPs, related to cotton fiber quality, in the public cotton marker database (https://www.cottongen.org/find/qtl). In two major cotton producing countries in the world, the United States and China, GAB made significant contribution toward improved seed cotton yield with desirable fiber quality standards.

Due to climate change, a number of cotton producing countries in the world are faced with high temperature conditions during cotton growing season. High temperature deteriorates fiber quality adversely (Zahid et al. 2016). Conventional breeding approaches are inadequate to meet elite cultivar development needs of post-climate change scenario. GAB has greater potential to meet this challenge. Therefore, a number of cotton producing countries are now attempting to incorporate GAB in routine cotton breeding programs to improve fiber quality, a complex attribute in cotton breeding.

References

Abdurakhmonov IY, Kohel RJ, Yu JZ, Pepper AE, Abdullaev AA, Kushanov FN, Salakhutdinov IB, Buriev ZT, Saha S, Scheffler BE, Jenkins JN, Abdukarimov A (2008) Molecular diversity and association mapping of fiber quality traits in exotic *G. hirsutum* L. germplasm. *Genomics* 92:478–487

Abdurakhmonov IY, Saha S, Jenkins JN, Buriev ZT, Shermatov SE, Scheffler BE, Pepper AE, Yu JZ, Kohel RJ, Abdukarimov A (2009) Linkage disequilibrium based association mapping of fiber quality traits in *G. hirsutum* L. variety germplasm. *Genetica* 136:401–417

Ballesta P, Mora F, Del Pozo A (2020) Association mapping of drought tolerance indices in wheat: QTL-rich regions on chromosome 4A. *Sci Agric* 77. https://doi.org/10.1590/1678-992x-2018-0153

Chen H, Khan MKR, Zhou Z, Wang X, Cai X, Ilyas MK, Wang C, Wang Y, Li Y, Liu F, Wang K (2015) A high-density SSR genetic map constructed from a F_2 population of *Gossypium hirsutum* and *Gossypium darwinii*. *Gene* 574(2):273–286. https://doi.org/10.1016/j.gene.2015.08.022

Cobb JN, Biswas PS, Platten JD (2019) Back to the future: revisiting MAS as a tool for modern plant breeding. *Theor Appl Genet* 132:647–667

Cooper M, Messina CD, Podlich D, Totir LR, Baumgarten A, Hausmann NJ, Wright D, Graham G (2014) Predicting the future of plant breeding: complementing empirical evaluation with genetic prediction. *Crop Pasture Sci* 65:311–336

Diouf L, Magwanga RO, Gong W, He S, Pan Z, Jia YH et al (2018) QTL mapping of fiber quality and yield-related traits in an intra-specific upland cotton using genotype by sequencing (GBS). *Int J Mol Sci* 19(2):441

Fan L, Wang L, Wang X, Zhang H, Zhu Y, Guo J et al (2018) A high-density genetic map of extra-long staple cotton (*Gossypium barbadense*) constructed using genotyping-by-sequencing based single nucleotide polymorphic markers and identification of fiber traits-related QTL in a recombinant inbred line population. *BMC Genomics* 19(1):489. https://doi.org/10.1186/s12864-018-4890-8

Fang L, Wang Q, Hu Y, Jia Y, Chen J, Liu B, Zhang Z, Guan X, Chen S, Zhou B, Mei G, Sun J, Pan Z, He S, Xiao S, Shi W, Gong W, Liu J, Ma J, Cai C, Zhu X, Guo W, Du X, Zhang T (2017) Genomic analyses in cotton identify signatures of selection and loci associated with fiber quality and yield traits. *Nat Genet* 49(7):1089–1098. https://doi.org/10.1038/ng.3887

Gore MA, Fang DD, Poland JA, Zhang J, Percy RG, Cantrell RG, Thyssen G, Lipka AE (2014) Linkage map construction and quantitative trait locus analysis of agronomic and fiber quality traits in cotton. *Plant Genome* 7(1). https://doi.org/10.3835/plantgenome2013.07.0023

Grover CE, Yoo MJ, Gore MA, Harker DB, Byers RL, Lipka AE, Hu G, Yuan D, Conover JL, Udall JA, Paterson AH, Gore MA, Wendel JF (2020) Genetic analysis of the transition from wild to domesticated cotton (*G. hirsutum*). *G3 (Bethesda)* 10(2):731–754. https://doi.org/10.1534/g3.119.400909

Gupta PK, Balyan HS, Sharma PC, Ramesh B (1996) Microsatellites in plants: a new class of molecular markers. *Curr Sci* 70:45–54

Han ZG, Guo WZ, Song XL, Zhang TZ (2004) Genetic mapping of EST-derived microsatellites from the diploid *Gossypium arboreum* in allotetraploid cotton. *Mol Gen Genomics* 272:308–327

Han ZG, Wang CB, Song XL, Guo WZ, Gou J, Li CH, Chen X, Zhang TZ (2006) Characteristics, development and mapping of *Gossypium hirsutum* derived EST-SSRs in allotetraploid cotton. *Theor Appl Genet* 112:430–439

Huang C, Nie X, Shen C, You C, Li W, Zhao W, Zhang X, Lin Z (2017) Population structure and genetic basis of the agronomic traits of upland cotton in China revealed by a genome-wide association study using high-density SNPs. *Plant Biotechnol J* 15(11):1374–1386

Hulse-Kemp AM, Lemm J, Plieske J, Ashrafi H, Buyyarapu R, Fang DD, Frelichowski J, Giband M, Hague S, Hinze LL, Kochan KJ, Riggs PK, Scheffler JA, Udall JA, Ulloa M, Wang SS, Zhu QH, Bag SK, Bhardwaj A, Burke JJ, Byers RL, Claverie M, Gore MA, Harker DB, Islam MS, Jenkins JN, Jones DC, Lacape JM, Llewellyn DJ, Percy RG, Pepper AE, Poland JA, Mohan Rai K, Sawant SV, Singh SK, Spriggs A, Taylor JM, Wang F, Yourstone SM, Zheng X, Lawley CT, Ganal MW, Van Deynze A, Wilson IW, Stelly DM (2015) Development of a 63K SNP array for cotton and high-density mapping of intraspecific and interspecific populations of *Gossypium* spp. *G3: Genes, Genomes, Genetics* 5(6):1187–1209

Iqbal MA, Rahman MU (2017) Identification of marker-trait associations for lint traits in cotton. *Front Plant Sci* 8:86. https://doi.org/10.3389/fpls.2017.00086

Islam MS, Thyssen GN, Jenkins JN, Fang DD (2015) Detection, validation and application of genotyping-by-sequencing based single nucleotide polymorphisms in upland cotton (*Gossypium hirsutum* L.). *Plant Genome* 8:1–10

Kantartzi SK, Stewart JM (2008) Association analysis of fiber traits in *Gossypium arboreum* accessions. *Plant Breed* 127:173–179

Khan MKR, Chen H, Zhou Z, Ilyas MK, Wang X, Cai X, Wang C, Liu F, Wang K (2016) Genome wide SSR high density genetic map construction from an interspecific cross of *Gossypium hirsutum* × *Gossypium tomentosum*. *Front Plant Sci* 7:436. https://doi.org/10.3389/fpls.2016.00436

Kole C, Muthamilarasan M, Henry R, Edwards D, Sharma R, Abberton M, Batley J, Bentley A, Blakeney M, Bryant J, Cai H, Cakir M, Cseke LJ, Cockram J, de Oliveira AC, De Pace C, Dempewolf H, Ellison S, Gepts P, Greenland A, Hall A, Hori K, Hughes S, Humphreys MW, Iorizzo M, Ismail AM, Marshall A, Mayes S, Nguyen HT, Ogbonnaya FC, Ortiz R, Paterson AH, Simon PW, Tohme J, Tuberosa R, Valliyodan B, Varshney RK, Wullschleger SD, Yano M, Prasad M (2015) Application of genomics-assisted breeding for generation of climate resilient crops: progress and prospects. *Front Plant Sci* 6:563. https://doi.org/10.3389/fpls.2015.00563

Kumar P, Singh R, Lubbers EL, Shen X, Paterson AH, Campbell BT, Jones DC, Chee PW (2019) Genetic evaluation of exotic chromatins from two obsolete interspecific introgression lines of upland cotton for fiber quality improvement. *Crop Sci* 59(3):1073–1084. https://doi.org/10.2135/cropsci2018.12.0745

Lander ES, Green P, Abrahamson J, Barlow A, Daly MJ, Lincoln SE, Newburg L (1987) Mapmaker: an interactive computer package for constructing primary genetic linkage maps of experimental and natural population. *Genomics* 1:174–181

Li B, Tian L, Zhang JY, Huang L, Han FX, Yan SR, Wang LZ, Zheng HK, Sun JM (2014) Construction of a high-density genetic map based on large-scale markers developed by specific length amplified fragment sequencing (SLAF-seq) and its application to QTL analysis for isoflavone content in *Glycine max*. *BMC Genomics* 15:1086

Li C, Dong Y, Zhao T, Li L, Li C, Yu E, Mei L, Daud MK, He Q, Chen J, Zhu S (2016) Genome-wide SNP linkage mapping and QTL analysis for fiber quality and yield traits in the Upland cotton recombinant inbred lines population. *Front Plant Sci* 7:1356. https://doi.org/10.3389/fpls.2016.01356

Li C, Fu Y, Sun R, Wang Y, Wang Q (2018) Single-locus and multi-locus genome-wide association studies in the genetic dissection of fiber quality traits in Upland cotton (*Gossypium hirsutum* L.). *Front Plant Sci* 9:1083. https://doi.org/10.3389/fpls.2018.01083

Liu G, Pei W, Li D, Ma J, Cui Y, Wang N, Song J, Wu M, Li L, Zang X, Yu S, Zhang J, Yu J (2019) A targeted QTL analysis for fiber length using a genetic population between two introgressed backcrossed inbred lines in upland cotton (*Gossypium hirsutum*). *Crop J* 7:273–282

Luo Z, Pernell Tomasi P, Fahlgren N, Abdel-Haleem H (2019) Genome-wide association study (GWAS) of leaf cuticular wax components in *Camelina sativa* identifies genetic loci related to intracellular wax transport. *BMC Plant Biol* 19:187. https://doi.org/10.1186/s12870-019-1776-0

Ma Z, He S, Wang X, Sun J, Zhang Y, Zhang G, Wu L, Li Z, Liu Z, Sun G, Yan Y, Jia Y, Yang J, Pan Z, Gu Q, Li X, Sun Z, Dai P, Liu Z, Gong W, Wu J, Wang M, Liu H, Feng K, Ke H, Wang J, Lan H, Wang G, Peng J, Wang N, Wang L, Pang B, Peng Z, Li R, Tian S, Du X (2018) Resequencing a core collection of upland cotton identifies genomic variation and loci influencing fiber quality and yield. *Nat Genet* 50:803–813

Magwanga RO, Lu P, Kirungu JN, Diouf L, Dong Q, Hu Y, Cai X, Xu Y, Hou Y, Zhou Z, Wang X, Wang K, Liu F (2018) GBS mapping and analysis of genes conserved between *Gossypium tomentosum* and *Gossypium hirsutum* cotton cultivars that respond to drought stress at the seedling stage of the BC_2F_2generation. *Int J Mol Sci* 19:1614. https://doi.org/10.3390/ijms19061614

Majeed S, Rana IA, Atif RM, Ali Z, Hinze L, Azhar MT (2019) Role of SNPs in determining QTLs for major traits in cotton. *J Cotton Res* 2(1):5. https://doi.org/10.1186/s42397-019-0022-5

Nguyen TB, Giband M, Brottier P, Risterucci A-M, Lacape J-M (2004) Wide coverage of the tetraploid cotton genome using newly developed microsatellite markers. *Theor Appl Genet* 109:167–175

Paterson AH, Saranga Y, Menz M, Jiang CX, Wright R (2003) QTL analysis of genotype× environment interactions affecting cotton fiber quality. *Theor Appl Genet* 106(3):384–396

Pegadaraju V, Nipper R, Hulke B, Qi LL, Schultz Q (2013) De novo sequencing of sunflower genome for SNP discovery using RAD (restriction site associated DNA) approach. *BMC Genomics* 14:556

Pugh DA, Offler CE, Talbot MJ, Ruan YL (2010) Evidence for the role of transfer cells in the evolutionary increase in seed and fiber biomass yield in cotton. *Mol Plant* 3(6):1075–1086

Qi H, Wang N, Qiao W, Xu Q, Zhou H, Shi J, Yan G, Huang Q (2017) Construction of a high-density genetic map using genotyping by sequencing (GBS) for quantitative trait loci (QTL)analysis of three plant morphological traits in upland cotton (*Gossypium hirsutum* L.). *Euphytica* 213(83). https://doi.org/10.1007/s10681-017-1867-7

Qureshi SN, Saha S, Kantety RV, Jenkins JN (2004) EST-SSR: a new class of genetic markers in cotton. *J Cotton Sci* 8(1):1122–1123

Reddy UK, Nimmakayala P, Abburi VL, Reddy CVCM, Saminathan T, Percy RG, Yu JZ, Frelichowski J, Udall JA, Page JT, Zhang D, Shehzad T, Paterson AH (2017) Genome-wide divergence, haplotype distribution and population demographic histories for *Gossypium hirsutum* and *Gossypium barbadense* as revealed by genome-anchored SNPs. *Sci Rep* 7:41285

Reinisch AJ, Dong JM, Brubaker CL, Stelly DM, Wendel JF, Paterson AH (1994) A detailed RFLP map of cotton, *Gossypium hirsutum* x *Gossypium barbadense*: chromosome organization and evolution in a disomic polyploid genome. *Genetics* 138(3):829–847

Rong J, Abbey C, Bowers JE, Brubaker CL, Chang C, Chee PW, Delmonte TA, Ding X, Garza JJ, Marler BS, Park CH, Pierce GJ, Rainey KM, Rastogi VK, Schulze SR, Trolinder NL, Wendel JF, Wilkins TA, Williams-Coplin TD, Wing RA, Wright RJ, Zhao X, Zhu L, Paterson AH (2004) A 3347-locus genetic recombination map of sequence-tagged sites reveals features of genome organization, transmission and evolution of cotton (Gossypium). *Genetics* 166(1):389–417

Ruan YL, Llewellyn DJ, Furbank RT (2003) Suppression of sucrose synthase gene expression represses cotton fiber cell initiation, elongation, and seed development. *Plant Cell* 15(4):952–964

Saeed M, Guo WZ, Ullah I, Tabbasam N, Zafar Y, Rahman M, Zhang TZ (2011) QTL mapping for physiology, yield and plant architecture traits in cotton (*Gossypium hirsutum* L.) grown under well-watered versus water-stress conditions. *Electron J Biotechnol* 14(3). http://www.ejbiotechnology.info/index.php/ejbiotechnology/article/view/v14n3-3/1307

Saeed M, Guo WZ, Zhang TZ (2014) Association mapping for salinity tolerance in cotton (*Gossypium hirsutum* L.) germplasm from US and diverse regions of China. *Aust J Crop Sci* 8(3):338–346

Saranga Y, Menz M, Jiang C, Wright R, Yakir D, Paterson A (2001) Genomic dissection of genotype × environment interactions conferring adaptation of cotton to arid conditions. *Genome Res* 11:1988–1995

Shaheen T, Zafar Y, Rahman M (2013) QTL mapping of some productivity and fiber traits in *Gossypium arboreum*. *Turk J Bot* 37:802–810

Shi Y, Li W, Li A, Ge R, Zhang B, Li J, Liu G, Li J, Liu A, Shang H, Gong J, Gong W, Yang Z, Tang F, Liu Z, Zhu W, Jiang J, Yu X, Wang T, Wang W, Chen T, Wang K, Zhang Z, Yuan Y (2015) Constructing a high-density linkage map for *Gossypium hirsutum* × *Gossypium barbadense* and identifying QTLs for lint percentage. *J Integr Plant Biol* 57(5):450–467. https://doi.org/10.1111/jipb.12288

Shi Y, Liu A, Li J, Zhang J, Zhang B, Ge Q, Jamshed M, Lu Q, Li S, Xiang X, Gong J, Gong W, Shang H, Deng X, Pan J, Yuan Y (2019) Dissecting the genetic basis of fiber quality and yield traits in interspecific backcross populations of *Gossypium hirsutum* × *Gossypium barbadense*. *Mol Gen Genomics* 294:1385–1402. https://doi.org/10.1007/s00438-019-01582-8

Song C, Li W, Wang Z, Pei X, Liu Y, Ren Z, He K, Zhang F, Sun K, Zhou X, Ma X, Yang D (2018) Genome resequencing reveals genetic variation between the parents of an elite hybrid upland cotton. *Agronomy* 8:305. https://doi.org/10.3390/agronomy8120305

Soto JC, Ortiz JF, Perlaza-Jiménez L, Vasquez AX, Lopez-Lavalle LA, Mathew B, Léon J, Bernal AJ, Ballvora A, López CE (2015) A genetic map of cassava (*Manihot esculenta Crantz*) with integrated physical mapping of immunity-related genes. *BMC Genomics* 16:190

Stam P, Van Ooijen JW (1995) JoinMap (tm) version 2.0: software for the calculation of genetic linkage maps. CPRO-DLO, Wageningen

Sun Z, Wang X, Liu Z, Gu Q, Zhang Y, Li Z, Ke H, Yang J, Wu J, Wu L, Zhang G, Zhang C, Ma Z (2017) Genome-wide association study discovered genetic variation and candidate genes of fiber quality traits in *Gossypium hirsutum* L. *Plant Biotechnol J* 15(8):982–996

Tanksley SD, Ganal MW, Prince JP, De Vicente MC, Bonierbale MW, Broun P, Fulton TM, Giovannoni JJ, Grandillo S, Martin GB, Messeguer R, Miller JC, Miller L, Paterson AH, Pineda O, Roder MS, Wing RA, Wu W, Young ND (1992) High density molecular linkage maps of the tomato and potato genomes. *Genetics* 132(4):1141–1160

Tautz D, Schlötterer C (1994) Simple sequences. *Curr Opin Genet Dev* 4(6):832–837

Temnykh S, Park WD, Ayres N, Cartinhour S, Hauck N, Lipovich L, Cho YG, Ishii T, McCouch SR (2000) Mapping and genome organization of microsatellite sequences in rice (*Oryza sativa* L.). *Theor Appl Genetics* 100(5):697–712

Thiel T, Michalek W, Varshney R, Graner A (2003) Exploiting EST databases for the development and characterization of gene-derived SSR-markers in barley (*Hordeum vulgare* L.). *Theor Appl Genet* 106(3):411–422

Tomkins JP, Paterson DG, Yang TJ, Main D, Wilkins J, Paterson AH, Wing RA (2001) BAC library construction, preliminary STC analysis, and identification of clones associated with fiber development. *Mol Breed* 8:255–261

TorelloMarinoni D, Valentini N, Portis E, Acquadro A, Beltramo C, Mehlenbacher SA, Mockler TC, Rowley ER, Botta R (2018) High density SNP mapping and QTL analysis for time of leaf budburst in *Corylus avellana* L. *PLoS One* 13(4):e0195408. https://doi.org/10.1371/journal.pone.0195408

Udall JA, Swanson JM, Haller K (2006) A global assembly of cotton ESTs. *Genome Res* 16:441–450

Ulloa M, Meredith WR (2000) Genetic linkage map and QTL analysis of agronomic and fiber quality traits in an intraspecific population. *J Cotton Sci* 4:161–170

Ulloa M, Meredith WR, Shappley ZW, Kahler AL (2002) RFLP genetic linkage maps from four F2.3 populations and a consensus map of *Gossypium hirsutum* L. *Theor Appl Genet* 104:200–208

Ulloa M, Saha S, Jenkins JN, Meredith WR, McCarty JC, Stelly DM (2005) Chromosomal assignment of RFLP linkage groups harboring important QTLs on an intraspecific cotton (*Gossypium hirsutum* L.) consensus map. *J Hered* 96:132–144

Van Deynze A, Stoffel K, Lee M (2009) Sampling nucleotide diversity in cotton. *BMC Plant Biol* 9:125. https://doi.org/10.1186/1471-2229-9-125

Van Ooijen JW, Voorrips RE (2001) JoinMap® Version 3.0: software for the calculation of genetic linkage maps. Wageningen, CPRO-DLO, The Netherlands

Varshney RK et al (2012) Draft genome sequence of pigeonpea (*Cajanus cajan*), an orphan legume crop of resource-poor farmers. *Nat Biotechnol* 30:83–99

Varshney RK, Mohan SM, Gaur PM, Gangarao NVPR, Pandey MK, Bohra A, Sawargaonkar SL, Chitikineni A, Kimurto PK, Janila P, Saxena KB, Fikre A, Sharma M, Rathore A, Pratap A, Tripathi S, Datta S, Chaturvedi SK, Mallikarjuna N, Anuradha G, Babbar A, Choudhary AK, Mhase MB, Bharadwaj CH, Mannur DM, Harer PN, Guo B, Liang X, Nadarajan N, Gowda CL (2013) Achievements and prospects of genomics-assisted breeding in three legume crops of the semi-arid tropics. *Biotechnol Adv* 31(8):1120–1134

Varshney RK et al (2014) Marker assisted backcrossing to introgress resistance to Fusarium wilt (FW) race 1 and Ascochyta blight (AB) in C 214, an elite cultivar of chickpea. *Plant Genome* 7:11

Varshney RK, Singh VK, Hickey JM, Xun X, Marshall DF, Wang J, Edwards D, Ribaut JM (2016) Analytical and decision support tools for genomics-assisted breeding. *Trends Plant Sci* 21(4):354–363

Verma S, Gupta S, Bandhiwal N, Kumar T, Bharadwaj C, Bhatia S (2015) High-density linkage map construction and mapping of seed trait QTLs in chickpea (*Cicer arietinum* L.) using Genotyping-by-Sequencing (GBS). *Sci Rep* 5:17512. https://doi.org/10.1038/srep17512

Wang HT, Jin X, Zhang BB, Shen C, Lin ZX (2015a) Enrichment of an intraspecific genetic map of Upland cotton by developing markers using parental RAD sequencing. *DNA Res* 22(2):147–160. https://doi.org/10.1093/dnares/dsu047

Wang YK, Ning ZY, Hu Y, Chen JD, Zhao R, Chen H, Ai N, Guo W, Zhang T (2015b) Molecular mapping of restriction-site associated DNA markers in allotetraploid Upland cotton. *PLoS One* 10(4):e0124781. https://doi.org/10.1371/journal.pone.0124781

Wang W, Sun Y, Yang P, Cai X, Yang L, Ma J, Ou Y, Liu T, Ali I, Liu D, Zhang J, Teng Z, Guo K, Liu D, Liu F, Zhang Z (2019) A high density SLAF-seq SNP genetic map and QTL for seed size, oil and protein content in upland cotton. *BMC Genomics* 20(1):599. https://doi.org/10.1186/s12864-019-5819-6

Weber JL (1990) Informativeness of human (dC-dA)n.(dG-dT)n polymorphisms. *Genomics* 7(4):524–530

Xu Z, Kohel RJ, Song G, Cho J, Yu J, Yu S, Tomkins J, Yu JZ (2008) An integrated genetic and physical map of homoeologous chromosomes 12 and 26 in Upland cotton (*G. hirsutum* L.). *BMC Genomics* 9(1):108

Xu Y, Lu Y, Xie C, Gao S, Wan J, Prasanna BM (2012) Whole-genome strategies for marker-assisted plant breeding. *Mol Breed* 29:833–854

Xu X, Xu R, Zhu B, Yu T, Qu W, Lu L, Xu Q, Qi X, Chen X (2015) A high-density genetic map of cucumber derived from specific length amplified fragment sequencing (SLAF-seq). *Front Plant Sci* 5:768. https://doi.org/10.3389/fpls.2014.00768

Yadav S, Sandhu N, Singh VK, Catolos M, Kumar A (2019) Genotyping-by-sequencing based QTL mapping for rice grain yield under reproductive stage drought stress tolerance. *Sci Rep* 9:14326. https://doi.org/10.1038/s41598-019-50880-z

Yasir M, He S, Sun G, Geng X, Pan Z, Gong W, Jia Y, Du X (2019) A genome-wide association study revealed key SNPs/genes associated with salinity stress tolerance in upland cotton. *Genes (Basel)* 10(10):829

Yu J, Kohel RJ, Smith CW (2010) The construction of a tetraploid cotton genome wide comprehensive reference map. *Genomics* 95:230–240

Yu JZ, Kohel RJ, Fang DD, Cho J, Van Deynze A, Ulloa M, Hoffman SM, Pepper AE, Stelly DM, Jenkins JN, Saha S, Kumpatla SP, Shah MR, Hugie WV, Percy RG (2012) A high-density simple sequence repeat and single nucleotide polymorphism genetic map of the tetraploid cotton genome. *G3* 2:43–58. https://doi.org/10.1534/g3.111.001552

Yu J et al (2014) CottonGen: a genomics, genetics and breeding database for cotton research. *Nucleic Acids Res* 42:D1229–D1236

Yu X, Zhang M, Yu Z, Yang D, Li J, Wu G, Li J (2020) An SNP-based high-density genetic linkage map for tetraploid potato using specific length amplified fragment sequencing (SLAF-Seq) technology. *Agronomy* 10:114. https://doi.org/10.3390/agronomy10010114

Yuan Y, Zhang H, Wang L, Xing H, Mao L, Tao J, Wang X, Feng W, Wang Q, Wang H, Wei Z, Zhang G, Song X-L, Sun X-Z (2019) Candidate quantitative trait loci and genes for fiber quality in *Gossypium hirsutum* L. detected using single- and multi-locus association mapping. *Ind Crops Prod* 134:356–369

Zahid KR, Ali F, Shah F, Younas M, Shah T, Shahwar D, Hassan W, Ahmad Z, Qi C, Lu Y, Iqbal A, Wu W (2016) Response and tolerance mechanism of cotton *Gossypium hirsutum* L. to elevated temperature stress: a review. *Front Plant Sci* 7:937

Zaidi SSA, Mansoor S, Paterson A (2018) The rise of cotton genomics. *Trends Plant Sci* 23(11):953–955

Zeng L, Meredith WR, Gutiérrez OA, Boykin DL (2009) Identification of associations between SSR markers and fiber traits in an exotic germplasm derived from multiple crosses among *Gossypium* tetraploid species. *Theor Appl Genet* 119:93–103

Zhang T, Qian N, Zhu X, Chen H, Wang S, Mei H, Zhang Y (2013a) Variations and transmission of QTL alleles for yield and fiber qualities in upland cotton cultivars developed in China. *PLoS One* 8:e57220

Zhang YX, Wang LH, Xin HG, Li DH, Ma CX, Ding X, Hong W, Zhang X (2013b) Construction of a high-density genetic map for sesame based on large scale marker development by specific length amplified fragment (SLAF) sequencing. *BMC Plant Biol* 13:141

Zhang Z, Shang HH, Shi YZ, Huang L, Li JW, Ge Q, Gong J, Liu A, Chen T, Wang D, Wang Y, Palanga KK, Muhammad J, Li W, Lu Q, Deng X, Tan Y, Song W, Cai J, Li P, Rashid HO, Gong W, Yuan Y (2016) Construction of a high-density genetic map by specific locus amplified fragment sequencing (SLAF-seq) and its application to quantitative trait loci (QTL) analysis for boll weight in upland cotton (*Gossypium hirsutum.*). *BMC Plant Biol* 16:79

Zhang Z, Ge Q, Liu A, Li J, Gong J, Shang H, Shi Y, Chen T, Wang Y, Palanga KK, Jamshed M, Deng X, Tan Y, Liu R, Zou X, Rashid H, Iqbal MS, Gong W, Yuan Y (2017) Construction of a high-density genetic map and its application to QTL identification for fiber strength in upland cotton. *Crop Sci* 57:774–788

Zhang C, Li L, Liu Q, Gu L, Huang J, Wei H, Wang H, Yu S (2019a) Identification of loci and candidate genes responsible for fiber length in upland cotton (*Gossypium hirsutum* L.) via association mapping and linkage analyses. *Front Plant Sci* 10:53. https://doi.org/10.3389/fpls.2019.00053

Zhang S, Cai Y, Guo J, Li K, Peng R, Liu F, Roberts JA, Miao Y, Zhang X (2019b) Genotyping-by-sequencing of *Gossypium hirsutum* races and cultivars uncovers novel patterns of genetic relationships and domestication footprints. *Evol Bioinforma* 15:1176934319889948

Zhu G, Gao W, Song X, Sun F, Hou S, Liu N, Huang Y, Zhang D, Ni Z, Chen Q, Guo W (2020) Genome-wide association reveals genetic variation of lint yield components under salty field conditions in cotton (*Gossypium hirsutum* L.). *BMC Plant Biol* 20(1):23. https://doi.org/10.1186/s12870-019-2187-y

Chapter 8
Genomics of Naturally Colored Cotton: A Way Forward to Initiate Precision Breeding

Waqas Malik, Naveed Anjum, Muhammad Usman Khan, Muhammad Ali Abid, Javaria Ashraf, Rui Zhang, Chengzhen Liang, Mamoona Hanif, Abdul Qayyum, and Mehboob-ur-Rahman

Abbreviations

4CL	4-Coumarate CoA ligases
5GT	5-*O*-glucosyltransferase
AI	Acid invertase
ANR	Anthocyanidin reductase
ANS	Anthocyanidin synthase
CA	Caffeic acid
CHI	Chalcone isomerase
CHS	Chalcone synthase
Cox	Cytochrome oxidase
DFR	Dihydroflavonol 4-reductase
DPA	Days post anthesis
F3H	Flavonone-3-hydroxylase
IAAO	Indoleacetic acid oxidase
LAR	Leucoanthocyanidin reductase
NAR	Naringenin

W. Malik (✉) · N. Anjum · M. Usman Khan · J. Ashraf · A. Qayyum
Genomics Lab, Department of Plant Breeding and Genetics, Faculty of Agricultural Sciences and Technology, Bahauddin Zakariya University, Multan, Pakistan
e-mail: waqasmalik@bzu.edu.pk

M. A. Abid · R. Zhang · C. Liang
Biotechnology Research Institute, Chinese Academy of Agricultural Sciences, Beijing, China

M. Hanif
Cotton Research Institute, Multan, Pakistan

Mehboob-ur-Rahman
Agricultural Biotechnology Division, National Institute for Biotechnology & Genetic Engineering (NIBGE) College, Faisalabad, Pakistan

Pakistan Institute for Engineering and Applied Science (PIEAS), Islamabad, Pakistan

Mehboob-ur-Rahman et al. (eds.), *Cotton Precision Breeding*,
https://doi.org/10.1007/978-3-030-64504-5_8

PAL	Phenylalanine
PCBER	Phenylocoumaran benzylic ether reductase-like protein
POD	Peroxidase
PPOs	Polyphenol oxidases
SPS	Sucrose-phosphate synthase
SuSy	Sucrose synthase
UDPG	Uridine diphosphate glucose
UPF	Ultraviolet protection factor
UVA	Ultraviolet A
UVB	Ultraviolet B

8.1 Introduction

Dyeing is the most expensive step in fabric manufacturing, due to the high use of energy and water. Immense amount of synthetic dyes and chemicals is required to dye and bleach cotton fiber. These synthetic dyes add heavy metals in the environment and thus have bad effects on human and wild life (Weisburger 2002). Additionally, high price of dyes is also a big issue for the textile industry. High costs of dyes together with environmental issues are diverting the thoughts of environment-conscious people for colored cotton production. Requiring less or no dying process in the textile industry, therefore, naturally colored cotton, reduces the pollution in the environment (Vreeland Jr 1999; Yuan et al. 2012a). Naturally colored cotton is a mutant of white cotton (Kohel 1985) which has been cultivated since 2700 BC (Fox 1987). Natural fiber of cotton was found in different colors including purple, brown, green, blue, red, and bronze; a few of these colors still exist, while others have disappeared (Chaudhry 1992). However, colored cotton is generally inferior in quality to white cotton, containing low micronaire, short fiber length, and less fiber strength. Usually, a negative correlation has been found between fiber quality traits and color due to the pleiotropic effects of fiber color-related genes (Tu et al. 2014; Gong et al. 2014). Sunlight also affects the color of brown-colored cotton that becomes darken with light exposure, while green cotton fiber fades to tan when it is exposed to light (Dickerson et al. 1999). However, low yield and a limited number of color variants hinder the cultivation of colored cotton at large scale. In contrast to white cotton, naturally colored cotton is grown on small area around the world, with 0.1 million ton annual production (China.org.cn 2006). These facts are provoking cotton breeders to develop the breeding programs for high yield with colored lint. Critical understanding about color development and the role of different genes in color developing pathway can play a vital role in developing color variations and stability. Awareness about hazardous effects of synthetic dyes and eco-friendly nature of naturally colored cotton can stimulate the interest of farmers to grow eco-friendly colored cotton. This chapter encompasses the breeding efforts, genetic mechanisms for pigmentation, and course of color development in naturally colored cotton.

8.2 Pernicious Effects of Dyes

In 1856, a British teenager WH Perkins discovered the first synthetic dye; since then, numerous synthetic dyes have been manufactured (Saratale et al. 2011). Each dye contains two major parts, "chromophore" and "auxochrome." Chromophore is responsible for coloration of substrate which absorbs certain wavelengths of light from nearby ultraviolet regions, while auxochrome helps in making bond between chromophore and fabrics. There are some classes of these colorants, such as azo, anthraquinone, vat, indigo, carbonium, phthalocyanine, and nitro dyes (Sponza 2006). Azo dyes are mainly (60–70%) used to dye textile stuffs in textile industry (Carliell et al. 1995). Annual production of these dyes is about 800,000 tons (Palmieri et al. 2005; Revankar and Lele 2007), and almost 140,000 tons of synthetic dyes are drained into the environment (Zollinger 1987; Cooper 1995). These dye effluents have a drastic effect on the environment due to high pH value and high chemical oxygen demand. Further, water-soluble dyes have extreme effects on aquatic life, even in low concentrations (10–50 mg/L) (Demirci and Hamamci 2013), because these hinder sun beam to penetrate in water, resulting in less photosynthesis (Çiçek et al. 2007). Heavy metals such as lead, cadmium, cobalt, nickel, and copper and certain other chemicals like soap, acetic acid, nitrates, etc. make textile effluents more harmful for life (Kant 2012). Previous researches have revealed that heavy metals ions which are present in textile wastewater are overly accumulated by algae (Banat et al. 1997). Currently, there are nearly 10,000 synthetic dyes available in market (Bazin et al. 2012), which affect either directly owing to their chemical structure or indirectly by producing intermediate metabolites, such as naphthalene and benzidines (Osugi et al. 2006; Bafana et al. 2009). For instance, mutagenic and carcinogenic by-products are produced due to the breakage of azo bond in azo dyes (Rafii and Cerniglia 1995), which are more dangerous than the original dye (Wong and Yuen 1996). To disseminate the knowledge about toxicity of textile effluents, awareness programs have been initiated over the past few years, and it is admitted that natural dyes are barely used in textile industry nowadays (Ganglberger 2009). Chemistry of synthetic dyes reveals their toxicity, and when human beings wear synthetically dyed fabrics, there is a chance that these dyes may sweep in their body with the help of sweat (Hansen 2005). Synthetic dyes are nonbiodegradable, and they cause skin infections and affect the functioning of the kidney, brain, reproductive system, liver, and central nervous system (Hijazi et al. 2015). Previous reports also revealed that workers of textile industries are more prone to lung diseases than other people (Salvi and Barnes 2009). Moreover, these synthetic dyes are not only harmful, but they are expensive too.

8.3 Benefits of Naturally Colored Cotton

The world is faced with issues such as skin cancer, DNA damage, genetic changes, and immune system disorders because of overexposure to ultraviolet radiations (Narayanan et al. 2010). In green-colored cotton, caffeic acid is present which has the best protection against ultraviolet radiations because of its ability to absorb ultraviolet radiations (Schmutz et al. 1993, 1994; Ryser 1999). The brown color of naturally colored cotton is derived from tannin vacuoles in the lumen of fiber and is developed only when cotton bolls are exposed to abundant oxygen and sunlight (Zhang et al. 2011c). These natural pigments in colored cotton play a vital role in shielding or blocking ultraviolet rays (Pailthorpe 1994). Cotton fabrics are commonly used in summers because they have the ability to absorb sweat and have pores for aeration. Simple white cotton does not protect us from ultraviolet radiations because its ultraviolet protection factor (UPF) ranges from 3 to 5 only (Wang et al. 2001). A fabric is considered "sun protective" when its UPF value is ≥15 (American Society for Testing and Materials 2015). In this regard, naturally colored cotton can absorb more ultraviolet radiations; for example, 28 and 29 UPF have been reported for brown and green cotton, respectively (Parmar et al. 2006). Previously, Ma et al. (2013) reported that naturally colored cotton has resistance against bacteria and the brown cotton has more tendency to resist bacteria than green cotton due to the presence of condensed tannins/pigments, while flavonoid pigments are present in green cotton. Additionally, naturally colored cotton has more antioxidant capacity than white cotton (Ma et al. 2016). Also, the naturally colored cotton has high flame resistance due to presence of more metal ions, i.e., phosphorus and boron contents (Hinchliffe et al. 2015), and also shows high resistance to abiotic (salt and drought) and biotic (pathogens and insects) stresses (Fox 1987; Lee 1996; Vreeland Jr 1987). The naturally colored cotton requires 70% less amount of water for processing than that of white cotton (Silva et al. 2005; Galindo et al. 2001). These benefits of naturally colored cotton over the white cotton have drawn the heed of both manufacturers and consumers.

8.4 Conventional Breeding Efforts

In recent years, several breeding efforts have been made to improve the production and quality of naturally colored cotton by paying more emphasis on color stability, brightness, and fiber quality improvement (Dutt et al. 2004; Matusiak and Frydrych 2014; Cuming et al. 2015). Genome-wide association studies have been conducted recently using colored cotton accessions to identify QTLs related to plant architecture (Wen et al. 2018). Hybrid development, selection, and recurrent crossing in naturally colored cotton can be useful for its improvement (Khan et al. 2009). Initial efforts for producing colored cotton varieties were undertaken by Fryxell (1984) who listed the different colored cotton lines and their respective fiber color. Earlier works reported that green color of fiber is due to dominant mutant *Lg* in *G. hirsutum* (Ware 1932). Later on, it has been shown that green color is governed by a single nuclear gene that exhibits dominance, incomplete dominance, and recessive inheritance (Carvalho et al. 2014; Wang et al. 2012). Brown color also shows dominant inheritance and is governed by a single nuclear gene (Carvalho et al. 2014), which is controlled by six loci (*Lc1* through *Lc6*). Initially, it has been reported that brown color is controlled by two independent dominant loci, *Lc1* and *Lc2*, in *G. hirsutum* and wild *G. barbadense* (Harland 1935), and in the absence of any pair of dominant alleles, lint becomes white in color (Симонгулян 1984). Afterward, linkage analysis revealed that *Lc1* is present on chromosome 7, while *Lc2* is linked to chromosome 6 (Kohel 1985). Further, *Lc3* is a supplementary gene which gives a dark brown color to the lint by strengthening the expression of the *Lc1* and *Lc2*, while *Lc4*, *Lc5*, and *Lc6* control the light brown color of lint (Kohel 1985). Recently, Wen et al. (2018) revealed that *Lc1* is further dissected into two loci, qBF-A07-1 and qBF-A07-2. Among these, the former locus is responsible for initiation of brown fiber color development, and its interaction with the latter causes the development of different shades of brown. Brown fuzz and lint show genetic variation due to dominance and additive effect; however, additive effect is predominant due to which linkage and interaction have been found between brown fuzz and lint color inheritance in *G. hirsutum* (Feng et al. 2010). These all reported colors are controlled by one gene, and each color is partially dominant over white (Carvalho et al. 2014). Future mapping and subsequent cloning of genes would assist cotton researchers to explore the molecular mechanism underlying color development in cotton fiber which will open new avenues toward the development of colored cotton cultivars. In recent years, several attempts are made to improve the fiber quality of green- and brown-colored cotton germplasm (Table 8.1). Unfortunately, only brown

Table 8.1 Conventional studies for colored cotton breeding

Fiber quality related traits	Fiber colors	References
BW, FL, FS, FF, FM, FU, CC, FC	Brown and green	Malik et al. (2018)
FL, FS, FF, FM, IAA and ABA	Brown, green, and white	Zhang et al. (2017)
FC, CH, CC, POD, SuSy, IAAO and β-1,3-glucanase	Brown, green, and white	Qian et al. (2015)
FC	Brown	Carvalho et al. (2014)
BW, SWB HSW, SV, SD, NSB, SCS, LS, GOT, FL, FS, FF, FU, FE, FC	Brown, green, and white	Malik et al. (2013)
LC, SCY, BW, GOT, Mic	Brown	Maralappanavar et al. (2013)
NSB/P, NN/P, PH, BW, NB/P, LY, FL, FS, FF, FU, FE, CC, FC, Fiber pH value, fructose, glucose, sucrose, total sugars	Brown and green	Yuan et al. (2013)
FL, CC, FC, SuSy sucrose synthase, SPS (sucrose-phosphate synthase), AI acid invertase, Cox fructose, glucose, sucrose	Brown, green, and white	Yuan et al. (2012a, b)
CC, N, S, Ca, K, P, Mg, fructose, glucose, galactose, sucrose, cellulose, contents in fiber	Green, brown, and white	Zhang et al. (2012)
FC, LP, FL, FU, FF, FS	Brown and white	Feng et al. (2011)
FC	Brown, green, and off-white,	Han et al. (2011)
PH, NSB/P, SCY, FL, FS, FF, FU	Brown, green, and white	Efe et al. (2010)
WC, LP, FL, FS, FF, FU	Brown, green, and white	Pan et al. (2010)

BW boll weight, *FL* fiber length, *FF* fiber fineness, *FS* fiber strength, *FM* fiber maturity, *FU* fiber uniformity, *SWB* seed weight per boll, *HSW* hundred seed weight, *SV* seed volume, *SD* seed density, *NSB* number of seeds per boll, *SCS* seed cotton per seed, *LS* lint per seed, *GOT* ginning outturn, *SCY* seed cotton yield, *FE* fiber elongation, *LP* lint percentage, *LI* lint index, *LC* lint color, *Mic* micronaire value, *NSB/P* number of sympodial branches per plant, *NN/P* number of nodes per plant, *PH* Plant height, *NB/P* number of bolls per plant, *CC* cellulose contents, *FC* fiber color/flavonoid contents, *WC* wax contents, *IAA* indole-3-acetic acid, *ABA* abscisic acid, *CH* carbohydrate contents, *IAAO* indoleacetic acid oxidase, *POD* peroxidase, *SuSy* sucrose synthase, *SPS* sucrose-phosphate synthase, *AI* acid invertase, *CoX* cytochrome c oxidase, *N* nitrogen, *S* sulfur, *K* potassium, *P* phosphorus and green naturally colored cultivars are currently available in germplasm, restricting the development of naturally colored cotton for textile market. Conventional breeding methods alone cannot develop new naturally colored cotton cultivars because of the lack of sufficient germplasm resources for different colored cottons. Therefore, biotechnology coupled with molecular breeding programs might be helpful for producing new cultivars of naturally colored cotton. The prerequisite to use this program is an exploration and understanding of the molecular basis of synthesis and deposition of pigments in cotton fibers.

8.5 Flavonoid Biosynthetic Genes with Fiber Pigmentation and Quality

Flavonoids constitute a diverse group of aromatic molecules that are derived from phenylalanine and are secondary metabolites responsible for most of the pigmentation in plants (Landi et al. 2015). Flavonoids are widely distributed in different concentrations depending on growth conditions, type of organ, and plant species (Debeaujon et al. 2001). Some of these compounds can translocate from the site of their synthesis to other tissues of the plant (Buer et al. 2007). In naturally colored cotton, the pigmentations of brown and green cotton fibers might be affected by the structural genes of flavonoid biosynthetic pathway (Feng et al. 2013; Li et al. 2013). Several previous reports showed that the transcription level of flavonoid biosynthesis genes is higher in colored than in white cotton fiber (Mikhailova et al. 2019). For instance, the expression level of different structural flavonoid biosynthesis genes, i.e., *GhCHI*, *GhANS*, *GhF3H*, *GhANR*, *GhDFR* (Xiao et al. 2007; Malik et al. 2015), *GhCHS*, *GhC4H*, *GhF3'H*, *GhF3'5'H* (Feng et al. 2013), *GhANR*, *GhCHI*, *GhPAL*, *GhDFR*, and *Gh3GT* (Li et al. 2013), was higher in brown as compared to white cotton fiber. On the other hand, *Gh4CL2* gene encoding 4-coumarate CoA ligases (*4CL*) is involved in the metabolism of caffeic acid (CA) derivatives and is responsible for pigment development in green cotton fibers (Feng et al. 2017). Similarly, three genes including *GhCHS*, *GhANR*, and *GhLAR* were significantly expressed in developing colored cotton fibers and identified as key genes for fiber color formation in cotton (Gao et al. 2019). In addition, several studies also depicted the up- or downregulation of structural genes of flavonoid pathway during different fiber development stages of colored cotton (Table 8.2). For example, five structural flavonoid biosynthesis genes *GhCHI*, *GhF3H*, *GhDFR*, *GhANS*, and *GhANR* had the higher transcript level at 10 and 20 days post anthesis (DPA) in brown cotton fiber as compared to green and white cotton (Malik et al. 2015). In contrast, silencing of *F3H* gene significantly increased the naringenin (NAR) content of fiber cells, resulting in suppression of fiber growth (Tan et al. 2013). Similarly, transformation of F3H-RNA interference segment into the DNA of brown cotton fiber has revealed fiber shortening (Gong et al. 2014). On the other hand, overexpression of the *F3H* gene has no effect on cotton fiber length, showing association of flavonoid metabolism with fiber pigmentation and quality (Hua et al. 2007; Tan et al. 2013).

Table 8.2 Fiber color development related genes in cotton

Fiber color	Genes related with fiber color	Tissue	References
Brown	GhCHS, GhANR, GhLAR	Fiber	Gao et al. (2019)
Brown, green, white	Gh3GT, GhCHI-1, GhCHI-2, Gh3GT, At3GT	Fiber	Liu et al. (2018)
Green	GhPAL,GhCHS,GhF3H,GhDFR, GhANS, GhANR	Fiber	Qian et al. (2016)
Green, white	Gh4CL1,Gh4CL2,Gh4CL3, Gh4CL4	Fiber	Feng et al. (2017)
Green, brown, white	GhCHI,GhF3H,GhDFR,GhANS, GhANR	Fiber	Malik et al. (2015)
Brown, white	GhLAR,GhANS,GhMATE1a,GhMATE1b	Fiber	Feng et al. (2014)
Brown, white	GhPAL,GhC4H,Gh4CL,GhChS,GhCHI,GhF3H,GhF3'5'H, GhDFR, GhLAR, GhANS, GhANR	Fiber	Xiao et al. (2007)
White	GhPAL,GhC4H,Gh4CL,GhCHS,GhCHI,GhF3H,GhDFR, GhFLS, GhANS, GhANR	Fiber	Yoo and Wendel (2014)
Brown, white	GhC4H,GhCHS, GhF3H,GhF3'5'H	Fiber	Feng et al. (2013)
White	GhCHS, GhCHI,GhF3'H,GhF3'5'H, GhDFR, GhANS, GhANR	Fiber	Tan et al. (2013)
Brown, white	GhPAL, GhCHI, GhDFR, GhANS,	Fiber	Li et al. (2013)
White	GhCHI,GhF3'H,GhDFR, GhANS, GhANR	Petal	Barkeer et al. (2012)
Brown	GhCHI, GhF3'H, GhDFR, GhANS, GhANR	Fiber	Xiao et al. (2007)

8.5.1 Exon/Intron Structure and Chromosomal Location of Structural Flavonoid Biosynthetic Genes

Online genomics tools were used for downstream analyses (Fig. 8.1). Several genes involved in flavonoid biosynthesis pathway were blasted using Cotton Functional Genomics Database (https://cottonfgd.org/) (Zhu et al. 2017) to identify the chromosomal location of each structural gene and then confirmed by BLASTN searches against *G. hirsutum* NAU assembly. The location of each flavonoid pathway structural gene was mapped to the upland cotton genome using the MapChart software (Voorrips 2002) (Fig. 8.2). Out of the 53 flavonoid biosynthesis pathway genes, 28 were found in At sub-genome, while 25 were located in Dt sub-genome. These genes appeared to be nonrandomly distributed across all chromosomes except Gh_A03, Gh_A06, Gh_D01, Gh_D02, and Gh_D04. A high density was observed in chromosomes Gh_A05, Gh_D05, and Gh_D12 which contained five genes, while low density was found in Gh_A01, Gh_A08, Gh_D06, and Gh_D08 chromosomes.

The diversity in gene exon and intron length and their numbers is considered as a potential reason for the evolution of multigene families. Thus, intron/exon number and organization were analyzed using cDNA and genomic sequences of structural genes to get further insights into structural diversity. The cDNA and genomic sequences of structural genes in cotton were downloaded from the Cotton Functional

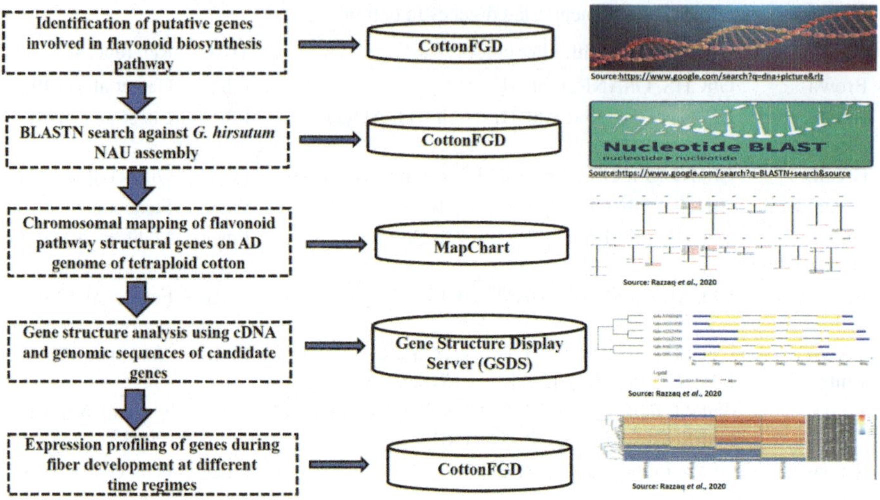

Fig. 8.1 Scheme for use of online genomics tools

Genomics Database. To observe the gene structure, the Gene Structure Display Server (http://gsds.cbi.pku.edu.cn/index.php) was used by comparing individual cDNA sequencing with its corresponding genomic sequence (Hu et al. 2015). A comprehensive analysis of the exon and intron length of structural genes is presented in Fig. 8.3. The most closely related members in each transcription factor family have a similar gene structure in terms of intron/exon length and their numbers. For example, flavonoid biosynthesis pathway genes belonging to PAL family had only one intron. However, current knowledge is limited in the application of flavonoid biosynthetic pathway to influence pigment deposition in cotton fiber.

8.5.2 Differential Expression Pattern of Structural Genes Regulating Flavonoid Biosynthetic Pathway

RNA-seq data of structural genes regulating biosynthetic pathway was downloaded from the Cotton Functional Genomic Database at 5 days post anthesis (DPA), 10 DPA, 20 DPA, and 25 DPA to determine their expression level. The transcript abundance of each gene was identified by fragments per kilobase of exon per million fragments mapped (FPKM). A fold change of at least 1 and cutoff q value of no more than 0.05 were used for identifying differentially expressed genes (Trapnell et al. 2013). Genes were upregulated and downregulated in response to different time regimes including 5 DPA, 10 DPA, 20 DPA, and 25 DPA (Fig. 8.4). This heat map clearly showed that 24 genes were downregulated in almost all developmental stages in white-colored fiber, and therefore, it was expected as these genes might be involved in the development of pigments in naturally colored cotton.

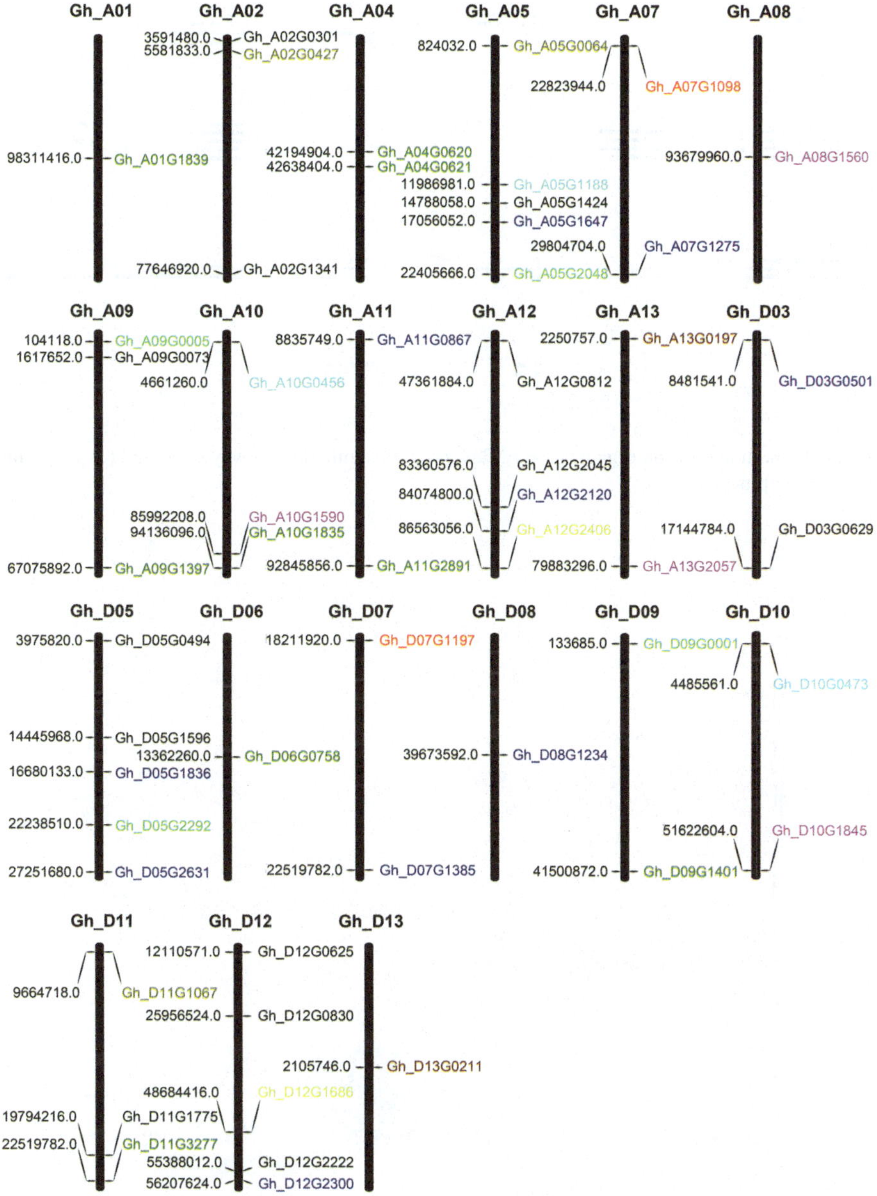

Fig. 8.2 Chromosomal position of flavonoid biosynthetic pathway genes expressed during fiber development in upland cotton. Structural genes of flavonoid biosynthetic pathway expressed during fiber color development were mapped on different chromosomes in the upland cotton. Genes belonging to different families have been represented in different colors

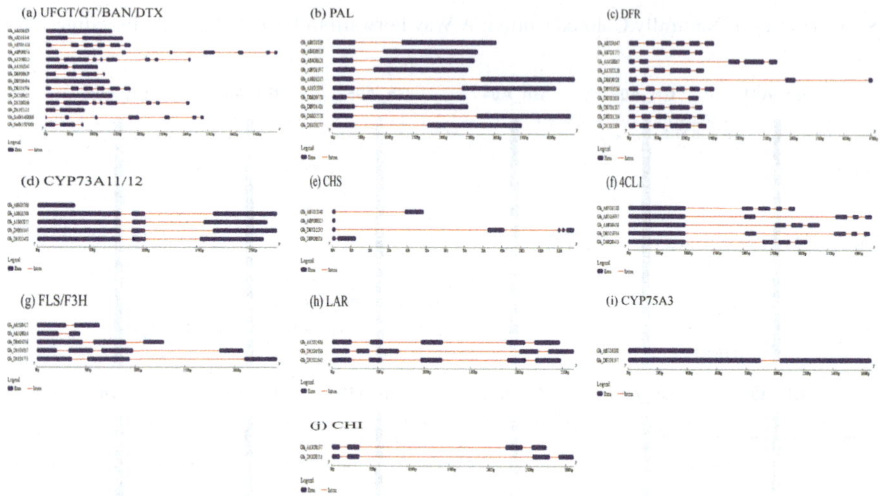

Fig. 8.3 Exon and intron organization of flavonoid biosynthetic pathway genes expressed during fiber development

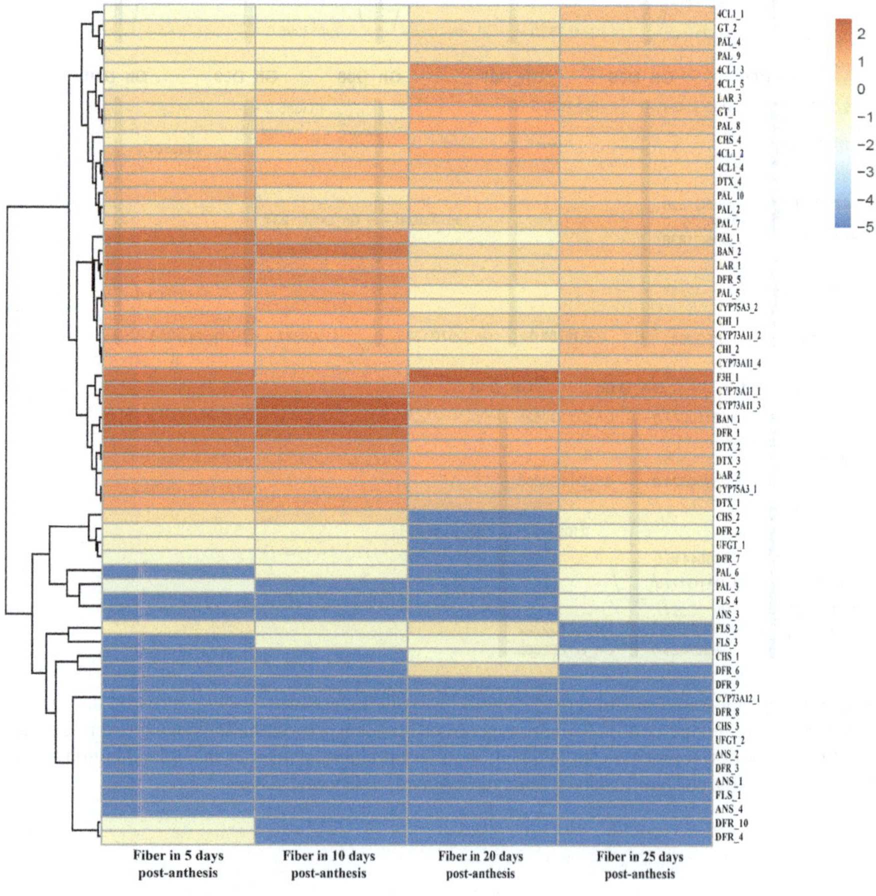

Fig. 8.4 Heat map representing the expression pattern of flavonoid biosynthetic pathway genes during different stages of fiber development. Heat maps of structural genes illustrating the change in expression levels at different fiber development stages, i.e., 5 DPA, 10 DPA, 20 DPA, and 25 DPA. *Blue* indicates decreased expression, whereas *red* represents upregulated transcript level during respective treatments

8.6 Role of Flavonoid Biosynthetic Pathway in Fiber Color Development

Flavonoid biosynthesis pathway is catalyzed by different enzymes encoded by structural and regulatory groups of genes and some derivatives of carbohydrates which act as precursor molecules like malonyl-CoA and 4-coumaroyl-CoA (Forkmann and Heller 1999; Takos et al. 2006). Schematic diagram of the flavonoid biosynthetic pathway encoding different colors in developing lint (Fig. 8.5) showed that structural genes *GhCHI*, *GhF3H*, *GhDFR*, *GhANS*, and *GhANR* encode chalcone isomerase (CHI), flavonone-3-hydroxylase (F3H), dihydroflavonol 4-reductase (DFR), anthocyanidin synthase (ANS), and anthocyanidin reductase (ANR) enzymes, respectively, which act on different intermediate precursors formed during flavonoid biosynthesis pathway (Takos et al. 2006). Additionally, different regulators, such as phytochromes, cryptochrome, phototropin, and UV-B photoreceptors, are also involved (Gonzalez et al. 2015; Zoratti et al. 2014). The flavonoid biosynthesis pathway is initiated by precursor phenylalanine (PAL) which is then converted into chalcone via chalcone synthase (CHS) and further catalyzed by CHI leading to production of naringenin (a flavanone). Further regulation of reaction is catalyzed by several intermediate enzymes which produce several classes of flavonoids (Schijlen et al. 2004). The flavonoid biosynthesis-related genes *GhCHS*, *GhCHI*, *GhF'3H*, and *GhF'3'5H* have been cloned from cotton, and the expression analysis of these genes depicted their involvement in pigmentation of brown cotton fiber (Feng et al. 2014). It was shown that the silencing of *GhCHI-1* in brown cotton fiber resulted in the genotypes with three different fiber colors including brown, green, and white cotton (Liu et al. 2018). It was also observed that overexpression of *Gh3GT* in brown cotton yielded green cotton fiber. The pigment biosynthesis pathway is catalyzed by flavonone-3-hydroxylase (F3'H) and DFR enzymes which act on naringenin and dihydroflavonols, respectively, to produce the colorless compound leucoanthocyanidin (flavan-3,4-diol) (Malik et al. 2015). Moreover, ANS instantly converts the colorless leucoanthocyanidins into colored compound called anthocyanidins. Sunlight works as signal to increase the accumulation of anthocyanidins in cotton fiber (Fan et al. 2016). While in white and green cotton leucoanthocyanidins are not converted into proanthocyanidin, in brown cotton, the reaction is catalyzed by ANR that leads to production of proanthocyanidins (Takos et al. 2006; Nakatsuka et al. 2005). Another reason for green color development could be phenylcoumaran benzylic ether reductase-like protein (PCBER) which is the prime lignin biosynthesis enzyme that may be responsible for green fiber color development as it is only found in green-colored fiber (Li et al. 2018). Four isomers, *Gh4CL1*, *Gh4CL2*, *Gh4CL3*, and *Gh4CL4*, are also involved in lignin biosynthesis pathway and formed different biochemical products. Higher activity was observed for *Gh4CL2* which showed higher metabolic efficiency for caffeate and ferulate. Two important caffeoyl derivatives like 22-*O*-caffeoyl-22-hydroxymonodocosanoin and 22-*O*-caffeoyl-22-hydroxydocosanoic acid were identified, whose concentrations were positively correlated with the degree of green color fiber. It is suggested

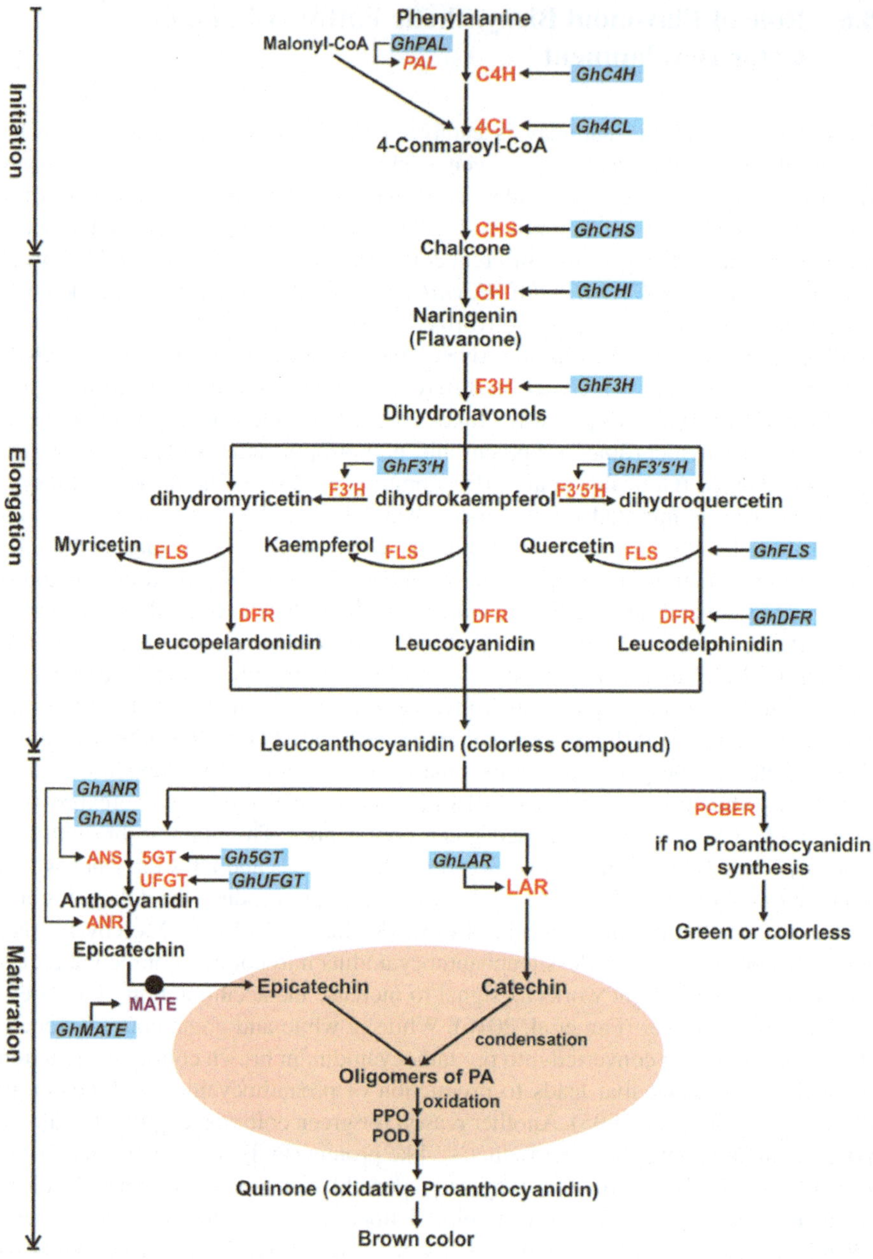

Fig. 8.5 A schematic presentation of the flavonoid biosynthetic pathway leading to green, brown, and white color development in cotton fiber. Enzyme abbreviations: *PAL* phenylalanine ammonia-lyase, *C4H* cinnamic acid 4-hydroxylase, *4CL* 4-coumarate CoA ligase, *CHS* chalcone synthase, *CHI* chalcone isomerase, *F3H* flavanone 3-hydroxylase, *F30H* flavonoid 30-hydroxylase, *F3'5'H* flavonoid 3050-hydroxylase, *FLS* flavonol synthase, *DFR* dihydroflavonol 4-reductase, *LAR* leucoanthocyanidin reductase, *ANR* anthocyanidin reductase, *ANS* anthocyanidin synthase, *UFGT* uridine diphosphate glucose:flavonoid

that *Gh4CL2* may be involved in the pigmentation of green color fiber (Feng et al. 2017). Proanthocyanidins (condensed tannins) are an important end product of phenylpropanoid pathway and are involved in the development of pigmentation and regulation of seed coat color in various plants (Matus-Cádiz et al. 2008), i.e., in pigmentation of *Arabidopsis* seed coat (Dixon et al. 2005), brown fiber in colored cotton (Li and Wang 2002), and seeds, leaves, fruits, flowers, and bark of different plant species (Dixon et al. 2005; Gabetta et al. 2000; Gu et al. 2004).

Leucoanthocyanidin is an important precursor catalyzed by ANS and leucoanthocyanidin reductase (LAR), both involved in the production of 3-OH-anthocyanidins and gallocatechin/catechin (2,3-*trans*-flavan-3-ols), respectively (Tanner et al. 2003; Xie et al. 2003). Also, it results in the production of anthocyanins by glycosylation and esterification under the control of uridine diphosphate glucose, flavonol 3-*O*-glucosyltransferase, and 5-*O*-glucosyltransferase, which are then transport to the plant vacuole with the help of transporters (MATE) where they undergo a series of oxidative polymerization. This oxidative polymerization is carried out by three candidate condensing enzymes, i.e., plant polyphenol oxidases (PPOs), plant laccases, and plant peroxidases. PPO exhibits dual activity: first, its monophenolase activity results in the production of *o*-diphenols through the hydroxylation of monophenols, and second, the diphenolase activity converts this *o*-diphenols into quinone (Chazarra et al. 2001; Dixon et al. 2005; Xie et al. 2003). During polymerization through covalent interaction, the quinone, an electrophilic molecule, combines with catechin and epicatechin which are nucleophilic molecules for rapid polymerization. This rapid polymerization produces red, black, and brown pigmentation in plant seeds and fruits (Sun et al. 2006). Quinone that is an oxidative product of proanthocyanidin is a crucial constituent for the brown color of cotton fiber (Feng et al. 2014). Quinone contents are more in mature brown fiber, and with fiber development, this decreases due to gradual oxidation of proanthocyanidin into quinone. Current studies have revealed that 5GT, through modification in the metabolic pathway, could make anthocyanin more stable, and lack of glucose at the 5 position of anthocyanidin could lead to brown fiber color variation due to interaction between oxidized tannins and glycosylated anthocyanin (Yamazaki et al. 1999; Nishizaki et al. 2011). Transcription analysis also revealed that *GhANR* and *GhANS* genes might involve in the alteration of white fiber to brown fiber, while *GhANS* gene is found to be strongly associated with brown color development (Malik et al. 2015).

8.7 Pattern for Color Development in Brown and Green Cotton

The biosynthesis of pigments (flavonoid) increases with the development of growing lint fiber; however, timing of pigmentation varies in brown and green cotton (Liang et al. 2011). Basically, structural genes *GhCHI*, *GhC4H*, *GhCHS*, *GhF3'H*,

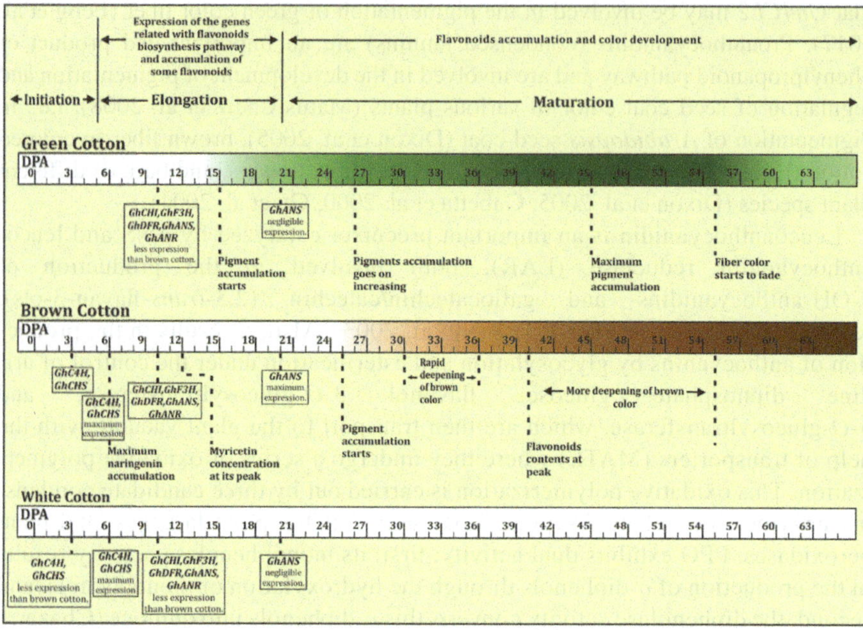

Fig. 8.6 Role of the flavonoid biosynthetic pathway genes during fiber initiation, elongation, and maturation in green, brown, and white colored cotton. Expression of genes during initiation, elongation, and maturation of fiber showed that in green-colored cotton, maximum accumulation of flavonoid deposition was found at 45 DPA, while for brown cotton fiber, maximum deposition of flavonoid pigments was observed between 40 and 55 DPA. In white-colored cotton, *C4H*, *CHS*, *CHI*, *F3H*, *ANS*, *ANR*, and *DFR* structural genes were downregulated during the course of fiber development

GhF3'5'H, *GhDFR*, *GhANS*, and *GhANR* play a vital role in flavonoid biosynthesis, and their expression levels fluctuate at different developmental stages in green, brown, and white fiber cotton (Indrais et al. 2011; Feng et al. 2013). The maximum pigment synthesis in colored cotton is between 5 DPA and 10 DPA (Fig. 8.6) and then keeps on declining with the advancement of different fiber developmental stages (Dutt et al. 2004; Hua et al. 2007; Yuan et al. 2012a, 2013).

8.7.1 Initiation

Fiber initiation phase starts from 0 DPA and ends at 5 DPA (Chen and Guan 2011). Almost the same amount of flavonoids has been found in green and brown fiber at 0 DPA, while white fiber accumulates more flavonoids than green and brown fiber (Yuan et al. 2012a). However, after 0 DPA, flavonoid content in white color fiber declines, while that in green and brown fiber shows an increasing trend. Transcriptome analysis revealed that there is a higher expression of *GhC4H* and *GhCHS* genes in brown cotton than that of white cotton at 3 DPA (Feng et al. 2013). At 5 DPA, the maximum amount of flavonoids was found in green and brown color fiber than in white color fiber (Malik et al. 2015).

8.7.2 Elongation

During the fiber elongation phase (6–20 DPA) (Chen and Guan 2011), expression level of *GhC4H* and *GhCHS* genes reaches its peak in brown and white cotton fiber. Transcript level of *GhC4H* was 2.5 times higher in brown than in white cotton (Feng et al. 2013). Moreover, transcript level of *F3'5'H* gene reaches its peak during 9–12 DPA in brown and white cotton. Naringenin, a substrate of *GhF3'H*, reaches to the maximum concentration at 6 DPA in brown and white cotton, but brown has more concentration than white. This naringenin becomes undetectable after 21 DPA (Feng et al. 2013). Transcript analysis of different structural genes at 10 DPA revealed that transcript level of *GhCHI*, *GhF3'H*, *GhDFR*, *GhANS*, and *GhANR* was higher in brown than in green and white fiber cotton. However, when transcript level of these quoted genes was measured at 20 DPA, all these genes except *GhANS* had less expression level than at 10 DPA. The transcript accumulation of *GhANS* was negligible in green and white fiber cotton while higher in brown fiber cotton. It has been reported that *GhANR* and *GhANS* are key genes involved in proanthocyanidins synthesis. This compound is important for producing the brown color pigment (Malik et al. 2015).

Concentration of flavonoid contents also varies during the fiber elongation phase. For instance, naringenin, a substrate of *GhF3'H*, accumulates in abundance at 6 DPA in brown fiber than that of white cotton fiber. However, afterward, it starts declining, and after 21 DPA, it becomes undetectable. Quercetin concentration dramatically increases between 15 and 18 DPA in brown fiber cotton. In contrast, kaempferol concentration was low in brown cotton fiber than that of the white fiber during the whole development period. Level of myricetin used to be high in brown cotton fiber from 6 to 27 DPA and myricetin concentration starts increasing from 6 DPA and reaches its peak at 12 DPA (Feng et al. 2013).

8.7.3 Fiber Maturation

Flavonoid biosynthesis genes express during fiber elongation. These genes encode different contents of flavonoids which accumulate in fiber during maturation phase for imparting color. These pigments accumulate in the lumen of brown fiber, while in green color fiber, these are found in the lumen and secondary cell wall (Li and Wang 2002; Zhang et al. 2011c). In the presence of sunlight, color begins to develop after boll opening, and then it starts fading with the continuous exposure of sunlight. During cell wall synthesis, color of fiber appears white before the primary cell wall stretching and then appears very light in color until secondary cell wall thickening (Qiu et al. 2002). There is a significant difference for the timing of color development in brown and green cotton. In brown cotton, flavonoid content increases gradually during fiber color development, and this process starts before 30 DPA. It is also observed that brown fibers significantly get deeper in color between 30 and 35 DPA (Wang and Li 2002). Brown fiber color rapidly becomes deeper from 35 to 40 DPA with the maximum flavonoid content at 40 DPA; this deepening of color continues until 55 DPA. In green fiber cotton, coloring of fiber starts from 15 DPA to 25

DPA (Dutt et al. 2004), which keeps on increasing between 25 and 45 DPA and reaches its peak at 45 DPA, while it keeps fading away under sunlight from 55 DPA till maturation (Zhang et al. 2011a).

8.8 Biochemical and Physiological Basis of Inferior Fiber Quality in Colored Cotton

Various changes occur at physiological and biochemical levels throughout the course of fiber development (Zhu et al. 2008). Fiber quality is primarily determined by cellulose contents (Yuan et al. 2012a) as these are vital components of secondary cell wall of a developing fiber (Haigler et al. 2005). Cellulose biosynthesis is distressed by different external factors, agronomic practices, and biosynthesis and accumulation of pigments in colored cotton (Dong et al. 2006; Dai and Dong 2014). Sucrose concentration is responsible for producing a high-quality fiber, and its low concentration leads to poor quality of a fiber (Pettigrew 2001). Earlier it was shown that less accumulation of cellulose in colored cotton is due to limited sugar content (Zhang et al. 2012). Carbohydrate metabolism involves various enzymes and coenzymes including sucrose synthase (SuSy), β-1,3-glucanase, sucrose-phosphate synthase (SPS), acid invertase (AI), indoleacetic acid oxidase (IAAO), peroxidase (POD), cytochrome oxidase (Cox), NAD+/NADH, and NADP+/NADPH (Zhang et al. 2017; Yuan et al. 2012a, b). Sucrose acts as a substrate for SuSy or AI and degrades into UDPG and fructose (Pettigrew 2001; Alonso et al. 2007; Coleman et al. 2009), which are further utilized in cellulose synthesis. β-1,3-glucanase also transfers the β-1,3-dextran and hydrolyses callose into UDPG which is a prime substrate required for cellulose synthesis. β-1,3-glucanase exhibits higher expression in colored cotton at 10, 20, 25, and 30 DPA, but its reduced activity is observed at 40 DPA (Qian et al. 2015). Activity of SuSy enzyme rapidly declines in white cotton from 15 to 20 DPA which is in contrast to the colored cotton where it is elevated at 15 DPA (Zhang et al. 2017) and then starts declining during 15–20 DPA (Qian et al. 2015). Several other studies reported the low concentration of sucrose in colored cotton fiber from 0 to 20 DPA which may limit the cellulose synthesis (Ruan and Chourey 1998; Haigler et al. 2001; Ruan et al. 2003; Ruan 2005). Further, it is confirmed that carbohydrate metabolism is involved not only in cellulose biosynthesis but also in flavonoid biosynthesis. The two enzymes, SuSy and AI, convert sucrose into glucose which is subsequently used for flavonoid and cellulose biosynthesis simultaneously (Yuan et al. 2012a). In addition to this, it is also reported that when the activities of SPS and SuSy were high, the sucrose content of cotton fiber was found low, suggesting their important role in fiber development in cotton (Feng et al. 2009). Another study depicted that the activity of β-1,3-glucanase was very low at 8 DPA; however a rapid increase in flavonoid and cellulose contents was observed at 10 DPA. This indicated that β-1,3-glucanase might serve as a precursor for cellulose and flavonoid synthesis (Qian et al. 2015).

Continuous supply of energy in the form of ATP is required for the completion of different biochemical pathways during cellulose synthesis, and it is provided by respiration in mitochondria. Respiration rate is positively correlated with carbohydrate concentration (Al-Ghazi et al. 2009). Cytochrome oxidase catalyzes the electron transport in mitochondria between cytochrome c and molecular oxygen. More Cox activity has been observed in brown than in white and green fiber cotton (Yuan et al. 2012b). Other coenzymes like NAD+, NADH, NADP+, and NADPH involved in fiber development also exhibit their high contents in brown fiber cotton than in green and white fiber cotton (Yuan et al. 2012a). Colored cotton exhibits higher enzymatic activity at different stages, and some exhibit lower activity owing to lower contents of carbohydrates. Glucose is produced during sucrose metabolism which is subsequently used in cellulose synthesis as well as in flavonoid synthesis. More competition for glucose reduces its amount for cellulose synthesis which results in low carbohydrate accumulation and subsequently leads toward inferior fiber quality. Therefore, currently, there is a dire need to focus on cellulose synthesis pathway to improve the quality of colored cotton fiber in naturally colored cotton.

8.9 Conclusion and Future Perspectives

With the awareness of pernicious effects of dyes and benefits of naturally colored cotton over white, the environmentalists are urging to adopt the cultivation of naturally colored cotton. Still, area under its cultivation is negligible as compared to white cotton, and its production is only 0.2% of total white cotton's production. Critical understanding on biochemical and molecular mechanisms underlying fiber color development in cotton and flavonoid biosynthesis genes has shown color development mechanism for green and brown cotton. These strides explained the role of structural flavonoid biosynthetic pathway genes (*GhC4H*, *GhCHS,GhCHI*, *GhF3'H*, *GhDFR*, and *GhANR*) in brown color development. With the expanding sequencing data of cotton genomes, cloning and genome-wide association studies of the fiber color genes would help us to understand the complex molecular mechanism of color development in cotton fiber. Gene editing techniques, especially CRISPR-Cas9, will pave a better way to modulate color development pathway and the impact of different genes on various shades development in colored cotton. This technique has been used to alter flower colors in ornamental plants. Anthocyanin and flavanol share the same **phenylpropanoid** pathway; development of mapping population and use of DNA markers might help in the identification of candidate transcription factors (bHLH, MYB, and WD40) responsible for regulating this pathway. Altering the activity of such transcription factors might induce development of different shades in colored cotton. Additionally, inferior fiber quality is another hindrance to cultivate naturally colored cotton on a large scale. Current studies have shown the competition between flavonoid and cellulose biosynthesis pathway to use glucose as a common substrate, which leads toward low cellulose accumulation, resulting in the production of cotton fiber with poor quality. Overexpression of glucose synthesis genes is another option to compensate utilization of sugars as a common substrate. Alternatively, negative association between cellulose biosynthesis and flavonoid pathways can be broken by site-specific nuclease alteration using CRISPR-Cas9. Therefore, more comprehensive investigations of the relationships between flavonoid metabolism and fiber development are warranted. Improvement of fiber quality by altering the cellulose and flavonoid biosynthesis pathways will be a first step toward producing cotton fiber with diverse colors and good quality for textile industry.Author ContributionsWaqas Malik conceived and designed this chapter, Muhammad Ali Abid analyzed the data, and all the other authors contributed to write this paper.

Funding This work was supported by the departmental research grant for the Department of Plant Breeding and Genetics, Faculty of Agricultural Sciences and Technology, Bahauddin Zakariya University, Multan, Pakistan.

References

Al-Ghazi Y, Bourot S, Arioli T, Dennis E, Llewellyn D (2009) Transcript profiling during Fiber development identifies pathways in secondary metabolism and cell wall structure that may contribute to cotton Fiber quality. *Plant Cell Physiol* 50:1364–1381

Alonso A, Raymond P, Hernould M, Rondeau-Mouro C, de Graaf A, Chourey P, Lahaye M, Shachar-Hill Y, Rolin D, Dieuaide-Noubhani M (2007) A metabolic flux analysis to study the role of sucrose synthase in the regulation of the carbon partitioning in central metabolism in maize root tips. *Metab Eng* 9:419–432

Bafana A, Jain M, Agrawal G, Chakrabarti T (2009) Bacterial reduction in genotoxicity of Direct Red 28 dye. *Chemosphere* 74:1404–1406

Banat I, Nigam P, Singh D, Marchant R (1997) Microbial decolorization of textile-dye containing effluents: a review. *Bioresour Technol* 58:217–227

Barkeer SK, Burdekar VK, Sangannavar PA, Mohan TC, Vamadevaiah HM, Khadi BM, Katageri I (2012) Temporal and spatial expression of flavonoid biosynthetic genes in petals of cotton. *J Cell Tissue Res* 12:0000–0000

Bazin I, IbnHadjHassine A, Haj Hamouda Y, Mnif W, Bartegi A, Lopez-Ferber M, De Waard M, Gonzalez C (2012) Estrogenic and anti-estrogenic activity of 23 commercial textile dyes. *Ecotoxicol Environ Saf* 85:131–136

Buer C, Muday G, Djordjevic M (2007) Flavonoids are differentially taken up and transported long distances in Arabidopsis. *Plant Physiol* 145:478–490

Carliell C, Barclay S, Naidoo N, Buckley C, Mulholland D, Senior E (1995) Microbial decolourisation of a reactive azo dye under anaerobic conditions. *Water SA* 21(1):61–69

Carvalho L, Farias F, Lima M, Rodrigues J (2014) Inheritance of different fiber colors in cotton (*Gossypium barbadense* L.). *Crop Breed Appl Biotechnol* 14(4):256–260

Chaudhry MR (1992) Natural colors of cotton, vol 10. ICAC Recorder, Washington, pp 3–5

Chazarra S, García-Carmona F, Cabanes J (2001) Evidence for a tetrameric form of iceberg lettuce (Lactuca sativa L.) polyphenol oxidase: purification and characterization. *J Agric Food Chem* 49:4870–4875

Chen Z, Guan X (2011) Auxin boost for cotton. *Nat Biotechnol* 29:407–409

Çiçek F, Özer D, Özer A, Özer A (2007) Low cost removal of reactive dyes using wheat bran. *J Hazard Mater* 146:408–416

Coleman H, Yan J, Mansfield S (2009) Sucrose synthase affects carbon partitioning to increase cellulose production and altered cell wall ultrastructure. *Proc Natl Acad Sci* 106:13118–13123

Cooper P (1995) Color in dye house effluent. Bradford Society of Dyers and Colourists, Bradford, on behalf of the Dyers' Company Publications Trust

Cuming DS, Altan FH, Akdemir H, Tosun M, Gürel A, Tanyolaç B (2015) QTL analysis of fiber color and fiber quality in naturally green colored cotton (*Gossypium hirsutum* L.). *Turk J Field Crops* 20(1):49–58

Dai J, Dong H (2014) Intensive cotton farming technologies in China: achievements, challenges and countermeasures. *Field Crop Res* 155:99–110

Debeaujon I, Peeters A, Léon-Kloosterziel K, Koornneef M (2001) The TRANSPARENT TESTA12 gene of Arabidopsis encodes a multidrug secondary transporter-like protein required for flavonoid sequestration in vacuoles of the seed coat endothelium. *Plant Cell* 13:853–871

Demirci O, Hamamci D (2013) Antioxidant responses in *Phanerochaete chrysosporium* exposed to Astrazone Red FBL textile dye. *Cell Biochem Funct* 31(1):86–90

Dickerson D, Lane E, Rodriguez D (1999) Naturally colored cotton: resistance to changes in color and durability when refurbished with selected laundry aids. California Agricultural Technology Institute, California State University, Fresno, pp 1–42

Dixon R, Xie D, Sharma S (2005) Proanthocyanidins—a final frontier in flavonoid research. *New Phytol* 165:9–28

Dong H, Li W, Tang W, Li Z, Zhang D, Niu Y (2006) Yield, quality and leaf senescence of cotton grown at varying planting dates and plant densities in the Yellow River Valley of China. *Field Crop Res* 98:106–115

Dutt Y, Wang X, Zhu G, Li Y (2004) Breeding for high yield and fibre quality in coloured cotton. *Plant Breed* 123:145–151

Efe L, Mustafayev A, Killi F (2010) Agronomic, fiber and seed quality traits of naturally coloured cottons in East Mediterranean region of Turkey. *Pak J Bot* 42:3865–3873

Fan X, Fan B, Wang Y, Yang W (2016) Anthocyanin accumulation enhanced in Lc-transgenic cotton under light and increased resistance to bollworm. *Plant Biotechnol Rep* 10:1–11

Feng Y, Zhao X, Wang Y, Ma R, Zhou Z (2009) Responses of carbohydrate metabolism to nitrogen in cotton fiber development and its relationships with fiber strength formation. *Sci Agric Sin* 42:93–102

Feng H, Wang J, Sun J, Zhang X, Jia Y, Sun J (2010) Genetic effects of fiber color in brown cotton (*Gossypium hirsutum* L.). *Acta Agron Sin* 36:961–967

Feng H, Sun J, Wang J, Jia Y, Zhang X, Pang B, Sun J (2011) Genetic effects and heterosis of the fibre colour and quality of brown cotton (*Gossypium hirsutum*). *Plant Breed* 130:450–456

Feng H, Tian X, Liu Y, Li Y, Zhang X, Jones B, Sun Y, Sun J (2013) Analysis of flavonoids and the flavonoid structural genes in brown fiber of upland cotton. *PLoS One* 8(3):e58820

Feng H, Li Y, Wang S, Zhang L, Liu Y, Xue F, Sun Y, Wang Y, Sun J (2014) Molecular analysis of proanthocyanidins related to pigmentation in brown cotton fibre (*Gossypium hirsutum* L.). *J Exp Bot* 65:5759–5769

Feng H, Yang Y, Sun S, Li Y, Zhang L, Tian J, Zhu Q, Feng Z, Zhu H, Sun J (2017) Molecular analysis of caffeoyl residues related to pigmentation in green cotton fibers. *J Exp Bot* 68:4559–4569

Forkmann G, Heller W (1999) Biosynthesis of flavonoids. In: Sankawa U (ed) Comprehensive natural products chemistry. Elsevier, Amsterdam, pp 713–748

Fox S (1987) Naturally colored cottons. *Spin Off* 46:29–31

Fryxell P (1984) Taxonomy and germplasm resources. *Cotton* 27–57

Gabetta B, Fuzzati N, Griffini A, Lolla E, Pace R, Ruffilli T, Peterlongo F (2000) Characterization of proanthocyanidins from grape seeds. *Fitoterapia* 71:162–175

Galindo C, Jacques P, Kalt A (2001) Photooxidation of the phenylazonaphthol AO20 on TiO2 kinetic and mechanistic investigations. *Chemosphere* 45:997–1005

Ganglberger E (2009) Environmental aspects and sustainability. In: Handbook of natural colorants. pp 353–366

Gao J, Shen L, Yuan J et al (2019) Functional analysis of GhCHS, GhANR and GhLAR in colored fiber formation of *Gossypium hirsutum* L. *BMC Plant Biol* 19(1):455

Gong W, He S, Tian J, Sun J, Pan Z, Jia Y, Sun G, Du X (2014) Comparison of the transcriptome between two cotton lines of different fiber color and quality. *PLoS One* 9(11):e112966

Gonzalez C, Fanzone M, Cortés L, Bottini R, Lijavetzky D, Ballare C, Boccalandro H (2015) Fruit-localized photoreceptors increase phenolic compounds in berry skins of field-grown *Vitis vinifera* L. cv. Malbec. *Phytochemistry* 110:46–57

Gu L, Kelm M, Hammerstone J, Beecher G, Holden J, Haytowitz D, Gebhardt S, Prior R (2004) Concentrations of proanthocyanidins in common foods and estimations of normal consumption. *J Nutr* 134:613–617

Haigler C, Ivanova-Datcheva M, Hogan P, Salnikov V, Hwang S, Martin K, Delmer D (2001) Carbon partitioning to cellulose synthesis. *Plant Mol Biol* 47:29–51

Haigler C, Zhang D, Wilkerson C (2005) Biotechnological improvement of cotton fibre maturity. *Physiologiaplantarum* 124:285–294

Han A, Chae Y, Lee M, Cho G (2011) Effect of color changes of naturally colored organic cotton fibers on human sensory perception. *Fibers Polym* 12:939

Hansen O (2005) Screening for health effects from chemical substances in textile colorants. In: Survey of chemical substances in consumer products. p 57

Harland S (1935) The genetics of cotton Part XIV The inheritance of brown lint in new world cottons. *J Genet* 31:27–37

Hijazi A, Mcheik A, Rammal H, Rammal W, Annan H, Toufaily J, Hamieh T (2015) Biosorption of methylene blue from waste water using Lebanese Cymbopogon citratus (citronnelle). *Eur Sci J* 11(21):293–308

Hinchliffe D, Condon B, Delhom C, Chang S, Montalvo J, Madison C, Reynolds M, VonHoven T, Cintrón M (2015) Physical and combustion properties of nonwoven fabrics produced from conventional and naturally colored cottons. *Text Res J* 85:1666–1680

Hu B, Jin J, Guo AY, Zhang H, Luo J, Gao G (2015) GSDS 2.0: an upgraded gene features visualization server. *Bioinformatics* 31:1296–1297

Hua S, Wang X, Yuan S, Shao M, Zhao X, Zhu S, Jiang L (2007) Characterization of pigmentation and cellulose synthesis in colored cotton fibers. *Crop Sci* 47:1540–1546

Indrais E, Cheema H, Samad A, Bashir A (2011) Temporal expression analysis and cloning of cotton (*Gossypium hirsutum*) fiber genes. *Int J Agric Biol* 13(1):89–94

Kant R (2012) Textile dyeing industry an environmental hazard. *Natural Science* 4(1):22–26

Khan N, Hassan G, Kumbhar MB, Marwat KB, Khan MA, Parveen A, Aiman U, Saeed M (2009) Combining ability analysis to identify suitable parents for heterosis in seed cotton yield, its components and lint % in upland cotton. *Crop Prod* 29:108–115

Kohel R (1985) Genetic analysis of fiber color variants in cotton 1. *Crop Sci* 25:793–797

Landi M, Tattini M, Gould K (2015) Multiple functional roles of anthocyanins in plant-environment interactions. *Environ Exp Bot* 119:4–17

Lee J (1996) A new spin on naturally colored cottons. *Agric Res Mag* 44:20–21

Li YY, Wang XD (2002) Observation of ultrastructure of colored cotton fiber. *J Zhejiang Univ Agric Life Sci* 28:379–382

Li YJ, Zhang XY, Wang FX, Yang CL, Liu F, Xia GX, Sun J (2013) A comparative proteomic analysis provides insights into pigment biosynthesis in brown color fiber. *J Proteome* 78:374–388

Li Y, Sun S, Zhang X, Wang X, Liu Y, Fei X, Jie S (2018) New clues concerning pigment biosynthesis in green colored fiber provided by proteomics-based analysis. *J Integr Agric* 17:46–53

Liang M, Liu H, Lu X, Li Y, Song W, Lu C, Xiao X, Li X (2011) Cloning of flavonoid 3′-hydroxylase gene (GhF3′ H) and expressional characteristics of several genes associated with pigment synthesis in brown cotton (*Gossypium hirsutum* L.) fibers. *J Agric Biotechnol* 19:808–815

Liu H, Luo C, Song W, Shen H, Li G, He Z, Chen W, Cao Y, Huang F, Tang S, Hong P, Zhao E, Zhu J, He D, Wang S, Huo G, Liu H (2018) Flavonoid biosynthesis controls fiber color in naturally colored cotton. *PeerJ* 6:e4537

Ma M, Li R, Du Y, Tang Z, Zhou W (2013) Analysis of antibacterial properties of naturally colored cottons. *Text Res J* 83:462–470

Ma M, Luo S, Hu Z, Tang Z, Zhou W (2016) Antioxidant properties of naturally brown-colored cotton fibers. *Text Res J* 86:256–263

Malik W, Khan A, Sadia B (2013) In situ characterization of coloured cotton genotypes. *Aust J Crop Sci* 7:299–304

Malik W, Khan A, Cheema H, Aslam U, Iqbal M, Qayyum A, Yasmeen A, Bibi N (2015) Transcriptome analysis of pigment related genes in colored cotton. *Int J Agric Biol* 17:205–210

Malik W, Shah MSA, Abid MA, Qanmber G, Noor E, Qayyum A, Liang C, Guo S, Zhang R (2018) Genetic basis of variation for fiber quality and quality related biochemical traits in Bt and non-Bt colored cotton. *Int J Agric Biol* 20:2117–2124

Maralappanavar MS, Patil SS, Kulkarni VV, Somshekhar, Madhura C, Immadi S, Narayanan K, Sanapapamma KJ, Vastrad JV (2013) Genetic improvement and registration of naturally coloured *Gossypium hirsutum* cotton genotypes suitable for textile industry. *J Cotton Res Dev* 27:157–164

Matus-Cádiz MA, Daskalchuk T, Verma B, Puttick D, Chibbar RN, Gray GR, Perron C, Tyler R, Hucl P (2008) Phenolic compounds contribute to dark bran pigmentation in hard white wheat. *J Agric Food Chem* 56:1644–1653

Matusiak M, Frydrych I (2014) Investigation of naturally coloured cotton of different origin–analysis of fibre properties. In: Fibres and textiles in Eastern Europe, vol 22. pp 34–42

Mikhailova A, Strygina K, Khlestkina E (2019) The genes determining synthesis of pigments in cotton. *Biol Commun* 64(2):133–145

Nakatsuka T, Nishihara M, Mishiba K, Yamamura S (2005) Temporal expression of flavonoid biosynthesis-related genes regulates flower pigmentation in gentian plants. *Plant Sci* 168:1309–1318

Narayanan D, Saladi R, Fox J (2010) Review: ultraviolet radiation and skin cancer. *Int J Dermatol* 49:978–986

Nishizaki Y, Matsuba Y, Okamoto E, Okamura M, Ozeki Y, Sasaki N (2011) Structure of the acyl-glucose-dependent anthocyanin 5-O-glucosyltransferase gene in carnations and its disruption by transposable elements in some varieties. *Mol Gen Genomics* 286:383–394

Osugi M, Umbuzeiro G, De Castro F, Zanoni M (2006) Photoelectrocatalytic oxidation of remazol turquoise blue and toxicological assessment of its oxidation products. *J Hazard Mater* 137:871–877

Pailthorpe M (1994) Textiles and sun protection: the current situation. University of New South Wales, Sydney

Palmieri G, Cennamo G, Sannia G (2005) Remazol Brilliant Blue R decolourisation by the fungus *Pleurotus ostreatus* and its oxidative enzymatic system. *Enzym Microb Technol* 36:17–24

Pan Z, Sun D, Sun J, Zhou Z, Jia Y, Pang B, Ma Z, Du X (2010) Effects of fiber wax and cellulose content on colored cotton fiber quality. *Euphytica* 173:141–149

Parmar M, Giri C, Singh M, Chabbra J (2006) Development of UV and flame-resistant fabric from natural coloured cotton. *Colourage* 53:57–60

Pettigrew W (2001) Environmental effects on cotton fiber carbohydrate concentration and quality. *Crop Sci* 41:1108–1113

Qian S, Hong L, Xu M, Cai Y, Lin Y, Gao J (2015) Cellulose synthesis in coloured cotton. *Sci Asia* 41:180

Qian S, Hong L, Cai Y, Gao J, Lin Y (2016) Effects of light on in vitro fiber development and flavonoid biosynthesis in green cotton (*Gossypium hirsutum*). *Acta Soc Botan Pol* 85

Qiu XM, Zhou WL, Li MS, Ma YG (2002) Study on the genetics and production of fiber pigments and color deviation after wetting process of naturally colored cotton. *Sci Agric Sin* 35:610–615

Rafii F, Cerniglia C (1995) Reduction of azo dyes and nitroaromatic compounds by bacterial enzymes from the human intestinal tract. *Environ Health Perspect* 103:17–19

Razzaq A, Ashraf J, Malik W, Shaban M, Zhang R, Liang C, Hanif M, Abid MA, Qayyum A (2020) In silico analyses of TALE transcription factors revealed its potential role for organ development and abiotic stress tolerance in cotton. *Int J Agric Biol* 23(1083):1094

Revankar MS, Lele SS (2007) Synthetic dye decolorization by white rot fungus, *Ganoderma sp.*, WR-1. *Bioresour Technol* 98:775–780

Ruan Y (2005) The delayed initiation and slow elongation of fuzz-like short fibre cells in relation to altered patterns of sucrose synthase expression and plasmodesmata gating in a lintless mutant of cotton. *J Exp Bot* 56:977–984

Ruan Y, Chourey P (1998) A fiberless seed mutation in cotton is associated with lack of Fiber cell initiation in ovule epidermis and alterations in sucrose synthase expression and carbon partitioning in developing seeds. *Plant Physiol* 118:399–406

Ruan Y, Llewellyn D, Furbank R (2003) Suppression of sucrose synthase gene expression represses cotton fiber cell initiation, elongation, and seed development. *Plant Cell* 15:952–964

Ryser U (1999) Cotton fibre initiation and histodifferentiation. In: Cotton fibres. Hawthorn Press

Salvi S, Barnes P (2009) Chronic obstructive pulmonary disease in non-smokers. *Lancet* 374:733–743

Saratale R, Saratale G, Chang J, Govindwar S (2011) Bacterial decolorization and degradation of azo dyes: a review. *J Taiwan Inst Chem Eng* 42:138–157

Schijlen E, Ric de Vos C, van Tunen A, Bovy A (2004) Modification of flavonoid biosynthesis in crop plants. *Phytochemistry* 65:2631–2648

Schmutz A, Jenny T, Amrhein N, Ryser U (1993) Caffeic acid and glycerol are constituents of the suberin layers in green cotton fibres. *Planta* 189:453

Schmutz A, Jenny T, Ryser U (1994) A caffeoyl-fatty acid-glycerol ester from wax associated with green cotton fibre suberin. *Phytochemistry* 36:1343–1346

Silva M, Beltrão N, Cardoso G (2005) Adubação do algodãocolorido BRS 200 emsistemaorgânico no SeridóParaiAZbano. *Rev Brasil Engenharia Agríc Ambient* 9:222–228

Sponza D (2006) Toxicity studies in a chemical dye production industry in Turkey. *J Hazard Mater* 138:438–447

Sun J, Jiang YM, Wei XY, Shic J, Youa YL, Liua H, Kakudad Y, Zhao M (2006) Identification of (–)-epicatechin as the direct substrate for polyphenol oxidase isolated from litchi pericarp. *Food Res Int* 39:864–870

Takos A, Ubi B, Robinson S, Walker A (2006) Condensed tannin biosynthesis genes are regulated separately from other flavonoid biosynthesis genes in apple fruit skin. *Plant Sci* 170:487–499

Tan J, Tu L, Deng F, Hu H, Nie Y, Zhang X (2013) A genetic and metabolic analysis revealed that cotton fiber cell development was retarded by flavonoid naringenin. *Plant Physiol* 162:86–95

Tanner G, Francki K, Abrahams S, Watson J, Larkin P, Ashton A (2003) Proanthocyanidin biosynthesis in plants purification of legume leucoanthocyanidin reductase and molecular cloning of its cDNA. *J Biol Chem* 278:31647–31656

Trapnell C, Roberts A, Goff O, Pertea G, Kim D, Kelley DR, Pimentel H, Salzberg SL, Rinn JL, Pachter L (2013) Differential gene and transcript expression analysis of RNA-seq experiments with TopHat and cufflinks. *Nat Protoc* 7:562–578

Tu L, Tan J, Guo K, Li Z, Zhang X (2014) Flavonoid pathway in cotton fiber development. *Sci Sin Vitae* 44:758–765

Voorrips RE (2002) MapChart: software for the graphical presentation of linkage maps and QTLs. *J Hered* 93:77–78

Vreeland JM Jr (1987) Coloured cotton: return of the native. *Int Dev Res Center Rep* 10:4–5

Vreeland JM Jr (1999) The revival of colored cotton. *Sci Am* 280:112–119

Wang X, Li Y (2002) Study on characteristics of colored cotton fiber development. *J Zhejiang Univ* 28:237–242

Wang S, Kopf A, Marx J, Bogdan A, Polsky D, Bart R (2001) Reduction of ultraviolet transmission through cotton T-shirt fabrics with low ultraviolet protection by various laundering methods and dyeing: clinical implications. *J Am Acad Dermatol* 44:767–774

Wang L, Liu H, Xiao X, Pang Z, Song W, Lu C, Luo C, Liu G, Xu J, Li X, Li X (2012) Studies on hereditary properties of colored cotton in Xinjiang. *Agric Sci Technol* 13:541–582

Ware J (1932) Inheritance of lint colors in upland cotton. *Agron J* 24:550–562

Weisburger J (2002) Comments on the history and importance of aromatic and heterocyclic amines in public health. *Mutat Res* 506:9–20

Wen T, Wu M, Shen C, Gao B, Zhu D, Zhang X, You C, Lin Z (2018) Linkage and association mapping reveals the genetic basis of brown fibre *(Gossypium hirsutum)*. *Plant Biotechnol J* 16:1654–1666

Wong P, Yuen P (1996) Decolorization and biodegradation of methyl red by *Klebsiella pneumoniae* RS-13. *Water Res* 30:1736–1744

Xiao Y, Zhang Z, Yin M, Luo M, Li X, Hou L, Pei Y (2007) Cotton flavonoid structural genes related to the pigmentation in brown fibers. *Biochem Biophys Res Commun* 358:73–78

Xie D, Sharma S, Paiva N, Ferreira D, Dixon R (2003) Role of anthocyanidin reductase, encoded by BANYULS in plant flavonoid biosynthesis. *Science* 299:396–399

Yamazaki M, Gong Z, Fukuchi-Mizutani M, Fukui Y, Tanaka Y, Kusumi T, Saito K (1999) Molecular cloning and biochemical characterization of a novel anthocyanin 5-O-glucosyltransferase by mRNA differential display for plant forms regarding anthocyanin. *J Biol Chem* 274:7405–7411

Yoo M, Wendel J (2014) Comparative evolutionary and developmental dynamics of the cotton (*Gossypium hirsutum*) fiber transcriptome. *PLoS Genet* 10(1):1004073

Yuan S, Hua S, Malik W, Bibi N, Wang X (2012a) Physiological and biochemical dissection of fiber development in colored cotton. *Euphytica* 187:215–226

Yuan S, Waqas M, Hua S, Bibi N (2012b) In vitro inhibition of pigmentation and fiber development in colored cotton. *J Zhejiang Univ Sci B* 13:478–486

Yuan SN, Malik W, Bibi N, Wen GJ, Ni M, Wang XD (2013) Modulation of morphological and biochemical traits using heterosis breeding in coloured cotton. *J Agric Sci* 121:57–71

Zhang X, Liu X, Dong Z, Chen Y, Chen D (2011a) Physiological characteristics associated with fiber development in two different types of natural colored-cotton cultivars. *Acta Agron Sin* 37:489–495

Zhang M, Song X, Sun X, Wang Z, Zhao Q, Li Z, Ji H, Xu X (2011b) Patterns of color formation in different fibers during development of colored cotton *Gossypium hirsutum* Cultivars. *Aust J Crop Sci* 5:1796–1800

Zhang M, Song X, Sun X, Wang Z, Zhao Q, Li Z, Ji H, Xu X (2011c) Observation of differentiation and pigment deposition process in colored cotton fibers. *Acta Agron Sin* 37:1280–1288

Zhang M, Song X, Sun X, Wang Z, Li Z, Ji H, Xu X, Li J (2012) The relationship between cellulose content and the contents of sugars and minerals during fiber development in colored cotton cultivars. *Cellulose* 19:2003–2014

Zhang X, Hu D, Yuan L, Yuan C, Abidallha E, Dong Z, Chen D, Zhang L (2017) Developmental and hormonal regulation of fiber quality in two natural-colored cotton cultivars. *J Integr Agric* 16:1720–1729

Zhu W, Liu K, Wang X (2008) Heterosis in yield, fiber quality, and photosynthesis of okra leaf-oriented hybrid cotton (*Gossypium hirsutum* L). *Euphytica* 164:283

Zhu T, Liang C, Meng Z, Sun G, Meng Z, Guo S, Zhang R (2017) CottonFGD: an integrated functional genomics database for cotton. *BMC Plant Biol* 17(1):101

Zollinger H (ed) (1987) Azo dyes and pigments. Colour chemistry-synthesis, properties and applications of organic dyes and pigments. *Angew Chem Int Ed* 43(40):5291–5292

Zoratti L, Karppinen K, Luengo Escobar A, Häggman H, Jaakola L (2014) Light-controlled flavonoid biosynthesis in fruits. *Front Plant Sci* 5(534):534

Симонгулян Н (1984) Genetic analysis of fiber color in upland cotton. *Agron Abroad Cotton* 3:172–191

Chapter 9
Mutagenesis for Targeted Breeding in Cotton

Mehboob-ur-Rahman, Saba Zafar, Momina Hussain, Hassan Abbas, and Bradley J. Till

9.1 Introduction

Cotton is cultivated in >90 countries including the five major cotton-producing countries: India, China, the USA, Brazil, Pakistan, and Australia (USDA, 2019). Cotton provides natural fiber as well as seed for extraction of oil for edible purposes and cotton seed cake for animal cultivation (Shuli et al. 2018).

Out of 52 cotton species belonging to the genus *Gossypium*, two are cultivated tetraploids, *G. hirsutum* L. (AD1) and *G. barbadense* L. (AD2), while two are cultivated diploids, i.e., *G. herbaceum* (A1) and *G. arboreum* (A2). However, *G. hirsutum* has been widely adapted because of its high lint production with medium fiber quality. It contributes ~90% of the total cotton production, while *G. barbadense* has been known for producing exceptionally high-quality fibers—a desired feature for textile industry. Efforts toward improving cotton yield potential, lint quality, and

Mehboob-ur-Rahman (✉)
Agricultural Biotechnology Division , National Institute for Biotechnology & Genetic Engineering (NIBGE) College, Faisalabad, Pakistan

Pakistan Institute for Engineering and Applied Science (PIEAS), Islamabad, Pakistan
e-mail: mehboob@nibge.org

S. Zafar · M. Hussain
Plant Genomics and Mol. Breeding Lab, National Institute for Biotechnology and Genetic Engineering (NIBGE), Faisalabad, Pakistan

Department of Biotechnology, Pakistan Institute of Engineering and Applied Sciences (PIEAS), Nilore, Islamabad, Pakistan

H. Abbas
Plant Genomics and Mol. Breeding Lab, National Institute for Biotechnology and Genetic Engineering (NIBGE), Faisalabad, Pakistan

B. J. Till
Veterinary Genetics Laboratory, University of California, Davis, Davis, California, USA

Mehboob-ur-Rahman et al. (eds.), *Cotton Precision Breeding*,
https://doi.org/10.1007/978-3-030-64504-5_9

capabilities for acclimatizing in changing environments have been made, but limited success was reported. Cotton like many other crop species has a narrow genetic base. This offers a limited scope for breeders to improve its genetic potential to combat biotic and abiotic stresses, yield, and quality features of the lint.

Development of genetically diverse cotton varieties has remained a major hallmark of most breeding programs. Secondly, breeding for genetically diverse cotton varieties may also help in sustaining cotton production under the changing environments (Arshad et al. 2009; Rahman 2002; Rahman and Zafar 2007, 2012). It was demonstrated that farmers should cultivate the genetically diverse cotton varieties as a strategy for sustaining cotton production (Rahman 2002). Similar recommendations were made for other crops where a narrow genetic base was made responsible for limiting retarding the breeding progress (Radhakrishnan 2017; Tester and Langridge 2010). Efforts were made to enhance the genetic diversity of several different crop species by deploying several procedures including conventional breeding (crossing two parent genotypes of the same species), introgression breeding, insertion of mutations through chemical or physical means, changing ploidy level, targeted editing of genomes, etc. The success of transferring desirable genes through crossing from untapped (wide) genetic resources may be limited due to linkage drag of unwanted traits. In this chapter, therefore, random mutagenesis procedures and their implications in targeted cotton breeding will be discussed.

In total, 3308 crop varieties have been developed using mutagens, according to publicly available information (https://mvd.iaea.org/). Out of these, 60% were released in Asia for general cultivation. In China alone, 25% of the reported mutant varieties have been released for cultivation. China, India, and Japan are the top three mutant variety-producing countries. These mutant varieties represent all major crops including cereals, legumes, fiber crops, etc. In cotton, in total 48 mutant varieties have been released till date. Out of 48 varieties, 12 varieties are released by Pakistan. In cotton, gamma-radiated mutant populations led to the development of two very potent cultivars; NIAB-78 was released in Pakistan in 1983, and Lumian No. 1 was released in China.

In mutagenesis experiments, usually a cultivated variety is exposed to a mutagen followed by identifying mutants for the trait of interest starting in the M_2 generation. Heritability of the mutant trait is further confirmed by advancing the generations. Numerous mutations are induced randomly throughout the genome, in contrast to site-directed mutagenesis approaches. In the past more than 80 years of mutation breeding, most varieties have been released without understanding the genes controlling the altered traits. This has changed in the past decades. Mutants carrying alerted traits can be exposed to several genomic screening procedures in order to identify alleles causative for novel phenotypes. These genes can be cloned for transformation purposes or to develop molecular markers developed to facilitate the introgression of mutant gene(s) into plants in segregating populations. The cotton plants containing the mutant allele/gene are selected by surveying their genomic DNA using specific primers and PCR-based genotyping assays that can be high throughput (Li et al. 2017a, b). These steps are required for initiating breeding by design approaches when the aim is to introduce the mutant allele into a cultivated variety. Also, the procedure for varietal development can be expedited using a greenhouse for advancing the generations. Most times, mutations are recessive in expression, and there are chances that these mutations are missed when exercising conventional phenotypic selection. Here the role of molecular markers is evident in that the mutant alleles can be selected in heterozygous plants containing the recessive mutations.

Several excellent reviews have been published which describe the methods, challenges, and potential of mutation breeding in crop improvement (Barkley and Wang 2008; Comai and Henikoff 2006; Jankowicz-Cieslak et al. 2017; Parry et al. 2009; Till et al. 2007). This article focuses on the substantial progress that has recently been made on the application of targeted mutation breeding to cotton. Targeted mutation breeding requires candidate gene sequence data, screening populations, and mutation screening techniques.

9.2 Mutagenesis

Mutations are changes or alterations in the genetic material. The concept of heritable mutations, those that are stably passed from generation to generation, dates to the early 1900s, decades before the establishment of DNA as the genetic material (Hershey and Chase 1952; de Vries 1901). While mutations in plants can be found in genomic DNA of nucleus, mitochondria, and chloroplast, nuclear DNA mutations are almost exclusively used for plant breeding and functional genomics. Mutation type varies from single-base substitutions to small insertions/deletions to larger chromosomal aberrations (Shu et al. 2012). In nature, mutations are "spontaneous" and occur during DNA replication and molecular damages caused by UV radiation and other environmental exposures. Genome sequencing experiments have established that the rate of spontaneous mutation accumulation can vary based on mutation type and position in the genome (e.g., Weng et al. 2019). Mutations can be induced in living organisms at much higher rates. This was first demonstrated by Muller in drosophila followed by Stadler in maize and barley by exposing cells to X-rays (Muller 1930; Stadler 1928).

Mutagens are of two broad types, physical or chemical. Both have been extensively used in inducing mutations in crop species including cotton, providing a means to generate novel variants (Andrews et al. 2005). A subset of mutations induced in a plant genome will affect gene function or regulation. These variations are responsible for altered traits which can be used in breeding new crop varieties including cotton. Strong correlation between mutation density and the rate of germination has been reported (Haughn and Gilchrist 2006; Parry et al. 2009).

9.2.1 *Physical Mutagens*

Physical agents, e.g., X-rays, UV rays, neutrons, and gamma radiations, have been used to induce mutations in genomes of several crop species (Kodym and Afza 2003). Genomic analysis of plants treated with fast neutrons, gamma irradiation, and carbon ion beams suggests that complex spectrum of mutations can be induced ranging from single-base substitutions to large chromosomal aberrations depending on the mutagen and dosage used (Li et al. 2016a, b, 2017a, b, 2019; Datta et al. 2018). Large genomic variations may be lethal and not meiotically heritable (Naito et al. 2005). While physical mutagens have been used for more than 80 years for crop improvement, little data exists regarding the genomic lesions conferring altered traits. It therefore remains possible that successful mutant varieties in polyploid crops such as wheat result from chromosomal alterations (e.g., large indels and rearrangements) that affect the dosage of many genes.

9.2.2 Chemical Mutagens

Chemical mutagens induce mutations randomly throughout the genome more frequently than the naturally occurring mutations (Greene et al. 2003). Chemical agents, including ethyl methanesulfonate (EMS) and *N*-ethyl-*N*-nitrosourea (ENU), have extensively been used for inducing point mutations in genomes (Hussain et al. 2018; Hussain and Rahman 2019). The EMS agent causes alkylation of guanine base that results in G/C-to-A/T transitions (Kaul and Bhan 1977; Kim et al. 2006; Witt et al. 2018). Thus these single-base changes in gene may lead to replace the amino acid—resultantly, the protein structure and function is altered, and most time gene activity can be knocked out. Mutagens such as EMS can induce mutations at a very high frequency, especially in polyploids (Till et al. 2018). This paves the way for generating several new alleles of a specific gene. It can be achieved by developing a small population irrespective of the genome size (Greene et al. 2003; Sabetta et al. 2011; Till et al. 2007). The induced mutations either by adding, replacing, or deleting a nucleotide may alter the function of a gene which is reflected in the form of altered phenotype in a mutant population (Aslam et al. 2013, 2016; Hussain et al. 2018; Hussain and Rahman 2019).

9.2.3 Determination of LD50 Value of a Mutagen

Maximum mutation density in a genome, and comparative effectiveness of mutagens can be achieved by exposing the biological material with optimal concentration of a mutagen. The definition of an effective mutagen dose relates with the production of relatively less biological damage (seedling injury, sterility, etc.) than that of induced mutations in the genome. Hence, choice of a proper mutagen and its effectiveness are important for inducing high frequency of mutations (Shah et al. 2008). In this regard, determination of an optimum concentration of a mutagen for inducing maximum mutations is prerequisite before launching a mutagenesis experiments for any crop variety. Mutagenesis of multicellular materials, such as seed, produces a genetic mosaic in the first (M_1) generation whereby different cells will accumulate different mutations. As such, direct measurement of DNA lesions to determine an optimum concentration is not advisable. As an alternative, different concentrations of chemical mutagen (or dosage of physical radiation) are applied, and a kill curve is developed that measures the percentage of lethality at different treatment conditions. Lethal dose (LD) values are often collected with mutagens such as EMS by measuring the percentage of germination that should be between 35 and 70% for carrying out forward and reverse genetic screens in various crop species (Sabetta et al. 2011; Caldwell et al. 2004; Chawade et al. 2010; Gottwald et al. 2009; Rawat et al. 2012; Jankowicz-Cieslak and Till 2016).

To determine the optimum concentration of a mutagen in cotton, seeds are exposed to different concentrations of physical and chemical mutagens at varying temperature and time periods. Ginned cotton seed are delinted with commercial grade concentrated H_2SO_4 and presoaked in sterile deionized water for 10 h in darkness at room temperature. For treating with physical mutagen, a gamma cell irradiator is used for inducing mutations in cotton. This instrument allows direct loading of the seed sample vertically as well as horizontally. The Cobalt-60 gamma source consists of multiple thin "pencils" distributed around a cylinder. This ensures a uniform application of gamma radiations. Often, 250 k rad (Cs 137) for 25 min at room temperature is used to treat the dry cotton seed. Seeds are then sterilized with 5% sodium hypochlorite and 70% ethanol and thereafter washing thrice for removing the mutagen residues (Balkan et al. 2019; El-Kameesy et al. 2019).

For EMS mutagenesis, different batches of 75 seeds pre-imbibed in water are made followed by exposing each batch with different EMS concentrations along with an untreated control. The seeds are further incubated at different temperatures for different time periods with mild shaking (45 rpm). After treating with EMS, the seeds are thoroughly washed thrice with running tap water followed by air-drying. The treated and untreated cotton seeds are sown in polythene bags (75 seeds/bag) in randomized 3 replicates of 25 seeds. Data is collected for germination percentage after 10 days. Then LD50 value is determined (the dose at which 45–55% seeds germinate). Guidelines should be followed for the safe handling of chemical mutagens as they are known carcinogens (Jankowicz-Cieslak and Till 2016).

9.2.4 Mutation Densities

Previously, mutation densities were estimated in cotton by counting the number of phenotypic mutants for various traits including plant height, boll size, fiber length, leaf color, flower color, sterility, etc. in M_2. The mutation frequency is expressed by the percentage of individual M_2 mutants out of the total population. Consequently, mutation densities differ among different traits and mutant populations resulting by exposing with different treatments. The density of induced mutation is calculated by using following formula:

$$d = (n \times 1000) / (i \times j)$$

where, "*i*" is size of the targeted gene in kilo bases (kb), "*n*" is the total number of mutant lines, "*j*" is the total number of mutations recognized in "*n*", and "*d*" is the mutation density. A wide range of mutation densities have been reported in various crop plants. In general, mutation densities in diploid are considered high if they range between 1 per 100 and 1 per 300 kb. Higher densities are reported for polyploids, e.g., 1/57 for triploids, 1/40 for tetraploids, and 1/24 for hexaploids (Till et al. 2018).

9.3 Mutation Screening Techniques

Mutations induced in a genome result in the substitution, addition, deletion, or positional rearrangement of one or more nucleotides. Recent developments in genomic science have empowered researchers for counting the number of mutations in genome much more accurately than that of conventional mutation detection systems which underestimate the number of mutations. Also, the genomic assays can screen mutants of a large populations derived by exposing with radiation and chemical mutagens in much less time period. These mutant alleles can be used in breeding program through tracking these alleles using specific primers. A number of techniques for mutation detection and screening that make use of target-specific PCR amplification have been used to identify sequence variation in plant genes (Table 9.1). These methods can be advantageous in that they can be very low cost and highly flexible.

9.3.1 Single-Strand Conformation Polymorphism (SSCP)

The precise analysis of genetic changes has main applications in most areas of agriculture research including the determination of unknown or known disease-causing mutations. For determining the variations in sequence of a mutant allele, single-strand conformation polymorphism (SSCP) coupled with PCR offers significant advantages over the other techniques. It is an effective and simple method for the determination of SNPs (Sheffield et al. 1993). In conventional gel electrophoresis, a single-nucleotide change in a gene is difficult to identify as both alleles have the same size. After amplifying the gene, denaturation of the amplified products is exercised. The single-stranded DNA may undergo a three-dimensional folding, and the resultant fragments may attain a unique conformational state. These fragments with different shapes travel with different pace in gel even if the size of the alleles is the same.

In this assay, genomic DNA is extracted. The target gene is amplified with the gene-specific primers. The amplicons are denatured followed by running on PAGE gels. For visualizing the fragments, there are several detection systems including through using radiolabeled primers, using florescent dyes, and or using

Table 9.1 Genetic techniques for mutation detection and screening

Sr. #	Methods	Abbreviations	Method	Usage
1	Single-strand conformation polymorphism	SSCP	Sequence variants are detected via analysis of DNA fragments using PAGE	It is used for identification of anonymous mutations Furthermore, it cannot be used for studying pooled DNA fragments
2	PCR-restriction fragment length polymorphism	PCR-RFLP	PCR products that are restricted via endonuclease restriction enzyme are analyzed using electrophoresis system	It is used for genotyping experiments It cannot be used for detection of anonymous SNPs
3	PCR-restriction fragment-SSCP	PCR-RF-SSCP	PCR products that are restricted via endonuclease restriction enzyme are analyzed using native PAGE system	It is highly used for detection of anonymous SNPs It cannot be used for studying pooled DNA fragments
4	Temperature gradient gel electrophoresis	TGGE	Heteroduplexed DNA is analyzed using temperature gradient gels	It can be used for the detection of changes in melting temperatures of the homoduplex of mutant DNA and wild type The difference in the electrophoresis resulted because of heteroduplex of these DNAs
5	Conformation-sensitive capillary electrophoresis	CSCE	The principle of CSCE is different movements of heteroduplex and homoduplex DNA during electrophoresis, when capillaries are filled with semi-denaturized polymers	It is used to detect mutation with known SNPs from a pooled set of DNAs
6	High-resolution melting curve analysis	HRM	HRM is a fast, high-throughput, advanced, and nonenzymatic method It describes DNA fragments through analysis. Of post-PCR melting curve	It is used to identify the difference between homozygous and heterozygous mutations It also often allows identification of mutation between homozygous wild types and mutants
7	Dot-blot SNP	–	It is analysis of SNPs through hybridization of labeled oligos to the dot-blotted PCR products—enhanced with reasonable hybridization	It is used to discriminate known induced genetic mutations It is also used for genotyping mutated genes But it cannot be used for analysis of unknown mutations

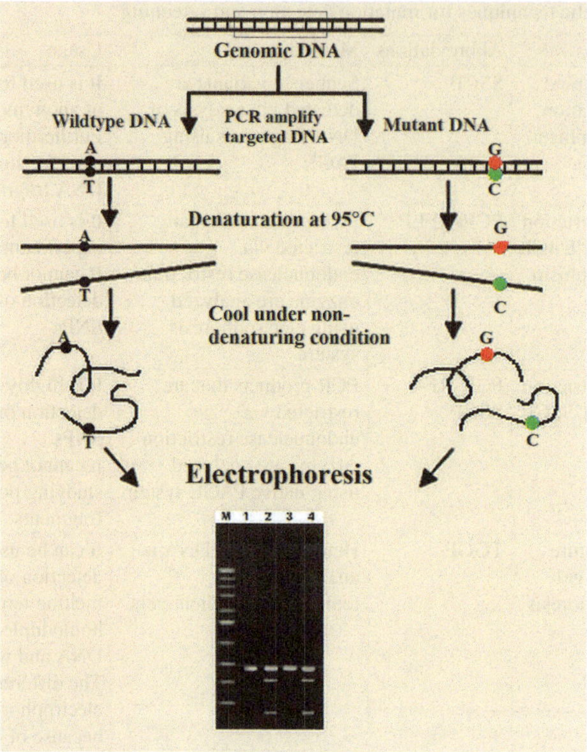

Fig. 9.1 Single-strand DNA conformation polymorphism method

conventional silver staining procedure (Fig. 9.1). It can identify SNPs for amplicon sizes of 400–500 bp. It typically takes 2 days to carry out the experiment. This is an accessible, sensitive but inexpensive, rapid, and potentially high-throughput platform (Gasser et al. 2006).

9.3.2 *Temperature Gradient Gel Electrophoresis (TGGE)*

In TGGE, temperature gradients are used to denature DNA molecules during their movement through either agarose gel or acrylamide gel. It is a form of electrophoresis and can analyze RNA, DNA, protein-DNA complexes, and often proteins. In TGGE, the dsDNA molecule's separation of depends upon melting of a DNA strands into two ssDNA molecules. That's why the movement of DNA molecules shows not only the size of the molecule but also the composition of its nucleotides, thus separating DNA molecules of similar size with different nucleotide composition. This assay can be completed in either a perpendicular or a parallel way, which is reliable on the relative direction of temperature gradient and electric field (Fig. 9.2). The parallel mode can analyze multiple samples in the same gel, while the perpendicular mode can do thorough analysis of a single sample (Viglasky 2013).

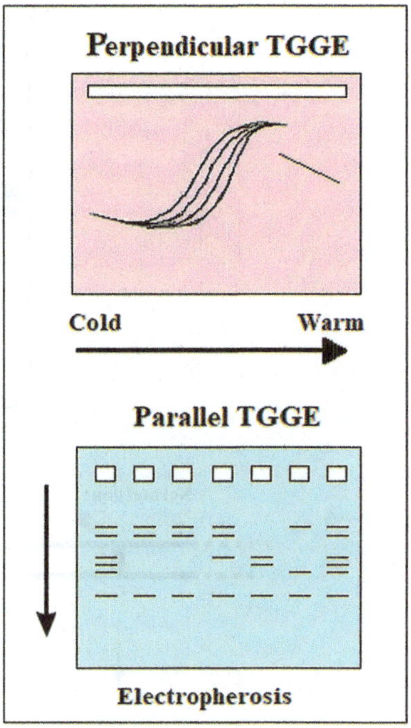

Fig. 9.2 Temperature gradient gel electrophoresis

9.3.3 Conformation-Sensitive Gel Electrophoresis (CSGE)

It is a fast and sensitive screening technique based on analysis of heteroduplexes and is used foridentification of DNA sequence mutation, especially point mutations or small deletions and insertions(Hill 2011) (Fig. 9.3). It has also been modified for use on a fluorescent platform that resulted in higher throughput and sensitivity (Ganguly 2002). It is automated and thus is useful in increasing sample throughput and sensitivity. However, it is a cheap, user-friendly, and effective technique for mutation identification that can be done with less specialist equipment.

9.3.4 Conformation-Sensitive Capillary Electrophoresis (CSCE)

It is an effective and high-throughput assay that can detect mutations in any part of the genome. This method is based on the principle of migration of DNA heteroduplexes and homoduplexes at different speed during electrophoresis at a bit denaturing conditions. The DNA fragments showing a changed peak morphology as compared to the wild type can be sequenced to obtain the exact position of a sequence variation (Ashton 2011).

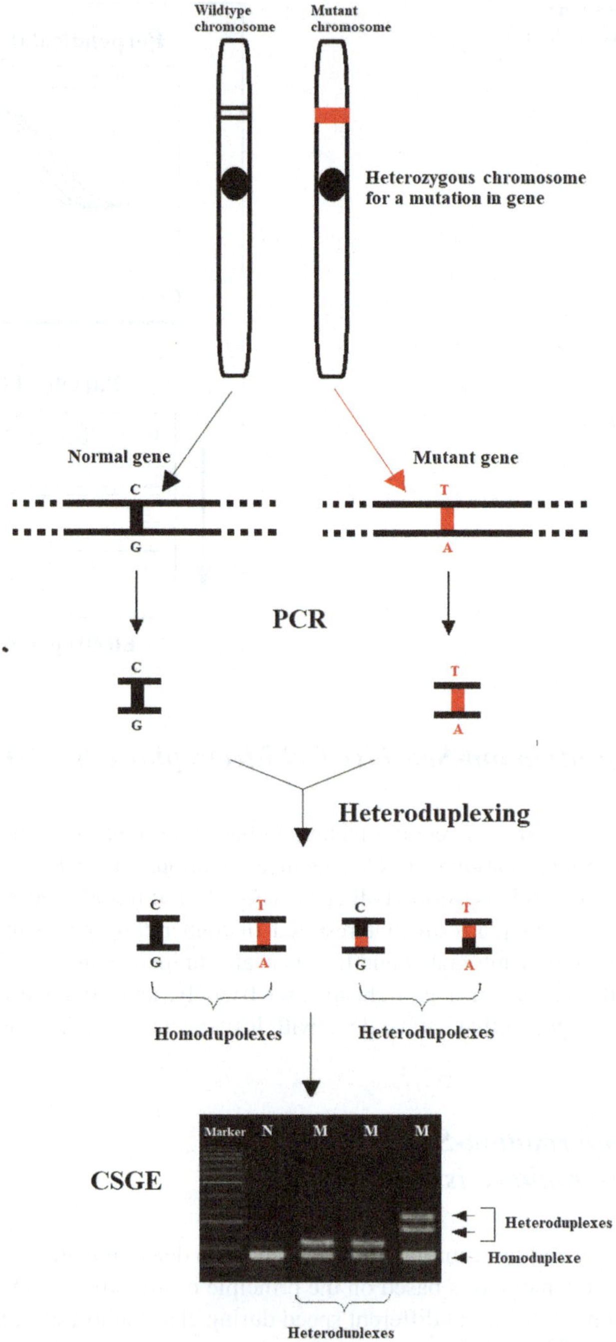

Fig. 9.3 Conformation-sensitive gel electrophoresis

9.3.5 High-Resolution DNA Melting Analysis

It is based on melting profile of a PCR product, which is determined by its GC content, heterozygosity, length, and sequence. These fragments can be traced with saturating dyes that shine in the presence of dsDNA. The melting temperature of the PCR products helps in genotyping of many mutations. Matching of sequence and mutation screening depends upon the differences in sequence that results in heteroduplexes, and ultimately form of the melting curve is changed. This technique has several advantages like low cost, fast, accurate, and homogeneous, over the other scanning and genotyping techniques (Reed et al. 2007).

9.3.6 Dot-Blot SNP Analysis

In this assay, genomic DNA is extracted from leaf tissues and is dissolved in 1 mL 0.1× TE buffer. DNA fragments covering the SNP sites are amplified with PCR, and the amplicons are denatured, and nylon membrane is used to dot-blot with multi-pin Blotter. The probe mixture containing oligonucleotide labeled with digoxigenin and unlabeled counterpart oligonucleotide is hybridized. After hybridization, the membranes are washed. Signals are detected (Shirasawa et al. 2006).

9.4 Applications Utilizing Next-Generation Sequencing

Different types of DNA mutation are caused with different mutagens resulting in different types of DNA damages in mutant populations. Therefore, different techniques are needed to identify different types of induced mutations. A variety of methods employing massively parallel sequencing provide high throughput of natural and induced mutations in cotton (Table 9.2).

9.4.1 Targeting Induced Local Lesions in Genome (TILLING)

This assay was described first time by Claire McCallum, a graduate student, and her colleagues from Fred Hutchinson Cancer Research Center in the late 1990s (McCallum et al. 2000). It is a reverse genetic technique whereby the DNAs from large mutant populations are screened to identify lesions in candidate genes. Plants harboring potentially interesting mutations are later phenotypically evaluated. A variety of different mutation discovery techniques can be used for TILLING. The first examples used denaturing HPLC. This was followed by high-throughput TILLING that utilized target-specific PCR with enzymatic mismatch cleavage

Table 9.2 Mutation screening using massively parallel DNA sequencing

Sr. #	Methods	Abbreviations	Method	Usage
1	Targeting induced local lesion in genomes (enzymatic mismatch cleavage, amplicon sequencing, exome capture)	TILLING	The DNA fragments of interest are amplified by PCR using gene-specific primers from 8× pooled DNA. Nicks are produced between a wild-type ssDNA and a mutant ssDNA These heteroduplexes are restricted using Cel1 endonuclease enzyme at a mismatch point After cleavage, PCR products are sequenced	This is a reverse-genetic method that allows recovery of deleterious alleles in specific genes
2	EcoTILLING	–	This procedure is used to detect spontaneous mutations (SNPs) in individuals or populations	It can be an important technique for the identification of SNPs in germplasm It can be used in assessing heterozygosity It can uncover mutants for disease resistance It can help in discovering the function of various genes It can regulate element by identifying natural variants It can be a good method for well-established population with thoroughly characterized morphological data
3	Sequencing of the genome	WGS, de novo assembly	Genomic DNA is fragmented, sequencing adapters ligated, and DNA sequenced. Bioinformatics tools are employed to assemble new genomes or map sequence to a reference genome. Tools are also applied to call nucleotide variations	It is used for sequencing and resequencing of genotypes for identifying SNPs associated with various traits of interest

(continued)

Table 9.2 (continued)

Sr. #	Methods	Abbreviations	Method	Usage
4	Exome capture assay	ECA	It is a genomic method used for the sequencing all of the exonic region of a genome It contains two main steps: 1. Selection of the subset of DNA that encodes proteins 2. Sequencing of the exonic DNA using any high-throughput NGS technique	This technique allows effective and relatively low-cost sequencing of hundreds to thousands of genes as compared to WGS
5	Genotype by sequencing	GBS	Restriction enzymes are used for chopping the genomic DNA for reducing the size of complex genome It can be used to genotype multiple DNA samples simultaneously After restriction, PCR is performed to maximize the fragments pool Afterward, GBS libraries are sequenced using NGS techniques resulting in about 100 bp single-end reads	It is used to identify SNPs in order to perform genotyping studies for undertaking genome-wide association studies (GWAS)

followed by fragment analysis. Limitations inherent to the sensitivity of DNA analysis meant that samples could be pooled up to eightfold to increase throughput (Till et al. 2003). TILLING assay was used first time in *Arabidopsis* TILLING Project (ATP) (Henikoff et al. 2004; Till et al. 2003). The same concept was successfully transferred on other crop species including cotton (Table 9.3). The generated mutants either can be released as a new variety or can be used in a breeding program. Also these mutations are heritable like the other traits (Dong et al. 2009; Kumar et al. 2013; Uauy et al. 2009). Also, through TILLING screening assay, recessive mutations present in heterozygous individuals can also be identified, otherwise not possible with conventional phenotyping procedures (Wilde et al. 2012). Mutations in individual homoeologous genes can be isolated independently and can be introgressed in the same line for knocking out the respective phenotype. Thus gene function can be studied (Kashtwari et al. 2019). This technique can be used to address many difficulties for genetics in diploids so that each homologue of a multigene family can be targeted independently as well combined into a single line using genetic crosses (Chen et al. 2017; Khan et al. 2018; Zafar et al. 2018; Aslam et al. 2013; Patel et al. 2014; Witt et al. 2018). TILLING requires several steps for detecting mutations (Fig. 9.4).

Table 9.3 Cotton TILLING populations

Verity	Mutagen dose	Generation	Pop size	Traits	SNPs	Reference
G. hirsutum var. Paymaster HS 200	3% EMS	M_3, M_4, and M_5	115 lines	Fiber quality and lint yield		Herring et al. (2004)
G. hirsutum var. (FiberMax 958, ACALA 1517-99, and TAM 94L-25) *G. barbadense* var. (Pima S7) *G. arboreum* var. (GA-TAMU, A2-60T, and A2-120W)	3% EMS, 0.05% Na azide and distilled H2O	M_4, M_5		Lint yields and lint percent, fiber quality		Lowery et al. (2007)
G. hirsutum var. Atlas, Tejas, SC 9023, Sphinx, Explorer, Holland 338, and Rocket	2.45% EMS	M_8 mutant lines		Oil seed contents, fiber traits including lint yield lint percentage, fibers/seed, fibers/mm^2, fiber quality, and spinning performance of lint		Bechere et al. (2009)
G. hirsutum var. MCU 5	10–50 Gy gamma rays and 1–5 mM EMS and SA	M_7	Selection of 19 mutant lines	Plant height, early flowering, bolls per plant, seed cotton yield, ginning out turn, seed index, harvest index, and fiber characters		Muthusamy and Jayabalan (2011)
G. hirsutum var. SC 9023	2.45% EMS	M_3	Stable line 9023 n^{t}_{4}	Agronomic and fiber traits		Bechere et al. (2012)
G. hirsutum var. PB-899, pb-900	0.2, 0.3% EMS			Elucidating gene functions		Aslam et al. (2013)

(continued)

Table 9.3 (continued)

Verity	Mutagen dose	Generation	Pop size	Traits	SNPs	Reference
G. hirsutum var. MCU 5 and MCU 11	Gamma irradiation (100–500 Gy), 1050 mM EMS and SA	M_2 generation for field trial while M_4 for in vivo and in vitro	Selection of 20 M_4 lines	Yield and protein content		
G. arboreum var. ANB-P, GOA-18, GOA-2, and C-118 *G. barbadense* var. PB-899, PB-900, CIM-496, and FH-113 *G. hirsutum* var. TADLA-16, TADLA-32, VPE-2 and PIMA-S2	EMS (0.1–0.8%) and two applications of γ-rays (100–800 Gy)	M_1 generation	Four genotypes from each of three species	Germination %age, plant height, number of bolls/plant, boll weight, lint yield, and ginning out turn percentage		Aslam et al. (2013)
G. hirsutum var. MD 15	3.2% EMS	M_4		Fiber strength, length, uniformity and micronaire, and seed index		Bechere et al. (2013)
G. hirsutum var. MD 15	3.2% EMS	M_4		Fiber traits, fiber length, strength, and uniformity		–
G. arboreum var. FDH786	EMS, DES, SA, and gamma rays	M_3	GaWM1, GaWM2, and GaWM3	Glossy leaves appearance, curved/wrinkled leaves, and flower size		Barozai and Husnain (2014)
G. hirsutum var. Gomal-93, Bt-131, Bt-121, and Bt-CIM-602	10, 15, 20, and 25 kR doses of gamma irradiation		Selection of 10 random plants	Plant height, # of branches per plant, # of bolls/plant, fiber micronaire value, and seed cotton yield		Khan et al. (2014)

(continued)

Table 9.3 (continued)

Verity	Mutagen dose	Generation	Pop size	Traits	SNPs	Reference
G. hirsutum lines, TAM 941-25, and Acala 151799	EMS	M_5	3164 lines	Fiber length, strength, uniformity, elongation, fineness, rd value, and lint percent		Patel et al. (2014)
G. hirsutum	5–50 Gray GR, 0.5–5.0 mM EMS and SA			Somaclonal variation		Muthusamy and Jayabalan (2014)
G. hirsutum var. Gomal-93, Bt-131, and Bt-CIM-602	0, 10, 15, 20, and 25 kR	M_2	Six genotypes	# of branches, leaves, # of flowers, bolls, seeds, and seed cotton weight plant^{-1}		Muhammad et al. (2015)
G. hirsutum var. Gomal-93, Bt-131 and Bt-CIM-602	Cobalt source at 0, 10, 15, 20, and 25 Krades	M_2	Six genotypes	Ginning out turn, and staple length, r fineness (micronaire), strength, and seed cotton yield		Aslam et al. (2016)
G. hirsutum var. PB 899 and PB-900	0.2 and 0.3% EMS	M_2	8000 plants	Morphology of leaf, pattern of branches, resistance to disease, photosynthetic lesions, and sterility of flowers	1/153 kb and 1/326 kb in var. "PB-899" and "PB-900"	Aslam et al. (2016)
G. hirsutum lines, TAM 94L25, and Acala 1517-99	EMS	M_5	3164 total lines, 106 lines with leaf and stem trichome and 55 control lines	Average boll weight (g), 1000 seed weight (g), ginning out turn, fiber upper half mean length, micronaire, bundle strength, uniformity index, elongation, reflectance, and yellowness.		Patel et al. (2016)

(continued)

Table 9.3 (continued)

Verity	Mutagen dose	Generation	Pop size	Traits	SNPs	Reference
G. hirsutum lines MD 15-Mut 13, MD 15-Mut 31, MD 15Mut 61, MD 15-Mut 89, MD 15-Mut 138	3.2% EMS			Fiber length, uniformity, and strength		Bechere et al. (2018)
G. hirsutum Raider 276 Tamcot Sphinx and TTU 774	EMS	M_4	Selection of 25 superior strains and 25 inferior strains	Plant height, # of nodes and internode, node of first fruiting branch, # of immature and mature bolls, length of taproot, # of lateral roots, distance to first and last lateral roots, seed cotton yield, ginning out turn, cotton seed yield, and lint yield		Witt et al. (2018)
G. hirsutum var. NIAB-777	Pollen irradiation method, 10 Gray	M_2	30 mutants	High yield, early maturity, resistance and/or tolerance to diseases		Aslam et al. (2018)

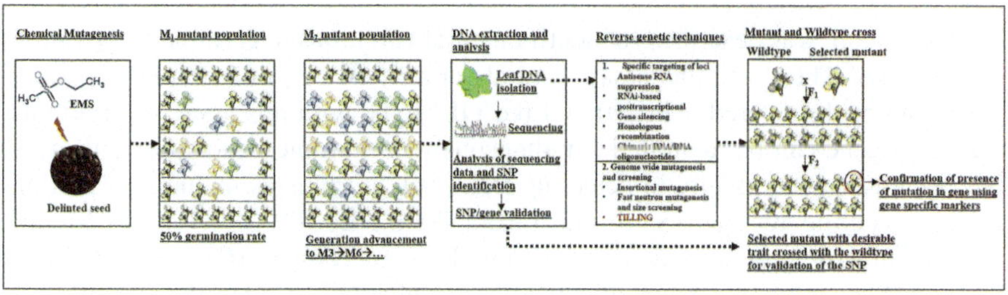

Fig. 9.4 Development of cotton mutant population for TILLING. Plants in blue, green, yellow, and orange color are mutant, while plants with no color are wild type

Basic steps involved in traditional TILLING have been described in several studies (McCallum et al. 2000; Perry et al. 2003; Till et al. 2003) and are described below:

1. Seeds are mutagenized with an optimum concentration of EMS, which causes G/C-to-A/T point mutations.
2. The M_1 population is raised. Each cotton plant is self-pollinated. From each boll, M_2 row is sown followed by scoring for the traits.
3. Leaf tissue is excised from each M_2 plant followed by extraction of genomic DNA by adopting an established protocol.
4. For screening, equal quantity of genomic DNA of eight plants is pooled for maximizing the efficiency of mutation detection.
5. Targeted genes are amplified using the end-labeled gene-specific primers. After excessive heat and cooling, heteroduplexes of the PCR products are formed.
6. These heteroduplexes showing mismatch can be cleaved with CEL I nuclease. This enzyme identifies the mismatch and chops down the genomic DNA from this particular mismatch. The resultant products can be fractionated on denaturing polyacrylamide gel electrophoresis for the detection of mutations.
7. The plants in each pool carrying the mutations are identified. These mutations are confirmed by sequencing followed by validating these mutations in F_2 population—developed by making cross between the mutant and wild type.

The advent of commercial next-generation sequencing has enabled direct sequencing of PCR products with much higher pooling, thus providing improvements to TILLING assay throughputs. This has been termed TILLING by sequencing and has been applied to a number of plant species (Gupta et al. 2017; Rigola et al. 2009; Tramontano et al. 2019; Tsai et al. 2015). Exome capture sequencing has also been applied for reverse genetics of mutant populations (Henry et al. 2014; Kettleborough et al. 2013). While the up-front costs are high, this approach is advantageous in that most coding sequence mutations can be recovered simultaneously, allowing the development of truly in silico TILLING resources (Till et al. 2018).

9.4.2 EcoTILLING

This procedure was described to identify natural variations in germplasm of various crop species. Thus with this assay, natural variants as well as function of putative genes can be discovered (Till 2014). Like TILLING, with this assay several mutations in a gene can be identified simultaneously in multiple genotypes. This assay has been used successfully in detecting DNA polymorphisms including variations in microsatellite regions (Sato and Tabata 2006). EcoTILLING was first time reported in *Arabidopsis* to identifying natural polymorphisms (Comai et al. 2004). Using this approach, a gene involved in fiber elongation was identified by silencing the GhHOX3. Suppression of this gene resulted in significant reduction of fiber length, while the overexpression of this gene conferred longer fiber length (Shan et al. 2014). In another study, EcoTILLING was used to identify SNPs in GhSus genes involved in conferring fiber and seed traits in *G. hirsutum* L. (Zeng et al. 2016). The information generated can be used for doing targeted breeding in cotton. As with ecoTILLING assays, NGS of highly pooled amplicons have been explained for evaluation of natural mutations in exomes of plants (Duitama et al. 2017; Marroni et al. 2011).

Following steps are involved intraditional EcoTILLING (Comai et al. 2004) (Fig. 9.5):

1. The genomic DNA of individual genotype is mixed with the reference DNA (1:1 ratio) followed by the amplification of around 1 kbp with asymmetrically labeled fluorescent primers.
2. After exposing the amplicon with heating and annealing, heteroduplexes are formed at the mismatched sites, and these heteroduplexes are recognized by the endonuclease CEL1 (restriction enzyme). The resultant restricted strands are visualized using Li-Cor gel analyzer.
3. The putative variations identified in one fluorescence channel which can be verified by the appearance of opposite restricted strand in the other channel.
4. The identified polymorphisms/variations in Li-Cor gel analyzer can be verified by sequencing the PCR product. Sequencing will help in identifying base pair changes, small insertions, and deletions.

9.4.3 Whole Genome Sequencing of Cotton Genome

The *Arabidopsis* genome was released in 2000, with priority given to it compared to other plants because of its small genome size and use as a model organism. Functional annotation of the complement of *Arabidopsis* genes continued after release of the reference sequence. This information opened a new gateway for making comparisons of the full or partial sequence information of several other eukaryotes; thus putative function for many genes was assigned. After sequencing the *Arabidopsis* genome, several other genomes including rice, papaya, poplar, grapes, cotton, etc. were sequenced using first-, second-, and third-generation sequencing technologies. Genome assemblies of *Gossypium raimondii* (0.88 Gb genome size), *Gossypium arboreum* (1.74 Gb), *Gossypium* hirsutum (2.3 Gb), and *Gossypium barbadense* (2.2 Gb) have been constructed (Li et al. 2014, 2015; Wang et al. 2019). Their sequence information resulted in estimating several protein coding genes, for example, 40,976, 41,333, 72,761, and 75,071 protein-coding genes were identified in *G. raimondii*, *G. arboreum*, *G. hirsutum*, and *G. barbadense,* respectively (Li et al. 2014, 2015; Wang et al. 2019). Recently, Hu and co-workers resequenced *G. barbadense* and *G. hirsutum* genome to look into the origin and evolution of allotetraploid cotton (Hu et al. 2019). Thus all these newly developed assemblies offer opportunities for resequencing cotton genotypes which can unravel new genes, alleles, etc. The cost of sequencing has been reduced substantially; thus one can

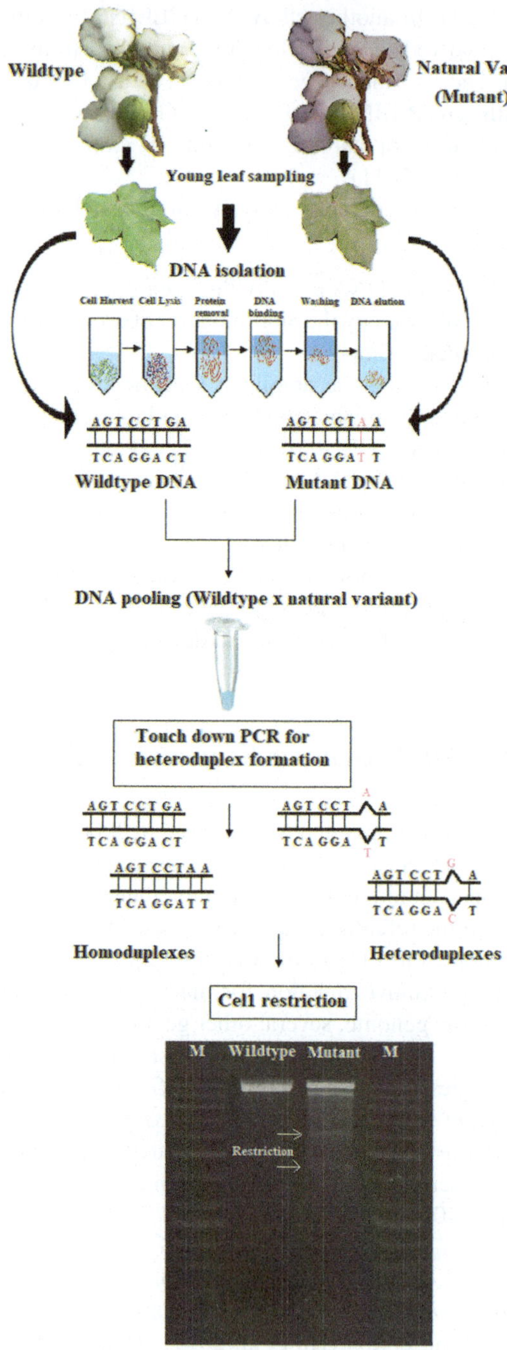

Fig. 9.5 EcoTILLING in cotton

sequence the genome of a mutant for identifying SNPs or new genes. After the identification of induced mutations (SNPs) in the genome, these SNPs can be filtered out to few regions especially the genic or expressed part of the genome for studying their impact on altering the amino acid sequences. Later on, these SNPs can be validated by developing an F_2 population of the mutant with its wild type. Once the allelic variations are confirmed, these can be exploited in molecular breeding programs by designing gene-specific primers for low-cost molecular markers. In cotton, bulked segregant analysis (BSA) based on next-generation sequencing was used and discovered a recessive nulliplex-branch gene that was mapped between two SNP markers which were ~600 kb apart (Chen et al. 2015). In another study, three QTLs conferring three fiber quality traits were validated and further fine-mapped with 27 SNP markers using the sequencing information of cotton genome (Islam et al. 2016b).

9.4.4 Exome Capture Assay/Targeted Sequence Capture

This assay enables efficient and relatively inexpensive sequencing of exons, the genomic sequences that translate into proteins, of all genes from many individuals (Fig. 9.6). The active part of the genome usually makes around 1–2% of the total genome. This assay has also been deployed for capturing and sequencing both members of each gene pair (homeologs) of wild and its derived mutants of upland cotton. In this particular study, 1000 genes (500 homeolog pairs) were targeted by hybridizing with probes—designed from the cotton transcriptomic data. These samples showed high coverage of the targeted genes, and substantial amount of flanking regions (untranslated regions and introns) were also captured along with the targeted exonic regions. In this study, low heterozygosity was found between the wild and cultivated upland cotton—owing to long inbreeding in natural *G. hirsutum*. Also, the extent of heterozygosity was found asymmetrical between the At and Dt of cultivated cotton (Salmon et al. 2012). This approach may be applied for targeting exonic regions of several variants of cotton species which would lead to identify new genes and their function. As mentioned above, exome capture sequencing can also be integrated into TILLING projects.

9.4.5 Genotype by Sequencing (GBS)

Several techniques have been established for the validation and determination of SNPs occurring naturally or induced through mutagens. Each technique has its own merits and demerits, but GBS is the most widely and commonly used technique in cotton. It is inexpensive and precise and requires less purification of DNA; that is why commonly used for the identification of SNPs (Davey et al. 2011). In GBS, genomic DNA is extracted, and after calculating the DNA concentration, it is

Fig. 9.6 Exome capture assay in cotton

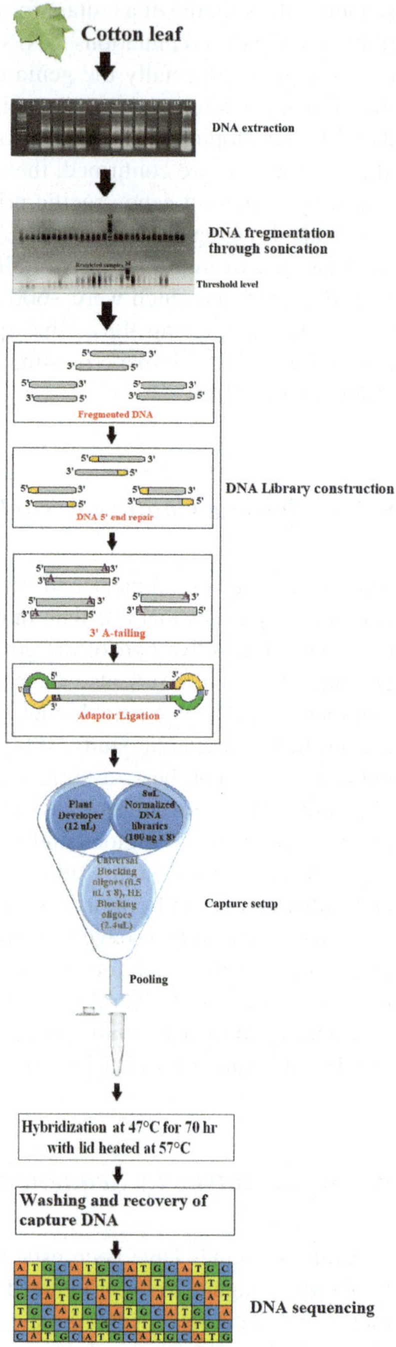

subjected for double digestion with two restriction enzymes. The selection of enzyme is very vital step in GBS. For the analysis of methylated DNA, methylation-sensitive restriction enzyme (MSRE) is used, and restricted DNA is ligated with PCR-specific adaptors. Multiple bioinformatics analyses are done on raw sequencing data for the identification and validation of SNPs, and these SNPs are further annotated for understanding their functions (Elshire et al. 2011). Only demerit of GBS is that it can miss few genomic regions as used restriction enzymes did not chop in these regions. Thus, the method is not ideal for gene-based approaches such as TILLING, but can be useful to establish the frequency of induced mutations in a genome (Monson-Miller et al. 2012; Kim et al. 2016).

In cotton, various SNPs related to fiber length and quality have been identified through GBS. In total, 71 QTLs associated with fiber quality traits were identified using cotton 63 K SNP array that were strongly associated with SNP markers. These 77 QTLs include 19 e-QTLs, 7 clusters of pleiotropic QTLs, 9 new QTLs, and 5 hotspots (Li et al. 2016a, b). Using these SNPs markers, the localization of chromosomes, linkage mapping, and characterization of phylogenomic of six MYB genes were carried out in four cotton species. These genes are vigorously involved in the development of cotton fiber. This helps in determining 108 SNPs for all of these genes (An et al. 2008). In total, 107 introgression lines developed through an interspecific cross between *G. hirsutum* and *G. tomentosum* were used for QTL mapping. In this study, in total 74 QTLs and also 5 clusters were identified, and these were found to be associated with various fiber quality traits (Keerio et al. 2018). In another study, 5617 SNPs have been validated and identified (Islam et al. 2015). This group has also described 6071 SNPs and 86 QTLs associated with GhRBB1_A07 gene. This experiment demonstrated the potential role of this gene in determining fiber quality. This gene was localized using a population developed via multi-parent advanced generation inter-cross (MAGIC) of *G. hirsutum* parents (Islam et al. 2016a).

The GBS technique has also been used on an interspecific population derived from a cross between *G. hirsutum* and *G. tomentosum*. In total 10,888 SNPs were mapped for identifying QTLs conferring drought tolerance. In total, 34,402 and 32,032 genes associated with drought tolerance were also identified in Dt and At sub-genomes using GBS, respectively (Magwanga et al. 2018). In another study, allelic diversity associated with leaf transcriptomes was identified using GBS technique in *G. barbadense* (Kottapalli et al. 2016). Similarly, many SNPs associated with various agronomic and biochemical traits in cotton were identified using GBS (Logan-Young et al. 2015).

9.5 Validation of Identified Mutations Through Sequencing

To validate the mutant alleles/genes, it is important to develop a mapping population by crossing mutant genotype with its wild-type version. After scoring the population for the mutant phenotype, plants are surveyed for mutant alleles/genes. This can be done on a candidate gene or whole genome level (Abe et al. 2012). Thereafter, if strong association between mutant phenotype and genotype exists, this suggests that mutant allele is responsible for mutant trait. Alternatively, quantitative reverse transcription PCR (qRT-PCR) was used to validate 12 genes involved in carbohydrate metabolism in wild-type cotton genotype TM-1 and its mutant version IM mutant (Wang et al. 2014).

9.6 Conclusions

Growing pressures on agriculture call for increased productivity in a changing environment. Genetic gain remains an important component for increasing productivity. Yet for many crops such as cotton, available genetic variation is limited or associated with unwanted linkage drag. Breeders are handicapped due to several inherent problems, for example, interaction of complex traits with the environment may lead to the selection of undesirable plants. Mutagenesis has proven to be a powerful solution for these problems. Novel genetic variation can be created orders of magnitude faster than spontaneous mutations. New mutations can be created in elite cultivars to avoid unwanted linked alleles. Advances in DNA sequencing suggest the induction and use of induced mutations will be much more efficient than in the past. Mutagenic treatments can now be fine-tuned. For example, mutagenic conditions favoring alleles causing major phenotypic variations can be easily identified using automated field phenomics screens. The resulting alleles from forward genetic screens can be efficiently combined and desired genetic backgrounds selected using tools such as genotyping by sequencing. Concomitantly, mutant populations designed for highly efficient reverse genetics can also be prepared. Together these resources will provide validated gene functions and alleles that can be directly incorporated into breeding programs. The tools exist for application in cotton breeding. A future can be envisioned where a database of new alleles exists with annotated biological functions such as drought tolerance, disease resistance, and fiber quality that can be rapidly introgressed into elite germplasm to counteract the negative effects of climate change and variation.

Acknowledgments We acknowledged the support of International Atomic Energy Agency (IAEA), Vienna, Austria, through an umbrella project entitled "Developing Germplasm through TILLING in Crop Plants Using Mutation and Genomic Approaches" (PAK/5/047) for establishing the facility of mutagenesis in NIBGE.

References

Abe A, Kosugi S, Yoshida K, Natsume S, Takagi H, Kanzaki H, Matsumura H, Yoshida K, Mitsuoka C, Tamiru M et al (2012) Genome sequencing reveals agronomically important loci in rice using MutMap. *Nat Biotechnol* 30:174–178

An C, Saha S, Jenkins JN, Ma D-P, Scheffler BE, Kohel RJ, John ZY, Stelly DM (2008) Cotton (Gossypium spp.) R2R3-MYB transcription factors SNP identification, phylogenomic characterization, chromosome localization, and linkage mapping. *Theor Appl Genet* 116:1015–1026

Andrews JH, Briand LC, Labiche Y (2005) Is mutation an appropriate tool for testing experiments? In: Proceedings of the 27th international conference on Software engineering, ACM, pp 402–411

Arshad M, Khan MI, Ali C, Afzal M, Rahman M-u (2009) Registration of 'CIM-496' cotton. *J Plant Regist* 3:231–235

Ashton EJ (2011) Conformation-sensitive capillary electrophoresis. In: Theophilus BDM, Rapley R (eds) PCR mutation detection protocols. Humana Press, Totowa, pp 1–6

Aslam U, Ali Khan A, Naseer Cheema HM, Imtiaz F, Malik W (2013) Kill curve analysis and response of ethyl methanesulfonate and γ-rays in diploid and tetraploid cotton. *Int J Agric Biol* 15:11–18

Aslam U, Cheema H, Ahmad S, Khan IA, Malik W, Khan AA (2016) COTIP: cotton TILLING platform, a resource for plant improvement and reverse genetic studies. *Front Plant Sci* 7:1863

Aslam M, Haq M, Bandesha A, Haidar S (2018) NIAB-777: an early maturing, high yielding and better quality cotton mutant developed through pollen irradiation technique-suitable for high density planting. *J Anim Plant Sci* 28:636–646

Balkan A, Bilgin O, Başer İ, Göçmen DB, Demirkan AK, Deviren B (2019) Improvement of grain yield and yield associated traits in bread wheat (Triticum aestivum L.) genotypes through mutation breeding Using gamma irradiation. *Tekirdağ Ziraat Fakültesi Dergisi* 16:103–111

Barkley N, Wang M (2008) Application of TILLING and EcoTILLING as reverse genetic approaches to elucidate the function of genes in plants and animals. *Curr Genomics* 9:212–226

Barozai MYK, Husnain T (2014) Development and characterization of the asiatic desi cotton (Gossypium arboreum L.) leaf epicuticular wax mutants. *Pak J Bot* 46:639–643

Bechere E, Auld D, Hequet E (2009) Development of 'naked-tufted' seed coat mutants for potential use in cotton production. *Euphytica* 167:333–339

Bechere E, Turley RB, Auld DL, Zeng L (2012) A new fuzzless seed locus in an upland cotton (Gossypium hirsutum L.) mutant. *Am J Plant Sci* 3:799

Bechere E, Meredith WR, Boykin JC (2013) Registration of mutant population MD 15 M4 Gossypium hirsutum L. with enhanced fiber quality. *J Plant Regist* 7:216–219

Bechere E, Zeng L, Auld D (2018) Registration of five upland cotton mutant germplasm lines with superior fiber length, strength, and uniformity. *J Plant Regist* 12:107–111

Caldwell DG, McCallum N, Shaw P, Muehlbauer GJ, Marshall DF, Waugh R (2004) A structured mutant population for forward and reverse genetics in Barley (Hordeum vulgare L.). *Plant J* 40:143–150

Chawade A, Sikora P, Bräutigam M, Larsson M, Vivekanand V, Nakash MA, Chen T, Olsson O (2010) Development and characterization of an oat TILLING-population and identification of mutations in lignin and β-glucan biosynthesis genes. *BMC Plant Biol* 10:86

Chen W, Yao J, Chu L, Yuan Z, Li Y, Zhang Y (2015) Genetic mapping of the nulliplex-branch gene (gb_nb1) in cotton using next-generation sequencing. *Theor Appl Genet* 128:539–547

Chen X, Lu X, Shu N, Wang S, Wang J, Wang D, Guo L, Ye W (2017) Targeted mutagenesis in cotton (Gossypium hirsutum L.) using the CRISPR/Cas9 system. *Sci Rep* 7:44304

Comai L, Henikoff S (2006) TILLING: practical single-nucleotide mutation discovery. *Plant J* 45:684–694

Comai L, Young K, Till BJ, Reynolds SH, Greene EA, Codomo CA, Enns LC, Johnson JE, Burtner C, Odden AR (2004) Efficient discovery of DNA polymorphisms in natural populations by Ecotilling. *Plant J* 37:778–786

Datta S, Jankowicz-Cieslak J, Nielen S, Ingelbrecht I, Till BJ (2018) Induction and recovery of copy number variation in banana through gamma irradiation and low-coverage whole-genome sequencing. *Plant Biotechnol J* 16(9):1644–1653

Davey JW, Hohenlohe PA, Etter PD, Boone JQ, Catchen JM, Blaxter ML (2011) Genome-wide genetic marker discovery and genotyping using next-generation sequencing. *Nat Rev Genet* 12:499

de Vries H (1901) Die mutationstheorie. Versuche und beobachtungen über die entstehung von arten im pflanzenreich. Veit & Comp., Leipzig

Dong C, Dalton-Morgan J, Vincent K, Sharp P (2009) A modified TILLING method for wheat breeding. *Plant Genome* 2:39–47

Duitama J, Kafuri L, Tello D, Leiva AM, Hofinger B, Datta S, Lentini Z, Aranzales E, Till B, Ceballos H (2017) Deep assessment of genomic diversity in cassava for herbicide tolerance and starch biosynthesis. *Comput Struct Biotechnol J* 15:185–194

El-Kameesy S, Salama E, Ghannam M, Roshdy S (2019) The influence of gamma irradiation doses on the morphological and physical properties of wheat. *J Phys Conf Ser* 1253:012031. IOP Publishing

Elshire RJ, Glaubitz JC, Sun Q, Poland JA, Kawamoto K, Buckler ES, Mitchell SE (2011) A robust, simple genotyping-by-sequencing (GBS) approach for high diversity species. *PLoS One* 6:e19379

Ganguly A (2002) An update on conformation sensitive gel electrophoresis. *Hum Mutat* 19:334–342

Gasser RB, Hu M, Chilton NB, Campbell BE, Jex AJ, Otranto D, Cafarchia C, Beveridge I, Zhu X (2006) Single-strand conformation polymorphism (SSCP) for the analysis of genetic variation. *Nat Protoc* 1:3121–3128

Gottwald S, Bauer P, Komatsuda T, Lundqvist U, Stein N (2009) TILLING in the two-rowed barley cultivar "Barke" reveals preferred sites of functional diversity in the gene HvHox1. BMC Res. Notes 2, 258. In: Gilchrist E, Haughn G (eds) 2013. Gene identification: reverse genetics, diagnostics in plant breeding. Springer, pp 61–89

Greene EA, Codomo CA, Taylor NE, Henikoff JG, Till BJ, Reynolds SH, Enns LC, Burtner C, Johnson JE, Odden AR (2003) Spectrum of chemically induced mutations from a large-scale reverse-genetic screen in Arabidopsis. *Genetics* 164:731–740

Gupta P, Reddaiah B, Salava H, Upadhyaya P, Tyagi K, Sarma S, Datta S, Malhotra B, Thomas S, Sunkum A et al (2017) Next-generation sequencing (NGS)-based identification of induced mutations in a doubly mutagenized tomato (Solanum lycopersicum) population. *Plant J Cell Mol Biol* 92:495–508

Haughn GW, Gilchrist EJ (2006) TILLING in the Botanical Garden: a reverse genetic technique feasible for all plant species. Floriculture, ornamental and plant biotechnology: advances and topical issues, 1st edn. Global Science Books, London, pp 476–482

Henikoff S, Till BJ, Comai L (2004) TILLING. Traditional mutagenesis meets functional genomics. *Plant Physiol* 135:630–636

Henry IM, Nagalakshmi U, Lieberman MC, Ngo KJ, Krasileva KV, Vasquez-Gross H, Akhunova A, Akhunov E, Dubcovsky J, Tai TH et al (2014) Efficient genome-wide detection and cataloging of EMS-induced mutations using exome capture and next-generation sequencing. *Plant Cell* 26:1382–1397

Herring AD, Auld DL, Ethridge MD, Hequet EF, Bechere E, Green CJ, Cantrell RG (2004) Inheritance of fiber quality and lint yield in a chemically mutated population of cotton. *Euphytica* 136:333–339

Hershey AD, Chase M (1952) Independent functions of viral protein and nucleic acid in growth of bacteriophage. *J Gen Physiol* 36:39–56

Hill M (2011) Conformation sensitive gel electrophoresis. In: Theophilus BDM, Rapley R (eds) PCR mutation detection protocols. Humana Press, Totowa, pp 7–16

Hu Y, Chen J, Fang L, Zhang Z, Ma W, Niu Y, Ju L, Deng J, Zhao T, Lian J, Baruch K, Fang D, Liu X, Ruan Y-l, Rahman M-u, Han J, Wang K, Wang Q, Wu H, Mei G, Zang Y, Han Z, Xu C, Shen W, Yang D, Si Z, Dai F, Zou L, Huang F, Bai Y, Zhang Y, Brodt A, Ben-Hamo H, Zhu X, Zhou B, Guan X, Zhu S, Chen X, Zhang T (2019) Gossypium barbadense and Gossypium hirsutum genomes provide insights into the origin and evolution of allotetraploid cotton. *Nat Genet* 51:739–748

Hussain M, Rahman M (2019) Registration of PGMB-15-30 Spring Wheat. *J Plant Regist* 13(2): 245–250

Hussain M, Iqbal MA, Till BJ, Rahman M (2018) Identification of induced mutations in hexaploid wheat genome using exome capture assay. *PLoS One* 13:e0201918

Islam MS, Thyssen GN, Jenkins JN, Fang DD (2015) Detection, validation, and application of genotyping-by-sequencing based single nucleotide polymorphisms in upland cotton. *Plant Genome* 8(1):1–10

Islam MS, Thyssen GN, Jenkins JN, Zeng L, Delhom CD, McCarty JC, Deng DD, Hinchliffe DJ, Jones DC, Fang DD (2016a) A MAGIC population-based genome-wide association study reveals functional association of GhRBB1_A07 gene with superior fiber quality in cotton. *BMC Genomics* 17:903

Islam MS, Zeng L, Thyssen GN, Delhom CD, Kim HJ, Li P, Fang DD (2016b) Mapping by sequencing in cotton (Gossypium hirsutum) line MD52ne identified candidate genes for fiber strength and its related quality attributes. *Theor Appl Genet* 129:1071–1086

Jankowicz-Cieslak J, Till BJ (2016) Chemical mutagenesis of seed and vegetatively propagated plants using EMS. *Curr Protoc Plant Biol* 1:617–635

Jankowicz-Cieslak J, Mba C, Till BJ (2017) Mutagenesis for crop breeding and functional genomics. In: Biotechnologies for plant mutation breeding. Springer, pp 3–18

Kashtwari M, Wani AA, Rather RN (2019) TILLING: an alternative path for crop improvement. *J Crop Improv* 33:83–109

Kaul M, Bhan A (1977) Mutagenic effectiveness and efficiency of EMS, DES and gamma-rays in rice. *Theor Appl Genet* 50:241–246

Keerio A, Shen C, Nie Y, Ahmed M, Zhang X, Lin Z (2018) QTL mapping for fiber quality and yield traits based on introgression lines derived from Gossypium hirsutum× G. tomentosum. *Int J Mol Sci* 19:243

Kettleborough RNW, Busch-Nentwich EM, Harvey SA, Dooley CM, de Bruijn E, van Eeden F, Sealy I, White RJ, Herd C, Nijman IJ et al (2013) A systematic genome-wide analysis of zebrafish protein-coding gene function. *Nature* 496:494–497

Khan S, Khan MI, Muhammad A, Khan F, Khan RA (2014) Gamma rays induced variations in some cotton (Gossypium hirsutum L.) genotypes and their evaluation in the environment of Bannu. *Int J* 2:512–518

Khan Z, Khan SH, Mubarik MS, Ahmad A (2018) Targeted genome editing for cotton improvement, past, present and future trends in cotton breeding. IntechOpen

Kim Y, Schumaker KS, Zhu J-K (2006) EMS mutagenesis of Arabidopsis. In: Arabidopsis protocols, pp 101–103

Kim C, Guo H, Kong W, Chandnani R, Shuang L-S, Paterson AH (2016) Application of genotyping by sequencing technology to a variety of crop breeding programs. *Plant Sci* 242:14–22

Kodym A, Afza R (2003) Physical and chemical mutagenesis, Plant functional genomics. Springer, pp 189–203

Kottapalli P, Ulloa M, Kottapalli KR, Payton P, Burke J (2016) SNP marker discovery in Pima cotton (Gossypium barbadense L.) leaf transcriptomes. Genomics insights 9, GEI. S40377

Kumar AP, Boualem A, Bhattacharya A, Parikh S, Desai N, Zambelli A, Leon A, Chatterjee M, Bendahmane A (2013) SMART–sunflower mutant population and reverse genetic tool for crop improvement. *BMC Plant Biol* 13:38

Li F, Fan G, Wang K, Sun F, Yuan Y, Song G, Li Q, Ma Z, Lu C, Zou C (2014) Genome sequence of the cultivated cotton Gossypium arboreum. *Nat Genet* 46:567

Li F, Fan G, Lu C, Xiao G, Zou C, Kohel RJ, Ma Z, Shang H, Ma X, Wu J (2015) Genome sequence of cultivated Upland cotton (Gossypium hirsutum TM-1) provides insights into genome evolution. *Nat Biotechnol* 33:524

Li C, Dong Y, Zhao T, Li L, Li C, Yu E, Mei L, Daud M, He Q, Chen J (2016a) Genome-wide SNP linkage mapping and QTL analysis for fiber quality and yield traits in the upland cotton recombinant inbred lines population. *Front Plant Sci* 7:1356

Li S, Zheng Y, Cui H, Fu H, Shu Q, Huang J (2016b) Frequency and type of inheritable mutations induced by γ rays in rice as revealed by whole genome sequencing. *J Zhejiang Univ Sci B* 17:905–915

Li G, Jain R, Chern M, Pham NT, Martin JA, Wei T, Schackwitz WS, Lipzen AM, Duong PQ, Jones KC et al (2017a) The sequences of 1504 mutants in the model rice variety Kitaake facilitate rapid functional genomic studies. *Plant Cell* 29:1218–1231
Li C, Unver T, Zhang B (2017b) A high-efficiency CRISPR/Cas9 system for targeted mutagenesis in Cotton (Gossypium hirsutum L.). *Sci Rep* 7:43902
Li F, Shimizu A, Nishio T, Tsutsumi N, Kato H (2019) Comparison and characterization of mutations induced by gamma-ray and carbon-ion irradiation in rice (Oryza sativa L.) using whole-genome resequencing. *G3 (Bethesda)* 9(11):3743–3751
Logan-Young CJ, Yu JZ, Verma SK, Percy RG, Pepper AE (2015) SNP discovery in complex allotetraploid genomes (Gossypium spp., Malvaceae) using genotyping by sequencing. *Appl Plant Sci* 3:1400077
Lowery CC, Auld D, Bechere E, Wright R, Hequet E, Abidi N, Smith C (2007) Use of chemical mutagenesis in improving upland cotton. In: World cotton research conference-4
Magwanga R, Lu P, Kirungu J, Diouf L, Dong Q, Hu Y, Cai X, Xu Y, Hou Y, Zhou Z (2018) GBS mapping and analysis of genes conserved between Gossypium tomentosum and Gossypium hirsutum cotton cultivars that respond to drought stress at the seedling stage of the BC2F2 generation. *Int J Mol Sci* 19:1614
Marroni F, Pinosio S, Di Centa E, Jurman I, Boerjan W, Felice N, Cattonaro F, Morgante M (2011) Large-scale detection of rare variants via pooled multiplexed next-generation sequencing: towards next-generation Ecotilling. *Plant J Cell Mol Biol* 67:736–745
McCallum CM, Comai L, Greene EA, Henikoff S (2000) Targeting induced local lesions in genomes (TILLING) for plant functional genomics. *Plant Physiol* 123:439–442
Monson-Miller J, Sanchez-Mendez DC, Fass J, Henry IM, Tai TH, Comai L (2012) Reference genome-independent assessment of mutation density using restriction enzyme-phased sequencing. *BMC Genomics* 13:72
Muhammad A, Wazir SM, Ullah H, Ullah R, Afridi S (2015) Performance of selected irradiated cotton varieties during M2 generation. *Am Eurasian J Agric Environ Sci* 15:154–160
Muller HJ (1930) Types of visible variations induced by X-rays in Drosophila. *J Genet* 22:299–334
Muthusamy A, Jayabalan N (2011) In vitro induction of mutation in cotton (Gossypium hirsutum L.) and isolation of mutants with improved yield and fiber characters. *Acta Physiol Plant* 33:1793–1801
Muthusamy A, Jayabalan N (2014) Radiation and chemical mutagen induced somaclonal variations through in vitro organogenesis of cotton (Gossypium hirsutum L.). *Int J Radiat Biol* 90:1229–1239
Naito K, Kusaba M, Shikazono N, Takano T, Tanaka A, Tanisaka T, Nishimura M (2005) Transmissible and nontransmissible mutations induced by irradiating Arabidopsis thaliana pollen with gamma-rays and carbon ions. *Genetics* 169:881–889
Parry MA, Madgwick PJ, Bayon C, Tearall K, Hernandez-Lopez A, Baudo M, Rakszegi M, Hamada W, Al-Yassin A, Ouabbou H (2009) Mutation discovery for crop improvement. *J Exp Bot* 60:2817–2825
Patel JD, Wright RJ, Auld D, Chandnani R, Goff VH, Ingles J, Pierce GJ, Torres MJ, Paterson AH (2014) Alleles conferring improved fiber quality from EMS mutagenesis of elite cotton genotypes. *Theor Appl Genet* 127:821–830
Patel JD, Wright RJ, Chandnani R, Goff VH, Ingles J, Paterson AH (2016) EMS-mutated cotton populations suggest overlapping genetic control of trichome and lint fiber variation. *Euphytica* 208:597–608
Perry JA, Wang TL, Welham TJ, Gardner S, Pike JM, Yoshida S, Parniske M (2003) A TILLING reverse genetics tool and a web-accessible collection of mutants of the legume Lotus japonicus. *Plant Physiol* 131:866–871
Radhakrishnan S (2017) Sustainable cotton production, sustainable fibres and textiles. Elsevier, pp 21–67
Rahman M-U (2002) DNA marker studies in cotton. University of Agriculture, Faisalabad
Rahman M-u, Zafar Y (2007) Registration of NIBGE-115 cotton. *J Plant Regist* 1:51–52
Rahman M-u, Zafar Y (2012) Registration of NN-3 cotton. *J Plant Regist* 6:342–347

Rawat N, Sehgal SK, Joshi A, Rothe N, Wilson DL, McGraw N, Vadlani PV, Li W, Gill BS (2012) A diploid wheat TILLING resource for wheat functional genomics. *BMC Plant Biol* 12:205

Reed GH, Kent JO, Wittwer CT (2007) High-resolution DNA melting analysis for simple and efficient molecular diagnostics. *Pharmacogenomics* 8(6):597–608

Rigola D, van Oeveren J, Janssen A, Bonné A, Schneiders H, van der Poel HJA, van Orsouw NJ, Hogers RCJ, de Both MTJ, van Eijk MJT (2009) High-throughput detection of induced mutations and natural variation using KeyPoint technology. *PLoS One* 4:e4761

Sabetta W, Alba V, Blanco A, Montemurro C (2011) sunTILL: a TILLING resource for gene function analysis in sunflower. *Plant Methods* 7:1

Salmon A, Udall JA, Jeddeloh JA, Wendel J (2012) Targeted capture of homoeologous coding and noncoding sequence in polyploid cotton. *G3: Genes, Genomes, Genetics* 2:921–930

Sato S, Tabata S (2006) Lotus japonicus as a platform for legume research. *Curr Opin Plant Biol* 9:128–132

Shah TM, Mirza JI, Haq MA, Atta BM (2008) Induced genetic variability in chickpea (Cicer arietinum L.). II. Comparative mutagenic effectiveness and efficiency of physical and chemical mutagens. *Pak. J Bot* 40:605–613

Shan C-M, Shangguan X-X, Zhao B, Zhang X-F, Chao L-m, Yang C-Q, Wang L-J, Zhu H-Y, Zeng Y-D, Guo W-Z (2014) Control of cotton fibre elongation by a homeodomain transcription factor GhHOX3. *Nat Commun* 5:5519

Sheffield VC, Beck JS, Kwitek AE, Sandstrom DW, Stone EM (1993) The sensitivity of single-strand conformation polymorphism analysis for the detection of single base substitutions. *Genomics* 16:325–332

Shirasawa K, Shiokai S, Yamaguchi M, Kishitani S, Nishio T (2006) Dot-blot-SNP analysis for practical plant breeding and cultivar identification in rice. *Theor Appl Genet* 113:147–155

Shu Q-Y, Forster BP, Nakagawa H, Nakagawa H (2012) Plant mutation breeding and biotechnology. CABI

Shuli F, Jarwar AH, Wang X, Wang L, Ma Q (2018) Overview of the cotton in Pakistan and its future prospects. *Pak J Agric Res* 31:291–418

Stadler LJ (1928) Genetic effects of X-rays in maize. *Proc Natl Acad Sci* 14:69–75

Tester M, Langridge P (2010) Breeding technologies to increase crop production in a changing world. *Science* 327:818–822

Till BJ (2014) Mining genetic resources via Ecotilling. In: Tuberosa R, Graner A, Frison E (eds) Genomics of plant genetic resources, vol 1. Managing, sequencing and mining genetic resources. Springer, Dordrecht, pp 349–365

Till BJ, Reynolds SH, Greene EA, Codomo CA, Enns LC, Johnson JE, Burtner C, Odden AR, Young K, Taylor NE (2003) Large-scale discovery of induced point mutations with high-throughput TILLING. *Genome Res* 13:524–530

Till BJ, Cooper J, Tai TH, Colowit P, Greene EA, Henikoff S, Comai L (2007) Discovery of chemically induced mutations in rice by TILLING. *BMC Plant Biol* 7:1

Till BJ, Datta S, Jankowicz-Cieslak J (2018) TILLING: the next generation. *Adv Biochem Eng Biotechnol* 164:139–160

Tramontano A, Jarc L, Jankowicz-Cieslak J, Hofinger BJ, Gajek K, Szurman-Zubrzycka M, Szarejko I, Ingelbrecht I, Till BJ (2019) Fragmentation of pooled PCR products for highly multiplexed TILLING. *G3 Genes Genomes Genetics* 9:2657–2666

Tsai H, Ngo K, Lieberman M, Missirian V, Comai L (2015) Tilling by sequencing. *Methods Mol Biol* 1284:359–380. Clifton NJ

Uauy C, Paraiso F, Colasuonno P, Tran RK, Tsai H, Berardi S, Comai L, Dubcovsky J (2009) A modified TILLING approach to detect induced mutations in tetraploid and hexaploid wheat. *BMC Plant Biol* 9:1

Viglasky V (2013) Polyacrylamide temperature gradient gel electrophoresis. In: Makovets S (ed) DNA electrophoresis: methods and protocols. Humana Press, Totowa, pp 159–171

Wang C, Lv Y, Xu W, Zhang T, Guo W (2014) Aberrant phenotype and transcriptome expression during fiber cell wall thickening caused by the mutation of the Im gene in immature fiber (im) mutant in Gossypium hirsutum L. *BMC Genomics* 15:94

Wang M, Tu L, Yuan D, Zhu D, Shen C, Li J, Liu F, Pei L, Wang P, Zhao G (2019) Reference genome sequences of two cultivated allotetraploid cottons, Gossypium hirsutum and Gossypium barbadense. *Nat Genet* 51:224

Weng M-L, Becker C, Hildebrandt J, Neumann M, Rutter MT, Shaw RG, Weigel D, Fenster CB (2019) Fine-grained analysis of spontaneous mutation spectrum and frequency in Arabidopsis thaliana. *Genetics* 211:703–714

Wilde HD, Chen Y, Jiang P, Bhattacharya A (2012) Targeted mutation breeding of horticultural plants. *Emirates J Food Agric* 24:31

Witt TW, Ulloa M, Pelletier MG, Mendu V, Ritchie GL (2018) Exploring ethyl methanesulfonate (EMS) treated cotton (Gossypium hirsutum L.) to improve drought tolerance. *Euphytica* 214:123

Zafar SA, Noor MA, Waqas MA, Wang X, Shaheen T, Raza M (2018) Temperature extremes in cotton production and mitigation strategies, past, present and future trends in cotton breeding. IntechOpen

Zeng Y-D, Sun J-L, Bu S-H, Deng K-S, Tao T, Zhang Y-M, Zhang T-Z, Du X-M, Zhou B-L (2016) EcoTILLING revealed SNPs in GhSus genes that are associated with fiber-and seed-related traits in upland cotton. *Sci Rep* 6:29250

Part III
Past, Present and Future Use of GM Cotton

Chapter 10
First-Generation Transgenic Cotton Crops

Aftab Ahmad, Muhammad Zubair Ghouri, Amer Jamil, Sultan Habibullah Khan, Niaz Ahmad, and Mehboob-ur-Rahman

10.1 Need for the Bt-Cotton

Cotton is cultivated in more than 80 countries on ~32.80 million hectares with an average production of 83.7 million bales (Shodhganga 2019). Cotton is a valuable fiber crop with high economic worth as it contributes ~500 billion US$ (Rahman et al. 2012). Cotton is also used in cotton buds, currency notes, and X-rays. Additionally, cottonseed is used for the extraction of oil for edible purposes, while the seed cake is a major component of livestock feed. Major cotton-producing countries are China, the USA, India, Pakistan, Uzbekistan, Australia, Brazil, Greece, Argentina, and Egypt. These countries contribute more than 85% of the overall cotton production (Campbell et al. 2010).

Aftab Ahmad and Muhammad Zubair Ghouri contributed equally with all other contributors.

A. Ahmad (✉)
Center for Advanced Studies in Agriculture and Food Security (CAS-AFS), University of Agriculture, Faisalabad, Pakistan

Department of Biochemistry, University of Agriculture, Faisalabad, Pakistan
e-mail: aftab.ahmad@uaf.edu.pk

M. Z. Ghouri · S. H. Khan
Center for Advanced Studies in Agriculture and Food Security (CAS-AFS), University of Agriculture, Faisalabad, Pakistan

Center for Agricultural Biochemistry and Biotechnology (CABB), University of Agriculture, Faisalabad, Pakistan

A. Jamil
Department of Biochemistry, University of Agriculture, Faisalabad, Pakistan

N. Ahmad · Mehboob-ur-Rahman
Agricultural Biotechnology Division, National Institute for Biotechnology & Genetic Engineering (NIBGE) College, Faisalabad, Pakistan

Pakistan Institute for Engineering and Applied Science (PIEAS), Islamabad, Pakistan

Mehboob-ur-Rahman et al. (eds.), *Cotton Precision Breeding*,
https://doi.org/10.1007/978-3-030-64504-5_10

Cotton production is threatened by more than 1300 plant-feeding insects and pests causing significant yield losses. Among these, a limited number of insects are common inhabitants, and only a few of these have economic value. Insect/pests of cotton are classified into sap-sucking insects (aphids, jassids, and whitefly) and chewing insects (bollworms, leaf-eating caterpillars). Bollworms are highly damaging because they feed on the most important tissue of the cotton plant (Naranjo 2010).

Bollworm complex is comprised of American bollworm (*Helicoverpa armigera*), pink bollworm (*Pectinophora gossypiella*), and spotted bollworm (*Earias vittella*) (Popp et al. 2013). In the recent years, American bollworm and pink bollworm have emerged as the most devastating pests causing up to 80% yield losses (Dhaliwal et al. 2010). Different strategies (pesticides, IPM, and host plant) have been used to control insect pests. Pesticides are most widely used to control cotton pests; however, the excessive use of these chemicals causes the resurgence of secondary pests, resistance to pesticides, killing of nontargets, and environmental contamination due to indiscriminate use (Gill and Garg 2014; Özkara et al. 2016).

With the advent of recombinant DNA tools, it was realized that the insect toxins could be produced inside the plant. The idea was fascinating as only those insects feeding on the crop would be targeted and appeared environmentally friendly as it would circumvent the release of harmful chemicals into the environment. Development of Bt-cotton (containing genes derived from *Bacillus thuringiensis*) has been adopted as one of the cost-effective and environmentally safe strategies to control the lepidopteran insect pests. These genes were taken from a soil bacterium *Bacillus thuringiensis.* These genes encode crystal endotoxin proteins which are detrimental to lepidopteran insect pests. These genes are inserted into cotton for producing these toxic proteins. When the insect started feeding on cotton expressing Bt genes, it binds to specific receptors in the insect gut. These toxins will help in dissolving the gut lining allowing the gut bacteria and spores to enter the body which ultimately leads to insect death. The commercial cultivation of transgenic Bt-cotton was approved in the USA in 1995 as it successfully controlled the lepidopteran insect population feeding on cotton. Different types of Cry toxins with different protein sizes are active against various insect classes summarized in Table 10.1 (Kumar et al. 2008). The development of Bt-cotton has particularly helped resource poor farming community for sustaining their livelihood through

Table 10.1 Types of Bt toxin genes, protein size, and shape of crystals

Gene	Protein size (kDa)	Crystal shape
Cry I A(a), A(b), A(c), B, C, D, E, F, G	130–138	Bipyramidal
Cry II (subgroup) A, B, C	69–71	Cuboidal
Cry III (subgroup) A, B, C	73–74	Flat irregular
Cry IV (subgroup) A, B, C, D	73–134	Bipyrimidal
Cry V–IX	35–129	Various

Source: Agritech (2019)

increased cotton production as well as reduction in the input cost. Various studies show that the cultivation of of Bt-cotton led to significant reduction in environmental pollution, minimizing the potential threats to farmers' health (Huang et al. 2002; Morse et al. 2006; Kouser et al. 2019). This book chapter summarizes the research work made toward the development of various generations of Bt-cotton and gives snapshot of the mechanism of action of the Bt toxin while emphasizing field-evolved resistance in insects, along with alternative strategies for controlling the insect pests. In addition, the risk assessment of Bt-cotton along with the challenges, threats, and the future prospects of Bt cotton are also discussed.

10.2 Timeline of Bt Cotton

Before the development of Bt-cotton, application of Bt toxins was applied on several crops including cotton as spray but were not successful in killing the chewing insect pests. It was largely due to unfavorable atmospheric conditions, i.e., high temperature, solar UV rays, washing due to rains, etc. Moreover, a limited number of bacterial strains were available for deriving Bt toxin, thus reducing the choice for killing the insects. For example, until 1977, there were only 13 Bt strains, and all of them were toxic to the lepidopteran species (Sanahuja et al. 2011). The first species toxic to the dipteran insects was discovered in 1977 (Ben-Dov 2014) followed by the discovery of another strain in 1983 that was toxic to coleopteran insects (Domínguez-Arrizabalaga et al. 2019). In addition, Bt spray was not reachable to all parts of the plant equally; thus the insects feeding on lower parts could continue their survival. The limited efficacy of the Bt toxins surged the use of synthetic sprays, which further limited application of Bt technology. With the discovery that the main insecticidal activity of Bt protein against insects was due to the parasporal crystals (Table 10.1), researchers diverted their attention for understanding the mechanism pertaining to the action of Bt toxin in insects (Sanahuja et al. 2011).

Bt was first produced in 1907 and has been used for insect control since 1983. During the 1980s, environmentalists found that the increasing use of synthetic insecticides was degrading the environment and the insects were also becoming resistant to these chemicals. Therefore, the use of Bt was encouraged after the problems encountered with the synthetic sprays. Bt-based biopesticide was first registered in the USA during the 1960s and later also used in India for IPM programs (Kumar et al. 2008). Thus, Bt has created a history of toxicity against insects and safety from nontarget organisms.

Today, there are more than 60 Bt subspecies which have been found producing insecticidal proteins (Sharma 2014). Bt strains mainly produce three types of toxins, viz., crystal (Cry), cytolytic (Cyt), and vegetative insecticidal protein (Vip) toxins, that are highly specific for different insect species (Osman et al. 2015; Palma et al. 2014). So far, a total of 229 Cry, 11 Cyt, and 102 Vip toxins have been reported (Tabashnik and Carrière 2017). At present, 342 *Bt* genes encoding different insect toxins have been discovered which can be exploited to develop insect-resistant crops (Bravo et al. 2018). After the adoption of Bt cotton in 1996, there has been a significant decrease in chemical pesticides to about 60–70% in China and Argentina, respectively (ISAAA 2018). Moreover, the yield generated an average increment of $65.05/ha in Argentina (ISAAA 2018). Bt-cotton has been very effective against cotton bollworms (*Helicoverpa armigera* and *Pectinophora gossypiella*) which are very dangerous pests worldwide (Rocha-Munive et al. 2018). After the emergence of resistance against the first-generation Bt cotton, second- and third-generation Bt cotton has been developed which carries a combination of two and three Bt genes, respectively (Brévault et al. 2013).

10.3 Success Stories of Bt Cotton

Globally, Bt cotton cultivation has resulted in significant economic and production advantages. Substantially, Bt cotton can reduce the pesticide sprays and improve quality of life of farmers by increasing incomes. Today, all major cotton-producing countries are benefiting from cultivation of Bt cotton (Purcell and Perlak 2004). In 2005, about 80% area of cotton in Australia, 76% in China, and 54% in the USA was grown with "single" or "double" Bt gene technology. The world's third largest cotton grower, India, has grown 1.36 million acres of Bt cotton crops (Purcell and Perlak 2004).

10.3.1 Adoption of Bt-Cotton in the USA

Cotton has been an important economic crop for the USA. Almost 75 million hectares of transgenic crops were planted in total out of which 4.58 million hectares were of Bt-cotton (ISAAA 2017a). This increase in Bt-cotton planted area was due to the adoption of transgenic technology which increased to 96% (an overall 2% increase from the year 2016). Bt-cotton was commercialized in the USA around 21 years back, and since then, the USA had acquired lots of economic benefits from it with more than 420,000 farmers cultivating transgenic crops including cotton (James, 2012; ISAAA 2017a; Fleming et al. 2018). Accordingly, in order to reduce production costs and output losses, cotton grower's adopted Bollgard®, Roundup Ready®, and stacked Bollgard/Roundup Ready® (Monsanto) (Witjaksono et al. 2014). In an empirical study in the USA from 1999 to 2006, one of the most important factors in farmer's adoption decision was the effectiveness of Bollgard® to target pests than the conventional practices (Suntornpithug and Kalaitzandonakes 2009).In another study, Sankula et al. (2005) reported an increase of 90 kg/ha for Bollgard II® over Bollgard® in the USA. Similarly, an average yield increase of 9% for Bollgard® and 11% increase for Bollgard II® was reported by Brookes and Barfoot (2008, 2009) in the USA. Additionally, several surveys have demonstrated that Bt-cotton growers are achieving higher yields and revenue compared to those growing non-Bt varieties (Witjaksono et al. 2014). In the USA, Monsanto's BXN plus Bollgard® cotton, Bollgard® and Bollgard II® expressing Cry1Ac, and Cry2Ab2 and Syngenta's VIPCOT® cotton expressing Vip3A (vegetative insecticidal protein 3A) are commercially and easily available in the market.

10.3.2 Adoption of Bt-Cotton in China

Similar to the USA, cotton is of prime importance to China as well with more than ten million workers in the cotton textile industry (Zhao and Tisdell 2009). Since 1997, China has been the leader in planting insect-resistant cotton. In the last 21 years, insect-resistant cotton in China was recorded highest in 2013 at a total of 4.2 million hectares. In 2017, 95% of the total cotton area of 2.8 million hectares were Bt transgenic harboring resistance to lepidopteran pests of cotton. However, a slow decrease was observed in the current area of Bt-cotton which was 2.8 million hectares in 2017 (ISAAA 2017a). The adoption rate of Bt-cotton in the year 2017 followed a similar trend as of year 2016 and remained ~95% of the total under cultivation of cotton (ISAAA 2017a). The adoption of Bt-cotton in China has benefited farmers by helping to reduce pesticides as well as the labor costs (Yang et al. 2005). Surveys demonstrated that small farmers have gained much benefit in terms of yield increase than the wealthier farmers in China (Pray et al. 2002). Particularly, this was due to the reduced insecticide applications in case of transgenic cotton for control of bollworms. Additionally, China has reduced (60%) insecticide usage—a primary benefit of transgenic cotton over conventional cotton. A series of surveys has been conducted from 1999 to 2001 in five villages (Hebei, Shandong, Henan, Anhui, and Jiangsu) of China to assess impact of GM cotton which has shown higher yields than its counterparts (non-Bt) (Huang et al. 2002). Currently, commercial cultivation of transgenic cotton in China includes SGK321 (Stacked Cry1Ac+ CpTI; developed in 1999 by the Chinese Academy of Agricultural Sciences) (Guo et al. 1999), Monsanto's Roundup Ready® cotton, Bollgard®, Bollgard II®, Roundup Ready Flex®, and Syngenta's VIPCOT® cotton (Peter 2017; ISAAA 2017b).

10.3.3 Adoption of Bt-Cotton in India

For the sustainable economy of India, cotton is of immense importance to the Indian farming community as they rely heavily on it. With a quarter share in the market of global cotton production, India has achieved a greater stride in production over the years. In 2002, Bt-cotton was introduced by a joint venture between Monsanto and Mahyco (Maharashtra Hybrid Seeds Co.) in India (Kazmin 2016). In 2016, the area under cultivation of Bt-cotton was 10.8 million hectares, which increased to 11.4 million hectares in 2017–2018, a 6% increase in the area over the previous year. Studies demonstrated that significant increase in Bt-cotton area was due to the reduced insecticide use and supportive weather conditions in 2017. In addition, a large number of farmers used to plant unauthorized stacked traits of insect-resistant cotton in the Central and Southern Zones of India in 2017 (ISAAA 2017a). Moreover, unlicensed Bt-cotton varieties were grown on more than 10,000 hectares in Gujarat, or seeds were multiplied and sold on a growing black market under different names in many Indian states (Murugkar et al. 2007). However, to achieve a yield level equal to global average cotton production, there is a need to introduce new-generation traits such as smart agronomic, stacked traits and high-yielding cotton varieties. Similarly, to maintain current yield levels, rigorous resistance management strategies are needed to be implemented (Saravanan and Mohanasundaram 2016; ISAAA 2017a).

Cotton farmers in Maharashtra had suffered an unusual outbreak of pink bollworm during the cotton seasons of 2017–2018 (Menon 2017). The evolution of resistance in pink bollworm was due to the mismanagement of the insect-resistant cotton technology. Therefore, to control pink bollworm infestation and use of unauthorized seed, cultivation of hybrids for long duration must be tackled with a high level of supervision (ISAAA 2017a). Bt technology has proved to be a success story in India, with farmers benefiting from it by reduced cost, chemical sprays, and increased yield (Manjunath 2011). Bt technology has increased the yield from 30 to 40% by reducing the pesticide use to about 50% generating an income of US$156 per hectare (Subramanian and Qaim 2009). The benefits provided by Bt are stable and indicate that they have increased over time. Bt-cotton has provided sustainable benefits contributing to the socioeconomic development in India. Till date, India has pioneered in cultivating Bt-cotton developed by Monsanto events Ingard® and Bollgard II®.

10.3.4 Adoption of Bt-Cotton in Pakistan

Bt-cotton in Pakistan was introduced a bit late in year 2003–2004 upon the release of several Bt-cotton strains including IR-FH-901 (approved in 2011 in Sindh), IR-NIBGE-2 (approved as IR-NIBGE-1524 in 2010), IR-CIM-448 (approved as IR-NIBGE-3701 in 2010), and IR-NIBGE-CIM-443 for conducting local field trials throughout the Pakistan. These varieties showed excellent performance in the field and consequently cultivated throughout Pakistan. Approximately 0.20 million ha was captured by these lines in 2005–2006 (Rao 2007), and thereafter the area under Bt-cotton has kept on increasing. At the moment, ~95% area is under Bt-cotton in Pakistan (ISAAA 2017a). Significant varieties were IR-NIBGE-1524, IR-NIBGE-3701, IR-NIBGE-901 (also called as IR-901), Bt-121, MNH-886, FH-142, and IUB-13. During the first decade of this century, Bt-cotton was used to moderate the infestations of lepidopteran pests. In Pakistan, Sindh was the first province region to adopt Bt-cotton, followed by Punjab, Khyber Pakhtunkhwa, and Baluchistan, respectively (Bakhsh et al. 2016). At present, cultivation of Bt-cotton in Pakistan is predominately Monsanto's Bollgard®, the first generation of Bt-cotton (Ali et al. 2010; Shahid 2015). However, a shift from first-generation to second- and third-generation Bt-cottons is expected soon (ISAAA 2017a).

10.3.5 Adoption of Bt-Cotton in Australia

Australia was among the first six countries to commercialize Bt-cotton in 1996. A total of 8% increase in the total area of Bt-cotton was observed from 2016 to 2017. The total area under Bt-cotton was 2.54 million hectares in 2017 (ISAAA 2017a). Australia has approved ~26 cotton events for general cultivation (ISAAA 2017a). In Australia, conventional cotton has been replaced by Bollgard II® varieties. In a report, Pyke (2000) demonstrated that adoption of Ingard and Bollgard II has reduced the insecticide sprays by two-fifths to four-fifths. Bollgard II has been proved to be very effective to control a range of sucking type of insects, while Ingard was not too effective in controlling *Helicoverpa armigera*. The majority of cotton farmers have realized economic benefit from Bt-cotton (Pyke 2000). However, performance varies with the climate and environmental differences across locations. In the growing season 2003–2004, 84% of paired comparisons of Bt with conventional cottons showed net profit, while in the growing season (2004–2005), 66% of 50 paired comparisons showed net profit. As of now, Australia has commercially cultivated Monsanto's Bollgard®, BXN plus Bollgard®, Bollgard II®, Roundup Ready®, Bollgard III®, Roundup Ready Flex®, Roundup Ready × Bollgard®, Roundup Ready Flex × Bollgard II®, Roundup Ready Flex × Bollgard III®, Syngenta's VIPCOT®, and Dow ArgoSciences' WideStrike® (ISAAA 2017a).

10.4 Development of Transgenic Cotton

Bt-cotton was developed with an aim to reduce the amount of cotton chemical pesticide applications and cost of cotton production and reduce the environmental impacts of pesticides. Monsanto developed Bt-cotton in 1996 by introducing Bt gene (Cry1Ac) in cotton, and after its commercial release, it was widely adopted in the USA, China, India, Australia, and Pakistan. Since its introduction in 1996, the acreage of Bt-cotton is growing every year (AgBioWorld 2011). Insects feeding on *Bt-cotton* expressing single Bt gene have developed resistance, and therefore scientists move toward stacked events (Gatehouse et al. 2011). Bt-cotton with multiple genes has further augmented single-gene Bt-cotton and was also found more effective against insects (Arshad et al. 2018). In addition to transgenic Bt-cotton, today, herbicide-tolerant transgenic cotton is also commercially available in the market (Awan et al. 2015). Commercially available herbicide-tolerant cotton includes Roundup Ready, Roundup Ready Flex, bromoxynil (BXNR), and glufosinate ammonium (LibertyLink R) (Awan et al. 2015), while insect-resistant genetically modified cotton comprises first-generation (Bollgard or Ingard), second-generation (Bollgard II, WideStrike Cotton, TwinLink), and third-generation Bt-cotton (Bollgard III, WideStrike III, TwinLink Plus) (Luo et al. 2015; Mall et al. 2019) (see Table 10.2 for details).

Table 10.2 Various generations of insect-resistant cotton

Generations	Bt technologies	Protein expressed	Insect activity	References
First-generation Bt cotton	Ingard	Cry1Ac	Lepidopteran larvae	Torres et al. (2009)
Second-generation Bt cotton	Bollgard 2	Cry1Ac + Cry2Ab	Lepidopteran and dipteran	Torres et al. (2009)
	WideStrike	Cry1Ac + CryIF	Caterpillar pests	
	TwinLink	Cry1Ab + Cry2Ae	Lepidopteran pests	
Third-generation Bt cotton	WideStrike 3	Cry1F + Cry1Ac + Vip3A	Lepidopteran larvae	Arya and Shrivastav (2015)
	Bollgard 3	Cry1Ac + Cry2Ab + Vip3A	Lepidopteran and herbicide tolerance	Monsanto (2019)
	TwinLink Plus	Cry1Ab + Cry2Ae + Vip3Aa19	Lepidopteran pests	BASF (2019)

In addition to these, cotton with a combination of both the insect resistance (Bt genes) and glyphosate tolerance genes has also been developed, for example, Roundup Ready and Cowpea Trypsin Inhibitor (CpTI) genes have been stacked with Cry genes (Castle et al. 2006; Ni et al. 2017). Recently, transgenic cotton lines expressing vegetative proteins (Vip) and genes conferring tolerance to 2-4-D have also been developed (Mall et al. 2019).

10.5 Mechanism of Action of Bt Toxin

Bt protein family originated from a soil bacterium *Bacillus thuringiensis* and comprises more than 200 members and are highly specific in their activity. Each Bt toxin is active against a particular insect type. The Cry endotoxin is present as insoluble protein of 130–138 kDa at low pH. Inside the insect midgut, it is converted into prototoxin (activated toxin ~60 kDa) when exposed to pH higher than 9 (Naimov et al. 2008; Bravo et al. 2007). The activated toxin destroys the midgut of the insect either through forming the pores or signal transduction. According to the pore formation model, activated Cry toxin binds to the cadherin receptors of the brush lining of midgut epithelial cells. As the Cry toxin binds the cadherin receptors of epithelial cells of insect midgut, it forms ion channel or pore formation from cells.

Pore formation in brush lining creates gaps, which allows bacterial infection and Bt spores to enter into the body cavity; thus insect dies from internal bacterial spore infection (Bravo et al. 2007; Jurat-Fuentes and Adang 2006; Naimov et al. 2008)

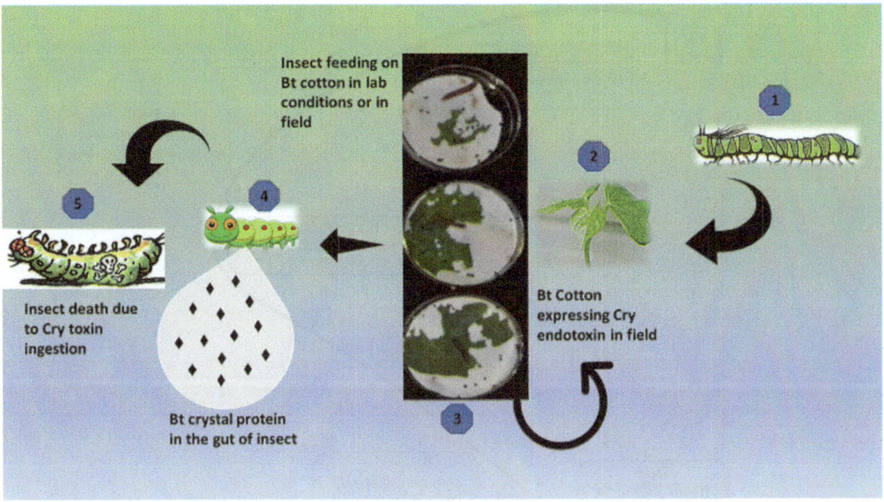

Fig. 10.1 Mechanism of action of Bt cotton. (1–3) Insects feeding on Bt crop ingest crystal toxins produced by Bt. (4) Cry toxins get activated by midgut proteases of insects, and they bind to different insect receptors on apical brush border membranes (BBM) of their midgut cells. (5) Toxins enter the membranes causing oligomerization and pore formation leading to cell breakdown and insect death

(Fig. 10.1). As a result, the midgut endothelial cells develop nonspecific pores causing swelling of the midgut, and larvae stops feeding. Ultimately, the pH of the midgut drops and bacterial spores start causing the death of the larvae which is called as *Septicaemia.* The first step in signal transduction model is similar to pore formation. By contrast, signal transduction method proposes that binding of cry toxin with cadherin receptor triggers stimulation of G protein and adenylate cyclase to increase cAMP which activates protein kinase A (Zhang et al. 2012). Activation of protein kinase A leads to oncotic cell death. Studies performed in different insect orders support pore formation model as compared to the signaling model which works better only with cell lines.

10.6 Bt Traits Available in Cotton

As already discussed, Bt-cotton was first introduced in 1996 in the USA, and so far, technology has been transferred from single gene to multiple genes. First-generation cotton, Bollgard, had only a single gene, while second-generation Bt crop, Bollgard II, TwinLink, and WideStrike had two genes. Most recent addition, the third-generation Bt crop comprises three genes, Bollgard III, WideStrike III, and TwinLink Plus. A timeline of Bt-cotton is given in Fig. 10.2.

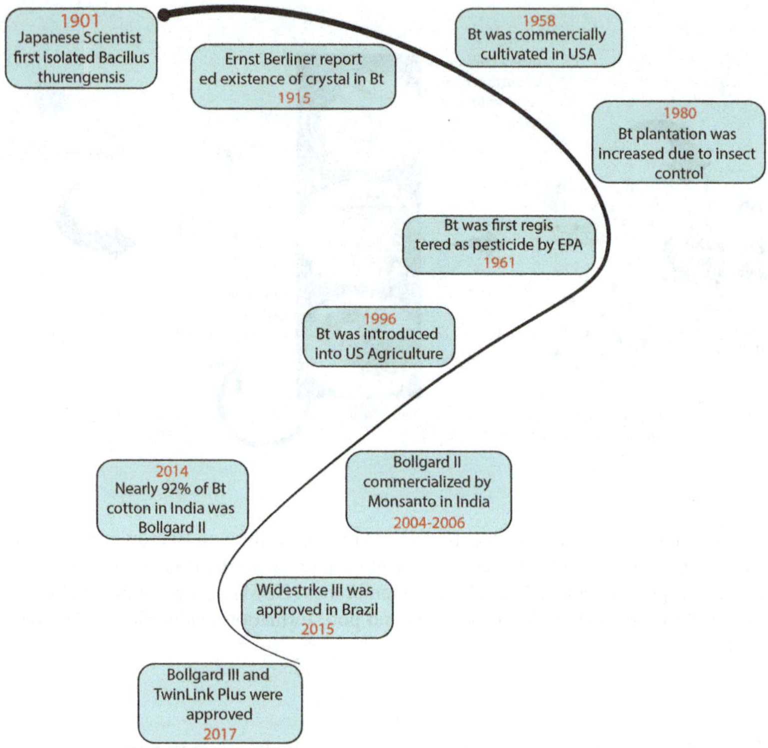

Fig. 10.2 Timeline of Bt cotton Source: Plewis (2014)

10.6.1 First-Generation Bt-Cotton (Bollgard or Ingard)

Since its first release in 1996, Bt-cotton has advanced from single-gene traits to multigenic packages (Ijaz, 2019). First-generation Bollgard had a single-gene Cry1Ac and was developed through *Agrobacterium* transformation by transforming *Cry1Ac* in Coker-312 (Downes et al. 2010; Hwang et al., 2015). Coker was transformed due to its higher regeneration potential compared to elite cultivars which are highly recalcitrant to current regeneration protocols. The initial transformation of cotton with native Bt gene resulted in lower expression of the protein, due to high "AT" content. Consequently, the codon usage of *Cry* gene was synthetically adjusted which resulted in 1000-fold higher expression compared with the native gene (Torres et al. 2009; Zhang, 2013). Similarly, codon optimization in *Cry1Ac* also resulted in higher expression in Bollgard (Torres et al. 2009). Full-length Bt protein in Bollgard was a hybrid molecule with amino acid 1–466 from Cry1Ab (*B. thuringiensis* subspecies kurstaki (Btk) strain HD-1) and 467–1178 Cry1Ac (Btk HD-73). So, Cry1Ac present in Bollgard was slightly different from native Cry1Ac. Cry1Ac gene in Bollgard was expressed under control of CaMV35S promoter with two copies of enhancer region and poly A signal from the beta-conglycinin gene of soybean (Torres et al. 2009).

10.6.2 Second-Generation Bt-Cotton

10.6.2.1 Bollgard II

Bollgard II was transformed with two Bt genes to improve resistance and develop a broad range of lepidopteran insects. Cry2Ab gene along with uidA (Gus) was transformed through the biolistic method in Bollgard which was already containing the Cry1A gene. Both *cry1Ab* and *uidA* genes were expressed under the CaMV 35S promoter (Torres et al. 2009). GUS was used as a screenable marker for transformation of the Cry1Ab gene in Bollgard II (Schünmann et al. 2003). Stacking Cry1Ab with Cry1Ac in Bollgard II increased the spectrum of resistance without compromising individual expression of both the genes and also throughout the growing season. The overall expression of Cry2Ab was higher than Cry1Ac in Bollgard II. Cotton lines expressing single Cry2Ab or stacked Cry2Ab and Cry1Ac showed better resistance against *Helicoverpa zea*, *Spodoptera frugiperda*, *Spodoptera exigua*, and *Pseudoplusia includens*. This broad-spectrum resistance is mainly due to the wider activity and higher expression of Cry2Ab (Adamczyk Jr et al. 2001; Greenplate et al. 2003). Cry2Ab had a uniform higher expression throughout the plant tissues except in the case of nectar and pollen.

10.6.2.2 WideStrike Cotton

WideStrike cotton (event 281-24-236), developed by Dow AgroSciences in 2005, contains two stacked genes, Cry1F and Cry1Ac (event 3006-210-23) (Robinson 2018). WideStrike contains modified Cry1F2a gene expressing Cry1F protein from Bt subspecies aizawai (Bta) strain PS811 and a modified Cry1Ac1 gene from Btk-HD-73 encoding Cry1Ac protein. To develop WideStrike, two different plasmids containing genes coding for Cry1F and Cry1C were separately transformed in cotton, which was then backcrossed with elite variety PSC355 followed by one self-pollination. Both backcrosses were then inter-crossed and self-pollinated to produce cotton variety 281-24234/3006-210-23 (Baeumler et al. 2006). The stacked cotton was effective against a wide range of lepidopteran insects.

10.6.2.3 TwinLink Bt-Cotton

TwinLink technology combines insect resistance Bt with weed management technology. TwinLink provides effective protection against major lepidopteran pests such as cotton bollworms, beet armyworm, fall armyworm, and pink bollworm and also full tolerance to LibertyLink herbicide used against a wide range of broad leaf and grass weeds. TwinLink is available as Glytol LibertyLink, TwinLink (GLT) trait package containing multiple Bt genes for high-level caterpillar control along with broad-spectrum tolerance to Liberty herbicide and glyphosate. GLT provides farmers a flexible choice of herbicide applications that suit the growing conditions. In addition to this, farmers may rotate herbicides to control weed effectively and reduce the risk of weed resistance. TwinLink contains two Cry proteins, Cry1Ab and Cry2Ae, with independent actions for broad-spectrum and effective control of leaf- and fruit-feeding cotton pests, thus contributing in maximum yield and fiber quality (Sawazaki et al. 2015).

10.6.3 Third-Generation Bt-Cotton

10.6.3.1 Bollgard III

The release of Bollgard III was a new addition in the Bt technology with new and improved benefits and effective strategy against resistance to pests as compared to Bollgard I and II. Bollgard III provided farmers a flexible control in the management of crops as it contains two *cry* genes (*Cry1Ac* and *Cry2Ab*) and one *Vip* gene (Vip3A) (Monsanto 2019). All these three proteins have a different mode of action which contributes to the longevity of the Bt technology besides being helpful in killing insects in three different ways (Monsanto 2019). In addition, Bollgard III also contributes significantly toward the sustainability of the Bt-cotton as it will be difficult for *Helicoverpa* spp. to develop simultaneous resistance against three layers of resistance. Different generations of Bt-cotton along with proteins expressed and their insect activity are summarized in Table 10.2.

10.6.3.2 WideStrike III

Like Bollgard III, WideStrike III developed by Dow AgroSciences contains three genes or multiple traits to provide a broad range of resistance against insects (Dow AgroSciences 2013). WideStrike III has been developed by crossing WideStrike cotton with Syngenta event Cot102. WideStrike III contains three genes Cry1F, Cry1Ac, and Vip3A and provides a high level of protection against several economically important cotton pests like tobacco budworm, pink bollworm, fall armyworm, and several other species. Therefore, it helps farmers to increase yield by inputting the same labor, pesticides, fuel, and equipment. Compared to Bollgard I and II, WideStrike will help in delaying breakdown of resistance in insects (Arya and Shrivastav 2015).

10.6.3.3 TwinLink Plus

TwinLink Plus was developed by Bayer in 2017 (Steadman et al. 2018) and provides a triple gene upgraded season-long protection to cotton against major lepidopteran insects. TwinLink plus technology combines two genes (Cry1Ab and Cry2Ae) present in TwinLink with Vip3Aa19 to create trademark TwinLink Plus. The enhanced action of TwinLink Plus against bollworms and armyworms decreases the cost of production as well as requirements of supplemental actions to control bollworms. Thus, TwinLink Plus helps farmers to achieve cotton production at its maximum yield. Early results from the new varieties show that TwinLink Plus has shown a remarkable increase in yield with outstanding fiber quality which was anticipated to release in January 2017.

10.6.3.4 Herbicide-Tolerant Cotton

Herbicide tolerance is one of the important traits addressing problems of cotton bollworms and weed competition with reduced risk and high yield. Herbicide-tolerant cotton can efficiently manage pests as well as reduce tillage practices ultimately contributing to the soil conservation. Commercially, Roundup Ready, Roundup Ready Flex (herbicide glyphosate-tolerant), bromoxynil (BXNR), and LibertyLink (glufosinate ammonium tolerant) cotton varieties are available in the market (Norsworthy et al. 2012). Among these four varieties, Roundup Ready was broadly used for weed control which is tolerant to glyphosate *N*-phosphornomethyl/glycine (Smyth et al. 2017). Roundup Ready cotton was made tolerant to glyphosate using *Agrobacterium*-mediated transformation in Coker-312 containing ESPS gene which encodes 5-enolpyruvylshikimate-3-phosphate (ESPS) synthase enzyme from *Agrobacterium* strain CD4. Roundup Ready cotton has limitations, i.e., it can only be applied at four-leaf stage or early developmental stages of plants, but its misapplication leads to fruit abortion, yield reduction, and nonviable pollen production. In 1997, Roundup Ready cotton was released for the first time which resulted in fewer herbicide applications than conventional weed control programs generating an equal net revenue (Culpepper and York 1999; Torres et al. 2009; Congreve 2015).

Upon limitations or concerns shown by Roundup Ready, a new transformation event based on reproductive growth was developed referred to as Roundup Ready Flex cotton which allows the tropical applications of glyphosate throughout seasons till harvesting. Unlike Roundup Ready cotton, Roundup Ready Flex cotton allows efficient weed control beyond four-leaf stage with multiple glyphosate applications. Roundup Ready Flex cotton was commercially planted in the USA, Australia, and China for the first time in the year 2006 (Torres et al. 2009). LibertyLink cotton developed by Bayer Crop Science in 2004 also contains herbicide-tolerant trait (glufosinate herbicide) for efficient weed control management.

In 1997, bromoxynil (BXNR) was another herbicide-tolerant cotton developed by Calgene and approved in 2005 in the USA. BXNR cotton varieties are tolerant to herbicide bromoxynil. It belongs to the oxynil family of herbicides and is active against dicot plants by blocking electron flow during photosynthesis. This type of cotton was also developed by *Agrobacterium*-mediated transformation containing a gene encoding nitrilase enzyme from a soil bacterium. Commercially available varieties of BXNR are BXN-47 and BXN-49B (Blair-Kerth et al. 2001; Torres et al. 2009). Dow AgroSciences has developed 2,4-D called 2,4-D choline which was significantly less volatile than previous salt formulations (Anonymous 2017).

The 2,4-D is a synthetic auxin herbicide that offers selective control of broadleaf weeds like Palmer amaranth (Craigmyle et al. 2013). The 2,4-D-tolerant cotton contains a 2,4-dichlorophenoxyacetate (2,4-D)/α-ketoglutarate dioxygenase (*tfdA gene)* which degrades 2,4-D into inactive compound glyoxylate. This modified 2,4-D-tolerant cotton was three times tolerant to recommended field concentrations for broadleaf weed control in wheat, sorghum, corn, and pastures (Egan et al. 2014). For grasses, the 2,4-D is a selective herbicide and produces strong vapor drift. However, it causes phytotoxicity to cotton when applied nearby. Therefore, 2-4-D-tolerant cotton, besides minimizing toxicity from drift, offers another option for directly applying herbicides to control dicotyledonous weeds.

10.7 Failure of Bt Technology

Since the deployment of Bt transgenic technology, Bt-cotton has transformed its production with manifold benefits to the productive farmers. Bt crops were first released in the the USA, Canada, Mexico, Argentina, China, and Australia in 1996, and in 2016, approximately 26 countries in the world had grown Bt crops (Seetharaman 2018). With the passage of the time, insect pests have been reported to evolve resistance against different Bt toxins. In India, pink bollworm infestation has caused a loss of about 90% cotton prior to the use of pesticides and introduction of transgenic cotton (Seetharaman 2018). Monsanto introduced Bollgard in 1996, and Bollgard II (second generation of Bt-cotton) in 2006, but later on it was reported that pink bollworm has evolved resistance against BG-I and BG-II (Seetharaman 2018). By the year 2012–2013, India grew more hectares of Bt-cotton than other countries in the world, and farmers preferred Bt pyramid cotton producing two or more toxins (Brévault et al. 2015).

There are reports that *H. armigera*, one of the devastating pests of cotton, has rapidly developed resistance against Bt-cotton in Australia, China, and Africa (Downes and Mahon 2012). Moreover, about 70% of the resistant larvae of *Helicoverpa* sp. were able to survive on Bt-cotton expressing Cry1Ac. The main reason of Bt failure was development of resistance in insects due to mutations in their receptors (detailed mechanism has been discussed in next section).

Although Bt pyramids have shown promise in controlling cotton pests, efficacy and durability of such pyramids are reduced by cross-resistance (Carrière et al. 2015). Some of the cotton insects such as *Heliothis virescens*, *Helicoverpa armigera*, *Pectinophora gossypiella*, *Plodia interpunctella*, *Ostrinia nubilalis*, *Spodoptera exigua*, and *Plutella xylostella* have shown resistance in laboratory conditions

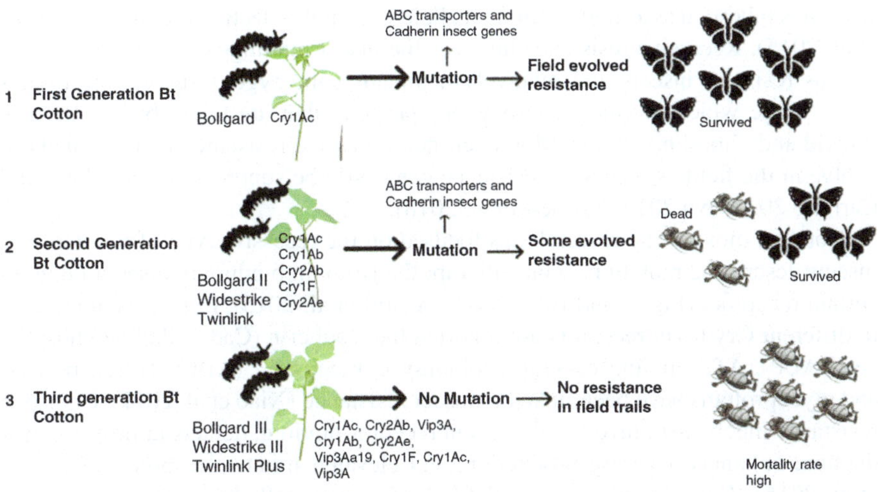

Fig. 10.3 Field-evolved resistance in insects. (1) Insects evolve resistance against Bollgard I because of mutation in ABC transporters and cadherin receptors. (2) Insects also evolve resistance against Bollgard II (Kurmanath 2015), WideStrike, and TwinLink (Francis 2019). (3) The third generation of Bt cotton is efficient in insect control, and up till now there is no mutation or resistance in insects

against Cry toxins, and *P. xylostella*, *P. gossypiella*, *Trichoplusia ni*, and *Spodoptera frugiperda* have evolved resistance in the field (Fig. 10.3) (Downes et al. 2010; Gassmann et al. 2011; Tabashnik et al. 2008; Jin et al. 2015; Zhang et al. 2012). Consequently, scientists have been urging to develop alternate tools to counter insect resistance even before the advent of Bt-cotton (Head and Greenplate 2012; Malone et al. 2008).

10.8 Development of Resistance in Insects Against the Bt Gene

Bt-cotton produces selective insecticidal toxins which are efficient in suppressing major pests, enhanced pest control by natural enemies, reduced insecticidal chemical spray applications and it also increases crop yield and grower's profits. However, the evolution of resistance in insets is overshadowing these potential benefits of Bt crops (Jin et al. 2018). Development of resistance in insects is due to the survival of some resistant insects over extensive pesticides applications. These resistant populations of insect survive and multiply, ultimately resulting in the entire generation to become resistant (Siegwart et al. 2015).

Resistance may involve two variables on which it depends: mechanism of inheritance (recessive, dominant, codominant, pleiotropic, or polygenic) and severity of the selection pressure. Generally, resistance in a population may spread rapidly when it is inherited as a single dominant allele with high selection pressure (Siegwart et al. 2015). Recessive resistance involves the insect mattings between rare homozygous resistant insects and relatively abundant homozygous susceptible insects from refuge which produce heterozygous progeny that are killed by the Bt crop (Gould and Tabashnik 1998). Moreover, non-recessive resistance is more likely to evolve in the field as recessive resistance can easily be suppressed (Tabashnik and Carrière 2017; Wu 2014; Mathew et al. 2018).

Notably, most of the research has focused on the recessive type of resistance in insects describing mutations which disrupt the Bt toxin binding in insect midgut to certain receptors (Pigott and Ellar 2007; Ocelotl et al. 2015). In insects, mutations in different Cry toxin receptors are reported like cadherin (Cad), alkaline phosphatase (ALP), ATP-binding cassette subfamily C member 2 (ABCC2) transporters, and glycoproteins have been linked to insect resistance (Xiao et al. 2017). Cry toxin resistance mechanism involves defects in receptor binding, defects in protease production, enhanced esterase production, and elevated immune response (Siegwart et al. 2015). *Heliothis virescens* and *Plodia interpunctella* have shown resistance due to defects in protease activities of midgut which affect the Cry1Ac protoxins (Bravo and Soberón 2008). Moreover, *H. virescens*, *Pectinophora gossypiella,* and *H. armigera* are the most devastating pests of cotton that have shown resistance due to mutations in receptor genes like cadherin and ABC transporters (Tay et al. 2015). Furthermore, mutations which lead to early stop codons that affect Cry1Ac binding causing a reduction in the expression of receptors have been reportedly associated with insect resistance (Xiao et al. 2016; Guo et al. 2015).

Most of the studies have been focused on recessive resistance where insects had mutations in receptors like cadherin, ALP (alkaline phosphatase), ABC transporters, and glycoproteins, while a little is known about the genetic basis of the dominant type of resistance in insects. Recently, many of the tetraspanin genes have been identified which play an important role in cell to cell communication and are also important in dominant resistance. Jin et al. (2018) reported a point mutation in a tetraspanin gene which confers dominant resistance to Cry1Ac in *Helicoverpa armigera*. In addition, CRISPR/Cas9 system-based complete knockout of HaTSPAN1 gene resulted in susceptibility of insects to Bt-cotton (Jin et al. 2018).

Different strategies for recessive resistance have been reported so far for delayed evolution of resistance. One of the important strategies is planting refuge crops (non-Bt host) along with Bt crops. Several countries have used non-Bt-cotton as a refuge, while in China, farmers have relied on natural refuges other than non-Bt-cotton. Refuge strategy was designed for control of *Helicoverpa armigera* (cotton bollworm) as Bt-cotton is its primary target and was not applied to *Pectinophora gossypiella,* which feeds entirely on cotton. So, refuge strategies are recommended to delay the evolution of resistance in insects. Similarly, second-generation Bt crops expressing two or more toxins have also been used for delaying the evolution of resistance in insects, while other conventional technologies have failed to control insects in cotton (Tabashnik et al. 2012).

High-dose strategy was also important and gained a relative amount of success in mitigating the insect resistance. This strategy utilizes a high dose of Bt with refuges to provide enough susceptible adult insects. This strategy is a single locus based and works against the recessive type of resistance. Incorporating high level of insecticidal toxins along with refuges host plant has been found to be effective in delaying resistance in insects (Sheikh et al. 2017). Recently, a new and innovative strategy was reported to counter insect resistance. This study involved hybridization of transgenic cotton varieties with conventional non-transgenic cotton varieties (Wan et al. 2017). Hybridization technology resulted in a reduced resistance against pink bollworm which is a voracious pest of cotton worldwide. China utilized this strategy in interbreeding Bt-cotton with non-Bt and which resulted in first-generation hybrid and planting second-generation hybrid seeds. These crosses generated a random mixture of 25% non-Bt-cotton and 75% Bt-cotton within the field (Wan et al. 2017). In another report, Xiao et al. (2017) performed multiple backcrosses from a BtR (Cry1Ac-resistant strain) from the cotton bollworm (*H armigera*) and successfully isolated a 516-fold Cry1Ac-resistant strain (96CAD). They also proposed that 96CAD had a tight linkage with mutant cadherin allele (mHaCad) containing 35 amino acids as compared to HaCad from the susceptible strain. They also observed significantly reduced levels of mHaCad proteins on midgut epithelium in 96CAD as compared to the control (96S).

Gene pyramiding combines two or more toxins with different modes of action against the insects in crops. In this case, it is difficult to evolve resistance against two toxins (Cry1Ac+ Cry1Ab) as they have different modes of action, and even though they bind to different receptors in insects, the insects still need multiple mutations to counteract these toxins. The first transgenic Bt-cotton expressing dual toxins (Cry1Ac and Cry2Ab) was tested in 2003 (Monsanto's Bollgard II) which proved to be highly efficient against lepidopteran pests (Manyangarirwa et al. 2006; Torres et al. 2009; Dhanaraj et al. 2019).

10.9 Cross-Resistance and Its Management Strategies

Cross-resistance occurs in insect populations when resistance to one pesticide also develops resistance to other pesticides because of common detoxification pathways or similar binding sites. For example, resistance in *H. armigera* against Cry1Ac caused cross-resistance to Cry1Aa and Cry1Ab through mutations in cadherin receptors. Cross-resistance arises frequently in insect populations to develop resistance against organophosphates and carbamates. Field-evolved resistance has been reported in insects against Bt crops in various countries which can be delayed by pyramiding two or more different Bt proteins in a single crop (Carrière et al. 2015; Yang et al. 2017). However, the durability of gene pyramiding can be compromised by evolution of cross-resistance. Therefore, understanding the patterns of cross-resistance in insects against different Bt proteins is vital for resistance management strategies (Carrière et al. 2015; Yang et al. 2017). In some cases, insect develop resistance to more than one class of pesticides called as multiple resistance. Such resistance may arise from the use of different class of pesticides (Xu et al. 2005).

For example, a strain of *P. xylostella* (NO-095) showed moderate level of resistance to Cry1C and high level of resistance to Cry1A, when selected under Bt formulation containing Cry1A and Cry1C in field and further selected with Cry1C in laboratory (Liu and Tabashnik 1997). Different resistance management strategies have been reported to encounter resistant insect populations such as saturation, moderation, and multiple attack. Management through saturation suppresses any detoxification and involves high dose of pesticides leaving no survivors behind. However, it works effectively only when resistance is dominant and target population is present in a contained area like green house. Moderation strategy involves application of low rate, short residual compounds with infrequent applications to ensure that susceptible gene will not be eliminated (ENT 2016). It uses minimal control to reduce population and works effectively when susceptible trait dominates over resistant one. Multiple attack strategy involves use of multiple control methods having different mechanisms, for example, rotating insecticides having different modes of action or by alternative use of chemical and nonchemical control strategies. Therefore, selection pressure will change from generation to generation for pests (ENT 2016).

10.10 Alternative Strategies to Bt Technology

Even though Bt-cotton has been performing well since 1996, researchers are trying to develop alternate strategies to control cotton bollworms due to the evolution of resistance. Alternate strategies for control of insect resistance include the expression of other toxins such as receptor proteins and engineering plants with proteases and expression of dsRNA. As an alternative technology to Bt toxins, RNA interference has great potential in insect/pest control. RNAi utilizes dsRNA for suppression of targeted gene expression in insects (Ni et al. 2017). For this, the focus point is to reduce the expression of genes essential to the pests and not to the nontarget organisms. It has been reported that juvenile hormones and cytochrome P450 monooxygenase genes (CYPAE14 and CYP6B6) are vital for insect functions and, hence, are potential targets for RNAi. Small dsRNA expression of CYP450 (P450 monooxygenase family) in bacteria and its feeding or bacterial spray (dsRNA) to cotton bollworms have shown promising results against the cotton bollworms (Mao et al. 2007, 2011).

Proteases from a variety of sources (viruses, bacteria, fungi, plant, and insects) exert insecticidal effect when overexpressed (Jouanin et al. 1998). As an alternative strategy, proteinase inhibitors can also be used against insect pest. Proteinase inhibitors are the plant defense proteins produced by plants when attacked by insects (Broadway 1995). Chitin is an insoluble structural polysaccharide that occurs in the exoskeletal and gut lining of insects. Because of the critical function of chitin, it has been considered as a potential target for insecticidal proteins. Dissolving of chitin with chitinases has been known to cause perforations in insect's exoskeleton, which has a serious effect on growth and molting of insect. Similarly, use of bacterial cholesterol oxidase has been proposed as an alternative strategy which has an insecticidal activity analogous to *Bt* toxins, dependent on its enzyme activity, which promotes membrane destabilization. Expression constructs of cholesterol oxidase containing a partial or complete sequence of proteins produce active enzymes in transgenic plants (Corbin et al. 2001). In an interesting study, binding of Cry1Ac to cadherin facilitates the proteolytic removal of the helix a-1 of the toxin, thereby inducing toxin oligomerization and pore formation (Ocelotl et al. 2015). In accordance with this observation, modified Cry1Ab and Cry1Ac toxins, which lacked helix a-1 (i.e., Cry1AbMod and Cry1AcMod), formed oligomers in the absence of cadherin. Moreover, in *M. sexta* insects by using these modified toxins, cadherin protein was silenced through RNAi (RNA interference) and which resulted in intolerance of higher levels of Cry1Ab. The insects which have shown Cry1Ac resistance have been killed by these modified toxins (Bravo and Soberón 2008).

10.11 Risk Assessment of Transgenic Cotton

Risk assessment is an integral part of an occupational health and safety management plan to create awareness of hazards and risks associated with the product of concern. Safety assessment relevant to Bt-cotton is focused on the biology of cotton and uses of products derived from Bt-cotton and biochemical characterization of introduced protein. An assessment of the biology of Bt plants for pest or weediness potential, relative to conventional cotton, includes the potential for cross-pollination to weedy relatives, dormancy and germination changes, and overwintering potential (Mayee et al. 2002; Bawa and Anilakumar 2013). A detailed review of risk assessment involves a history of safe consumption by animals including humans, toxicity testing of proteins on animals, safety results from lab and field for allergic effect, digestibility of proteins, and dietary consumption of cotton products by animals or humans.

In cotton testing, multiple traits of agronomic characteristics, fiber quality, plant morphology, and nutritional components (proximates, fatty acid spectrum, amino acid spectrum, and gossypol) of cottonseed oil and cottonseed meal must be evaluated. In Mexico, risk assessment studies of GM cotton were performed which included the evaluation of the risk of gene flow to wild relatives, its potential effects on nontarget organisms, risk of resistant weeds, and evaluation of resistance to Cry proteins by insect/pests (Rocha-Munive et al. 2018). Results demonstrated that there were no insect resistance in Mexico. Addditionally, it was also demonstrated that there were no effects on the non-target organisms, herbicides resistant weeds growth was still low (i.e.,) transgenic cotton managed herbicide resistance very well, and farmer's have achieved high incomes by growing GM cotton with reduced insecticide applications overall. Risk assessment of Bt-cotton on human health has also been evaluated and approved by the US Food and Drug Administration (FDA) in the USA. Feeding studies of cottonseed and cottonseed meal to rats and some other animals were also evaluated for health hazards and behavioral effects (Gadelha et al. 2014; Singla and Garg 2013). In addition, cotton-derived products used in medical, food, and personal hygiene products were also tested for associated risks (Mayee et al. 2002; Bawa and Anilakumar 2013).

Review of all safety information indicated that Bt transgenic cotton does not pose any potential damage to animals or humans. The expression of recombinant proteins into cotton has not shown any visible negative impacts on host plant fitness. Tolerance level set by the Environmental Protection Agency (EPA-USA) includes safe limits of pesticides in food (cottonseed oil) and feed (cottonseed, cottonseed meal, cottonseed hulls). Most of the countries have decided maximum safe limits for pesticide residues in food to protect consumer rights. Therefore, trade limitations may arise in case of any difference in those permissible limits. The Codex Committee on Pesticide Residues (CCPR) is responsible for establishing Codex Maximum Residue Limits (MRLs) for pesticide residues in specific food items or in groups of food or feed that move in international trade. Accordingly, biosafety studies carried out on Bt-cotton demonstrated that it did not pose any risk to animal or human health (Koch et al. 2015). Additional approvals of Bt-cotton have been obtained following the scientific reviews in Japan, Australia, Argentina, South Africa, Mexico, Canada, and China. Moreover, scoured and bleached cotton used for medical and hygiene products as well as for chemical products does not contain any DNA or protein of transgenic plant.

The US Department of Agriculture (USDA) evaluates whether the technology could pose a threat to animal or plant. The regulatory authority of the US EPA evaluates the Bt crop pesticidal properties, environmental impacts, and impact on nontarget insects and other organisms. Other countries and regulatory authorities do similar type of exercises prior to approval. Agronomic characteristics of all cotton varieties were evaluated in the field such as yield, lint quality, plant growth, and susceptibility to diseases and insects (Mayee et al. 2002; Koch et al. 2015). These factors were remained unaffected in Bt-cotton and thus considered safe for use against insects. Similarly, testing and assessment were also conducted to evaluate the impacts of Bt proteins on nontarget insects. Therefore, large-scale testing of Bt spray demonstrates that it is considered safe and has no adverse effects on nontarget organisms. Additionally, a population increase was found in nontarget organisms due to the reduced load of chemical pesticides (Akhtar et al. 2009). Due to the reduced load of chemical pesticides, it was also found that persistence of Cry toxins in root exudates of Bt-cotton and Bt corn has no effect on soil microbes, earthworms, and nematodes (Saxena et al. 2002). Furthermore, the impact of Bt-cotton on the environment based on a low level of exposure of Bt proteins has demonstrated no adverse effects on the environment since its introduction in 1996. Indeed, Bt-cotton has provided farmers with lots of benefits over conventional cotton.

10.12 Prospects and Conclusion

Bt-cotton has shown promise to control cotton bollworms. First-generation Bollgard I or Ingard had control over major lepidopteran insects. After this, Bollgard II cotton remarkably benefited the farmers by reducing chemical pesticide load and reducing pest attack. Unfortunately, field-evolved resistance posed more damage to the existing Bt-cotton, and therefore cotton bollworms (*Helicoverpa* and *Pectinophora*) are still the most devastating ones. Development of cotton with multiple layers of resistance such as Bollgard III expressing three Bt genes is viewed as an immediate strategy to delay the onset of resistance with the provision of a higher level of protection. In a different strategy, combining different herbicide tolerance genes with different Cry endotoxins has also proven to be an effective insect and weed control strategy (Lombardo et al. 2016).

Although Bt-cotton has offered many benefits to the farmers, evolution of resistance in insects has become a significant issue. Therefore, scientists are pursuing to develop alternate strategies to control insect problems in cotton. One of these potential strategies includes the use of dsRNA (RNAi) approach to control insects. For effective RNAi against pests, identification of the gene targets is not only the key aspect, but the expression and delivery of bulk amount of dsRNA to target pest is also important as well. Although dsRNA-mediated silencing in plants and insects has been as an initial proof of concepts, commercial application of dsRNA must be cost-effective comparable to the traditional chemical pesticides. Additionally, CRISPR-based gene drives have also been proposed to control insects and vector-borne diseases in agriculturally important crops. Gene drive system enables biased inheritance of a desired trait in offsprings with more than 50% chances of inheritance. Proof of concept studies have been reported in mosquitoes to control Zika, dengue, and malaria; however, no commercial gene drives have been approved so far in the world.

References

Adamczyk JJ Jr, Adams LC, Hardee DD (2001) Field efficacy and seasonal expression profiles for terminal leaves of single and double *Bacillus thuringiensis* toxin cotton genotypes. *J Econ Entomol* 94(6):1589–1593

AgBioWorld (2011). http://www.agbioworld.org/biotech-info/articles/biotech-art/safety-bt-cotton.html. Accessed 13 Dec 2019

Agritech (2019) TNAU Agritech portal: biotechnology. http://agritech.tnau.ac.in/bio-tech/biotech btcotton_env.html

Akhtar W, Sengupta D, Chowdhury A (2009) Impact of pesticides used in agriculture: their benefits and hazards. *Interdiscip Toxicol* 2(1):1–12

Ali S, Hameed S, Masood SHAHID, Ali GM, Zafar Y (2010) Status of Bt cotton cultivation in major growing areas of Pakistan. *Pak J Bot* 42(3):1583–1594

Anonymous (2017) Enlist Duo herbicide product label. Dow AgroSciences Publication No. D02-407-003. Dow AgroSciences, Indianapolis, 7 p

Arshad M, Khan RR, Aslam A, Akbar W (2018) Transgenic Bt cotton: effects on target and non-target insect diversity. In: Past, present and future trends in cotton breeding. IntechOpen

Arya S, Shrivastav S (2015) Seeds of doubt: Monsanto never had Bt cotton patent. https://timesofindia.indiatimes.com/india/Seeds-of-doubt-Monsanto-never-had-Bt-cotton-patent/articleshow/47578304.cms

Awan MF, Abbas MA, Muzaffar A, Ali A, Tabassum B, Rao AQ et al (2015) Transformation of insect and herbicide resistance genes in cotton (*Gossypium hirsutum* L.). *J Agric Sci Technol* 17(2):287–298

Baeumler S, Wulff D, Tagliani L, Song P (2006) A real-time quantitative PCR detection method is specific to widestrike transgenic cotton (event 281-24-236/3006-210-23). *J Agric Food Chem* 54(18):6527–6534

Bakhsh K, Akram W, Jahanzeb A, Khan M (2016) Estimating the productivity of Bt cotton and its impact on pesticide use in Punjab (Pakistan). *Pak Econ Soc Rev* 54(1):15

BASF (2019) Insect Resistance Management (IRM) Guide. https://industries.basf.com/assets/northamerica/us/en/Agriculture/Crop%20Protection/BCS/InsectResistanceManagementGuide.pdf

Bawa AS, Anilakumar KR (2013) Genetically modified foods: safety, risks and public concerns—a review. *J Food Sci Technol* 50(6):1035–1046

Ben-Dov E (2014) *Bacillus thuringiensis* subsp. israelensis and its dipteran-specific toxins. *Toxins* 6(4):1222–1243

Blair-Kerth LK, Dotray PA, Keeling JW, Gannaway JR, Oliver MJ, Quisenberry JE (2001) Tolerance of transformed cotton to glufosinate. *Weed Sci* 49(3):375–380

Bravo A, Soberón M (2008) How to cope with insect resistance to Bt toxins? *Trends Biotechnol* 26(10):573–579

Bravo A, Gill SS, Soberon M (2007) Mode of action of *Bacillus thuringiensis* Cry and Cyt toxins and their potential for insect control. *Toxicon* 49(4):423–435

Bravo A, Gill SS, Soberón M (2018) Bacillus Thuringiensis: mechanisms and USE. Reference module in life sciences. https://doi.org/10.1016/b978-0-12-809633-8.04071-1

Brévault T, Heuberger S, Zhang M, Ellers-Kirk C, Ni X, Masson L et al (2013) The potential shortfall of pyramided transgenic cotton for insect resistance management. *Proc Natl Acad Sci* 110(15):5806–5811

Brévault T, Tabashnik BE, Carrière Y (2015) A seed mixture increases the dominance of resistance to Bt cotton in Helicoverpa zea. *Sci Rep* 5:9807

Broadway RM (1995) Are insects resistant to plant proteinase inhibitors? *J Insect Physiol* 41(2):107–116

Brookes G, Barfoot P (2008) GM crops: global socio-economic and environmental impacts 1996-2006. PG Economics Ltd, Dorchester

Brookes G, Barfoot P (2009) Global impact of biotech crops: socio-economic & environmental effects 1996–2007. *Outlooks Pest Manag* 20(6):258–264

Campbell BT, Saha S, Percy R, Frelichowski J, Jenkins JN, Park W et al (2010) Status of the global cotton germplasm resources. *Crop Sci* 50(4):1161–1179

Carrière Y, Crickmore N, Tabashnik BE (2015) Optimizing pyramided transgenic Bt crops for sustainable pest management. *Nat Biotechnol* 33(2):161

Castle LA, Wu G, McElroy D (2006) Agricultural input traits: past, present, and future. *Curr Opin Biotechnol* 17(2):105–112

Congreve M (2015) Roundup ready flex cotton—friend or foe for grass weed control? https://grdc.com.au/news-and-media/news-and-media-releases/north/2015/09/roundup-ready-flex-cotton-and-grass-weed-control

Corbin DR, Grebenok RJ, Ohnmeiss TE, Greenplate JT, Purcell JP (2001) Expression and chloroplast targeting of cholesterol oxidase in transgenic tobacco plants. *Plant Physiol* 126(3):1116–1128

Craigmyle BD, Ellis JM, Bradley KW (2013) Influence of herbicide programs on weed management in soybean with resistance to glufosinate and 2, 4-D. *Weed Technol* 27(1):78–84

Culpepper AS, York AC (1999) Weed management and net returns with transgenic, herbicide-resistant, and nontransgenic cotton (*Gossypium hirsutum*). *Weed Technol* 13(2):411–420

Dhaliwal GS, Jindal V, Dhawan AK (2010) Insect pest problems and crop losses: changing trends. *Indian J Ecol* 37(1):1–7

Dhanaraj AL, Willse AR, Kamath SP (2019) Stability of expression of Cry1Ac and Cry2Ab2 proteins in Bollgard-II hybrids at different stages of crop growth in different genotypes across cropping seasons and multiple geographies. *Transgenic Res* 28(1):33–50

Domínguez-Arrizabalaga M, Villanueva M, Fernandez AB, Caballero P (2019) A strain of *Bacillus thuringiensis* containing a novel cry7Aa2 gene that is toxic to *Leptinotarsa decemlineata* (Say) (Coleoptera: Chrysomelidae). *Insects* 10(9):259

Dow AgroSciences (2013). http://www.plantmanagementnetwork.org/pub/php/news/2013/ThreeGeneInsectCotton/

Downes S, Mahon R (2012) Evolution, ecology, and management of resistance in Helicoverpa spp. to Bt cotton in Australia. *J Invertebr Pathol* 110(3):281–286

Downes S, Mahon RJ, Rossiter L, Kauter G, Leven T, Fitt G, Baker G (2010) Adaptive management of pest resistance by Helicoverpa species (Noctuidae) in Australia to the Cry2Ab Bt-toxin in Bollgard II® cotton. *Evol Appl* 3(5–6):574–584

Egan JF, Barlow KM, Mortensen DA (2014) A meta-analysis on the effects of 2, 4-D and dicamba drift on soybean and cotton. *Weed Sci* 62(1):193–206

ENT (2016) General Entomology | Resource Library (Tutorials). (2016, March 26). Retrieved from https://projects.ncsu.edu/cals/course/ent425/library/tutorials/applied_entomology/resistance.html

Fleming D, Musser F, Reisig D, Greene J, Taylor S, Parajulee M et al (2018) Effects of transgenic *Bacillus thuringiensis* cotton on insecticide use, heliothine counts, plant damage, and cotton yield: a meta-analysis, 1996-2015. *PLoS One* 13(7):e0200131

Francis L (2019) Developing resistance to Bt genes in cotton bollworm. https://agrilife.org/texasrowcrops/2019/03/18/developing-resistance-to-bt-genes-in-cotton-bollworm/

Gadelha ICN, Fonseca NBS, Oloris SCS, Melo MM, Soto-Blanco B (2014) Gossypol toxicity from cottonseed products. *ScientificWorldJournal* 2014:231635

Gassmann AJ, Petzold-Maxwell JL, Keweshan RS, Dunbar MW (2011) Field-evolved resistance to Bt maize by western corn rootworm. *PLoS One* 6(7):e22629

Gatehouse AMR, Ferry N, Edwards MG, Bell HA (2011) Insect-resistant biotech crops and their impacts on beneficial arthropods. *Philos Trans R Soc B Biol Sci* 366(1569):1438–1452

Gill HK, Garg H (2014) Pesticides: environmental impacts and management strategies. In: Pesticides-toxic aspects. IntechOpen

Gould F, Tabashnik B (1998) Bt-cotton resistance management. In: Now or never: serious new plans to save a natural Pest control. pp 65–105

Greenplate JT, Mullins JW, Penn SR, Dahm A, Reich BJ, Osborn JA et al (2003) Partial characterization of cotton plants expressing two toxin proteins from *Bacillus thuringiensis*: relative toxin contribution, toxin interaction, and resistance management. *J Appl Entomol* 127(6):340–347

Guo S.D., Cui H., Xia L., Wu D., Ni W., Zhang Z., Zhang B., Xu Y. (1999) Development of bivalent insect-resistant transgenic cotton plants. *Scientia Agricultura Sinica*, 32: 1–7.

Guo Z, Kang S, Chen D, Wu Q, Wang S, Xie W et al (2015) MAPK signaling pathway alters the expression of midgut ALP and ABCC genes and causes resistance to *Bacillus thuringiensis* Cry1Ac toxin in diamondback moth. *PLoS Genet* 11(4):e1005124

Head GP, Greenplate J (2012) The design and implementation of insect resistance management programs for Bt crops. *GM Crops Food* 3(3):144–153

Huang J, Rozelle S, Pray C, Wang Q (2002) Plant biotechnology in China. *Science* 295(5555):674–676

Hwang HH, Gelvin SB, Lai EM (2015) Agrobacterium biology and its application to transgenic plant production. *Front Plant Sci* 6:265

Ijaz (2019) First Bt cotton grown in Pakistan. http://www.pakissan.com/english/advisory/biotechnology/first.bt.cotton.grown.in.pakistan.shtml. Accessed 17 Dec 2019

ISAAA (2017a) Global status of commercialized biotech/GM crops in 2017: biotech crop adoption surges as economic benefits accumulate in 22 years. ISAAA Brief No. 53, ISAAA, Ithaca

ISAAA (2017b) GM crop events approved in China. https://www.isaaa.org/gmapprovaldatabase/approvedeventsin/default.asp?CountryID=CN. Accessed 17 Dec 2019

ISAAA (2018 Oct) Pocket K No. 6: Bt Insect Resistant Technology. https://www.isaaa.org/resources/publications/pocketk/6/default.asp

James C (2012) Global status of commercialized biotech/GM crops. ISAAA Brief No. 44, International Service for the Acquisition of Agri-biotech Application, ISAAA, Ithaca

Jin L, Zhang H, Lu Y, Yang Y, Wu K, Tabashnik BE, Wu Y (2015) A large-scale test of the natural refuge strategy for delaying insect resistance to transgenic Bt crops. *Nat Biotechnol* 33(2):169

Jin L, Wang J, Guan F, Zhang J, Yu S, Liu S et al (2018) A dominant point mutation in a tetraspanin gene associated with field-evolved resistance of cotton bollworm to transgenic Bt cotton. *Proc Natl Acad Sci* 115(46):11760–11765

Jouanin L, Bonadé-Bottino M, Girard C, Morrot G, Giband M (1998) Transgenic plants for insect resistance. *Plant Sci* 131(1):1–11

Jurat-Fuentes JL, Adang MJ (2006) Cry toxin mode of action in susceptible and resistant *Heliothis virescens* larvae. *J Invertebr Pathol* 92(3):166–171

Kazmin A (2016 Oct 18) Monsanto faces growing troubles in India. Financial Times. Retrieved 8 June 2016

Koch MS, Ward JM, Levine SL, Baum JA, Vicini JL, Hammond BG (2015) The food and environmental safety of Bt crops. *Front Plant Sci* 6:283

Kouser S, Spielman DJ, Qaim M (2019) Transgenic cotton and farmers' health in Pakistan. *PLoS One* 14(10):e0222617

Kumar S, Chandra A, Pandey KC (2008) *Bacillus thuringiensis* (Bt) transgenic crop: an environment-friendly insect-pest management strategy. *J Environ Biol* 29(5):641–653

Kurmanath KV (2015) Wily Pink Bollworm Survives Monsanto's Bollgard-II. http://www.thehindubusinessline.com/economy/agri-business/wily-pink-bollworm-survives-monsantos-bollgardii/article7814810.ece

Liu Y, Tabashnik BE (1997) Inheritance of resistance to the *Bacillus thuringiensis* toxin Cry1C in the diamondback moth. *Appl Environ Microbiol* 63(6):2218–2223

Lombardo L, Coppola G, Zelasco S (2016) New technologies for insect-resistant and herbicide-tolerant plants. *Trends Biotechnol* 34(1):49–57

Luo JY, Zhang S, Lv LM, Wang CY, Zhu XZ, Cui JJ (2015) The effects of transgenic Cry1Ac+ Cry2Ab cotton on cotton bollworm control and functional response of predators on whitefly. *Yi chuan=. Hereditas* 37(6):575–581

Mall T, Gupta M, Dhadialla TS, Rodrigo S (2019) Overview of biotechnology-derived herbicide tolerance and insect resistance traits in plant agriculture. In: Transgenic plants. Humana Press, New York, pp 313–342

Malone LA, Gatehouse AM, Barratt BI (2008) Beyond Bt: alternative strategies for insect-resistant genetically modified crops. In: Integration of insect-resistant genetically modified crops within IPM programs. Springer, Dordrecht, pp 357–417

Manjunath TM (2011) Question Answer on BT_cotton in India. Association of a biotechnology Led Enterprises

Manyangarirwa W, Turnbull M, McCutcheon GS, Smith JP (2006) Gene pyramiding as a Bt resistance management strategy: how sustainable is this strategy? *Afr J Biotechnol* 5(10):781–785

Mao YB, Cai WJ, Wang JW, Hong GJ, Tao XY, Wang LJ et al (2007) Silencing a cotton bollworm P450 monooxygenase gene by plant-mediated RNAi impairs larval tolerance of gossypol. *Nat Biotechnol* 25(11):1307

Mao YB, Tao XY, Xue XY, Wang LJ, Chen XY (2011) Cotton plants expressing CYP6AE14 double-stranded RNA show enhanced resistance to bollworms. *Transgenic Res* 20(3):665–673

Mathew LG, Ponnuraj J, Mallappa B, Chowdary LR, Zhang J, Tay WT et al (2018) ABC transporter mis-splicing associated with resistance to Bt toxin Cry2Ab in laboratory-and field-selected pink bollworm. *Sci Rep* 8(1):13531

Mayee CD, P Singh AB Dongre MRK Rao, S Raj (2002) Transgenic Bt cotton. Technical Bulletin 22, Central Institute For Cotton Research, Nagpur. http://www.cicr.org.in/pdf/transgenic_bt_cotton.pdf

Mehboob-ur-Rahman, Tayyaba Shaheen, Nabila Tabbasam, Muhammad Atif Iqbal, Muhammad Ashraf, Yusuf Zafar, Andrew H. Paterson, (2012) Cotton genetic resources. A review. *Agronomy for Sustainable Development* 32 (2):419–432

Menon M (2017) A frayed history: the journey of cotton in India. Oxford University Press, Oxford

Monsanto (2019) Bollgard 3. http://www.monsantoglobal.com/global/au/products/pages/bollgard3.aspx

Morse S, Bennett R, Ismael Y (2006) Environmental impact of genetically modified cotton in South Africa. *Agric Ecosyst Environ* 117(4):277–289

Murugkar M, Ramaswami B, Shelar M (2007) Competition and monopoly in the Indian cotton seed market* May 2007

Naimov S, Boncheva R, Karlova R, Dukiandjiev S, Minkov I, de Maagd RA (2008) Solubilization, activation, and insecticidal activity of *Bacillus thuringiensis* serovar thompsoni HD542 crystal proteins. *Appl Environ Microbiol* 74(23):7145–7151

Naranjo SE (2010) Impacts of Bt transgenic cotton on integrated pest management. *J Agric Food Chem* 59(11):5842–5851

Ni M, Ma W, Wang X, Gao M, Dai Y, Wei X et al (2017) Next-generation transgenic cotton: pyramiding RNAi and Bt counters insect resistance. *Plant Biotechnol J* 15(9):1204–1213

Norsworthy JK, Ward SM, Shaw DR, Llewellyn RS, Nichols RL, Webster TM et al (2012) Reducing the risks of herbicide resistance: best management practices and recommendations. *Weed Sci* 60(SP1):31–62

Ocelotl J, Sánchez J, Arroyo R, García-Gómez BI, Gómez I, Unnithan GC et al (2015) Binding and oligomerization of modified and native Bt toxins in resistant and susceptible Pink Bollworm. *PLoS One* 10(12):e0144086

Osman GEH, Already R, Assaeedi ASA, Organji SR, El-Ghareeb D, Abulreesh HH, Althubiani AS (2015) Bioinsecticide *Bacillus thuringiensis* a comprehensive review. *Egypt J Biol Pest Control* 25(1):271

Özkara A, Akyıl D, Konuk M (2016) Pesticides, environmental pollution, and health. In: Environmental health risk-hazardous factors to living species. IntechOpen

Palma L, Muñoz D, Berry C, Murillo J, Caballero P (2014) *Bacillus thuringiensis* toxins: an overview of their biocidal activity. *Toxins* 6(12):3296–3325

Peter (2017 Apr 17) Monsanto's Bollgard 3 not registered for cottonseed exports to China. https://www.weeklytimesnow.com.au/agribusiness/monsantos-bollgard-3-not-registered-for-cottonseed-exports-to-china/news-story/48ca44f7af5ef38f4c0f1958 2ca569c9

Pigott CR, Ellar DJ (2007) Role of receptors in *Bacillus thuringiensis* crystal toxin activity. *Microbiol Mol Biol Rev* 71(2):255–281

Plewis I (2014) Indian farmer suicides: is GM cotton to blame? *Significance* 11(1):14–18

Popp J, Pető K, Nagy J (2013) Pesticide productivity and food security. A review. *Agron Sustain Dev* 33(1):243–255

Pray CE, Huang J, Hu R, Rozelle S (2002) Five years of Bt cotton in China–the benefits continue. *Plant J* 31(4):423–430

Purcell JP, Perlak FJ (2004) Global impact of insect-resistant (Bt) cotton. *AgBioForum* 7(1&2):27–30. Available on the World Wide Web: http://www.agbioforum.org

Pyke B (2000) The performance of INGARD® cotton in Australia in the 1998/99 season. Cotton R&D Corporation Occasional Paper, Narrabri New South Wales

Rao IA (2007) Why not GM crops [Internet]. http://www.pakistan.com/english/advisory/biotechnology/why-not.gm.crops.shtml. Accessed 2008-08-05

Robinson E (2018 Dec 7) What will second generation Bt cotton contribute? https://www.farmprogress.com/what-will-second-generation-bt-cotton-contribute

Rocha-Munive MG, Eguiarte LE, Soberón M, Castañeda S, Niaves-Nava E, Scheinvar E et al (2018) Evaluation of the impact of genetically modified cotton after 20 years of cultivation in Mexico. *Front Bioeng Biotechnol* 6:82

Sanahuja G, Banakar R, Twyman RM, Capell T, Christou P (2011) *Bacillus thuringiensis*: a century of research, development and commercial applications. *Plant Biotechnol J* 9(3):283–300

Sankula S, Marmon G, Blumenthal E (2005) Biotechnology-derived crops planted in 2004: impacts on US agriculture. National Center for Food and Agricultural Policy, Washington, DC

Saravanan, S., and V. Mohanasundaram (2016) Development and adoption of Bt cotton in India: Economic, environmental and health issues. 1–19.

Sawazaki, H.E., Duarte, A.P., Fuzatto, M.G., Sawazaki, E., Grandi, S.H.R., de Ponte, J.F. and Nogueira, L. (2015) Identification and Quantification of Corn, Soybean and Cotton Genetically Modified by Real Time PCR. *American Journal of Molecular Biology*, 5, 84–93. http://dx.doi.org/10.4236/ajmb.2015.53007

Saxena D, Flores S, Stotzky G (2002) Bt toxin is released in root exudates from 12 transgenic corn hybrids representing three transformation events. *Soil Biol Biochem* 34(1):133–137

Schünmann PH, Llewellyn DJ, Surin B, Boevink P, De Feyter RC, Waterhouse PM (2003) A suite of novel promoters and terminators for plant biotechnology. *Funct Plant Biol* 30(4):443–452

Seetharaman G (2018 Jan 21) These two issues could put the brakes on the Bt cotton story. https://economictimes.indiatimes.com/news/economy/agriculture/the-brakes-are-applied-on-the-bt-cotton-story/articleshow/62583116.cms?from=mdr

Shahid J (2015 Dec 28) Punjab reports low cotton production. https://www.dawn.com/news/1229109

Sharma PD (2014) Environmental biology & toxicology. Rastogi Publications

Sheikh AA, Wani MA, Bano P, Un S, Nabi TAB, Bhat MA, Dar MS (2017) An overview of resistance of insect pests against Bt crops. *J Entomol Zool Stud* 5:941–948

Shodhganga (2019). https://shodhganga.inflibnet.ac.in/bitstream/10603/142626/9/09_chapter3.pdf. Accessed 14 Dec 2019

Siegwart M, Graillot B, Blachere Lopez C, Besse S, Bardin M, Nicot PC, Lopez-Ferber M (2015) Resistance to bio-insecticides or how to enhance their sustainability: a review. *Front Plant Sci* 6:381

Singla N, Garg M (2013) Effect of crude cottonseed oil containing gossypol on the fertility of male and estrous cycle of female Bandicota bengalensis Gray and Hardwicke. *J Appl Anim Res* 41(2):156–165

Smyth, S., Andrew, AgSciGuy, Jc, Rauschenberger, G., John, … Kms (2017, October 1). Why Roundup Ready Crops Have Lost their Allure. Retrieved from http://sitn.hms.harvard.edu/flash/2015/roundup-ready-crops/

Steadman J, Steadman J, Ramkumar S, Thompson C, Grower C (2018 Apr 25) Bayer Adds TwinLink Plus Pest Protection for 2017. https://www.cottongrower.com/cotton-news/bayer-adds-twinlink-plus-pest-protection-for-2017/

Subramanian A, Qaim M (2009) Village-wide effects of agricultural biotechnology: the case of Bt cotton in India. *World Dev* 37(1):256–267

Suntornpithug P, Kalaitzandonakes NG (2009) Understanding the adoption of cotton biotechnologies in the US: firm level evidence. *Agric Econ Rev* 10(389-2016-23316):80–96

Tabashnik BE, Carrière Y (2017) The surge in insect resistance to transgenic crops and prospects for sustainability. *Nat Biotechnol* 35(10):926

Tabashnik BE, Gassmann AJ, Crowder DW, Carrière Y (2008) Insect resistance to Bt crops: evidence versus theory. *Nat Biotechnol* 26(2):199

Tabashnik BE, Wu K, Wu Y (2012) Early detection of field-evolved resistance to Bt cotton in China: cotton bollworm and pink bollworm. *J Invertebr Pathol* 110(3):301–306

Tay WT, Mahon RJ, Heckel DG, Walsh TK, Downes S, James WJ et al (2015) Insect resistance to *Bacillus thuringiensis* toxin Cry2Ab is conferred by mutations in an ABC transporter subfamily A protein. *PLoS Genet* 11(11):e1005534

Torres JB, Ruberson JR, Whitehouse M (2009) Transgenic cotton for sustainable pest management: a review. In: Organic farming, pest control and remediation of soil pollutants. Springer, Dordrecht, pp 15–53

Wan P, Xu D, Cong S, Jiang Y, Huang Y, Wang J et al (2017) Hybridizing transgenic Bt cotton with non-Bt cotton counters resistance in pink bollworm. *Proc Natl Acad Sci* 114(21):5413–5418

Witjaksono J, Wei X, Mao S, Gong W, Li Y, Yuan Y (2014) Yield and economic performance of the use of GM cotton worldwide over time: a review and meta-analysis. *China Agric Econ Rev* 6(4):616–643

Wu Y (2014) Detection and mechanisms of resistance evolved in insects to Cry toxins from *Bacillus thuringiensis*. In: Advances in insect physiology, vol 47. Academic, pp 297–342

Xiao Y, Liu K, Zhang D, Gong L, He F, Soberón M et al (2016) Resistance to *Bacillus thuringiensis* mediated by an ABC transporter mutation increases susceptibility to toxins from other bacteria in an invasive insect. *PLoS Pathog* 12(2):e1005450

Xiao Y, Dai Q, Hu R, Pacheco S, Yang Y, Liang G et al (2017) A single point mutation resulting in cadherin mislocalization underpins resistance against *Bacillus thuringiensis* toxin in cotton bollworm. *J Biol Chem* 292(7):2933–2943

Xu X, Yu L, Wu Y (2005) Disruption of a cadherin gene associated with resistance to Cry1Ac δ-endotoxin of *Bacillus thuringiensis* in Helicoverpa armigera. *Appl Environ Microbiol* 71(2):948–954

Yang P, Iles M, Yan S, Jolliffe F (2005) Farmers' knowledge, perceptions and practices in transgenic Bt cotton in small producer systems in Northern China. *Crop Prot* 24(3):229–239

Yang F, Kerns DL, Head GP, Price P, Huang F (2017) Cross-resistance to purified Bt proteins, Bt corn and Bt cotton in a Cry2Ab2-corn resistant strain of Spodoptera frugiperda. *Pest Manag Sci* 73(12):2495–2503

Zhang B (2013) Transgenic cotton: from biotransformation methods to agricultural application. In: Transgenic cotton. Humana Press, Totowa, pp 3–15

Zhang H, Tian W, Zhao J, Jin L, Yang J, Liu C et al (2012) The diverse genetic basis of field-evolved resistance to Bt cotton in cotton bollworm from China. *Proc Natl Acad Sci* 109(26):10275–10280

Zhao X, Tisdell CA (2009) The sustainability of cotton production in China and in Australia: Comparative economic and environmental issues (No. 1741-2016-140564).

Chapter 11
GM Cotton for Stress Environments

Nasir Ahmad Saeed, Moddassir Ahmad, and Zahid Mukhtar

11.1 Introduction

Cotton is the most important world-leading natural fiber and second largest oilseed crop. It is grown in more than 30 countries with major share from China, India, the USA, Brazil, and Pakistan. It is predominantly cultivated in areas with warmer climates (Riaz et al. 2013). Population growth stimulates a growing demand for textile fiber that is met either by natural fiber (cotton) or by the synthetic fibers. The annual growth in the demand for total textile fiber is 3–4%. Cotton fiber accounts for only 35% of the total demand, while the rest is fulfilled by the synthetic man-made fiber. Contrary to natural fiber production from cotton plants grown in soil, synthetic fiber production can be increased by simply setting up more factories. Since the synthetic fiber is made from petroleum products, the real challenge is to produce more natural fiber from the available land resources. The land area under cotton production is limited to 32 million hectares (Salman et al. 2016). Over-farming, land degradation, and a number of other environmental factors are constant threats to existing cotton production in certain areas of cotton-growing regions. Major cotton-growing areas are rain-fed and constantly under threats of water shortage and desiccation. Thus modern technology inputs are required to maximize production per hectare from marginal lands in stress environments. In the past, growers realized the impact of biotech GM cotton. The reductions in inputs were accompanied by positive impact on cotton sustainability, production, and profitability. These insect-resistant (IR) and herbicide-tolerant (HT) biotech GM cotton was developed and commercialized more than two decades ago. In 2017, GM cotton was planted on 24.8 million hectares (ISAAA 2018), and this area is increasing day by day. In the future, there will be more demand for abiotic stress-tolerant GM crops.

N. A. Saeed · M. Ahmad · Z. Mukhtar (✉)
Agricultural Biotechnology Division, National Institute for Biotechnology and Genetic Engineering (NIBGE), Faisalabad, Pakistan

Mehboob-ur-Rahman et al. (eds.), *Cotton Precision Breeding*,
https://doi.org/10.1007/978-3-030-64504-5_11

Environmental or abiotic stresses are a major global problem limiting crop productivity from 50 to 70% in many parts of the world. Global climate change is further aggravating the situation, and the challenge to feed the growing population is getting more serious (Gill et al. 2013). Plants are sessile in nature and react to internal and external signals by modulating their growth and development. In nature, plants are not totally free from stresses. Certain environmental factors may have potential adverse effect on crop plant's growth, survival, development, and reproductive processes. These environmental stresses such as salinity, drought, heat, heavy metals, nutrient starvation, waterlogging, flooding, light intensity and photoperiod, strong winds, cold and freezing, pollution, increased level of greenhouse gasses, etc. can have a devastating effect on major agricultural crops including cotton (Sett 2017). Drought and salinity are likely to cause serious salinization of more than 50% of all the arable lands by 2050 as they are already widespread in many regions (Ashraf 2002). Among the abiotic stresses, drought, soil salinity, and low soil fertility are the most significant problems in developing countries. In nature, plants that cannot tolerate an environmental stress either will perish or will be outcompeted by other plant species and become extinct. In the recent scenario of climate change, environmental stresses are important factors in determining the geographic ranges of crop plants. In this chapter, some of the common abiotic stresses that affect cotton crop yield and their possible remedies through biotechnology approaches will be examined.

11.2 Cotton Crop Improvement Strategies for Abiotic Stressed Environments

Although a variety of breeding, hybridization, mutations, and other genetic techniques have been used for the improvement of crops, the overall annual genetic gain have not exceeded 0.8–1.2% in crop productivity due to low genetic diversity. Traits regulating abiotic stresses are highly complicated and quantitative in nature and therefore need much more consideration (Bakhsh and Hussain 2015). Even by adopting modern genetic modifications and genome editing tools, the target of accomplishing 2% genetic gain is a tough errand. The crop yields have reached to a plateau in some parts of the world due to narrow genetic base, and further increase through conventional means is not feasible under extreme weather and environmental conditions. The introduction of new genes from diverse sources or overexpression of existing genes, selection accuracy, and efficiency through global partnership may overcome these stress-related problems and improve crop yields.

11.3 Types of Abiotic Stresses, Molecular Mechanisms, and Pathways Controlling Abiotic Stress in Cotton

There are a number of abiotic or environmental stresses affecting cotton crop in the field (Fig. 11.1). Plants detect changes in the surrounding environment and initiate growth responses. Two major classes of stress-related genes have been identified (a)

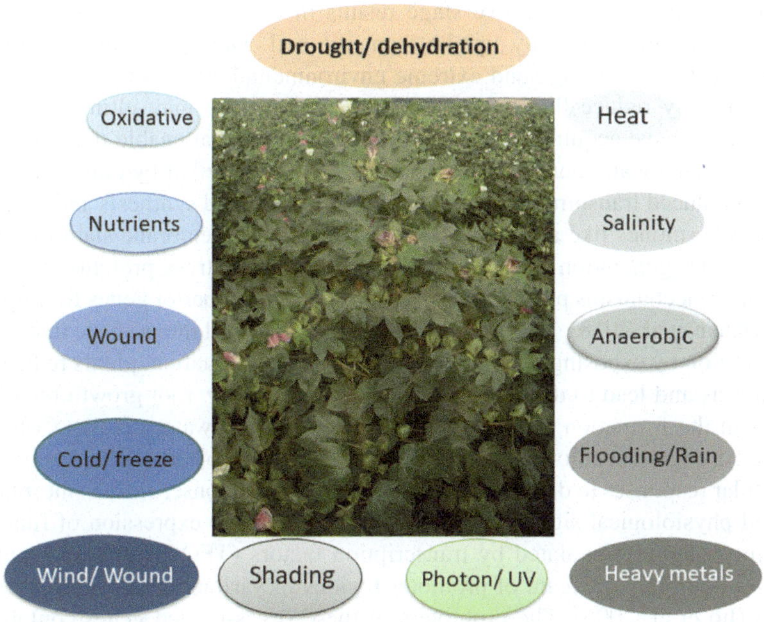

Fig. 11.1 Types of abiotic stresses affecting cotton crop in the field

genes controlling functions and (b) genes controlling regulatory pathways. The corresponding genes include ions and water channels, osmoprotectants, dehydrins, transcription factors, cell signaling and transduction pathways, ROS scavenging mechanisms, Ca^{+} and nitric oxide (NO) pathways, heat shock proteins, chaperones, etc. (Karan and Subudhi 2012). A large number of stress-responsive genes have been identified using differential hybridization, RNA sequencing, high-throughput genome sequencing and microarrays, etc. (Li et al. 2018). These genes under suitable stress-inducible promoters play a major role in controlling environmental stresses in cotton and other crop plants. Fine-tuning of selected genes under suitable promoters is required to develop abiotic stress-tolerant cotton.

11.4 Drought Stress

Cotton is predominantly a drought-tolerant crop because of its deep root system. Drought or water deficit is the single most important yield-limiting factor (Khan et al. 2018). Drought may occur at any time of growth cycle, but the timing and gravity of stress may significantly affect cotton crop production. Among the more pronounced effects on plants are the reduction of plant growth, photosynthesis, carbon fixation, leaf expansion, transpiration, cell growth, cell wall synthesis, protein synthesis, etc. So water availability is crucial for sustainable cotton fiber production (Khan et al. 2018). But a significant part of cotton production comes from rain-fed areas of the world. Drought during the early stages reduces vegetative growth whereas at reproductive stage results in boll drop. Plants lose water by transpiration and restore by uptake from the soil. During summers, prolonged drought or water shortage and extreme environmental conditions can stress crop and ultimately reduce fiber yield. Acute water shortage may ultimately kill the plant. Plants have acquired some control mechanisms that enable them to survive in less extreme water deficits. Plants respond to water deficit by conserving water through reduced transpiration, stomatal closure, increased synthesis and regulation of stress hormones like abscisic acid, maintenance of membrane stability, carbon fixation rate, generation of antioxidants, induction of stress proteins, etc. These molecular mechanisms play a vital role in discovering superior genes for important traits (Khan et al. 2018). Leaves of many plant species roll into a shape that reduces transpiration by exposing less leaf surface. Under this situation, plants reduce photosynthesis and lead to diminishing crop yield. Similarly, root growth retards and results in the formation of shallow roots, thus limiting water uptake. Cotton has evolved drought-specific and some general morphophysiological, biochemical, and molecular responses to drought stress. In this whole response, various morphological and physiological signal pathways are involved. The expression of functional proteins is largely regulated by transcription factors (TFs). These are regulatory proteins and bind upstream of promoter regions of various stress-related genes in plants (Ito et al. 2006). The promoters of these stress-related genes contain conserved cis-domain PyCCGACAT, named as DRE/CRT, which is involved in regulation of stress response (Yamaguchi-Shinozaki and Shinozaki 2005). With the advent of recombinant DNA technology, plenty of genes have been cloned and assessed. Over-expression of *DREB1A* under *rd29* promoter in wheat improved drought tolerance in wheat and produced 25% more grain yield than that of the best available check (personal experiments by Nasir A. Saeed). Expression of *AtDREB2A* driven by *rd29* promoter increased total root volume, surface area, and total root length without affecting shoot morphometric growth parameters (Lisei-de-Sa et al. 2017). The plant-specific transcription factors (TFs) play essential roles in regulating development and stress. Plant-specific NAC proteins comprise one of the largest transcription factor families in plants. Over-expression of *Arabidopsis* transcription factors such as AtRAV1/2 and AtAB15 (Mittal et al. 2014) and rice NAC gene *SNAC1*improves salt and drought tolerance by enhancing root development and reducing transpiration rate in transgenic cotton (Liu et al. 2014). Similarly, expression of *GhNAC2* (derived from *G. herbaceum*) improves root growth and imparts tolerance to drought in transgenic cotton (Gunapati et al. 2015).Over-expression of *GhNAC79* resulted in early flowering phenotype in *Arabidopsis* and cotton. Furthermore, VIGS-induced silencing of *GhNAC79* in cotton led to a drought-sensitive phenotype (Guo et al. 2017). *GhNAC79* positively regulates drought stress, and it also responds to ethylene and methyl jasmonate (MeJA) treatments, making it a good candidate gene for cotton improvement. Regulatory genes such as *LOS5* (encodes a molybdenum cofactor sulfurase) responsible for aldehyde oxidase activity (Yue et al. 2012) and isopentenyl transferase (a rate-limiting enzyme for cytokinin biosynthesis) (Kuppu et al. 2013) have also been used to improve drought tolerance.

Cotton plants over-expressing choline dehydrogenase (*betA*) gene derived from *E. coli* (Lv et al. 2007; Zhang et al. 2011) and choline monooxygenase (*Ah-CMO*) gene derived from *Atriplex hortensis* (Zhang et al. 2009) showed elevated levels of glycine betaine and were found tolerant to cold and drought stress conditions. Aldehyde dehydrogenase (ALDH) is essential for scavenging aldehydes when plants are exposed to stress. Ectopic expression of *ScALDH21* (isolated from *Syntrichia caninervis*, which is an extremely tolerant moss) improved drought tolerance in cotton (*G. hirsutum*). Plants accumulated more proline and had higher peroxidase activity. Cotton plants showed greater plant height, larger bolls, and improved cotton fiber (Yang et al. 2016).

In cotton, expression of *Arabidopsis* 14-3-3 protein (GF141), tobacco osmotin (Nt Osm), and *G. arboretum* heat shock protein (Ghsp26) genes resulted in increased water-stress tolerance and reduced leaf-wilting under drought stress (Yan et al. 2004; Parkhi et al. 2009; Maqbool et al. 2010). Transgenic cotton, expressing H^+-pyrophosphatase (H^+-PPase) genes from *Thellungiella halophila* (*Ts-VP*) and *Arabidopsis thaliana* (*AVP1*), conferred increased salinity tolerance up to 250 mM of NaCl as well as improved photosynthetic ability under drought. Transgenic plants could produce ~40% enhanced seed yield than the untransformed control (Lv et al. 2008, 2009; Pasapula et al. 2011). Furthermore, these transgenic plants could accumulate 20% more proline and 25% more abscisic acid and produced enhanced biomass after exposure to drought conditions (Yue et al. 2012).

Dehydrin (DHN) is a class of hydrophilic proteins widely found in plants. These proteins belong to second member of late embryogenesis abundant protein family. LEA gene *HVA1* has been transformed into wheat, which exhibited high tolerance to drought and salinity. Barley dehydrin *dhn3* and *dhn4* greatly improved the osmotic stress tolerance of *Arabidopsis thaliana* seedlings (Park et al. 2006). *Saussurea involucrata SiDhn2* gene encodes a dehydrin protein that belongs to the KS subtype of dehydrin family. When transformed, transgenic cotton plants exhibited remarkable decrease in boll abscission rate and highly increased seed yield (Liu et al. 2017). When cotton LEA2 genes, *Cot_AD24498*, *CotAD_20020*, *CotAD_21924*, and *CotAD_59405*, were expressed in *Arabidopsis*, these resulted in increased antioxidants, catalase (CAT), peroxidase (POD) and superoxide dismutase (SOD) accumulation, increased root length and significantly reduced oxidants like hydrogen peroxide and malondialdehyde (MDA) concentrations in leaves (Magwanga et al. 2018). The key physiological responses against drought stress in cotton, including stomatal closure, root development, cellular adaptations, reduced photosynthesis, enhanced production of abscisic acid (ABA) and jasmonic acid (JA), and scavenging of reactive oxygen species (ROS), have been identified (Ullah et al. 2017). Drought stress induces stress-related transcription factors, genes, and signaling pathways in cotton. Park et al. (2012) identified 519 differentially expressed transcript derived fragments. Nearly 70% of these transcripts belong to four major categories: (1) unclassified, (2) stress/defense, (3) metabolism, and (4) gene regulation. They found that heat shock protein-related and reactive oxygen species-related transcripts were induced by water-deficit stress. TaMnSOD plays a crucial role as a peroxidation scavenger when expressed in cotton (Zhang et al.

2014). When water was withheld for 18 days, plants accumulated more proline and soluble sugar than wild-type cotton with increased biomass under drought stress. Transformation of GbMYB5 transcription factor improves drought tolerance in cotton and tobacco. Virus-induced gene silencing compromised tolerance to drought and reduced post-rewatering water recovery survival rate to 50% as compared to 90% in wild-type. Silencing GbMYB5 decreased proline content, antioxidant enzyme activities and increased malondialdehyde content in cotton, while the expression of drought-inducible genes *NCED3*, *RD22*, and *RD26* were not affected (Chen et al. 2015).

To improve stress tolerance, Mishra et al. (2017) transformed Rice SUMO E3 ligase gene *OsSIZ1* in cotton. This enzyme is involved in a sumoylation reaction, a widely conserved post-translational modification that comprises the triggering of SUMO's C-terminus by the E1 activating enzyme, its subsequent transfer through the E2 conjugating enzyme, and ultimately the joining up of SUMO protein to the target protein substrates via the E3 ligase. These cotton plants showed higher photosynthesis and growth under drought and thermal stress. This stress tolerance is correlated with higher fiber yield in both controlled environment and field trials.

11.5 Salinity Stress

Globally, one billion hectares of arid and semi-arid land is salt affected (FAO 2005; Dagar and Minhas 2016). Soil salinity will become progressively more severe over time due to climatic changes, non-scientific irrigation practices, and excessive use of fertilizers (Han et al. 2015). It is predicted that soil salinity will affect more than 50% of arable land by 2050 (Ashraf and Wu 1994; Demiral and Türkan 2006). Although cotton is considered as moderately salt-tolerant crop, excess of salts in the soil hinders plant growth and reduces yield. These salts lower the water potential of soil solution and cause water deficit in plants even though soil has plenty of water. Sodium chloride is abundantly found in saline soils that contributes Na^+ ions (Munns and Tester 2008). More salts present in soil results in building up of negative water potential and thus lowers the water potential gradient from soil to roots and thereby reduces uptake. The Na^+ and other ions when present in higher concentrations are toxic to plants that may reduce acquiring of solute-rich water from soil. In sodic soils, plants show reduced dry mass and lint yield and can quickly take up toxic Na^+ ions with limited concentrations of essential nutrients like phosphorus, potassium, and micro-nutrients in tissues. Early indicators of soil salinity include poor crop growth. Severe symptoms can include bare, salt-encrusted soil surfaces and greasy-looking black patches due to dispersion of organic matter and total crop failure. Plants respond to moderate levels of salinity by different mechanisms. Some solutes are produced that help tolerate salinity. Most crops are susceptible to high salinity (except halophytes) and give reduced yield at farmer's fields. Salt-tolerant plants have mechanisms like salt glands that pump salts across the leaf epidermis. These mechanisms are absent in major crops like wheat, rice, maize, soybean, cotton, etc.

A number of genes have been identified that are responsible for salinity tolerance, and their manipulation and expression may be useful for developing salt-tolerant crops.

Screening of cotton varieties and germplasm is of great importance (Ashraf and Wu 1994; Ahmad et al. 2002). Some wild upland cotton and wild species exhibit relatively high tolerance to salinity, but Ye and Liu (1998) postulated that the probability of success by direct selection or screening could be low particularly in commercial varieties. Little success has been achieved in developing considerable salt-tolerant varieties with the conventional methods due to narrow germplasm resources available and particularly due to the complexity of the tolerance mechanism in cotton (Ma et al. 2011). Under stress, plants temporarily adapt saline conditions using K^+, Na^+, and osmoprotectants for osmoregulation (Munns and Tester 2008). Based on earlier studies, Ashraf (2002) reported that plasma membrane-associated ATPase responds to increasing supply of Na^+ in the growth medium, but the activity of transport proteins in plasma membrane alone was insufficient to regulate intracellular Na^+ levels. This needs to be empirically tested to verify individual genes in cotton. Vacuolar-ATPase is also not responsive to increased external Na^+. The inability of V-ATPase to respond to Na^+ is an indication of the lack of effective driving force for compartmentalization of Na^+ in cotton. To counter ion toxicity, plants have devised mechanisms to compartmentalize Na^+ in vacuoles and remove the excess Na^+ from cytosol to the apoplast (Munns et al. 2012). Only few studies addressing salt tolerance traits are available in cotton (Tiwari et al. 2013; Oluoch et al. 2016). Wei et al. (2017) performed transcriptomic analysis of cotton species *G. klotzschianum* under salinity stress to elucidate the mechanism of salt tolerance. Firstly they detected changes in hormones like H_2O_2 and glutathione (GSSH and GSH) and later explored the gene expression pattern of leaves and roots treated with 300 mM NaCl for different time regimes including 0, 3, 12, and 48 h along with respective controls by RNA-seq using Illumina platform. They found significant increase in ABA and H_2O_2 while a decrease in GSH content and increase in the GSSH content at 48 h, respectively, under 300 mM NaCl stress. Recent genome-wide association study (GWAS) using Illumina Infinium CottonSNP63K array identified 23 SNPs. Most of these genes were involved in encoding transcription factors, transporters, or enzymes (Sun et al. 2018) and were previously reported as being involved in plant salt tolerance such as NAC, MYB, NHX, WD40, VP, CDPK, LEA, and CIPK. The differential expression further paved the way for identification and cloning of stress-responsive genes. Previous studies have shown that ectopic expression of cotton CBL-interacting protein kinase gene (*GhCIPK16*) and *SnRK2* could enhance abiotic stress tolerance (He et al. 2013; Bello et al. 2014). Over-expression of vacuolar *AtNHX1* gene in cotton can also improve salt tolerance (He et al. 2005). Transcription factors GhWRKY39 and GhDREB1 (Shi et al. 2014; Huang et al. 2009) and over-expression of rice NAC gene (*SNAC1*) could also improve salt tolerance in transgenic cotton (Liu et al. 2014). Over-expression of ROS scavengers such as *GhSOD1*, *GhCAT1*, and *GhMT3a* showed high salt tolerance in cotton (Luo et al. 2013). These and many other genes have been characterized for developing abiotic stress-tolerant cotton.

11.6 Heat Stress

According to the temperature analysis studies conducted by scientists at the NASA's Goddard Institute for Space Studies (GISS), the average global temperature on earth has risen slightly more than 1° Celsius since 1880, and two-thirds of this has arisen since 1975, at a rate of about 0.15–0.2 °C per decade. This is because of increasing concentration of atmospheric CO_2 and other greenhouse gases. Government approaches right now submit us to surface warming of 3–4°C above pre-industrial levels by 2100, which will boost ice sheet melt (Golledge et al. 2019). Professor Adam Scaife (Head of Long Range Prediction at Met Office, Devon UK) forecasted that global average temperature for 2019 will be 1.10 °C above the pre-industrial average period from 1850 to 1900. In their simulations, future ice sheet melt will enhance global temperature variability and contributes up to 25 cm to sea level increase by the year 2100. The increased temperature causes heat stress and reduces the productivity of major crops with dramatic consequences, if urgent action is not taken. Climate change is a huge and growing threat to nature and the people. In agriculture sector, loss of 4.2 billion dollars was recorded due to excessive heat and drought stress (Pachauri et al. 2014). Many of the countries of Asia, Africa, and Middle East are located in very hot desert climate where temperature raises up to 50 °C in summer. This situation either renders poor growth of crops, or otherwise a heavy supply of irrigation water is required to cope with transpiration loss and normal growth of crops. In these harsh conditions, most crops cannot be grown. This excessive heat harms and kills the plant by denaturing its enzymes and damaging its metabolism and inhibits photosynthesis by deactivation of Rubisco. Sometimes, drought and heat come together and make the situation worst for plant growth. Hot dry weather tends to dehydrate many crop plants and allows closing of stomata and sacrifices evaporative cooling. Above 40 °C, plants start producing heat shock proteins which prevent denaturation and help protecting cell machinery from further damage. Cotton is a perennial shrub and grows in semi-desert warm climate. Temperature is a primary environmental factor controlling growth, development, and adaptation of cotton plant (Reddy et al. 1991). Despite its origin, cotton does not necessarily yield best at excessively high temperature, and a negative correlation has been reported. When temperature increases to more than optimum levels (day/night temperature of 30/22 °C to 35/27 °C), the cotton plant suffers. High temperature (35/27 °C to 40/32 °C) damages photosynthesis activity, and even short-term exposure to very high temperature (>40/32 °C) has a pronounced effect on growth and reduces yield (Azhar et al. 2009). Cotton grows in regions of Asia and Africa where summer approaches 50 °C that reduces crop yield in terms of low plant population per unit area and reduces fiber yield and quality. It has been estimated that cotton crop produces about 25% of its potential yield because of the damaging effects of environmental stresses (Boyer 1982) including heat. Heat stress can affect at any stage/time and causes boll shedding and pollen sterility. Every degree of maximum daily temperature over 30 °C in the month of July decreases lint yield to the extent of 50 kg per hectare (Oosterhuis 2002; Khan et al. 2017) and also reduces the fiber quality.

Relative cell injury level from leaf disks at high temperature has been suggested as one of the screening techniques for heat tolerance in plants (Sullivian 1972). This technique is simpler, quicker, and less expensive than the whole plant screen and can be used during early growth stages. Cotton plant has been improved using several indices including a visual index, phenotypic index and fruit height response index (Singh et al. 2007), high stomatal conductance, okra-type leaf, thicker leaves, earliness, and cell membrane stability. However, low level of genetic variability is present in domesticated upland cotton for improvement. Only a limited number of heat-tolerant lines are available for breeding. Cotton plant can withstand adverse environmental conditions in several phases, and accumulation of chemicals is extremely vital. Calcium (Ca^{++}), kinases, ROS, carbohydrates, transcription factors, regulation of gene expression, second messenger, and phytohormone signaling pathways play a major role in gene activation (Zahid et al. 2016). Heat stress is a complex trait, and the underlying mechanism how the plants respond is partially known. Recently, global analysis revealed as many as 575 genes respond to heat stress. Genes encoding heat shock proteins (HSPs), transcription factors (TFs), and protein cleavage enzymes were induced, whereas genes encoding proteins associated with electron flow, photosynthesis, glycolysis, cell wall synthesis, and secondary metabolism were generally repressed under heat stress (Cottee et al. 2014). Out of 61 abiotic stress-related genes, 41 were specifically associated with heat stress mediation. Over-expression of some key genes improves heat tolerance in cotton. Among the downstream processes, protection against oxidative damage and protein aggregation are crucial for maintaining cellular membrane integrity and photosynthesis. Therefore, over-expression of heat shock proteins (HSPs) initially appeared to be a promising approach for enhancing heat tolerance in cotton (Zhang et al. 2016b). HSPs are important type of stress-induced proteins that are produced in plants in response to external stresses. Heat shock transcription factors bind to the heat shock elements in the upstream region of HSPs to increase their expression. Increased expression improves the ability of plant to resist various abiotic stresses and act as a molecular chaperone that has a role in plant stress physiology. HSPs can be classified as high molecular mass proteins (e.g., HSP100, HSP90, HSP70/DnaK, and HSP60/GroE), low molecular mass proteins (e.g., HSP20) and small heat shock proteins (sHSP) (Ma et al. 2016). HSP20 are the most abundant heat shock proteins in plants and provide temporary protection. HSP20 proteins (14 sub-families) are encoded by nuclear multigene families and are localized in different cellular compartments. To enhance heat tolerance, these individual HSPs have been transformed into plants.

11.7 Waterlogging, Submergence, and Flooding Stress

Excess water is a problem for growing plants. During heavy rains, flooding, or soil waterlogging conditions, roots cannot breathe properly due to lack of oxygen for respirations and leads to suffocation and cell death. The situation varies with different crop plants and climatic changes. Rice is naturally adapted to submerged conditions, while many other crops like cotton cannot survive in standing warm water for more than 2 days. Oxygen deprivation (hypoxia) stimulates the production of ethylene and carbon dioxide, impairs root growth; reduces photosynthesis, leaf area, nutrient uptake and triggers apoptosis. Waterlogging reduces cotton yield and dry matter (due to low radiation use efficiency) by causing nutrient loss, decrease in stomatal conductance and leaf water potential, decline in soil structure, fruit shedding, and slowing growth of new fruiting sites.

According to FAO (2007), 20–30 million hectares of land is affected by soil waterlogging mainly because of heavy rainfall. In nature, cotton is poorly adapted to waterlogged conditions that may reduce cotton yield up to 10% (Conaty et al. 2008) under short-term waterlogging. Yield reduction can reach up to 30% under 9 days of waterlogging. In the USA, losses in crop production due to flooding were second to drought in many years of the past 12 years (Bailey-Serres et al. 2012) and accounted for more than 70% of the reduction in harvests in 2011. There is limited information available on genetic variation for waterlogging. However, some physiological studies indicated resistance in some cotton varieties which were originated on heavy clay soils. Cotton yield is severely affected due to waterlogging conditions because of its sensitivity at flowering and boll-setting stages. Bange et al. (2004) found that lint yield was reduced by 48 kg/ha for every day when the soil was low in oxygen.

Plants have evolved adaptive mechanisms that enable them to survive short periods of low oxygen supply. The molecular basis of adaptation to transient low oxygen conditions has not been completely characterized (Millar and Dennis 1996; Dennis et al. 2000). Genes responsible for submergence tolerance, promoter elements, and transcription factors have been identified in few plant species. These genes can be used in future for improving resistance to waterlogging in cotton. Alcohol dehydrogenase (*Adh*) gene family of *Gossypium hirsutum* was expressed under CaMV35S promoter. Christianson et al. (2010) assayed global gene transcription response in roots and leaf tissues of partially submerged plants. Waterlogging caused significant reduction in stem elongation, shoot and root biomass, leaf number and altered the expression of 1012 genes in root tissue. Many of these were associated with cell wall modifications and growth pathways, glycolysis, fermentation, mitochondrial electron transport, and nitrogen metabolism. Waterlogging of roots also altered global gene expression of leaf tissues, significantly changing the expression of 1305 genes after 24 h of flooding. Genes affected were associated with cell wall growth and modification, tetrapyrrole synthesis, hormone response, starch metabolism, nitrogen metabolism, etc. Similarly, Qi et al. (2012) and others revealed complex responses to low O_2 in 5–10% of all the genes assayed. In cucumber, these genes are linked to carbon metabolism, photosynthesis, ROS generation/scavenging, and hormone synthesis. So water stress should be seen as a compound stress. Rice handles submergence stress by internal aeration and growth controls. A quiescence strategy based on Submergence-1A (SUB1A) or an escape strategy based on SNORKEL1 (SK1) and SNORKEL2 (SK2) is used for controlling the growth. Rice controls waterlogging by forming lysigenous aerenchyma and a barrier to radial O_2 loss (ROL) in roots to supply O_2 to the root tip (Nishiuchi et al.

2012). With iTRAQ labeling with LC-MS/MS, Li et al. (2014) identified 169 differentially expressed proteins including 116 upregulated and 53 downregulated proteins. Go, COG, and KEGG pathway analysis revealed that 12% of these proteins were involved in water stress response of cotton. Zhang et al. (2015) reported that waterlogging reduced the photosynthetic rate by 16.9% and NO concentration by 17.5% and increased malondialdehyde (MDA) accumulation by 22.2%. Waterlogging regulated the expression of a set of genes associated with leaf photosynthesis, ROS scavenging, anaerobic metabolism, or cell growth like LHCB, CSD, ACS6, ADH, PDC, ERFs, XTHs, and EXPAs (Zhang et al. 2015).Waterlogging accelerates abscission of young fruits/leaves, and this process is potentially regulated by high ethylene production. Application of an anti-ethylene agent, aminoethoxyvinylglycine (AVG), blocked ethylene accumulation and increased growth and fruit retention. Yield losses can be minimized by downregulation of genes controlling ethylene biosynthesis (Najeeb et al. 2015). Waterlogging increased the expression of anaerobic fermentation-related genes, such as alcohol dehydrogenase (ADH), suggesting that ethylene may play a key role in the survival of cotton under waterlogging stress (Zhang et al. 2017). Furthermore, in okra and maize crops, ethylene priming improved their tolerance to waterlogging stress (Vwioka et al. 2017).

Pathways of anoxia tolerance are also a potential tool toward developing waterlogging tolerance in cotton genotypes. Under stress, plants produce reactive oxygen species. Plants have scavenging mechanisms to cope with these stresses. Superoxide dismutase (SOD) plays a critical function. Zhang et al. (2016a) performed microarray-based expression analysis and found ten SOD genes in cotton. These include Cu-Zn-SOD, Fe-SODs, and Mn-SODs etc. Waterlogged conditions adversely affect cotton fiber quality. Sucrose phosphate synthase and fiber invertase are most sensitive (downregulated), whereas expansin, β-1,4-glucanase, and endoglucan transferase are unregulated under waterlogged conditions. Exposure to waterlogged conditions for more than 6 days limited fiber development (Kuai et al. 2016). RNA-seq-based studies in cucumber revealed de novo adventitious root primordial initiation and broadened the understanding of mechanism of waterlogging in plants. More than 27,000 transcripts were detected from which 1494 genes were differentially expressed 2 days after exposure to waterlogging. Formation of adventitious roots is an adaptive trait for waterlogging tolerance. It allows the submerged tissue to obtain oxygen directly from air (Xu et al. 2017). So the mechanisms of tolerance have to be understood. This requires global analysis of cotton plant at molecular level by deploying transcriptomics, metabolomics, and proteomics techniques. Secondly, a large number of cotton varieties have been developed around the world in the past. These breeding efforts mainly focused on high lint yield and quality concomitant with loss of genetic diversity and stress resistance. Exploring the gene pools for tolerance to waterlogging and incorporation through molecular markers is possible (Mustroph 2018). Zhang et al. (2019) tested 104 entries from 81 genotypes and identified 12 waterlogging-tolerant genotypes (7 New Mexico cotton Acala lines, 3 glandless lines, and 2 commercial cultivars). Inheritance studies on hypoxia tolerance are very scanty. Predominance of additive genetic effects and high heritability along with partial dominance suggest that selection for waterlogging in early generation can be useful (Hussain et al. 2019). Thirdly, known flooding resistance genes (as given above) could be transformed into cotton varieties. Lastly, based on the information from global analysis, downregulation of key genes can be done through genome editing CRISPR-Cas9 technology. Targeted genome editing with engineered nucleases can be used for disruption of undesirable genes or metabolic pathways including abiotic stresses. In maize, the best candidate gene *GRMZM2G110141* is identified that enhanced waterlogging tolerance and can be transferred through marker-assisted breeding (Yu et al. 2018) and genetic engineering. Homologues of this gene can be identified in cotton. In *Chrysanthemum* genome-wide association studies, 14 SNPs were identified. Four putative candidate genes for waterlogging tolerance have been identified (Su et al. 2019). In waterlogged field, oxygen-deficient conditions hamper plant growth (Biswas and Kalara 2018). The ability to detoxify the adverse effects of ROS by producing different types of antioxidants is the preferred method for cotton improvement.

11.8 Cold and Freezing Stress

At low temperature, movement of fluid in cell membranes fluctuates substantially. A biological membrane is a fluid mosaic with proteins and lipids which move in plane of the membrane. In cold conditions, lipids and proteins lose their fluidity. Low temperature adversely affects the functions of membrane proteins and alters solute transport across the membranes. Under cold stress, plants alter lipid composition by altering proportion of unsaturated fatty acids and impede crystal formation. In colder regions, cells of many frost-tolerant plant species increase cytoplasmic levels of specific solutes such as sugars that together help reduce loss of water from cell during extracellular freezing. The unsaturation of membrane lipids also increases and maintains proper membrane fluidity.

Cotton is a perennial plant. It originated in hot climatic regions and prefers to achieve maximum production. In nature, cotton has certain resilience to high temperature but no resilience to low temperature. With time, cotton is now modified and adapted to grow in diverse environmental conditions prevalent in many countries around the world. Temperature affects all stages of cotton growth including germination, fruiting, growth rate, photosynthesis and fiber quality. Generally, cotton plant performs best with long and hot season. Plant grows faster in hot environment and gives higher economic yield. The optimum temperature for growth and development of cotton plant is considered to be around 28 °C with a range of 23–32 °C. This is considered to be a non-stressful condition (Burke et al. 1988). The temperatures experienced by cotton during the growth season potentially range from a minimum of 0 °C at the time of planting and/or harvest to a maximum of ~45 °C. Below 0 °C is considered as frost/freezing temperature. Cold-sensitive species like rice, cotton, tomato, and maize stop growing below this temperature and reduce fertility and thus yield (Holaday et al. 2016). The most typical symptoms are poor germination, stunted seedlings, yellowing of leaves (chlorosis), reduced leaf expansion, withering, and wilting which may lead to plant death. This is caused by over-efflux of ions and metabolites through the plasma membrane, e.g., K+ ions, amino acids, and sugars.

Cotton is produced across a wide range of environments and management conditions from hot humid sub-tropical to semi-arid environments. It is clear that the physiological resilience to cold and other abiotic stresses is considerable. Temperature below and above the optimum affects metabolism differently. As the temperature declines below optimum, reaction rates decrease in a largely reversible manner. Below a certain limit, oxidative stress and changes in membrane structure can result in non-reversible damage.

The expressions of many genes are upregulated by cold such as lipid transfer proteins, late embryogenesis proteins, alcohol dehydrogenase, dehydrins (Dhn), and translation elongation factors (Nishida and Murata 1996; Lee et al. 1999). ABA (Rab genes) has long been believed to be involved in cold signaling and has an important role in cold acclimation. Tolerance to freezing in cotton can be increased after treating with ABA. Cold acclimation is a process by which plants acquire freezing tolerance upon their prior exposure to low non-freezing temperature. Unfortunately, cotton is incapable of cold acclimation and cannot tolerate ice formation in its tissues. *Gossypium barbadense* genotypes are generally more tolerant to low temperature than that of the other species. It was reported that soluble sugars (Fan et al. 1995), peroxidase (Fan et al. 1995), peroxidase isoenzyme (Guo et al. 1991), cytochrome oxidase (Fan et al. 1989), non-protein nitrogen (Li et al. 1996), etc. are involved in chilling tolerance in cotton seedlings. Acquisition of freezing tolerance is a phenomenon which depends on multiple pathways with the C-repeat/dehydration-responsive binding factor CBF pathway, responsible for cold-regulated (COR) gene regulation which is well-known and widely characterized. *GhDREB1* plays an important role in improving cold tolerance and affects plant growth and development (Shan et al. 2007). Liu et al. (2014) reported that P-type ATPases translocate ions across membranes and are involved in the transport of phospholipids. *GbPATP* in cotton was induced at low temperatures. *GbPATP*-silenced cotton plants were more sensitive to low temperatures and exhibited greater malondialdehyde (MDA) content and lower catalase (CAT) activity. *GbPATP* transgenic tobacco plants showed better chilling tolerance. Over-expression of AmDUF1517 (isolated from *Ammopiptanthus mongolicus*) in cotton had stronger resistance to cold and other stresses. Further analysis showed that trans-AmDUF1517 cotton displayed significantly higher antioxidant enzymes and lower ROS accumulation. This suggests that over-expression of AmDUF1517 can improve cotton resistance to stress by not only maintaining ROS homeostasis but also by alleviating cell membrane injury (Yu-qiang et al. 2018). From a normalized cDNA library of *Gossypium barbadense*, a number of genes such as *MYB*-related, *C2H2*, *FAR1*, *bHLH*, *bZIP*, *MADS*, and *mTERF* were identified (Zhou et al. 2016). Upon cold treatment, a total of 2782 and 1430 differentially expressed genes were identified in leaves and roots (Wang et al. 2017). They found five aquaporins and CBF1s genes. Another class of endogenous non-coding RNAs called microRNAs (miRNAs) modulate the expression of target genes by degrading mRNA or repressing translation of genes involved in plant development and stress tolerance. Wang et al. (2016) used small RNA and mRNA degradome sequencing to identify low- and high-temperature stress-responsive miRNAs and their targets in cotton (*G. hirsutum* L.). Majority of the miRNAs identified were from genes likely to be involved in response to hormonal stimulus, oxidation-reduction reaction, photosynthesis, plant-pathogen interaction, and plant hormone signaling in cotton (Wang et al. 2016). These results improve our understanding of cold stress tolerance in cotton.

11.9 Oxidative Stress

Oxidative stress is a complex chemical and physiological phenomenon. This is caused by excessive production of reactive oxygen species (ROS) including superoxide radicals, singlet oxygen, and hydrogen peroxide and hydroxyl radicals. ROS are produced in high amount under both biotic and abiotic conditions in the field. Cotton is grown in hot environments where plant faces heat, drought, wind, and other abiotic stresses. ROS are produced by NADPH oxidases, peroxisomes, chloroplast, apoplast, plasma membrane, cell walls, endoplasmic reticulum, and mitochondria (Das and Roychoudhury 2014). ROS target lipids, proteins, and DNA. Excessive ROS damage all parts of cotton plant and reduce crop yield. To counter the deleterious effects of ROS, plants have acquired antioxidant mechanisms (enzymatic and non-enzymatic) that scavenge ROS. Enzymatic antioxidants include superoxide dismutase (*SOD*), catalases (*CAT*), ascorbate peroxidase (*APX*), glutathione peroxidase, glutathione S-transferases, glutathione reductase (*GR*), monodehydroascorbate reductase (*MDHAR*), dehydroascorbate reductase (*DHAR*), guaiacol peroxidase (*GPX*), and peroxiredoxin (*POX*), while non-enzymatic antioxidants include ascorbate, glutathione, proline, tocopherol, flavonoids, proline, and carotenoids. When the balance of ROS production and scavenging is disturbed, the cell faces risk of oxidative stress. Glutathione reductase (GR) activity significantly increases with high temperature. Protection of photosynthesis in cotton from the effects of chilling and high photon flux density can be enhanced by increasing the activities of antioxidant enzymes particularly those of APX and GR (Payton et al. 1999). Enhanced chloroplastic GR activity in transgenic plants results in the increased protection against oxidative stress (Pilon-Smits et al. 2000). Hediye et al. (2014) found that drought- and heat-tolerant cotton lines maintain constitutive activities of superoxide dismutase, ascorbate peroxidase and induced CAT and POX. In another study, accumulation of proline, soluble proteins, soluble sugars, hydrogen peroxide, and superoxide radicals increased significantly in TM-1 cotton line. The relative expression of drought-responsive genes including coding for transcription factors and other regulatory proteins or enzymes controlling genes (ERF, ERFB, DREB, WRKY6, ZFP1, FeSOD, CuZnSOD, MAPKKK17, P5CR, and PRP5) were higher in TM-1 under drought (Hassan et al. 2018). These and many other genes have significant role in protection of cotton against oxidative abiotic stresses.

11.10 Heavy Metal Stress

Heavy toxic metals stress is becoming a major threat to plant growth and development. Increased industrialization, urbanization, and pollution have exacerbated the contamination of soils especially around the cities due to inadequate treatment of wastewater and spill out from wastewater channels. This results in accumulation of heavy metals in soil that are toxic and hazardous to the living organisms. While plants require some heavy metals like cobalt (Co), copper (Cu), iron (Fe), manganese (Mn), molybdenum (Mo), nickel (Ni), vanadium (V), and zinc (Zn) in very small concentration, higher levels can be toxic for plant growth. Certain heavy metals like lead (Pb), cadmium (Cd), mercury (Hg), and arsenic (As) are considered highly toxic and pose a threat to crop production (Chibulke and Obiora 2014). Due to scarcity of irrigation water and population pressure, farmers are using brackish and contaminated water for growing crops. Increased metal concentration in the soil leads to the competition between toxic metals and essential nutrients. Direct effect includes inhibition of cellular functions and oxidative stress and ion homeostasis in plants. Plants control these stresses by overproduction of proline, antioxidant phytochelatins, metallothioneins, and metal transporters (Hayat et al. 2012; Nikalje and Suprasanna 2018). Plants accumulate proline when exposed to heavy metals. Transcriptomic studies revealed genes responsible for salt tolerance mechanism in halophytes. These include phytochelatins (PCS), metallothionein (MT), plasma membrane ATPases (PM H+ -ATPase), pyrroline-5-caboxylate synthase (*P5CS*), and catalase (CAT). Luo et al. (2017) showed induction of transcription factors like *MYB*, *AP2-EREBP*, *bZIP*, and *NAC* which may interact with their downstream targets to activate the tolerance mechanism. To reduce the damage by heavy metals, higher peroxidase activities contribute to heavy metal tolerance in cotton. Aluminum-activated malate transporters are essential for aluminum tolerance, ferritin (iron storage) increase tolerance to oxidative stress caused by excess Fe and over-expression of arsC are essential for Cd tolerance (Ray et al. 2011).

11.11 Nutrient Deficiency

Globally transgenic cotton (*Gossypium hirsutum* L.) varieties are being introduced since the last two decades, covering around 77% cotton-growing area. Most of the transgenic cotton varieties were developed to generate insect pest resistance. However, there are still more opportunities to increase its lint production by improving nutrient use efficiency. Short supplies of fertilizers contribute to nutrient deficiency and stress. So there is considerable interest to improve nutrient use efficiency in cotton to increase its productivity. Nutrient use efficiency in cotton can be enhanced by adapting genetic, molecular, microbial, and management practices. In this section, genes involved in nutrient use efficiency, their associated pathways, and prospects for developing genetically engineered cotton for stress environments are presented.

The rate of suitable fertilizer application depends upon the type of soil, presence of organic matter, availability of various nutrients, and acquisition of available nutrients by crop plants and analysis of plants for uptake of various nutrients. Nonetheless, this information could only be acquired through determination of rate of nutrients uptake at various phenological stages of crop development (Augostinho et al. 2008). The comprehension of soil nutrients dynamics, rate of nutrient uptake, nutrient loss potential through microbes, leaching or precipitation, and nutrient use efficiency of cotton plant would help to devise best strategies for optimal rate, time, and mode of fertilizer application to achieve optimization of fertilizer use efficiency in cotton (Bruulsema et al. 2012).

Nitrogen (N) For a large number of crops including cotton, there is a genetic variability for N absorption and utilization efficiency. In the past, most crop varieties were developed in the presence of high mineral fertilization inputs, thus missing to exploit the genetic differences under a low level of fertilizers. It is necessary to increase the knowledge of cotton gene expression and regulation under N-limited conditions to understand the response of this crop to different N regimes. Such information is vital and useful to clarify the signal transduction pathways and the mechanism that regulate the N-uptake, assimilation, and remobilization pathways. Digital gene expression (DGE) is a useful technique to study gene expression under N-limited conditions. By using this technique, a number of soybean genes were differentially expressed between the low-N-tolerant and low-N-sensitive varieties under N-limited conditions. Some of these genes may be candidate for improving NUE in crops (Hao et al. 2011). Transgenic modifications of N transporters are mostly achieved in rice. Expression of a barley alanine aminotransferase (*AlaAT*) gene under tissue-specific promoter improved biomass and grain yield in rice by increasing nitrogen uptake (Shrawat et al. 2008). Root system architecture and nutrient uptake can play a significant role in next green revolution. Low N availability can stimulate lateral root branching. Uptake and redistribution of nutrients are mediated by a number of transporters. Nitrogen is most often taken up by plants as NO^{3-} and NH^{4+}. There are two types of nitrate transporters: NPFs and NRT2s. NH_4^+ is carried out by plasma membrane-based *AMT/MEP/Rh* transporters. The N assimilation requires reduction of NO^{3-} to NH^{4+} followed by NH^{4+} conversion to amino acids. There are a number of known genes for improvement of root growth (*TAR2*, *TNOD1*, *DRO1*, *MADS25*, *NAC2-5A*, *EXPB2*, *EXPB23*, *PSTOL1*, *NFYA-B1*, and *VP*) and nutrient use efficiency (*AMT*, *NRT*, *HAK*) (Wan et al. 2017) that can be used for improvement of cotton.

Phosphorus (P) P is an essential macronutrient for cotton for performing basic biological functions. Plant absorbs orthophosphate (Pi) either as $H_2PO_4^-$ or HPO_4^2 depending upon soil pH. Phosphorus deficiency is critical in tropical and subtropical soils. Plants respond to Pi starvation. For example, in *Arabidopsis*, the coordinate induction of more than 600 genes under conditions of Pi starvation has been reported (Mission et al. 2005). Increased expression of these genes enhances both the phosphorus uptake and cycling. The upregulation of vacuolar H^+-PPase has been documented in *Brassica napus* (Shrawat et al. 2008; Gaxiola et al. 2011), and similarly rice transcription factor (*OsPTF1*) has been reported to enhance rice tolerance to Pi starvation (Yi et al. 2005). To date, many phosphate transporters (PHT1s) have been identified in many plant species (Wan et al. 2017). The rice genome contains 13 PHT1 genes. In cotton, a total of 17 *GhPT1* genes were identified in upland cotton genome (Maoni et al. 2017) suggesting that *OsPHT1.6* and others have the potential to increase productivity of crops by using lesser amount Pi fertilizer. *TaPHT1.2* and *TaPHT1.4* were expressed in wheat; both of these increased P uptake and plant growth. The *GhPT* expression patterns revealed that GhPT6 and GhPT14 were highly expressed in roots. Several transcription factors have been found and showed potential in improving nutrient use efficiency. These include *NAC2-5A*, *DEP1*, *MADS25*, *NAP*, *Dof1*, *NLP7*, *HY5*, *NFYA-B1*, *PTF1*, *PHR1-A1*, *PHR2*, *PHR3*, *MYB2P-1*, *WRKY45*, *PHO2-1A*, *CIPK23*, *DDF2*, *LJO*, *TIII_A*, *bHLH121*, and *ARF2* (Wan et al. 2017). The expressions of many of these genes and regulatory elements were upregulated under low-phosphorus and low-potassium conditions. At present, most of the transgenic plants were developed by upregulation. Further improvement in NUE is possible through genome editing (downregulation) of some of these genes. These findings could be used to improve P uptake and nutrient deficiency in this important fiber crop.

11.12 Conclusions and Future Prospects

Conventional breeding has been playing a major role in developing crops with improved traits for centuries. However, the identification and selection of useful stress-responsive genes and their subsequent introgression into crop cultivars through conventional breeding approaches are labor-intensive and time-consuming. On the other hand, biotechnology offers great opportunity to identify and isolate useful genes from diverse organisms and their introduction into crops cultivars. Biotechnology has played a very vital role for the development of transgenic plants especially for those traits which were thought difficult to be achieved through conventional means. Till date, a wide majority of commercialized biotech crops harbor genes for insect resistance and herbicide tolerance. On the other hand, the commercialization of GM crops bearing other useful traits like those conferring abiotic stress tolerance continues to be challenging mainly because of complex regulatory mechanism and public acceptability.

A large number of stress-responsive genes have been identified and successfully introduced into crop plants to produce transgenic crops with improved characteristics. Recent advances in our understanding of the mechanisms involved in crop abiotic stress tolerance and the development of molecular genetic methods for tailoring transgenic plants have allowed us to address these issues much more efficiently than in the past. Transgenic plants have been developed with improved resistance to drought, salinity, and extreme temperature through expression/overexpression of genes regulating specific proteins, osmolytes, antioxidants, ion homeostasis, transcription factors, etc. Almost all countries have regulatory bodies which monitor and approve the commercialization of GM crops with the aim of protecting environment and human/animal health from the unintended effects of GM crops. Regulatory approval is the real bottleneck in the commercialization of GM crops as it takes much longer than the time devoted to the development of GM crops. The currently available genome editing technologies like CRISPR-Cas9, VIGS, and other gene editing and silencing technologies are invaluable in improving the required traits of important food crops. The major advantage of these technologies is that genome-edited crops does not require stringent regulatory approvals that are required for transgenic crops. Further developments in these biotechnological approaches will better equip our future crops with new means to survive in hostile environments.

References

Ahmad S, Khan NI, Iqbal MZ, Hussain A, Hassan M (2002) Salt tolerance of cotton (*Gossypium hirsutum* L.). *Asian J. Plant Sci* 1(6):715–719

Ashraf M (2002) Salt tolerance of cotton: some new advances. *Crit Rev Plant Sci* 21(1):1030. CRC Press LLC

Ashraf M, Wu L (1994) Breeding for salinity tolerance in plants. *Crit Rev Plant Sci* 13:17–42. https://doi.org/10.1080/07352689409701906

Augostinho LMD, Prado RM, Rozane DE, Freitas N (2008) Dry mass accumulation and nutrient absorption march in 'Pedro Sato' guava seedlings. *Bragantia* 67(3):577–585

Azhar FM, Ali Z, Akhtar MM, Khan A, Trethowan R (2009) Genetic variability of heat tolerance, and its effect on yield an fibre quality traits in upland cotton (*Gossypium hirsutum* L.). *Plant Breed* 128:356–362

Bailey-Serres J, Lee SC, Brinton E (2012) Waterproofing crops: effective flooding survival strategies. *Plant Physiol* 160:1698–1709

Bakhsh A, Hussain T (2015) Engineering crop plants against abiotic stress: current achievements and prospects. *Emir J Food Agric* 27(1):24–39

Bange MP, Milroy SP, Thongbai P (2004) Growth and yield of cotton in response to water logging. *Field Crop Res* 88(2):129–142

Bello B, Zhang X, Liu C, Yang Z, Yang Z, Wang Q (2014) Cloning of *Gossypium hirsutum* sucrose non-fermenting 1-related protein kinase 2 gene (*GhSnRK2*) and its overexpression in transgenic *Arabidopsis* escalates drought and low temperature tolerance. *PLoS One* 9:e112269. https://doi.org/10.1371/journal.pone.0112269. PMID: 25393623

Biaswas J, Kalara N (2018) Effect of waterlogging and submergence on crop physiology and growth of different crops and its remedies: Bangladesh perspective. *Saudi J Eng Technol.* https://doi.org/10.21276/sjeat.2018.3.6.1

Boyer JS (1982) Plant productivity and environment. *Science* 218:443–448

Bruulsema TW, Mullen R, O'Halloran IP, Watters H (2012) Reducing loss of fertilizer phosphorus to Lake Erie with the 4Rs. IPNI Insights, International Plant Nutrition Institute, Norcross

Burke JJ, Mahan JR, Hatfield JL (1988) A relationship between crop specific thermal kinetic windows and biomass production. *Agron J* 80:553–556

Chen T, Li W, Hu X, Guo J, Liu A, Zhang B (2015) A cotton MYB transcription factor, GbMYB5, is positively involved in plant adaptive response to drought stress. *Plant Cell Physiol* 56(5):917–929

Chibulke GU, Obiora SC (2014) Heavy metal polluted soils: effect on plants and bioremediation methods. *Appl Environ Soil Sci* 2014:752708

Christianson JA, Llewellyn DJ, Dennis ES, Wilson IW (2010) Global gene expression response to waterlogging and roots and leaves of cotton (*Gossypium hirsutum* L.). *Plant Cell Physiol* 51(1):21–37

Conaty WC, Tan DKY, Constable GA, Sutton BG, Field DJ, Mamum EA (2008) Genetic variation for water logging tolerance in cotton. *J Cotton Sci* 12:53–61

Cottee NS, Wilson IW, Tan DKY, Bange MO (2014) Understanding the molecular events underpinning cultivar differences in the physiological performance and heat tolerance of cotton (*Gossypium hirsutum*). *Funct Plant Biol* 41:56–67

Dagar JC, Minhas PS (2016) Global perspectives on agroforestry for the management of salt-affected soils. In: Dagar JC, Minhas PS (eds) Agroforestry for the management of waterlogged saline soils and poor-quality waters. Springer, New Delhi, pp 5–32. https://doi.org/10.1007/978-81-322-2659-8_2

Das K, Roychoudhury A (2014) Reactive oxygen species (ROS) and response of antioxidants as ROS-scavenging during environmental stress in plants. *Front Environ Sci* 2(53):1–13

Demiral T, Türkan I (2006) Exogenous glycinebetaine affects growth and proline accumulation and retards senescence in two rice cultivars under NaCl stress. *Environ Exp Bot* 56:72–79. https://doi.org/10.1016/j.envexpbot.2005.01.00

Dennis ES, Delferous R, Ellis M, Rehman M, Wu Y, Hoeren FU, Grover A, Ismond KP, Good AG, Peacock WJ (2000) Molecular strategies for improving waterlogging tolerance in plants. *J Exp Bot* 51:342

Fan Y, Pan D, Ji M, Guo D (1989) The relationship between the cytochrome oxidase activity and chilling resistance in cotton seedling. *China Cotton* 6:11–13

Fan Y, Zhang S, Li S (1995) Study on relationship between cotton chilling seedling resistance and peroxidase activity variation. *Cotton Sci* 7:172–174

FAO (2005) Global network on integrated soil management for sustainable use of salt-affected soils. FAO Land and Plant Nutrition Management Service, Rome. http://www.fao.org/ag/agl/agl/spush

FAO (2007) Agriculture and water quality interactions: a global overview. SOLAW Background Thematic Report—TR08. http://www.fao.org/3/a-bl092e.pdf

Gaxiola RA, Edwards M, Elser JJ (2011) A transgenic approach to enhance phosphorus use efficiency in crops as part of a comprehensive strategy for sustainable agriculture. *Chemosphere* 84:840–845

Gill SS, Gill R, Anjum NA, Tuteja N (2013) Transgenic approaches for abiotic stress tolerance in crop plants. *Plant Stress* 7:73–83

Golledge NR, Keller ED, Gomez N, Naughten KA, Bernales J, Trusel L, Edwards TL (2019) Global environmental consequences of twenty-first-century ice-sheet melt. *Nature* 566:65–72

Gunapati S, Naresh R, Ranjan S, Nigam D, Has A, Verma PC, Gadre R, Pathre UV, Sane AP, Sane VA (2015) Expression of GhNAC2 from *G. herbaceum*, improves root growth and imparts tolerance to drought in transgenic cotton and Arabidopsis. *Sci Rep* 6:24978

Guo D, Fan Y, Guo C, Pan D, Wan L, Xie G (1991) Peroxidase isoenzyme and chilling resistance of cotton seedling. *China Cotton* 18:11–13

Guo Y, Pang C, Jia X, Ma Q, Dou L, Zhao F, Gu L, Wei H, Wang H, Fan S, Su J, Yu S (2017) An NAM domain gene, *GhNAC79*, improves resistance to drought stress in upland cotton. *Front Plant Sci* 8:1657. https://doi.org/10.3389/fpls.2017.01657

Han J, Shi J, Zeng L, Xu J, Wu L (2015) Effects of nitrogen fertilization on the acidity and salinity of greenhouse soils. *Environ Sci Pollut Res* 22:2976–2986. https://doi.org/10.1007/s11356-014-3542-z

Hao QN, Zhou XA, Sha AH, Wang C, Zhou R, Chen S (2011) Identification of genes associated with nitrogen-use efficiency by genome-wide transcriptional analysis of two soybean genotypes. *BMC Genomics* 12:525

Hassan MMU, Ma F, Prodhan ZH, Li F, Shen H, Chen Y, Wang X (2018) Molecular and physio-biochemical characterization of cotton species for assessing drought stress tolerance. *Int J Mol Sci* 19:2636

Hayat S, Haya Q, Alyemeni MN, Wani AS, Pichtel J, Ahmad A (2012) Role of proline under changing environments. *Plant Signal Behav* 7(11):1456–1466

He C, Yan J, Shen G, Fu L, Holaday AS, Blumwald E, Zhang H (2005) Expression of an *Arabidopsis* vacuolar sodium/proton antiporter gene in cotton improves photosynthetic performance under salt conditions and increases fiber yield in the field. *Plant Cell Physiol* 46:1848±1854. https://doi.org/10.1093/pcp/pci201. PMID: 16179357

He L, Yang X, Wang L, Zhu L, Zhou T, Deng T, Zheng X (2013) Molecular cloning and functional characterization of a novel cotton CBL-interacting protein kinase gene (*GhCIPK6*) reveals its involvement in multiple abiotic stress tolerance in transgenic plants. *Biochem Biophys Res Commun* 435:209–215

Hediye A, Rengin S, Baris O, Turkan UI (2014) Reactive oxygen species scavenging capacities of cotton (*Gossypium hirsutum* L.) cultivars under combined drought and heat induced oxidative stress. *Environ Exp Bot* 99:141–149

Holaday AS, Mahan JR, Payton P (2016) Effects of chilling temperatures on photosynthesis. J *Cotton Sci* 20:220–231

Huang JG, Yang M, Liu P, Yang GD, Wu CA, Zheng CC (2009) *GhDREB1* enhances abiotic stress tolerance, delays GA-mediated development and represses cytokinin signalling in transgenic *Arabidopsis*. *Plant Cell Environ* 32:1132±1145

Hussain A, Zafar ZU, Athar H-R, Farooq J, Ahmad S, Nazeer W (2019) Assessing gene action for hypoxia tolerance in cotton (*Gossypium hirsutum* L.). *Agron Mesoam* 30(1):51–62

ISAAA (2018) Global status of commercialized biotech/GM crops in 2018:biotech crops continue to help meet the challenges of increased population and climate change. ISAAA Brief No. 54, ISAAA, Ithaca

Ito Y, Katsura K, Maruyama K, Taji T, Kobayashi M, Seki M, Shinozaki K, Yamaguchi-Shinozaki K (2006) Functional analysis of rice DREB1/CBF-type transcription factors involved in cold-responsive gene expression in transgenic rice. *Plant Cell Physiol* 47:141–153

Karan R, Subudhi P (2012) Approaches to increasing salt tolerance in plants. In: Ahmad P, Parasad MNV (eds) Abiotic stress responses in plants: metabolism, productivity and sustainability. Springer Science-Business Media, LLC

Khan N, Ahmad I, Azhar MT (2017) Genetic variation in relative cell injury for breeding upland cotton under high temperature stress. *Turk J Field Crops* 22(1):152–159

Khan A, Pan X, Najeeb U, Tan DKY, Fahad S, Zahoor A, Lou H (2018) Coping with drought: stress and adaptive mechanism, and management through cultural and molecular alternatives in cotton as vital constituents for plant stress resilience and fitness. *Biol Res* 51:47

Kuai J, Chen Y, Wang Y, Meng Y, Chen B, Zhao W, Zhou Z (2016) Effect of waterlogging on carbohydrate metabolism and the quality of fiber in cotton (*Gossypium hirsutum* L.). *Front. Plant Sci* 7:877

Kuppu S, Mishra N, Hu R, Sun L, Zhu X, Shen G et al (2013) Water deficit inducible expression of a cytokinin biosynthetic gene IPT improves drought tolerance in cotton. *PLoS One* 8:e64190

Lee H, Xiong L, Ishitani M, Stevenson B, Zhu J-K (1999) Cold-regulated gene expression and freezing tolerance in an *Arabidopsis thaliana* mutant. *Plant J* 17(3):301–308

Li S, Zhang J, Fan Y (1996) Study on relationship between cotton chilling seedling resistance and non-protein nitrogen content increase. *China Cotton* 23:11–12

Li W, Zhonghua Z, Shunil Z, Jinxiang C, Feng W, Ruilian L, Hao Z (2014) Primary identification of waterlogging tolerance genes in cotton (*Gossypium hirsutum* L) based on quantitative proteomic approach. *Cotton Sci* 4

Li H, Li K, Guo Y, Guo J, Miao K, Botella JR, Song C-P, Miao Y (2018) A transient transformation system for gene characterization in upland cotton (*Gossypium hirsutum*). *Plant Methods* 14:50

Lisei-de-Sa ME, Arraes FBM, Brito GG, Beneventi MA, Lourenco-Tessutti IT, Basso AMM, Amorim RMS, Silva MCM, Faheem M, Oliveira NG, Mizoi J, Yamaguchi-Shinozaki K, Grossi-de-Sa MF (2017) *AtDREB2A-CA*influences root architecture and increases drought tolerance in transgenic cotton. *Agric Sci* 8:1195–1225. https://doi.org/10.4236/as.2017.810087

Liu G, Li X, Jin S, Liu X, Zhu L, Nie Y, Zhang X (2014) Overexpression of rice NAC gene *SNAC1* improves drought and salt tolerance by enhancing root development and reducing transpiration rate in transgenic cotton. *PLoS One* 9(1):e86895

Liu B, Mu J, Zhu J, Liang Z, Zhang L (2017) *Saussurea involucrata* SiDgn2 gene confers tolerance to drought stress in upland cotton. *Pak J Bot* 49(2):465–473

Luo X, Wu J, Li Y, Nan Z, Guo X, Wang Y, Zhang A, Wang Z, Xia G, Tian Y (2013) Synergistic effects of GhSOD1 and GhCAT1 overexpression in cotton chloroplasts on enhancing tolerance to methyl viologen and salt stresses. *PLoS One* 8(1):e54002. https://doi.org/10.1371/journal.pone.0054002

Luo Y, Reid R, Freese D, Li C, Watkins J, Shi H (2017) Salt tolerance response revealed by RNA-Seq in a diploid halophyte wild relative of sweet potato. *Sci Rep* 7:1–13

Lv S, Yang A, Zhang K, Wang L, Zhang J (2007) Increase of glycinebetaine synthesis improves drought tolerance in cotton. *Mol Breed* 20:233–248

Lv S, Zhang K, Gao Q, Lian L, Song Y, Zhang J (2008) Overexpression of an H^+-PPase gene from Thellungiella halophila in cotton enhances salt tolerance and improves growth and photosynthetic performance. *Plant Cell Physiol* 49:1150–1164

Lv S, Lian LJ, Tao PL, Li ZX, Zhang KW, Zhang J (2009) Overexpression of *Thellungiella halophila* H ($^+$)-PPase (TsVP) in cotton enhances drought stress resistance of plants. *Planta* 229:899–910

Ma X, Dong H, Li W (2011) Genetic improvement of cotton tolerance to salinity stress. *Afr J Agric Res* 6(33):6798–6803

Ma W, Zhao T, Liu B, Fang L, Hu Y, Zhang T (2016) Identification and characterization of the GhHsp20 gene family in *Gossypium hirsutum*. *Sci Rep* 6:32517

Magwanga RO, Lu P, Kirungu JN, Dong Q, Hu Y, Zhou Z, Cai X, Wang X, Hou Y, Wang K, Liu F (2018) Cotton late embryogenesis abundant (LEA2) gene promotes root growth and confers drought stress tolerance in transgenic *Arabidopsis thaliana*. *G3 (Bethesda)* 8(8):2781–2803

Maoni C, Zhiyong Z, Haina S, Chengqi L, Xin Z, Genhai H, Jinbao Z, Qinglian W (2017) Genome-wide identification and expression analysis of Pht1 family genes in cotton (*Gossypium hirsutum* L.). *Cotton Sci* 29(1):59–69

Maqbool A, Abbas W, Rao AQ, Irfan M, Zahur M, Bakhsh A, Riazuddin S, Husnain T (2010) *Gossypium arboreum* GHSP26 enhances drought tolerance in *Gossypium hirsutum*. *Biotechnol Prog* 26:21–25

Millar AA, Dennis ES (1996) Protein synthesis during oxygen deprivation in cotton. *Aust J Plant Physiol* 23:341–348

Mishra N, Sun L, Zhu X, Smith J, Prakash Srivastava A, Yang X, Pehlivan N, Esmaeili N, Luo H, Shen G, Jones D, Auld D, Burke J, Payton P, Zhang H (2017) Overexpression of the rice SUMO E3 ligase gene OsSIZ1 in cotton enhances drought and heat tolerance, and substantially improves fiber yields in the field under reduced irrigation and rainfed conditions. *Plant Cell Physiol* 58(4):735–746. https://doi.org/10.1093/pcp/pcx032

Mission J, Raghothama KG, Jain A, Jouhet J, Block MA, Bligny R, Ortet P, Creff A, Stomerville S, Rolland N, Doumas P, Nacry P, Herrera-Estrella L, Nussaume L, Thibaud MC (2005) A genome-wide transcriptional analysis using Arabidopsis thaliana Affymetrix gene chips determined plant responses to phosphate deprivation. *Proc Natl Acad Sci U S A* 94:11934–11939

Mittal A, Gampala SS, Ritchie GL, Payton P, Burke JJ, Rock CD (2014) Related to ABA-Insensitive3 (ABI3)/Viviparous1 and AtABI5 transcription factor co-expression in cotton enhances drought stress adaptation. *Plant Biotechnol J* 12:578–589

Munns R, Tester M (2008) Mechanisms of salinity tolerance. *Annu Rev Plant Biol* 59:651–681. https://doi.org/10.1146/annurev.arplant.59.032607.092911

Munns R, James RA, Xu B, Athman A, Conn SJ, Jordans C, Byrt CS, Hare RA, Tyerman SD, Tester M, Plette D, Gilliham M (2012) Wheat grain yield on saline soils is improved by an ancestral NaC transporter gene. *Nat Biotechnol* 30:360–364. https://doi.org/10.1038/nbt.2120

Mustroph A (2018) Improving flooding tolerance of crop plants. *Agronomy* 8:160

Najeeb U, Bange MP, Tan DY, Atwell BJ (2015) Consequences of waterlogging in cotton and opportunities for mitigation of yield losses. *AoB Plants* 7:plv080

Nikalje G, Suprasanna P (2018) Coping with metal toxicity-Cues from halophytes. *Front Plant Sci* 9:777

Nishida I, Murata N (1996) Chilling sensitivity in plants and cyanobacteria: the crucial contribution of membrane lipids. *Annu Rev Plant Physiol Plant Mol Biol* 47:541–568

Nishiuchi S, Yamauchi T, Takahashi H, Kotula L, Nakazono M (2012) Mechanism for coping with submergence and waterlogging in rice. *Rice* 5:2

Oluoch G, Zheng J, Wang X, Khan MKR, Zhou Z, Cai X, Wang C, Wang Y, Li X, Wang H, Liu F, Wang K (2016) QTL mapping for salt tolerance at seedling stage in the interspecific cross of *Gossypium tomentosum* with *Gossypium hirsutum*. *Euphytica* 209:223–235. https://doi.org/10.1007/s10681-016-1674-6

Oosterhuis DM (2002) Day or night high temperature: a major cause of yield variability. *Cotton Grower* 46:8–9

Pachauri RK, Allen MR, Barros VR, Booome J, Cramer W, Christ R, Church JA, Clark L, Dahe Q, Dasgupta D (2014) Climate change 2014: synthesis report. In: Contribution of working groups I, II and III to the fifth assessment report of the intergovernmental panel on climate change, Switzerland

Park SY, Noh KJ, Yoo JH, Yu JW, Lee BW, Kim JG, Seo HS, Paek NC (2006) Rapid up-regulation of Dehyrin3 and Dehydrin4 in response to dehydration is a characteristic of drought-tolerant genotypes in barley. *J Plant Biol* 49:455–462

Park W, Scheffler BE, Bauer PJ, Campbell BT (2012) Genome-wide identification of differentially expressed genes under water deficit stress in upland cotton (*Gossypium hirsutum* L.). *BMC Plant Biol* 12:90

Parkhi V, Kumar V, Sunilkumar G, Campbell LM, Singh NK, Rathore KS (2009) Expression of apoplastically secreted tobacco osmotin in cotton confers drought tolerance. *Mol Breed* 23:625–639

Pasapula V, Shen G, Kuppu S, Paez-Valencia J, Mendoza M, Hou P, Chen J, Qiu X, Zhu L, Zhang X, Auld D, Blumwald E, Zhang H, Gaxiola R, Payton P (2011) Expression of an Arabidopsis vacuolar H+-pyrophosphatase gene (AVP1) in cotton improves drought- and salt tolerance and increases fiber yield in the field conditions. *Plant Biotechnol J* 9:88–99

Payton P, Holaday AS, Allen RD, Webb RP (1999) Response of transgenic cotton to oxidative stress. In: Proceedings of the beltwide cotton conference, vol 1. pp 523–524

Pilon-Smits EAH, Zhu Y, Sears T, Terry N (2000) Overexpression of glutathione reductase in Brassica juncea—effects on cadmium accumulation and tolerance. *Physiol Plant* 110:455–460

Qi X-H, Xu X-W, Lin X-J, Zhang W-J, Chen X-H (2012) Identification of differentially expressed genes in cucumber (Cucumis sativus L) root under waterlogging stress by digital gene expression profile. *Genomics* 99:160–168

Ray B, Noren SK, Mandal AB, Basu AK (2011) Genetic engineering for abiotic stress tolerance in agricultural crops. *Biotech* 10(1):1–22

Reddy VR, Baker DN, Hodges HF (1991) Temperature effects on cotton canopy growth, photosynthesis and respiration. *Agron J* 83:699–704

Riaz M, Farooq J, Sakhawat G, Mahmood A, Sadiq MA, Yaseen M (2013) Genotypic variability for root/shoot parameters under water stress in some advanced lines of cotton (*Gossypium hirsutum* L.). *Genet Mol Res* 12(1):552–561

Salman M, Khan AA, Rana IA, Maqsood RH, Azhar MT (2016) Genetic architecture of relative cell injury and some yield related parameters in *Gossypium hirsutum* L. *Turk J Field Crops* 21(2):246–253

Sett R (2017) Tolerance of plants in response to abiotic stress factors. Recent Adv Petrochem Sci 1(5). RAPSCI.MS.ID. 555573. Juniper Publishers. ISSN: 2575-8578

Shan D-P, Huang J-G, Yang Y-T, Guo Y-H, Wu C-A, Yang G-D, Gao Z, Zheng C-C (2007) Cotton GhDREB1 increases plant tolerance to low temperature and is negatively regulated by gibberellic acid. *New Phytol* 176:70–81

Shi W, Hao L, Li J, Liu D, Guo X, Li H (2014) The *Gossypium hirsutum* WRKY gene *GhWRKY39-1* promotes pathogen infection defense responses and mediates salt stress tolerance in transgenic *Nicotiana benthamiana*. *Plant Cell Rep* 33:483–498. https://doi.org/10.1007/s00299-013-1548-5. PMID: 24337818

Shrawat AK, Carroll RT, DePauw M, Taylor GT, Good AG (2008) Genetic engineering of improved nitrogen use efficiency in rice by the tissue-specific expression of alanine aminotransferase. *Plant Biotechnol J* 6:722–732

Singh RP, Parasad PVV, Sunjita K, Giri SN, Reddy KR (2007) Influence of high temperature and breeding for heat tolerance in cotton: a review. *Adv Agron* 93:313–385

Su J, Zhang F, Ching X, Song A, Guan Z, Fang W, Chen F (2019) Genome-wide association study identifies favorable SNP alleles and candidate genes for waterlogging tolerance in chrysanthemums. *Hortic Res* 6:21

Sullivian CY (1972) Mechanism of heat and drought tolerance in grain sorghum and methods of measurement. In: Rao NGP, House LR (eds) Sorghum in the seventies. Oxford and IBH Publishing Co., New Delhi

Sun Z, Li H, Zhang Y, Li Z, Ke H, Wu L, Zhnag G, Wang X, Ma Z (2018) Identification of SNPs and candidate genes association with salt tolerance at the seedling stage in cotton (*Gossypium hirsutum* L.). *Front Plant Sci* 9:1011

Tiwari RS, Picchioni GA, Steiner RL, Jones DC, Hughs SE, Zhang J (2013) Genetic variation in salt tolerance at the seedling stage in an interspecific backcross inbred line population of cultivated tetraploid cotton. *Euphytica* 194:1–11. https://doi.org/10.1007/s10681-013-0927-x

Ullah A, Sun H, Yang X, Zhang X (2017) Drought coping strategies in cotton: increased crop per drop. *Plant Biotechnol J* 15:271–284

Vwioka E, Adinkwu O, El-Esawi MA (2017) Comparative physiological, biochemical and genetic response to prolonged waterlogging stress in Okra and maize given exogenous ethylene priming. *Front Plant Sci* 8:632

Wan T, Xue H, Yi-ping T (2017) Transgenic approaches for improving use efficiency of nitrogen, phosphorus and potassium in crops. *J Integr Agric* 16(12):2657–2673

Wang Q, Liu N, Yang X, Tu L, Zhang X (2016) Small RNA-mediated responses to low- and high temperature stress in cotton. *Sci Rep* 6:35558

Wang J, Zhang Q, Cui F, Hou L, Zhao S, Xia H, Qiu J, Li T, Zhang Y, Wang X, Zhao C (2017) Genome wide analysis of gene expression provides new insights into cold responses in *Thellungiella salsuginea*. *Front Plant Sci* 8:713

Wei Y, Xu Y, Lu P, Wang X, Li Z, Cai XZ, Wang Y, Zhang Z, Lin Z, Liu F, Wang K (2017) Salt stress responsiveness of a wild cotton species (*Gossypium klotzschianum*) based on transcriptomic analysis. *PLoS One* 12(5):e0178313

Xu X, Chen M, Ji J, Xu Q, Qi X, Chen X (2017) Comparative RNA-seq based transcriptome profiling of waterlogging response in cucumber hypocotyls reveals novel insight into the de novo adventitious root primordial initiation. *BMC Plant Biol* 17:129

Yamaguchi-Shinozaki K, Shinozaki K (2005) Organization of cis-acting regulatory elements in osmotic and cold-stress-responsive promoters. *Plant J* 10:88–94

Yan J, He C, Wang J, Mao Z, Holaday SA, Allen RD, Zhang H (2004) Overexpression of the Arabidopsis 14-3-3 protein GF14 lambda in cotton leads to a "stay-green" phenotype and improves stress tolerance under moderate drought conditions. *Plant Cell Physiol* 45:1007–1014

Yang H, Zhang D, Li X, Li H, Zhang D, Lan H, Wood AJ, Wang J (2016) Overexpression of ScALDH21 gene in cotton improves drought tolerance and growth in greenhouse and field conditions. *Mol Breed* 36:34

Ye W, Liu J (1998) Identification technique and application of salt-tolerant germplasm resources in cotton. *China Cotton* 25(9):34–38

Yi K, Wu Z, Zhou J, Du L, Guo L, Wu Y, Wu P (2005) *OsPTF1*, a novel transcription factor involved in tolerance to phosphate starvation in rice. *Plant Physiol* 138:2087–2096

Yu F, Lian K, Zhang Z, Du D, Zhang X, Zhao H, Haq BU, Qiu F (2018) Dissecting the genetic architecture of waterlogging stress-related traits uncovers a key waterlogging tolerance gene in maize. *Theor Appl Genet* 131:12299–12310

Yue Y, Zhang M, Zhang J, Tian X, Duan L, Li Z (2012) Overexpression of the AtLOS5 gene increased abscisic acid level and drought tolerance in transgenic cotton. *J Exp Bot* 63:3741–3748

Yu-qiang H, Guo-qing L, Li-hua W, Chun-ling W, Hui-ming G, Yi-fei L, Hong-mei H (2018) Overexpression of AmDUF1517 enhanced tolerance to salinity, drought, and cold stress in transgenic cotton. *J Integr Agric* 17(10):2204–2214

Zahid KR, Ali F, Shah F, Younas M, Shah T, Shawar D, Hassan W, Ahmad Z, Qi C, Lu Y, Aqbal A, Wu W (2016) Response and tolerance mechanism of cotton *Gossypium hirsutum* L. to elevated temperature stress: a review front. *Plant Sci* 7:937

Zhang H, Dong H, Li W, Sun Y, Chen S, Kong X (2009) Increased glycine betaine synthesis and salinity tolerance in AhCMO transgenic cotton lines. *Mol Breed* 23:289–298

Zhang K, Guo N, Lian L, Wang J, Lv S, Zhang J (2011) Improved salt tolerance and seed cotton yield in cotton (*Gossypium hirsutum* L.) by transformation with *betA* gene for glycinebetaine synthesis. *Euphytica* 181:1–16

Zhang D-Y, Yang H-L, Li X-S, Li H-Y, Wang Y-C (2014) Overexpression of Tamarix albiflonum TaMnSOD increases drought tolerance in transgenic cotton. *Mol Breed* 34:1–11

Zhang Y, Song X, Yang G, Li Z, Li H, Kong X, Eneji AE, Dong H (2015) Physiological and molecular adjustment of cotton to waterlogging at peak-flowering in relation to growth and yield. *Field Crop Res* 179:164–172

Zhang J, Li B, Yang Y, Hu W, Chen F, Xie L, Fan L (2016a) Genome wide characterization and expression profiles of the superoxide dismutase gene family in Gossypium. *Int J Genomics* 2016:8740901

Zhang J, Srivastava V, Stewart JMD, Underwood J (2016b) Heat tolerance in cotton is correlated with induced overexpression of heat-shock factors, heat-shock proteins and general stress response gene. *J Cotton Sci* 20:253–262

Zhang Y, Kong X, Dai J, Luo Z, Li Z, Lu H, Xu S, Tang W, Zhang D, Li W, Xin C, Dong H (2017) Global gene expression in cotton (*Gossypium hirsutum* L.) leaves to waterlogging stress. *PLoS One* 12(9):e0185075

Zhang J, Abdelraheem A, Wedegaertner T (2019) Genetic variation of waterlogging tolerance in Pima (*Gossypium barbadense*) cotton and glanded and glandless upland cotton (*Gossypium hirsutum*) under field conditions. *Ind Crop Prod* 129:169–174

Zhou B, Zhang L, Ullah A, Jin X, Yang X, Zhang X (2016) Identification of multiple stress responsive genes by sequencing a normalized cDNA library from sea-land cotton (*Gossypium barbadense* L.). *PLoS One* 11(3):e0152927

Chapter 12
GM Technology and Fiber Traits

Ayesha Latif, Ammara Ahad, Mukhtar Ahmed, Sidra Akhtar, Aneela Yasmeen, Ahmad Ali Shahid, Abdul Qayyum Rao, Idrees Ahmad Nasir, and Tayyab Husnain

12.1 GM Technology

Genetic modification of plants is one of the major areas of biotechnology which deals with manipulation of genetic material in living organisms. The concept of genetic modification can be traced back to ancient times when artificial selection and selective breeding were practiced by human ancestors for getting plants with desirable traits. Current cultivars of edible corn, cotton, wheat, etc. have been developed through artificial selection to alter plant genetics from their wild progenitors; this is often referred to as the most dramatic alteration of plant genetics (Doebley et al. 2006). Similarly, current variants of apples, bananas, and broccoli which are desirable for human consumption have been obtained by the use of similar techniques.

12.2 Timeline of Genetic Modification

Revolution in genetic modification took place back in 1946 when studies revealed that genetic material can be transferred between different species. The research work done by Boyer and Cohen in 1973 proved to be the breakthrough in genetic engineering when the world's first GM organism, bacteria resistant to kanamycin, was created (Cohen et al. 1973). The first GM animal was a mouse created by Rudolf Jaenisch in 1974, while genetic modification in plants started in 1983 (Rangel 2015). The commercialization of GM plant was made for the first time by China which officially presented the case of virus resistant tobaccos in 1990. It took

A. Latif · A. Ahad · M. Ahmed · S. Akhtar · A. Yasmeen · A. A. Shahid · A. Q. Rao
I. A. Nasir (✉) · T. Husnain
Center of Excellence in Molecular Biology, University of the Punjab, Lahore, Pakistan
e-mail: idreesnasir.cemb@pu.edu.pk

Mehboob-ur-Rahman et al. (eds.), *Cotton Precision Breeding*,
https://doi.org/10.1007/978-3-030-64504-5_12

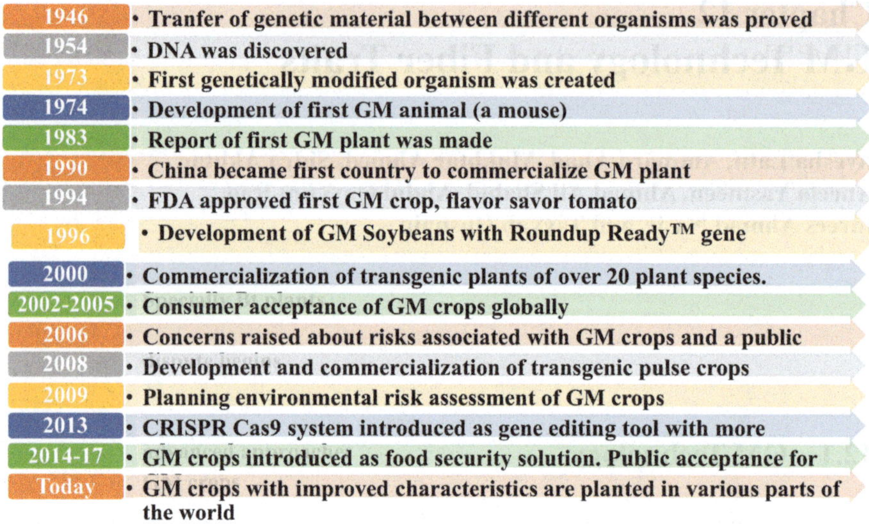

Fig. 12.1 Timeline of genetic modification

four more years when the Food and Drug Administration (FDA) approved flavor savor tomatoes for human consumption in 1994 (Bawa and Anilakumar 2013). With the progress in technology, need-based approvals were granted to many GM crops for plantation, i.e., corn/maize, potatoes, canola, cotton, and soybeans for large-scale human production. These days a large number of GM fruits, vegetables, and cereals are under cultivation (Fig. 12.1).

12.3 Transition from Genes to Fabrics

12.3.1 Cotton

Cotton being a major source of natural fiber is the most significant crop of the world (Fryxell 1992) having history of ancient times (Hassan and Militky 2012). A significant number of cotton growers are found in various countries including the USA, China, India, Pakistan, Australia, and the Middle East (Smith 1995). Although many other fibers are in use since ancient times, cotton fiber stands up among all due to its unique characteristics and has its own distinct superior qualities (Pillay and Myers 1999). It is made up of more than 95% cellulose (w/w); therefore it is ranked as the world's most important natural textile crop (Li et al. 2014).

Among four widely cultivated species of cotton, *G. hirsutum* L. and *G. barbadense* L. are tetraploid (Gilbert et al. 2008), and *G. herbaceum* L. and *G. arboretum* are diploid (Brubaker et al. 1999). *G. hirsutum* contributes to 95% of world cotton production which is grown on account of its yield including quality and quantity despite its susceptibility to many biotic and abiotic stresses. It is characterized by high yield and with moderate fiber quality, and these attributes enabled this crop as the main target for research (Rauf et al. 2015).

12.4 Peeking into the World of Cotton Fiber Structure

12.4.1 Structure of Cotton Fiber

Primary cell wall (PCW) together with several layers of secondary cell wall (SCW) constitutes the cotton fiber. The lumen is found in between these walls. Cellulosic and noncellulosic materials jointly make up the uniform primary cell wall (PCW) (Hearle 2007). Neighboring fibers are joint by the middle lamella, which is adjacent to PCW, into tissue-like structure (Singh et al. 2009). Existence of the middle lamella within cotton fiber was unexpected because at maturity, individual fibers are found. It was later reported that this layer is dissolved at the commencement of secondary cell wall synthesis by cell wall enzymes; therefore upon maturity single individual fibers are released. A winding layer is deposited next to PCW during transition phase followed by rearrangements of cytoskeleton microtubules in helical pattern occurs (Singh et al. 2009). Thickening and strengthening of cotton fiber is achieved by the deposition of secondary wall in form of multiple consecutive pure cellulose layers. Fiber cells remain alive until boll dehiscence; as soon as the boll splits and is exposed to air, these cells become dead, dried, and twisted (Abidi et al. 2010) (Fig. 12.2).

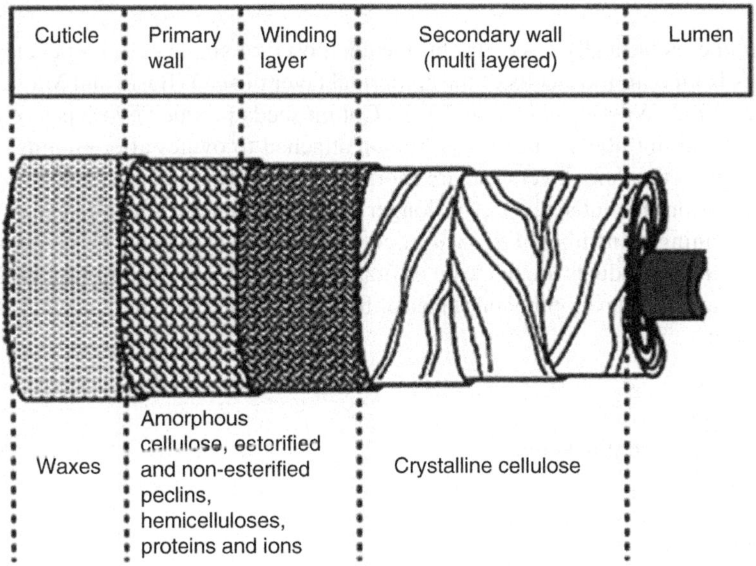

Fig. 12.2 Detailed structure of the cotton fiber (Kozlowski 2012). Moving from outside to inside, the outermost layer is of the waxy cuticle which surrounds the primary wall. Next to the primary wall is the winding layer which is followed by a multilayered cellulosic secondary wall and a lumen in the middle

12.5 Compositional Architecture of Cotton Fiber

Cellulose is the main component of cotton fiber constituting 88–96% of fiber and is present in SCW. On the other hand, PCW is comprised of pectin, which in turn is mixture of 0.9–1.2% polygalacturonic acid and magnesium salt. PCW is also composed of 1.1–1.9% proteins and 0.3–1.00% waxes. Mineral salts (i.e., calcium, magnesium, oxides of silicon, phosphates, potassium, and sulfates) and organic acids, e.g., L-maleic acid and citric acid, are also part of primary cell wall ranging between 0.5 and 1.0% and 0.7 and 1.6%, respectively. In addition, some sugars like fructose, glucose, galactose, and pentose; pigments, i.e., flavonoid; and vitamins are also present in cotton fibers (Hassan and Militky 2012).

The quantity of sugars, uronic acid, protein, and cellulose varies considerably at different developmental stages of cotton fiber. Protein contents decline from highest to lowest quantity when fiber development transition occurs from elongation phase to SCW synthesis. It again reaches to its highest at the end of SCW synthesis. Same is the case with sugar contents, while uronic acid decreases continuously (Meinert and Delmer 1977).

The components of primary cell wall are secreted by Golgi apparatus during elongation and PCW synthesis phase, while components of secondary cell wall are secreted during SCW deposition by endoplasmic reticulum and plasma membrane (Westafer and Brown 1976).

12.6 Unraveling Cotton Fiber Developmental Phases

Cotton fiber is basically a seed trichrome composed of single cell. It's development initiates from outermost cells of the epidermal layer of seed (Basra and Malik 1984; Li et al. 2009; Wilkins and Arpat 2005). Cotton seeds produce two types of fibers, i.e., fuzz and lint fibers. Fuzz fibers remain attached to ovules after ginning. These are present in the inner layers and are shorter in length. On the other hand, lint fibers are present on the outer layers, are longer in length, and are separated from seeds during ginning. Lint fibers are used in cotton yarn development, while rayon and other cellulose products are made up of fuzz fibers (Poehlman and Borthakur 1969).

There are four developmental stages of fiber as reported by Naithani et al. (1982). These are:

1. Initiation
2. Elongation
3. Secondary wall thickening
4. Maturation

12.6.1 Initiation

Cotton fiber development starts from the formation of round protrusions termed as fiber initials as a result of differentiation of epidermal cells of the seed. This stage is called fiber initiation. It lasts for 5 days starting 2 days prior anthesis to 3 days post anthesis (DPA). The fiber initiation results in formation of 16,000 fiber initials on each ovule (Seagull and Giavalis 2004). Fiber initials are enlarged as the result of turgor pressure generated by central vacuole which is formed with passage of time. At this stage nucleus in the growing fiber moves in the middle (Hu et al. 2019) (Fig. 12.3).

Different biochemical processes involved in fiber initiation have been reported, and certain molecules having their impact in the fiber development have been identified. Ruan et al. (2003) reported that SS3 suppression caused inhibition in fiber initiation and elongation in transgenic cotton. Reduction in expression of SuS protein in seed coat results in the production of fibreless or rudimentary fuzz-like fibers in transgenic cotton plants. Reduced SuS activity at initial stages of fiber development (from −2 to 5 DPA) led to fibreless seed phenotype (Ahmed et al. 2019). Reduction in hexoses, starch, and UDP-glucose production is also reported due to

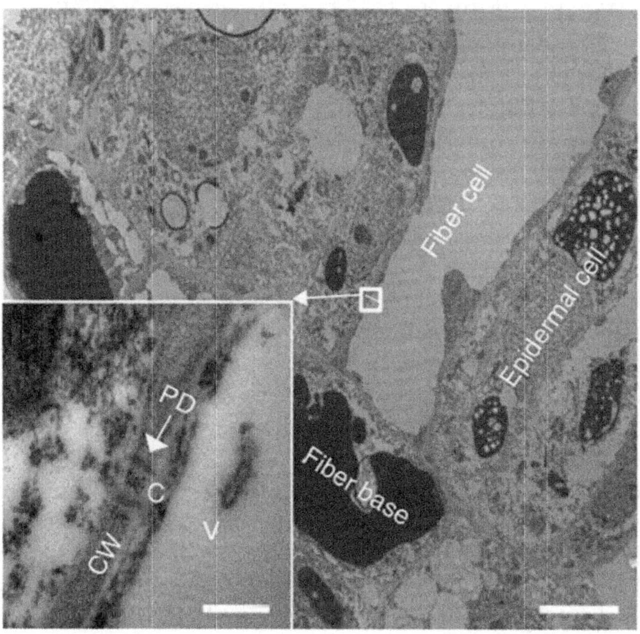

Fig. 12.3 Transmission electron microscopy (TEM) analysis of epidermal cell structure. Transmission electron microscopy (TEM) analysis of epidermal cell structure at 5-DPA ovule, showing the large central vacuole in the fiber cell compared with fiber base and epidermal cell. Scale bars, 10 μm in the figure and 100 nm in the inset. *PD* plasmodesmata, *C* cytosol, *V* vacuole, *CW* cell wall (Hu et al. 2019)

decrease in SuS activity. Reduced hexose levels ultimately result in reduced turgor and low osmotic pressure. In addition decreased UDP-glucose production also disturbs the hexose signaling pathway, which is also a substrate for cell wall synthesis (Stiff and Haigler 2012).

12.6.2 Elongation

Elongation is the second most important phase of fiber development. Fibers undergo extensive elongation and reach up to 2–3 cm length. According to Lee et al. (2007), primary cell wall (PCW) is formed from 3 to 20 DPA. PCW in cotton fiber is similar to that of dicot plants; it is mainly composed of cellulose and hemicelluloses and pectin. A waxy cuticle is formed outside the fiber cell wall (Doblin et al. 2002). Semicrystalline cellulose fibrils constitute about 22% of PCW. These fibrils are embedded in a matrix composed of xyloglucans and pectin (Singh et al. 2009). Initiation and elongation phases of fiber development in cotton are controlled by rapid primary cell wall formation.

Diffuse growth mechanism of leaf trichomes is also applicable to cotton fibers growth, but recent findings suggest that cotton fiber elongates through tip growth mechanism (Yu et al. 2019). Longitudinal growth of cotton fibers is promoted, instead of radial expansion, by hindrance in turgor pressure movement caused by components of cellulose fibrils cytoskeleton which are arranged transversely to the growing fiber axis (Haigler et al. 2001). Silencing of ACTIN1 gene resulted in reduced fiber growth, proving importance of cytoskeleton in fiber elongation. F-actin and microtubules in cytoskeleton aid in cell wall and membrane materials. Fiber elongation is mainly dependent upon plasmodesmatal regulation, turgor pressure, and vacuolar invertase (GhVINV). *Expansin* genes promote cell wall loosening (Ruan et al. 2004).

Fiber expansion and elongation is also dependent upon cell wall extensibility. Xyloglucans present in cell wall are chopped and rejoined by xyloglucan endotransglycosylases (XTHs). Its digestion eases transport of cellulose microfibrils for cell. Fiber length was reported to be increased by the overexpression of *GhXTH1* as compared to the non-transgenic controls with no adverse effects on fiber characteristics (Lee et al. 2010).

Fiber elongation is also affected by increase in level of K^+, sugars, and malate which play an important role in turgor pressure control. Increase in activity of GhPEPC1 at fiber elongation phase confirms the role of phopsphoenolpyruvate carboxylase (PEPC) in malate synthesis. Substrate for cellulose synthesis is UDP-glucose, which along with fructose is synthesized by reversible breakdown of sucrose. This step is catalyzed by SuS which is unregulated at fiber elongation phase (Ruan et al. 2003; Xu et al. 2012).

Many proteins are expressed during cotton fiber elongation including proteins related to plasma membrane, tonoplast, and some aquaporins (Ferguson et al. 1997). A decisive role is played by aquaporins for water transport across tonoplasts and plasma membrane of fiber cells (Arpat et al. 2004). Aquaporins were reported to be the most significant downregulated genes of developing fibers in ligon-lintless mutants (Li_1, Li_2) in RNA-seq analysis (Naoumkina et al. 2015).

12.6.2.1 The Transition Phase

Transition phase between PCW and SCW occurs during 16–20 DPA. Fibril-like structures are observed on cotton fibers at 17 DPA which shows the initiation of SCW synthesis. The fibers become flat and slightly thicker at 20 DPA suggesting switching of PCW to SCW synthesis. In total, 100-fold increases in rate of cellulose synthesis as compared to earlier stages of fiber development are reported during transition phase. Fiber elongation is limited by transverse orientation of cellulose microfibril changes to helical pattern. Thickening in cotton fiber cell wall might be favored by a significant increase in ratio of ABA and IAA during rapid cellulose synthesis at 30–40 DPA (You-Ming et al. 2001).

Sus is present in cotton fiber cells in its cytoplasm and on plasma membrane (Amor et al. 1995). Carbon is transported directly to form cotton fiber cellulose by a complex formed between cellulose synthase (CES) and Sus present on plasma membrane (Haigler et al. 2001). Interestingly Kor and Tubulin genes are reported to be upregulated during cellulose synthesis along with lipid transfer proteins which interact with GhCESA1. These deposited fatty acids within cotton fiber cell wall layers are found to be in alternating pattern with cellulose (Doblin et al. 2002).

12.6.3 Secondary Wall Synthesis and Maturation

Secondary cell wall synthesis of cotton fiber is laid down after elongation phase at 16–40 DPA (Lee et al. 2007). Deposition of pure cellulose in large quantity at this stage causes narrowing of fiber lumen. Mature fibers have almost 96% cellulose upon harvesting. Cellulose chains deposited during SCW synthesis are longer, about 14,000 units, than that of the chains deposited during PCW synthesis and contribute to fiber strength at this stage (Stewart et al. 2016).

Both micronaire and strength of cotton fiber are affected by the minimal stress that develops on cotton plant during SCW synthesis. Cellulose synthesis is increased during secondary wall synthesis due to stronger carbon sink as compared to primary wall. A large number of genes are involved in deposition of carbon irreversibly in secondary wall, mainly the ones coding cellulose synthase (Haigler et al. 2001).

Synchronization exists between synthesis of SCW and elongation of cotton fiber. Decrease in contents of living protoplast and shrinkage in space occupied by large vacuole is observed at this stage (Brill et al. 2011). Microfibrils are formed during cellulose biosynthesis. Cellulose molecules are accumulated in a precise pattern, while UDP-glucose serves as precursor at this stage (Brill et al. 2011).

Different factors, i.e., environmental stresses phytohormones, genes, and transcription factors, affect the maturity of cotton fiber and synthesis of secondary wall (Ayele et al. 2017). Manipulation of various genes and promoters involved in synthesis of SCW is involved in improvement of fiber (Qin and Zhu 2011). Important genes that play significant role in different stages of fiber development are elaborated in Table 12.1 along with their potential functions.

Table 12.1 Genes involved in fiber development and elongation (Ahmed et al. 2018)

Gene	Accession no	Potential function	Reference
GhCESA1	U58283	Upregulated at the onset of secondary wall synthesis	Pear et al. (1996)
GhE6	BM356398	Fiber protein E6, fiber elongation, and secondary wall biosynthesis	John and Keller (1996)
pGhEX1	AF043284	Found abundantly in cotton fiber cells and regulated during fiber elongation	Orford and Timmis (1998)
GhTUB1	AF487511	Plays a role in polar elongation of cotton fiber	Zhang et al. (2003)
GhGlcAT1	AY346330	Glucuronosyltransferase-like protein involved in the synthesis of noncellulosic cell wall components during fiber elongation	Wu et al. (2006)
CEL	AY574906	Endo 1,4-beta-glucanase, necessary for plant cellulose biosynthesis	Zhu et al. (2012)
CelA1	GHU58283	Cellulose synthase	Zhu et al. (2012)
CelA1	AF150630	Cellulose synthase catalytic subunit, cellulose biosynthesis in developing cotton fibers	Zhu et al. (2012)
Exp1	DQ204495	Alpha expansin1, cell wall extension and effect on length and quality of fiber	Zhu et al. (2012)
ACT 1	AY305723	Actin1, plays a major role in fiber elongation	Zhu et al. (2012)
BG	DQ103699	Beta 1,4-glucanase, loosening of primary wall and promotion of secondary cell wall synthesis	Zhu et al. (2012)
Pel	DQ073046	Pectate lyase, degradation of de-esterified pectin and helps in normal fiber elongation	Zhu et al. (2012)
SuS1	U73588	Sucrose synthase, plays an important role in cotton fiber initiation and elongation by influencing carbon partitioning to cellulose synthesis	Zhu et al. (2012)
LTP3	AF228333	Lipid transfer protein gene, cutin synthesis during fiber primary cell wall synthesis stage	Zhu et al. (2012)
WLIM1a	JX648310	Fiber elongation and secondary wall synthesis in developing fibers	Han et al. (2013)
GbEXPATR	DQ912951	Enhances cotton fiber elongation through reorganizing secondary cell wall synthesis	Li et al. (2016)

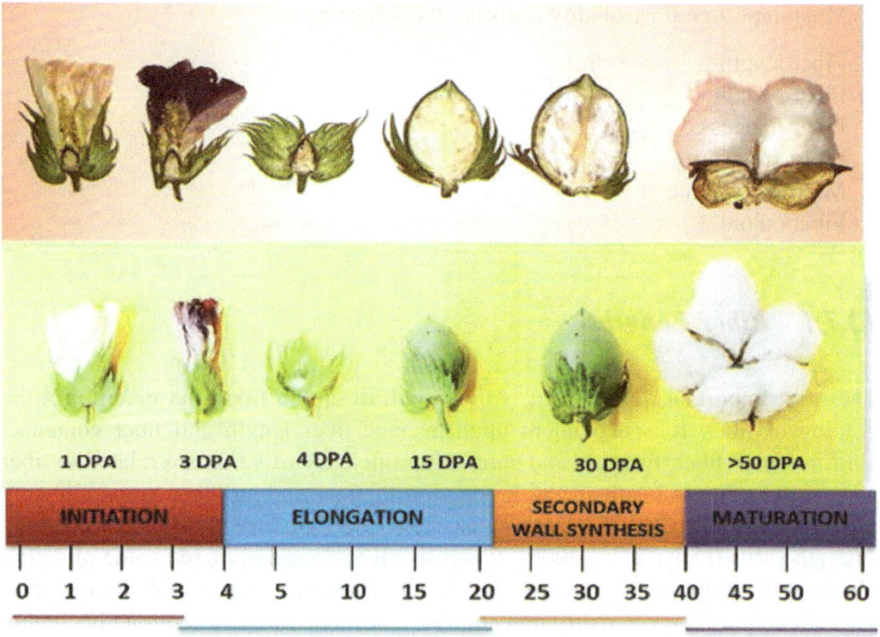

Fig. 12.4 Stages of cotton fiber development in *Gossypium hirsutum*. Initiation is the first step of cotton fiber development. It starts from −2 to 3 DPA. Elongation phase comes after initiation and has duration of 3 to 20 DPA. The secondary cell wall synthesis commences at 20 DPA till 40 DPA, and the last stage of fiber maturation begins at 40 DPA till >50 DPA (Lee et al. 2007)

12.6.4 Fiber Maturation

The fiber maturation occurs at 40–50 DPA (Qin and Zhu 2011). During this stage the cotton bolls open up, and moisture of fiber evaporates due to their exposure to atmosphere leading to its dryness. The evaporation also results in the collapse the lumen of fiber, thereby decreasing the length of fiber up to one-third of their original size, and introduces twists in fibers which are the most important aspects required for cotton yarn from lint via spinning. Typical white mass of open cotton boll is the result of fluffing of seeds fiber upon boll opening (Ruan et al. 2003) (Fig. 12.4).

12.7 Cotton Fiber Quality Traits

Physical properties of cotton define its fiber's quality. These properties have impact in textile performance due to effect these cause on fiber spinnability into yarn (Draye et al. 2005). Environmental condition, cotton cultivars, and crop management define fiber quality traits.

Most important fiber quality traits are the following:

1. Fiber length
2. Fiber strength
3. Fiber fineness
4. Fiber uniformity
5. Micronaire value
6. Fiber color

12.7.1 Fiber Length

The most important quality trait is the length of cotton fiber that determines the grading of fiber. It is dependent upon average fiber length and fiber contents., Uniformity of fiber, fineness and spinning proficiency of yarns are related to fiber length. Long fibers result in the production of finer and stronger yarns allowing more twist around each other. In contrast, short fibers produce poor-quality and low-strength yarns (Draye et al. 2005). It is difficult to define any fixed value of cotton fiber length within specific sample or genotype due to intrinsic variability of cotton fiber (Behery 1993). Yet in different countries, different ranges of cotton fiber length have been introduced to define fiber quality traits. Fiber length for Australian cotton cultivars was reported to be 29–32 mm by the Australian Cotton Shippers Association (ACSA). Four staple categories of cotton based on fiber length are reported in the USA, i.e., short fibers (<21 mm), medium fibers (22–25 mm), medium-long fibers (26–28 mm), and long fibers (29–34 mm) (Bradow et al. 1997), while the local cotton cultivars in Pakistan are divided as short staple (20.64 mm or less), medium staple (21.64–25.4 mm), medium-long staple (26.19–27.78 mm), long staple (28.58–33.34 mm), and extra-long staple (35.7–41.28 mm) (Malik and Ahsan 2016).

12.7.2 Fiber Strength

Strength of individual cotton fibers is responsible for the ultimate strength of yarn, while fiber strength itself is dependent upon the length of cellulose chains. Thus longer cellulose chains will generate stronger fibers which will result in good-quality yarn production (Pettigrew et al. 1996). A comparison of cottons' fiber strength range among different cultivars of USA and Pakistan is given in Table 12.2 (Malik and Ahsan 2016).

12.7.3 Fiber Fineness

Fiber fineness is determined by diameter and thickness of wall. Soft and silky touch of some fibers with coarse and hard of others is due to fiber fineness. Thickness of cell wall is increased as the fiber matures, and it becomes equal and sometimes more

Table 12.2 Comparison of fiber strength range among cotton cultivars in the USA and Pakistan

Sr. no	Category	Value in the USA	Value in Pakistan
1	Weak	23 g/tex or below	20 g/tex or below
2	Intermediate	24–25 g/tex	21–25 g/tex
3	Average	26–28 g/tex	26–29 g/tex
4	Strong	29–30 g/tex	30–32 g/tex
5	Above strong	31 g/tex or above	32 g/tex or above

than the diameter of the fiber cell lumen. Fiber fineness directly affects the fiber processing and is responsible for determining the quality of products manufactured (Ramey 1982).

12.7.4 Fiber Uniformity

Fiber uniformity index is defined as "ratio between mean and upper half of fiber's mean length." It is always less than 100 due to non-similarity of fibers length in any test sample or even within single boll or single seed and described in percentage. Cotton cultivar which shows low uniformity index resulted in high percentage of short fiber content ultimately resulting in low quality yarn production. It affects the strength and symmetry of yarn and also the spinning process. Upland cotton is divided into five categories based on fiber uniformity index as follows (Bradow et al. 1997).

1. Very low (77 or below)
2. Low (77–79)
3. Intermediate (80–82)
4. High (83–85)
5. Very high (85 or above)

12.7.5 Micronaire Value of Cotton Fiber

Airflow resistance by a specific volume of fiber sample is measured as micronaire value of the fiber. Extent of fiber lumen development is sometimes confused with micronaire value. Low micronaire value is considered an indication of fine fiber, but fiber immaturity can also result in low micronaire value which causes neps (creation of knots in yarn processing) and dyeing problems, thereby hindering yarn processing (Steadman 1997). The micronaire value is determined by SCW thickness and fiber diameter. It is a unit less trait with optimum range from 3.5 to 4.9 (Haigler et al. 2005).

On the other hand, the high micronaire value results in extra maturity of cotton fibers which do not take the bean shape upon drying and make weak yarn. Mic range of upland cotton is divided into a premium range (3.7–4.2), a base range (3.5–4.9), and a discount range (3.4 or less or 5 or above) (Bradow et al. 1997).

12.7.6 *Fiber Color*

Fiber color is directly influenced by the prevailing environmental conditions during fiber development. Creamy or bright white color is the characteristic of high-quality fiber, while the low-quality fiber is usually grayish in color or dull yellow. Low-quality fiber is usually due to harsh field conditions like drought frost, insect infestation, or impurities caused by foreign matter (Perkins et al. 1984).

12.8 Improvement of Fiber Traits Through Genetic Modification

Understanding the biochemistry of fiber quality and unraveling the fiber development mechanism have helped in understanding relationship between fiber quality and maturity. Identification of thousands of fiber characteristics regulating genes opened up new avenues of fiber improvements.

Fiber quality and yield have been reported to be improved through gene pool exchange among different cultivars in classical cotton breeding. But this fact is actually limited to a few characteristics, i.e., introduction/exchange of traits among interspecies of organisms through breeding (Ahmed et al. 2018). On the other hand, biotechnological approaches have the potential to introduce desirable target genes of synthetic in origin from other organisms into cotton to overcome the problems of conventional breeding (John and Keller 1996).

Low level of genetic diversity has been reported in *G. hirsutum* which can be enhanced through application of advanced approaches like germplasm introgression, mutagenesis, and genetic transformation (Lacape et al. 2010).

12.8.1 *Effect of Change in Genes' Expression*

Enzymes involved in sucrose metabolism and cellulose synthesis influenced the growth of cotton fiber. The characteristics of fibers have well been characterized upon overexpression and under expression of different traits by many researchers. Improvement of fiber traits through overexpression of different enzymes like sucrose synthase (SuS), invertase (INV), and sucrose phosphate synthase (SPS) has been reported (Shu et al. 2009). The overexpression of these traits has been found to result in improved fiber maturity and increased plant biomass through enhanced plant ability to withstand abiotic stresses.

Fiber characteristics can be altered by the production of aliphatic polyester poly-D-(2)-3-hydroxybutyrate (PHB) in cotton which is a thermoplastic polymer. PHB is produced from acetyl-CoA in a reaction catalyzed by enzymes acetyl-coA reductase (phaB), β-ketothiolase (phaA), and poly hydroxyl alkanoate synthase (phaC). Endogenous phaA activity is exhibited by cotton fibers. The transgenic cotton expressing phaB and phaC fused with GUS determined the expressed protein in fiber cells. Improved insulating properties were exhibited by cotton fibers having PHB genes. A high influence on the textile industry output can be achieved through successful implementation of the transgenic technology for modification of these fiber traits (John and Keller 1996).

High rates of cellulose deposition are reported to be the result of high GhGluc1, GhCeSA1, and GhCeSA2 expression during cell wall biosynthesis (Ruan et al. 2004). Increase in cell wall loosening and intracellular turgor pressure is caused by Ca^{+2} conductance which is regulated by ghFAnnxA during SCW synthesis (Qin and Zhu 2011). Downregulation of GhAnne due to decrease in Ca^{2+} flux at the cell apex resulted in inhibition of fiber length (Tang et al. 2014) (Fig. 12.5).

Improved fiber length is most desirable characteristics in fiber development process. Actin-binding proteins are known to raise fiber elongation level (Wang et al. 2010). Importance of F-actin arrays in staple elongation and root hair growth of *Arabidopsis* was reported. Fiber cell elongation is reported to be inhibited through silencing of GhACTIN1 due to reduction in quantity of F-actin (Qin and Zhu 2011). Fiber length can be modified through combined function of actin filaments and MTs linked by kinetin (Xu et al. 2009).

Overexpression of fiber-specific α-expansins GbEXPA2 and GbEXPATR genes of *G. barbadense* in *G. hirsutum* revealed increase in fiber length. Another important gene reported for fiber improvement is GhEXPA8. Major improvement in staple length and micronaire values in transgenic cotton plants was found in the combined data of research done on three generations of local cotton variety NIAB 846 (Bajwa et al. 2015).

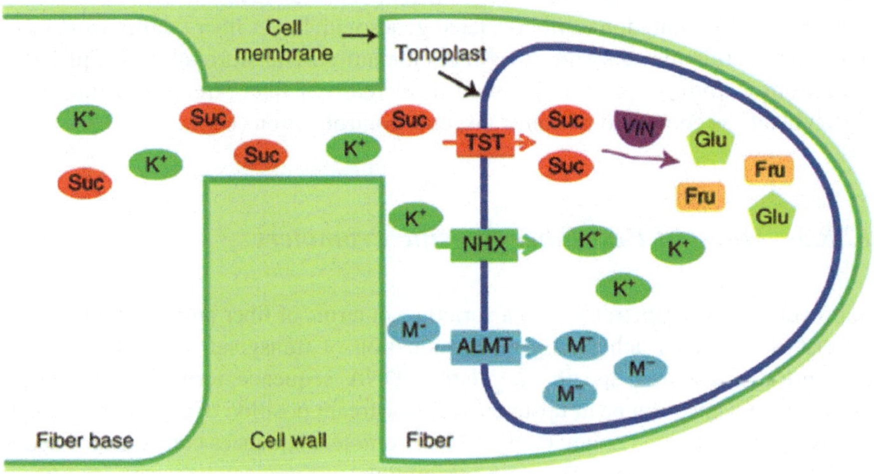

Fig. 12.5 Model depicting the roles of open PD and the genes related to the main osmotically active solutes in fiber elongation. Suc, sucrose; M, malate; Fru, fructose; Glu, glucose (Hu et al. 2019)

A novel link between fiber elongation and Ca^{2+} and K^+ was unrevealed by Guo et al. (2017). Potassium is the primary osmotic agent that increases cell turgor pressure and results in fiber elongation. Deficiency in potassium (K) results in reduction of fiber length in cotton (Yang et al. 2014). Regulation of plant's natural biological cycle mechanism is a new trend in research area of green revolution to drive desired traits.

12.8.2 Transgene Influence on Fiber Traits of Cotton

Transformation of foreign fiber-related genes in cotton plants has proved to be a successful strategy for improvement of cotton fiber trait. Introduction of foreign gene in cotton has never been an easy task. Cotton is the most difficult crop to be transformed with foreign genes. Initially coker was found to be an easy cotton genotype cultivar for transformation. So foreign genes for fiber quality improvement were introduced in coker and thereafter were backcrossed with the desired variety (Bayley et al. 1992). It is a time taking and much laborious attempt as it took years for the recovery of desirable traits of the adapted variety. New transformation technique has been demonstrated that involves the direct transformation of elite cotton cultivars.

Genes are directly transformed in pollen grains of the selected cotton variety. A significant increase in fiber length (15.7%), strength (up to 17.3%), and cellulose content of transgenic cotton plants as compared to non-transgenic plants was observed after introducing the bacterial cellulose synthesis genes acsA and acsB into cotton (Li et al. 2004). Thus introduction of bacterial acsA and B genes can improve cotton fiber length and strength (Zhang et al. 2011).

Genetic transformation of fiber-related genes with better fiber quality from other plants into cotton may also improve fiber elongation. Transformation of CpEXPA3 gene from *Calotropis procera* resulted in increase in fiber strength in transgenic cotton plants as compared to non-transgenic control cotton (Bajwa et al. 2013).

12.8.3 Quest of Best Fiber-Specific Promoters

Improvement in commercial cotton varieties, in terms of fiber and agronomic characteristics, has been achieved. The introduction of transgene in plants has been demonstrated under a specific regulatory DNA sequence termed as promoter. Promoter sequence can be at upsteam or downstream position of a gene, but mostly it is 1000 bp upstream (Zhang et al. 2011). Promoter sequences can be expressed in all tissues of the transgenic plant, i.e., ubiquitous, or these can be tissue specific.

CaMV35S is the most widely used promoter which showed strong constitutive expression in all dicots including cotton. But tissue-specific promoters are required for introduction of specific traits in cotton like fiber quality or yield. Despite the discovery of several tissue-specific promoters, characterization of these in cotton is a very laborious process; therefore it remained slow as compared to other plants (Wilkins and Arpat 2005). The issue was addressed through evaluation of tissue-specific promoters in other plants like *Arabidopsis* and tobacco. Introduction of two promoters of lipid transfer proteins in tobacco was done from cotton fiber. The results of this study showed that trichrome-specific expression in tobacco leaves can be triggered even with a promoter sequence of less than 614 nucleotides (Liu et al. 2000). Expression in leaf trichromes of *Arabidopsis* was obtained after introduction of promoter of CESA family from cotton fiber in it (Kim and Triplett 2001).

Later studies have shown that different regulatory mode of development is shared between leaf trichromes and cotton fibers. Therefore, it was not a worth working idea to introduce promoters in tobacco or *Arabidopsis* and translate the results to cotton fiber. Strong expression of seed coat and fiber-specific proteases (SCFP) was detected in cotton fiber only in a study done on both cotton fiber and tobacco trichrome (Hou et al. 2008).

Cloning and characterization of several fiber-specific promoters have been carried out in cotton. Higher and long-term expression of GhSCFP was reported when a comparison study was undertaken for cotton-specific promoters, i.e., GhSCFP, GhACT1, FbL2A, and E6It. These have shown expression as high as that of CaMV35 promoter (Li et al. 2005).

12.8.4 Genetic Modification in Transcription Factors (TFs)

Transcriptional factors also play an important role in modification of fiber traits. The three genes of GhHOX family, i.e., GhHOX1, GhHOX2, and GhHOX3, are reported to be lined with quantitative trait loci. These play role to refine fineness, length, and uniformity of cotton fiber. A significant role in production of fine quality and elongated fiber has been reported for GhHOX3 (Shah and Brown 2005).

Expression of various target genes can be regulated by single TFs by binding specifically to the cis-acting element in promoters of respective genes (Patil et al. 2016). Transcription factors like *differentiation* GhMYB25, GhMYB109, and GhMYB2 play a very important role in fiber development (Lu et al. 2002).

12.8.5 Phytohormonal Regulation of Fiber Cell Development

Plant hormones also play a distinct role in regulation of cotton fiber development. In vitro and in vivo studies have demonstrated the effect of hormones on different fiber development stages (Ahmed et al. 2018). Indole acetic acid (IAA) and gibberellic acid (GA) have important role in fiber development (Samuel Yang et al. 2006; Zhang et al. 2011). Increase in fiber initials and fiber lint was reported when IAA synthesizing gene was transformed in cotton. Xiao et al. (2010) have shown positive effect of GA synthesizing gene overexpression on increase in number of fiber initials and fiber elongation. An increase of 12.7% in culture ovules was reported by external application of brassinolide (BL) which enhanced expression of elongation-related genes EXP, XTH, AGP, and GhTUB1 (Xiao et al. 2007). The BL can retard the inhibition in fiber growth and elongation caused by brassinazole (BRZ) (Zhu et al. 2006). The BL biosynthesis genes, i.e., auxins and gibberellins, are expressed during different fiber growth stages which enhance fiber growth (Shi et al. 2006). Shi et al. (2006) demonstrated significant role of ethylene in fiber development. Increase in fiber growth was observed by external application of ethylene on ovule cultures, while AVG's (an ethylene inhibitor) application resulted in retarded fiber growth.

Abscisic acid inhibits fiber growth (Gokani and Thaker 2002). Cytokinins are also reported to inhibit fiber growth despite their role in boosting ovule cultures, cellular division, and vascular tissues development (Dhindsa et al. 1976).

12.9 Conclusion

Fiber quality traits are topic of significant importance specifically for textile industry. Genetic modification for improvement of fiber quality has been point of interest for scientists across world. Research work done so far has resulted in detection of genes important for different aspects of cotton fiber. Alteration in expression of fiber-related genes, transcription factors, and, genes expressing phytohormones has given quite encouraging results. Introduction of fiber-related foreign genes in cotton also proved to be fruitful for enhancing fiber traits. More research needs to be done. Work can be carried out for changing the color of cotton fiber to avoid losses and infections caused by dyeing industry. Detection of more foreign genes related to fiber synthesis and their introduction in fiber can enhance fiber qualities. Further studies need to be done on studying biochemical basis of different mechanisms involved in fiber development.

References

Abidi N, Hequet E, Cabrales L (2010) Changes in sugar composition and cellulose content during the secondary cell wall biogenesis in cotton fibers. *Cellulose* 17(1):153–160

Ahmed M, Shahid AA, Din SU, Akhtar S, Ahad A, Rao AQ et al (2018) An overview of genetic and hormonal control of cotton fiber development. *Pak J Bot* 50(1):433–443

Ahmed M, Akhtar S, Fanglu M, Hasan M, Shahid A, Yanang X et al (2019) Sucrose synthase (SuSy) gene expression: an indicator for cotton fiber initiation and early development. *Russ J Plant Physiol* 66:41–49

Amor Y, Haigler CH, Johnson S, Wainscott M, Delmer DP (1995) A membrane-associated form of sucrose synthase and its potential role in synthesis of cellulose and callose in plants. *Proc Natl Acad Sci* 92(20):9353–9357

Arpat A, Waugh M, Sullivan JP, Gonzales M, Frisch D, Main D et al (2004) Functional genomics of cell elongation in developing cotton fibers. *Plant Mol Biol* 54(6):911–929

Ayele A, Hequet E, Kelly B (2017) The impact of fiber maturity on estimating the number of cotton (*Gossypium hirsutum* L.) fibers per seed surface area. *Ind Crop Prod* 102:16–22

Bajwa KS, Shahid AA, Rao AQ, Kiani MS, Ashraf MA, Dahab AA et al (2013) Expression of Calotropis procera expansin gene CpEXPA3 enhances cotton fibre strength. *Aust J Crop Sci* 7(2):206

Bajwa KS, Shahid AA, Rao AQ, Bashir A, Aftab A, Husnain T (2015) Stable transformation and expression of GhEXPA8 fiber expansin gene to improve fiber length and micronaire value in cotton. *Front Plant Sci* 6:838

Basra AS, Malik C (1984) Development of the cotton fiber. *Int Rev Cytol* 89:65–113. Elsevier

Bawa A, Anilakumar K (2013) Genetically modified foods: safety, risks and public concerns—a review. *J Food Sci Technol* 50(6):1035–1046

Bayley C, Trolinder N, Ray C, Morgan M, Quisenberry J, Ow D (1992) Engineering 2, 4-D resistance into cotton. *Theor Appl Genet* 83(5):645–649

Behery HM (1993) Short fiber content and uniformity index in cotton. CAB International

Bradow JM, Wartelle LH, Bauer PJ, Sassenrath-Cole GF (1997) Small-sample cotton fiber quality quantitation. *J Cotton Sci* 1(1):48–60

Brill E, van Thournout M, White RG, Llewellyn D, Campbell PM, Engelen S et al (2011) A novel isoform of sucrose synthase is targeted to the cell wall during secondary cell wall synthesis in cotton fiber. *Plant Physiol* 157(1):40–54

Brubaker C, Paterson A, Wendel J (1999) Comparative genetic mapping of allotetraploid cotton and its diploid progenitors. *Genome* 42(2):184–203

Cohen SN, Chang ACY, Boyer HW, and Helling RB (1973) Construction of biologically functional bacterial plasmids in vitro. *Proc Natl Acad Sci USA* 70(11):3240–3244.

Dhindsa R, Beasley C, Ting I (1976) Effects of abscisic acid on in vitro growth of cotton fiber. *Planta* 130(2):197–201

Doblin MS, Kurek I, Jacob-Wilk D, Delmer DP (2002) Cellulose biosynthesis in plants: from genes to rosettes. *Plant Cell Physiol* 43(12):1407–1420

Doebley JF, Gaut BS, Smith BD (2006) The molecular genetics of crop domestication. *Cell* 127(7):1309–1321

Draye X, Chee P, Jiang C-X, Decanini L, Delmonte TA, Bredhauer R et al (2005) Molecular dissection of interspecific variation between *Gossypium hirsutum* and *G. barbadense* (cotton) by a backcross-self approach: II. Fiber fineness. *Theor Appl Genet* 111(4):764–771

Ferguson DL, Turley RB, Kloth RH (1997) Identification of a δ-TIP cDNA clone and determination of related A and D genome subfamilies in Gossypium species. *Plant Mol Biol* 34(1):111–118

Fryxell P (1992) A revised taxonomic interpretation of Gossypium L (Malvaceae). *Rheedea* 2:108–116

Gilbert M, Slingenbergh J, Xiao X (2008) Climate change and avian influenza. *Rev Sci Tech* 27(2):459–466

Gokani SJ, Thaker VS (2002) Physiological and biochemical changes associated with cotton fiber development: IX. Role of IAA and PAA. *Field Crops Res* 77(2–3):127–136

Haigler CH, Ivanova-Datcheva M, Hogan PS, Salnikov VV, Hwang S, Martin K, Delmer DP (2001) Carbon partitioning to cellulose synthesis. *Plant Cell Walls* 47:29–51. Springer

Haigler CH, Zhang D, Wilkerson CG (2005) Biotechnological improvement of cotton fibre maturity. *Physiol Plant* 124(3):285–294

Han L-B, Li Y-B, Wang H-Y, Wu X-M, Li C-L, Luo M et al (2013) The dual functions of WLIM1a in cell elongation and secondary wall formation in developing cotton fibers. *Plant Cell* 25(11):4421–4438

Hassan SZU, Militky J (2012) Acetylcholinesterase based detection of residual pesticides on cotton. *Am J Anal Chem* 3(02):93

Hearle JW (2007) Protein fibers: structural mechanics and future opportunities. *J Mater Sci* 42(19):8010–8019

Hou L, Liu H, Li J, Yang X, Xiao Y, Luo M et al (2008) SCFP, a novel fiber-specific promoter in cotton. *Chin Sci Bull* 53(17):2639–2645

Hu Y, Chen J, Fang L, Zhang Z, Ma W, Niu Y, Ju L, Deng J, Zhao T, Lian J, Baruch K, Fang D, Liu X, Ruan Y, Rahman M, Han J, Wang K, Tianzhen Zhang T (2019) Gossypium barbadense and Gossypium hirsutum genomes provide insights into the origin and evolution of allotetraploid cotton. *Nature Genetics* Vol 51 PP: 739–748

John ME, Keller G (1996) Metabolic pathway engineering in cotton: biosynthesis of polyhydroxybutyrate in fiber cells. *Proc Natl Acad Sci* 93(23):12768–12773

Kim HJ, Triplett BA (2001) Cotton fiber growth in planta and in vitro. Models for plant cell elongation and cell wall biogenesis. *Plant Physiol* 127(4):1361–1366

Kozlowski RM (2012) Handbook of natural fibres: types, properties and factors affecting breeding and cultivation. Elsevier

Lacape J-M, Llewellyn D, Jacobs J, Arioli T, Becker D, Calhoun S et al (2010) Meta-analysis of cotton fiber quality QTLs across diverse environments in a *Gossypium hirsutum x G. barbadense* RIL population. *BMC Plant Biol* 10(1):132

Lee JJ, Woodward AW, Chen ZJ (2007) Gene expression changes and early events in cotton fibre development. *Ann Bot* 100(7):1391–1401

Lee J, Burns TH, Light G, Sun Y, Fokar M, Kasukabe Y et al (2010) Xyloglucan endotransglycosylase/hydrolase genes in cotton and their role in fiber elongation. *Planta* 232(5):1191–1205

Li X, Wang X, Zhao X, Dutt Y (2004) Improvement of cotton fiber quality by transforming the acsA and acsB genes into *Gossypium hirsutum* L. by means of vacuum infiltration. *Plant Cell Rep* 22(9):691–697

Li X-B, Fan X-P, Wang X-L, Cai L, Yang W-C (2005) The cotton ACTIN1 gene is functionally expressed in fibers and participates in fiber elongation. *Plant Cell* 17(3):859–875

Li X-R, Wang L, Ruan Y-L (2009) Developmental and molecular physiological evidence for the role of phosphoenolpyruvate carboxylase in rapid cotton fibre elongation. *J Exp Bot* 61(1):287–295

Li CQ, Song L, Zhao HH, Wang QL, Fu YZ (2014) Identification of quantitative trait loci with main and epistatic effects for plant architecture traits in Upland cotton (*Gossypium hirsutum* L.). *Plant Breed* 133(3):390–400

Li Y, Tu L, Pettolino FA, Ji S, Hao J, Yuan D et al (2016) GbEXPATR, a species-specific expansin, enhances cotton fibre elongation through cell wall restructuring. *Plant Biotechnol J* 14(3):951–963

Liu H-C, Creech RG, Jenkins JN, Ma D-P (2000) Cloning and promoter analysis of the cotton lipid transfer protein gene Ltp3. *Biochim Biophys Acta* 1487(1):106–111

Lu Y, Wei G, Zhu Y (2002) Cloning whole cellulose-synthesizing operon (ayacs operon) from Acetobacter xylinum and transforming it into cultivated cotton plants. *Acta Bot Sin* 44(4):441–444

Malik TH, Ahsan MZ (2009) Review of the cotton market in Pakistan and its future prospects. *OCL* 2016: 23(6):A606. https://doi.org/10.1051/ocl/2016043

Meinert MC, Delmer DP (1977) Changes in biochemical composition of the cell wall of the cotton fiber during development. *Plant Physiol* 59(6):1088–1097

Naithani S, Rao NR, Singh Y (1982) Physiological and biochemical changes associated with cotton fibre development: I. Growth kinetics and auxin content. *Physiol Plant* 54(2):225–229

Naoumkina M, Thyssen GN, Fang DD (2015) RNA-seq analysis of short fiber mutants Ligon-lintless-1 (Li 1) and - 2 (Li 2) revealed important role of aquaporins in cotton (Gossypium hirsutum L.) fiber elongation" *BMC Plant Biol.* 15:65. https://doi.org/10.1186/s12870-015-0454-0

Orford SJ, Timmis JN (1998) Specific expression of an expansin gene during elongation of cotton fibres. *Biochim Biophys Acta* 1398(3):342–346

Patil G, Do, T, Vuong T D, Valliyodan B, Lee J, D, Chaudhary J. et al. (2016). Genomic-assisted haplotype analysis and the development of high-throughput SNP markers for salinity tolerance in soybean. *Sci. Rep.* 19:19199. https://doi.org/10.1038/srep19199

Pear JR, Kawagoe Y, Schreckengost WE, Delmer DP, Stalker DM (1996) Higher plants contain homologs of the bacterial celA genes encoding the catalytic subunit of cellulose synthase. *Proc Natl Acad Sci* 93(22):12637–12642

Perkins HH, Jr., Ethridge DE, Bragg CK (1984) Fiber. p. 437–509. In R.J. Kohel and C.F. Lewis (ed.) Cotton. Agron. Monogr. 24. ASA, CSSA, and SSSA, Madison, WI
Pettigrew WT, Heitholt JJ, Meredith WR (1996) Genotypic interactions with potassium and nitrogen in cotton of varied maturity. *Agron J* 88(1):89–93
Pillay M, Myers G (1999) Genetic diversity in cotton assessed by variation in ribosomal RNA genes and AFLP markers. *Crop Sci* 39(6):1881–1886
Poehlman JM and Borthakur DN (1969) Breeding Asian Field Crops. Oxford and IBH Publishing Co, New Delhi, pp 1–100.
Qin Y-M, Zhu Y-X (2011) How cotton fibers elongate: a tale of linear cell-growth mode. *Curr Opin Plant Biol* 14(1):106–111
Ramey H (1982) The meaning and assessment of cotton fibre fineness. International Institute for Cotton, Manchester
Rangel G (2015) From corgis to corn: a brief look at the long history of GMO technology. Blog, Special Edition on GMOs. Science in the News
Rauf MA, Zubair S, Azhar A (2015) Ligand docking and binding site analysis with pymol and autodock/vina. *Int J Basic Appl Sci* 4(2):168
Ruan Y-L, Llewellyn DJ, Furbank RT (2003) Suppression of sucrose synthase gene expression represses cotton fiber cell initiation, elongation, and seed development. *Plant Cell* 15(4):952–964
Ruan Y-L, Xu S-M, White R, Furbank RT (2004) Genotypic and developmental evidence for the role of plasmodesmatal regulation in cotton fiber elongation mediated by callose turnover. *Plant Physiol* 136(4):4104–4113
Samuel Yang S, Cheung F, Lee JJ, Ha M, Wei NE, Sze SH et al (2006) Accumulation of genome-specific transcripts, transcription factors and phytohormonal regulators during early stages of fiber cell development in allotetraploid cotton. *Plant J* 47(5):761–775
Seagull R, Giavalis S (2004) Pre-and post-anthesis application of exogenous hormones alters fiber production in *Gossypium hirsutum* L. cultivar Maxxa GTO. *J Cotton Sci* 8:105–111
Shah J, Brown RM (2005) Towards electronic paper displays made from microbial cellulose. *Appl Microbiol Biotechnol* 66(4):352–355
Shi Y-H, Zhu S-W, Mao X-Z, Feng J-X, Qin Y-M, Zhang L, Cheng J, Wei LP, Wang ZY, Zhu YX (2006) Transcriptome profiling, molecular biological, and physiological studies reveal a major role for ethylene in cotton fiber cell elongation. *Plant Cell* 18(3):651–664
Shu H, Zhou Z, Xu N, Wang Y, Zheng M (2009) Sucrose metabolism in cotton (*Gossypium hirsutum* L.) fibre under low temperature during fibre development. *Eur J Agron* 31(2):61–68
Singh B, Avci U, Inwood SEE, Grimson MJ, Landgraf J, Mohnen D et al (2009) A specialized outer layer of the primary cell wall joins elongating cotton fibers into tissue-like bundles. *Plant Physiol* 150(2):684–699
Smith L (1995) Cotton response to deep tillage with controlled traffic on clay. *Trans ASAE* 38(1):45–50
Steadman R (1997) Cotton testing. *Text Prog* 27(1):1–63
Stewart JM, Oosterhuis D, Heitholt JJ, Mauney JR (2016) Physiology of Cotton, Springer
Stiff MR, Haigler CH (2012) Recent advances in cotton fiber development. Flowering and fruiting in cotton. The Cotton Foundation, Tennessee, pp 163–192
Tang W, He Y, Tu L, Wang M, Li Y, Ruan Y-L, Zhang X (2014) Down-regulating annexin gene GhAnn2 inhibits cotton fiber elongation and decreases Ca 2+ influx at the cell apex. *Plant Mol Biol* 85(6):613–625
Wang L, Li X-R, Lian H, Ni D-A, He Y-k, Chen X-Y, Ruan Y-L (2010) Evidence that high activity of vacuolar invertase is required for cotton fiber and Arabidopsis root elongation through osmotic dependent and independent pathways, respectively. *Plant Physiol* 154(2):744–756
Westafer JM, Brown JR (1976) Electron microscopy of the cotton fibre: new observations on cell wall formation. *Cytobios* 15(58–59):111–138
Wilkins TA, Arpat AB (2005) The cotton fiber transcriptome. *Physiol Plant* 124(3):295–300
Wu Y, Machado AC, White RG, Llewellyn DJ, Dennis ES (2006) Expression profiling identifies genes expressed early during lint fibre initiation in cotton. *Plant Cell Physiol* 47(1):107–127

Xiao Y-H, Zhang Z-S, Yin M-H, Luo M, Li X-B, Hou L, Pei Y (2007) Cotton flavonoid structural genes related to the pigmentation in brown fibers. *Biochem Biophys Res Commun* 358(1):73–78

Xiao Y-H, Li D.-M., Yin M.-H, Li, X.-B, Zhang M, Wang Y.-J, Pei Y (2010). Gibberellin 20-oxidase promotes initiation and elongation of cotton fibers by regulating gibberellin synthesis. *Journal of Plant Physiology*, 167(10), 829–837. https://doi.org/10.1016/j.jplph.2010.01.003

Xu T, Qu Z, Yang X, Qin X, Xiong J, Wang Y et al (2009) A cotton kinesin GhKCH2 interacts with both microtubules and microfilaments. *Biochem J* 421(2):171–180

Xu S-M, Brill E, Llewellyn DJ, Furbank RT, Ruan Y-L (2012) Overexpression of a potato sucrose synthase gene in cotton accelerates leaf expansion, reduces seed abortion, and enhances fiber production. *Mol Plant* 5(2):430–441

Yang T, Zhang S, Hu Y, Wu F, Hu Q, Chen G et al (2014) The role of a potassium transporter OsHAK5 in potassium acquisition and transport from roots to shoots in rice at low potassium supply levels. *Plant Physiol* 166(2):945–959

You-Ming Y, Chu-Nian X, Bao-Min W, Jun-Zhen J (2001) Effects of plant growth regulators on secondary wall thickening of cotton fibres. *Plant Growth Regul* 35(3):233–237

Yu Y, Wu S, Nowak J, Wang G, Han L, Feng Z et al (2019) Live-cell imaging of the cytoskeleton in elongating cotton fibres. *Nat Plants* 5(5):498

Zhang T, Yuan Y, Yu J, Guo W, Kohel RJ (2003) Molecular tagging of a major QTL for fiber strength in Upland cotton and its marker-assisted selection. *Theor Appl Genet* 106(2):262–268

Zhang M, Zheng X, Song S, Zeng Q, Hou L, Li D et al (2011) Spatiotemporal manipulation of auxin biosynthesis in cotton ovule epidermal cells enhances fiber yield and quality. *Nat Biotechnol* 29(5):453

Zhu S-W, Gao P, Sun J-S, Wang H-H, Luo X-M, Jiao M-Y et al (2006) Genetic transformation of green-colored cotton. *In Vitro Cell Dev Biol Plant* 42(5):439–444

Zhu H, Lv J, Zhao L, Tong X, Zhou B, Zhang T, Guo W (2012) Molecular evolution and phylogenetic analysis of genes related to cotton fibers development from wild and domesticated cotton species in Gossypium. *Mol Phylogenet Evol* 63(3):589–597

Chapter 13
Second and Third Generations of Transgenic Cotton

Syed Shan-e-Ali Zaidi, Niaz Ahmad, and Mehboob-ur-Rahman

13.1 Introduction

Genetically modified (GM) crops are the crop varieties with modified DNA using genetic engineering methods, in most cases to introduce a trait of interest such as increased plant height or disease resistance. GM crops are dominating the agriculture industry worldwide because of their superior production. Cotton is one of the major fiber crops of global significance. It is cultivated in tropical and subtropical regions of more than 80 countries of the world occupying nearly 33 m ha with an annual production of around 25 million tones. China, the USA, India, Pakistan, Uzbekistan, Australia, Brazil, Greece, Argentina, and Egypt are major cotton-producing countries. These countries contribute nearly 85% of the total global cotton production. The first GM cotton plant was obtained in 1987 by two independent groups (Firoozabady et al. 1987; Umbeck et al. 1987), in which a reporter gene was transformed in cotton genome using *Agrobacterium*-mediated genetic transformation system.

GM cotton is now the third largest biotech crop in terms of area under cultivation. In 2014, GM cotton occupied 68% of the global cotton area, mostly involving insect-resistant Bt varieties (James 2014). Bt cotton is particularly popular in developing countries such as China, India, Pakistan, South Africa, Burkina Faso, and others. In these countries, Bt cotton is grown by over 15 million smallholder farmers, contributing to significant economic, social, and environmental benefits (Qaim 2016). In

S. Shan-e-Ali Zaidi (✉)
Plant Genetics, TERRA Teaching and Research Center, Gembloux Agro-Bio Tech, University of Liège, Gembloux, Belgium

N. Ahmad · Mehboob-ur-Rahman (✉)
Agricultural Biotechnology Division, National Institute for Biotechnology & Genetic Engineering (NIBGE) College, Faisalabad, Pakistan

Pakistan Institute for Engineering and Applied Science (PIEAS, Islamabad, Pakistan
e-mail: mehboob@nibge.org

Mehboob-ur-Rahman et al. (eds.), *Cotton Precision Breeding*,
https://doi.org/10.1007/978-3-030-64504-5_13

this chapter, we have discussed the ongoing development of GM cotton using different genetic assays, which has been classified into three so-called generations. After briefly discussing first-generation GM cotton, which has been detailed in another chapter, here we shall focus on second and third/next generation of GM cotton.

13.2 First Generation of GM Cotton

First-generation GM cotton would refer to the early GM cotton varieties (introduced in the mid-1990s) containing gene(s) for a single desired trait, for example, Cry1Ac for production of Bt cotton. First-generation GM cotton can be further classified into two distinct eras: (1) first-generation Bt cotton, where a single *Cry1Ac* gene has been used to introduce a single trait, lepidopteron insect resistance, and (2) second-generation Bt cotton, where a single trait is enhanced by using multiple genes at a time, for example, gene-stacking or developing hybrids carrying multiple Cry genes.

Bt cotton contains Cry genes isolated from the soil bacterium *Bacillus thuringiensis* (Bt). These Cry genes induce the plant to produce substances that are toxic to insect pests of the lepidopteron order, especially cotton bollworms. Bollworms infest 88% of the global cotton area and are accountable for large crop damage and intensive chemical pesticide applications (Zehr 2010). The company Monsanto instigated the commercialization of Bt cotton in the USA in the mid-1990s. In Pakistan, Bt cotton was officially approved for the first time in 2010 (Kouser and Qaim 2013); however, unofficial cultivation of Bt varieties had already commenced in 2002 (Ali and Abdulai 2010; Kouser et al. 2017). With 7.1 million acres of Bt cotton in 2014 (88% of the total national cotton area), Pakistan is now the country with the fourth largest GM cotton area in the world (James 2014).

13.3 Second-Generation GM Cotton

Second-generation GM cotton is referred to the introduction of improved GM cotton varieties, commercialized in the 2000s after the worldwide success of first-generation GM cotton varieties. In most cases, these varieties contained multiple desired traits in single variety or traits introduced using latest GM technologies such as RNA interference (RNAi). Second-generation GM cotton is sometimes also referred to as "third-generation Bt cotton," where the Bt trait (lepidopteron insect resistance introduced using Cry gene(s)) is combined with other traits of interest, for example, glyphosate resistance (using EPSPS gene(s)).

Glyphosate is an organophosphorus compound that acts as a broad-spectrum systemic herbicide. Glyphosate interferes with the shikimate pathway, which produces the aromatic amino acids phenylalanine, tyrosine, and tryptophan in plants and microorganisms. It blocks this pathway by inhibiting the enzyme 5-enolpyruvy lshikimate-3-phosphate synthase (EPSPS), which catalyzes the reaction of shikimate-3-phosphate (S3P) and phosphoenolpyruvate to form 5-enolpyruvylshikimate-3-phosphate (EPSP) (Steinrücken and Amrhein 1980). Inhibiting the enzyme causes shikimate to accumulate in plant tissues and diverts energy and resources away from other processes, eventually killing the plant. Therefore, glyphosate is used to kill weeds, especially annual broadleaf weeds and grasses that compete with crops. Second-generation GM cotton varieties with Bt-plus glyphosate-resistance stacked genes were introduced in the mid-1990s, and the acreage was dramatically increased afterward (Coupe and Capel 2016).

Invasive agricultural pests have been reported to rapidly evolve resistance to transgenic crops expressing only one Bt toxin. So the farmers had to switch to Bt crop "pyramids" producing two or more toxins that kill the same pest (Carriere et al. 2015; Ni et al. 2017). However, the efficacy and durability of such gene pyramids/stacks reduced by cross-resistance and antagonism between Bt toxins (Carriere et al. 2015, 2016). Therefore, an immediate urge for alternative management tactics paved the way for the adaptation of latest technologies. For example, transgenic cotton pyramids were developed that combine protection from Bt and RNAi against *Helicoverpa armigera*, one of the world's most destructive pests of cotton and many other crops (Ni et al. 2017). Computer modeling incorporating the data from bioassays shows that the pyramids can substantially delay evolution of resistance (Ni et al. 2017).

One of the bottlenecks in cotton production in the Indian subcontinent is the infestation of one of the world's most invasive insect pests whitefly (*Bemisia tabaci*), which is also the vector for most important virus disease of cotton, cotton leaf curl disease (CLCuD) (Briddon et al. 2014). RNAi has also been used to demonstrate engineered resistance against CLCuD in cultivated cotton by targeting genome of cotton leaf curl Burewala virus and cotton leaf curl Multan betasatellite (Ahmad et al. 2017). Moreover, the engineered resistance against whitefly has also been demonstrated in transgenic plants expressing RNAi constructs targeting important whitefly genes (Malik et al. 2016; Raza et al. 2016).

Cotton *Verticillium* wilt, called "cotton cancer," is a destructive disease, annually leading to 250–310 million US dollars in economic losses in China alone (Wang et al. 2017). RNAi and virus-induced gene silencing (VIGS) has been used to demonstrate the efficacy of several genes, including cotton 14-3-3c/d, NINJA, and CYP82D, to engineering *Verticillium* wilt resistance in cotton. These genes are directly involved in the infection of *Verticillium dahliae*, and their utilization against *V. dahliae* infestation has been confirmed by RNAi approaches (Gao et al. 2013; Sun et al. 2014; Wang et al. 2017) (Fig. 13.1).

Yield reduction was also noticed by the escalating temperature and limited water supply to cotton crop. Various drought-tolerant genes have been introduced in cotton, and modest level of drought resistance was claimed. Overexpressing of *TsVP*, an H^+-PPase-coding gene from *Thellungiella halophile*, in cotton improved shoot and root growth as compared to the wild type. These beneficial features enhanced drought tolerance in transgenic cotton, and seed cotton yield was 51% higher than wild-type cotton plants (Lv et al. 2009). In another study, a gene, *ScALDH21*, from *Syntrichia caninervis* was transformed into cotton, and transgenic lines showed greater plant height, larger bolls, and greater fiber yield than wild type during different treatments of drought stress (Yang et al. 2016).

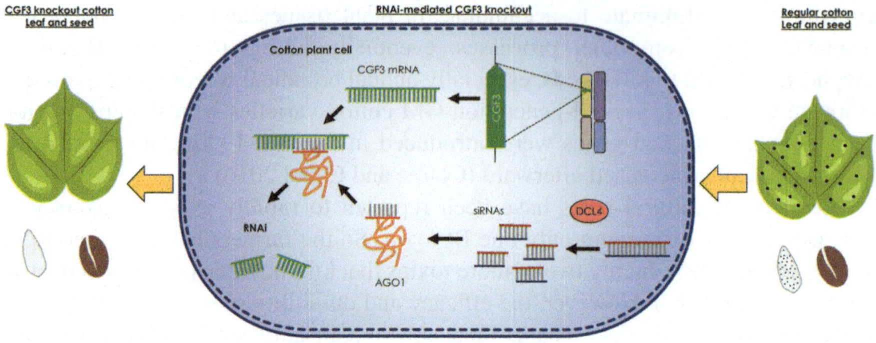

Fig. 13.1 Gene silencing of gossypol-containing glands (dark-colored dots) in the seed and leaf. Left top and bottom, normal cotton exhibits gossypol-containing glands (dark-colored dots) in the seed and leaf. Middle panel indicates the RNAi-mediated knockout of gossypol-related gene (e.g., Cotton Gland Formation gene 3, CGF3). Small interfering RNAs (siRNA) will bind the messenger RNA (mRNA) of CGF3, resulting in the mRNA cleavage ultimately downregulating the CGF3 expression. Right top and bottom indicate the resulting phenotype. Seed and leaf show the effect of CGF3 knockout

Also, a relatively high amount of protein (~23%) in seed cake and oil (~21%) is present in the cottonseed. Cottonseed is a fifth oil-producing crop in China and third in the USA (Small 2009), thus contributing substantial amount of the total domestic fat and oil supply. At the moment, quality of the oil is not of high standard owing to the presence of several antinutrients such as gossypol—naturally occurring terpenoid compound that is found in pigment glands present throughout the cotton plant. Gossypol is a double-edged sword. On the one hand, plants use gossypol as a defense compound such as insect deterrent, and on the other hand it acts as a cumulative toxin in simple-stomached animals. Plant biotechnology utilizing RNAi was used to create a plant that has no gossypol in the seed while retaining gossypol in all other plant tissues to prevent predation (Wedegaertner and Rathore 2015). Since cotton is drought and heat tolerant, it is well suited to serve as a source of both food and fiber as climate change puts other crops in jeopardy.

Secondly, accumulation of Bt protein has been reported in transgenic cottonseeds (Dong and Li 2007; Kranthi et al. 2005). Likewise, the number of constitutive promoters has been used to develop transgenic cotton; resultantly these promoters let the transgene to express in all parts of the plant including in the developing seed. For avoiding the expression of recombinant proteins in seed, it is important to use promoters, which can only express in green parts of the plants, for example, green-tissue-specific promoter PNZIP. This promoter was used for the expression Cry9C gene in vegetative parts of the cotton plant, while very faint expression of Cry protein, relatively 100 times lower, was recorded in reproductive organs including pollen, petals, and developing seed. The transgene was equally effective in controlling lepidopteron insect pests (Wang et al. 2016). Thus, safe use of cottonseed cake can be ensured in the future as well as addressing the consumers concerns.

Apart from the aforementioned traits, transgenic strategies including RNAi (see Fig. 13.1) and gene overexpression have also been demonstrated to introduce several important agronomic traits in cultivated cotton, for example, cotton fiber development (Tang et al. 2014; Wan et al. 2016) and stress responses (Li et al. 2016; Zahid et al. 2016). However, it must be noted here that for most of the transgenic strategies mentioned above, the data is mostly coming from lab and greenhouse experiments. These experiments are performed under controlled conditions and the applied efficacy in the field trials still need to be assessed. The other challenges toward sustaining fiber yield are the changing environments, shrinking cultivated land resources due to urbanization, and the competition offered by the other crops, and the decreasing trend in petroleum prices can be overcome by developing ideotype cotton plant which can produce more lint yield with improved quality. This can be achieved by introducing genes for improving the efficacy of various existing pathways involved in conferring lint quality and lint percentage and can be the best target of next-generation GM technologies. For all these, highly efficient plant regeneration protocols will be required for introducing multiple genes or new pathways in cotton.

For doing this task, parallel understanding of various molecular pathways will be required for producing ideotype cotton plant. For achieving this, integration of several other disciplines including automation for genotyping as well as phenotyping, role of bioinformatics, and functional and synthetic biology is needed. Together these fields can help understanding and improving the efficiency of pathways for important traits.

13.4 Third Generation of GM Cotton

With the advancement in plant biotechnology, numerous novel applications of GM crops have emerged. Thus, even while speculating, there are many ways to look at the third- or next-generation GM cotton. The third generation of GM crops promises for producing huge quantity of industrial products much more cheaply. These products are biosensors, industrial enzymes and epoxies, plastics and cosmetics, and pharmaceutical drugs such as vaccines, antibodies, and therapeutic proteins.

A major difference between first-/second- and third-generation GM cotton is the technology used for genetic modification. Most of the first-/second-generation GM cotton varieties have been developed using conventional T-DNA transformation approaches. However, with the rise of precise genome editing tools, mainly CRISPR-Cas9 (clustered regularly interspaced short palindromic repeats-CRISPR associated 9), several new protocols have been developed to introduce trait of interest (Butt et al. 2020; Khan et al. 2019; Zaidi et al. 2017, 2018b). Moreover, the development of transgene-free approaches can significantly decrease the cost and time for variety development (Zaidi et al. 2019). The utilization of CRISPR for developing third-generation GM cotton, including the transgene-free systems, has been discussed in another chapter.

Another way to look at the next-generation GM cotton can be the number of introduced genes and improved traits in single GM cotton variety. Cotton varieties with single or double genes for a number of distinct traits are already in use. Although we can only speculate about the number of genes controlling the development and architecture of cotton plants and the various physiological processes impacting yield at the moment, this knowledge is increasing exponentially with the whole genome sequencing being available for the various cotton cultivars from around the world (Zaidi et al. 2018a). It seems safe to assume, though, that genetically complex traits will require additional years of research to understand their molecular genetic circuits, before the development of commercial varieties. Nevertheless, complex traits, including those controlling adaptation to abiotic stresses, such as drought and salinity, flowering and reproduction, and hybrid vigor, are being actively investigated, and it would not be surprising if some of these could be regulated in GM cotton varieties within the next 5–10 years.

Various fungal, bacterial, and transgenic animal expression systems are available to produce recombinant proteins. Among all these, plants represent a highly cost-effective expression platform for commercial applications. They offer low production costs, improved safety, purity, ease of storage, and consistent and scalable production—all of which can be exploited to meet diverse demands and applications. "Pharmacrops" are plants genetically modified to produce pharmaceuticals, for example, vaccines, antibodies, and proteins to treat human or animal diseases. Maize engineered to express human gastric lipase, used to treat cystic fibrosis, is already in advanced clinical trials.

Plant foods could also be used as edible vaccines (Walmsley and Arntzen 2000). Many seeds contain proteins that are allergenic for certain people. When these allergenic proteins are digested, small fragments derived from them are absorbed into patches of cells on the small intestine that are part of the immune system. Antibodies are produced against these proteins, and this leads to an immune response in the individual, with potentially severe consequences after subsequent exposure to the allergenic proteins. This same process can be used to create immunity against common viral and bacterial pathogens by producing antigenic proteins derived from them in edible plant parts. Of particular interest are a group of pathogens—Norwalk virus, *Vibrio cholerae* (the cause of cholera), and enterotoxigenic *Escherichia coli* (a source of "traveler's diarrhea")—that cause the deaths of several million children each year, mainly in developing countries. Preliminary studies indicate that uncooked plant foods, such as potatoes or bananas, can be used to produce pathogen-derived proteins (such as virus coat proteins). These foods might then be used to inoculate children and adults against a variety of common diseases. Although a great deal of research remains to be done to demonstrate the efficacy and economic viability of this approach, results from preliminary experiments are promising (Takeyama et al. 2015).

There is also interest in using plants to produce human monoclonal antibodies. Preliminary research has demonstrated that several types of plant tissues, including seeds and leaves, have the capacity to express genes encoding the protein subunits of monoclonal antibodies and assemble them into functional complexes (Daniell et al. 2001). It remains to be seen whether cotton plants can produce these antibodies in sufficient quantities to meet therapeutic requirements. However, if it proves possible, the technology has tremendous potential because of the expense of producing monoclonal antibodies in mammalian tissue cultures (Tiwari et al. 2009). It would be essential to grow these plants in restricted locations, but the value of the products would easily be sufficient to offset the cost of growing the crop in isolation.

Other possibilities for sustaining cotton production is to bring the wild species under cultivation by introducing the domestication-related genes. The wild cotton species can tolerate stresses much effectively than that of the domesticated species. However, these species lack genes which produces spinnable fiber as well as high yield. Sometimes these species have genes but are not expressing. Thus after getting insight into the genetics of cotton fiber, various new emerging tools such as CRISPR-Cas can be used to induce modifications in these genes for making them functional. Alternatively, if the genes are missing, they can be introduced using different genetic tools. In the genus *Gossypium*, only few genotypes of *G. hirsutum* respond to transformation technology; thus accessions belonging to other cotton species are not amenable to induce targeted genetic changes using transformation protocols. In the future for bringing these species under cultivation, it is important to explore the genetic mechanism of totipotency in cotton followed by introduction of these genes or inducing genetic modifications in the already present genes for making the wild and/or other cultivated species or varieties amenable to transformation technologies.

Since, third generation of GM crops aims for the production of plant-made pharmaceuticals (PMPs) and plant-made industrial products (PMIPs) abundantly for economizing the price of the conventional pharmaceutical drugs as well as industrial products (plastics, cosmetics, enzymes, etc.). Such type of GM crops especially food crops may have poor acceptability among the masses owing to the presence of PMPs/PMIPs with the traditional food. Here, cotton largely grown for its lint can be utilized for producing PMPs/PMIPs as this crop is largely cultivated for natural fiber. However, cautionary measures should be taken that these traits should not appear in cottonseed oil (used for edible purpose) and seed cake (consumed by animals) as both are consumed for edible purpose in different parts of the world (Shanahan et al. 2001). Up till now, third-generation GM cotton has not been commercialized. However, the third-generation GM cotton would have to face the public reaction, and the stringent regulations required for conducting trials and commercial release are not clear yet.

One way to produce high-value targets could be the introduction of transgenes into plastid genome. Plastids have their own genome, which is quite small (160 kbp) and gene dense (around one gene per 1 kbp) and compact in its organizational architecture (for more details about the cotton chloroplast genome, please see Wu et al. (2018)). With few exceptions, its size remains almost uniform in the angiosperms. The chloroplast genome exists in a high copy number (up to 10,000 copies per plant cell). This high copy number ensures the conserved organizational arrangement of the genome to be preserved, resulting in relatively low mutational rates compared to the nucleus. It has been estimated that the chloroplast genome evolves at an ~10× lower rate than the nuclear genome (Rousseau-Gueutin et al. 2018; Wolfe et al. 1987). The highly conserved order, low mutation recombination rate, and low level of nucleotide substitution rate make it an excellent tool for phylogenetic and comparative genome evolution studies (Twyford and Ness 2017).

Plastids in cotton like many angiosperms are transmitted maternally. This maternal mode of inheritance provides a sort of natural gene containment. Engineering of the chloroplast genome therefore would allow the open-field cultivation of the crop, as plastid genomes do not fly with the pollen. In addition, chloroplast transformation allows targeted insertion of transgenes at a predetermined location on the genome using homologous recombination (Fig. 13.2).

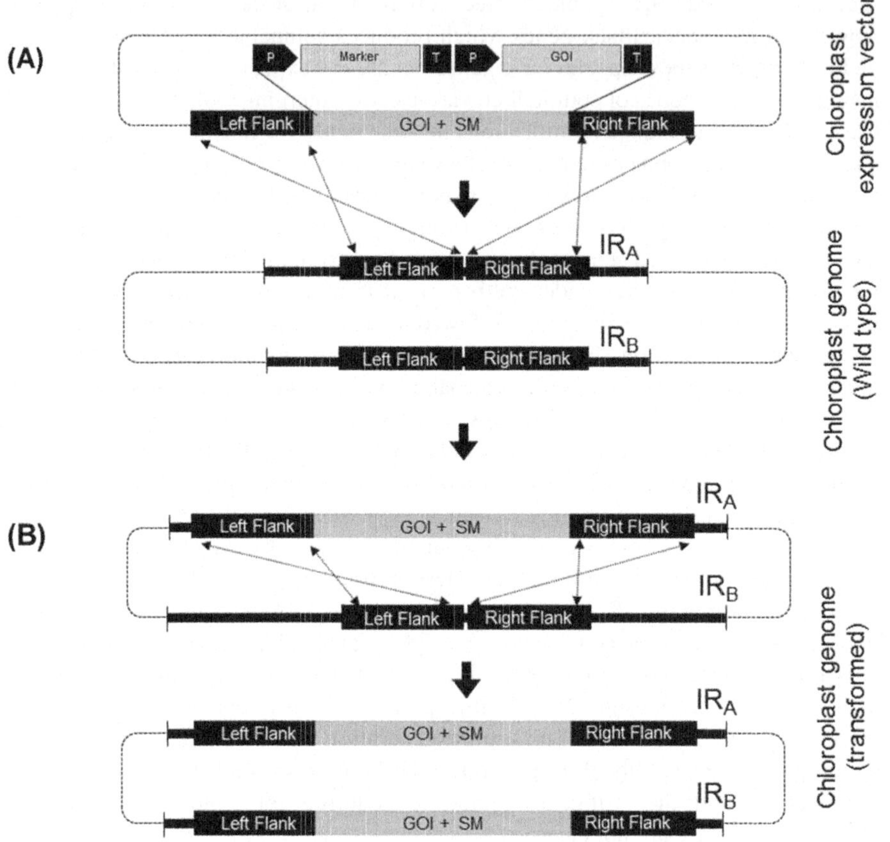

Fig. 13.2 Illustration of homologous recombination process for the integration of transgenes at a site-specific site in the plastid genome. (**a**) Expression vector showing the placement of expression cassette containing gene of interest (GOI) and selection cassette carrying the selection marker (SM) for their integration into plastid genome at a chosen site through homologous recombination, a hallmark of plastid genome engineering technology. Two flanking regions from plastid genome at which insertion is required are, therefore, always incorporated into the chloroplast expression vectors to allow a crossing event to occur for the delivery of transgenes at the intended site. The GOI and SM are always placed between these flanking regions. The inclusion of these flanking regions into chloroplast expression vectors makes this technology species-specific, which means vectors constructed for one species cannot be used to transform another (when used, both the transformation efficiency and expression of recombinant proteins were compromised (Ruhlman et al. 2010)). In addition, the strategy could be employed to study gene functions either reverse genetics or forward genetics or by carrying site-directed mutagenesis. (**b**) Shows the duplication of GOI and SM from IR_A into IR_B through a process known as copy-correction, also mediated by homologous recombination, which takes place between inverted repeat regions. Initially few copies of the plastome are transformed, and therefore the explant contains a mixture of both the transformed and untransformed copies, a state known as heteroplasmy. The wild-type copies are sorted out gradually by repeating few regeneration cycles against selection to reach homoplasmy, a state when all copies of the plastome are transformed. The dotted arrows show the regions which will undergo recombination, whereas the thick-black arrows show the progress of reaction to purify homoplasmic plant lines (*see* Ahmad and Mukhtar (2013) for further details). Abbreviations: *IR* inverted repeat, *GOI* gene of interest, *SM* selection marker

Being a nonfood crop, use of cotton as a host for the production of high-value targets would make it a good choice. At the moment, cotton is one of the most cultivated GM crops perhaps due to the reason that it's a non-feed crop. Its foremost product is fiber which is made up of cellulose. The expression of recombinant proteins in green plastids—chloroplasts—would therefore limit their accumulation in the leaves. Chloroplasts do not export proteins into the cytoplasm; therefore, any protein expressed in them will remain "compartmentalized" inside the cell. Transformation of the cotton chloroplasts for the production of high-value targets would provide several advantages, including high-level gene expression, site-specific integration of transgenes, and uniform gene expression due to the defiance of the Mendelian laws of inheritance. These features make cotton chloroplast transformation as an attractive tool for the development of next generation of environmentally friendly transgenic cotton. However, the biggest hurdle to realize the potential of the chloroplast genomes for developing biotech cotton is the incredibly lengthy time of recovering transgenic cotton plants via somatic embryogenesis. Somatic embryogenesis is the only viable approach at present through which stable transgenic plant lines can be recovered with relatively higher transformation efficiency compared to other approaches (Ahmad and Mukhtar 2017). However, somatic embryogenesis is much lengthy procedure. It can take somewhere between 1 and 2 years to convert cotton calli into somatic embryos and then another 6 months to obtain plantlets. Therefore, at present, the cotton chloroplast transformation is not a suitable platform to use as a biopharming crop. However, the development of shorter tissue culture protocols and the identification of tissue culture-responsive genotypes may alleviate these hurdles in future.

13.5 Conclusions and Prospects

The abovementioned modifications to crop plants raise a number of issues. There might be indirect human health risks mediated through the environment that would require new expert analysis. Nontarget risks associated with these plants with altered nutritional characteristics (both macronutrients and micronutrients), increased concentrations of "health-producing" compounds, or edible vaccines may be considerably more subtle than the direct mortality risks associated with plants producing insecticidal toxins, which are being evaluated presently. Moreover, concern over their release into the environment has led to high levels of conflict within the policy community and great concern among the interested public. This third generation can be seen as not only encapsulating the benefits of genetically modified plants through the promise of innovative production processes leading to better goods that are less expensive to produce, with the potential to enhance personal health and welfare, but also the inherent threats of a complex technology with the perceived potential for dreaded outcomes impacting the environment and public health.

More work on cotton plants is needed to link physiology and system biology with field performance. Understanding traits in cotton plants that are associated with root architecture, stomatal conductance, photosynthesis, and osmotic adjustment would help better understand biotic and abiotic stresses in cotton. It is important to enhance the stress tolerance capability of cotton, and there is still much work to be performed to secure future generations from the upcoming crisis. Cotton is a complex crop; however, rapid advances in the omics technologies will make it possible to use a system biology approach to understand cotton plants responses to various stresses and introduce respective tolerance/resistance in cotton.

References

Ahmad N, Mukhtar Z (2013) Green factories: plastids for the production of foreign proteins at high levels. *Gene Ther Mol Biol* 15:14–29

Ahmad N, Mukhtar Z (2017) Genetic manipulations in crops: challenges and opportunities. *Genomics* 109:494–505

Ahmad A, Zia-Ur-Rehman M, Hameed U, Qayyum Rao A, Ahad A, Yasmeen A, Akram F, Bajwa KS, Scheffler J, Nasir IA, Shahid AA, Iqbal MJ, Husnain T, Haider MS, Brown JK (2017) Engineered disease resistance in cotton using RNA-interference to knock down cotton leaf curl kokhran virus-burewala and cotton leaf curl multan betasatellite expression. *Viruses* 9:257

Ali A, Abdulai A (2010) The adoption of genetically modified cotton and poverty reduction in Pakistan. *J Agric Econ* 61:175–192

Briddon RW, Akbar F, Iqbal Z, Amrao L, Amin I, Saeed M, Mansoor S (2014) Effects of genetic changes to the begomovirus/betasatellite complex causing cotton leaf curl disease in South Asia post-resistance breaking. *Virus Res* 186:114–119

Butt H, Zaidi SS, Hassan N, Mahfouz M (2020) CRISPR-based directed evolution for crop improvement. *Trends Biotechnol* 38:236–240

Carriere Y, Crickmore N, Tabashnik BE (2015) Optimizing pyramided transgenic Bt crops for sustainable pest management. *Nat Biotechnol* 33:161–168

Carriere Y, Fabrick JA, Tabashnik BE (2016) Can pyramids and seed mixtures delay resistance to Bt crops? *Trends Biotechnol* 34:291–302

Coupe RH, Capel PD (2016) Trends in pesticide use on soybean, corn and cotton since the introduction of major genetically modified crops in the United States. *Pest Manag Sci* 72:1013–1022

Daniell H, Streatfield SJ, Wycoff K (2001) Medical molecular farming: production of antibodies, biopharmaceuticals and edible vaccines in plants. *Trends Plant Sci* 6:219–226

Dong HZ, Li WJ (2007) Variability of endotoxin expression in Bt transgenic cotton. *J Agron Crop Sci* 193:21–29

Firoozabady E, DeBoer DL, Merlo DJ, Halk EL, Amerson LN, Rashka KE, Murray EE (1987) Transformation of cotton (*Gossypium hirsutum* L.) by Agrobacterium tumefaciens and regeneration of transgenic plants. *Plant Mol Biol* 10:105–116

Gao W, Long L, Zhu LF, Xu L, Gao WH, Sun LQ, Liu LL, Zhang XL (2013) Proteomic and virus-induced gene silencing (VIGS) Analyses reveal that gossypol, brassinosteroids, and jasmonic acid contribute to the resistance of cotton to Verticillium dahliae. *Mol Cell Proteomics* 12:3690–3703

James C (2014) Global status of commercialized biotech/GM crops: 2014. International Service for the Acquisition of Agri-Biotech Applications

Khan MZ, Zaidi SS, Amin I, Mansoor S (2019) A CRISPR way for fast-forward crop domestication. *Trends Plant* Sci 24(4):293–296

Kouser S, Qaim M (2013) Bt cotton, damage control and optimal levels of pesticide use in Pakistan. *Environ Dev Econ* 19:704–723

Kouser S, Abedullah, Qaim M (2017) Bt cotton and employment effects for female agricultural laborers in Pakistan. *N Biotechnol* 34:40–46

Kranthi KR, Naidu S, Dhawad CS, Tatwawadi A, Mate K, Patil E, Bharose AA, Behere GT, Wadaskar RM, Kranthi S (2005) Temporal and intra-plant variability of Cry1Ac expression in Bt-cotton and its influence on the survival of the cotton bollworm, Helicoverpa armigera (Hübner) (Noctuidae: Lepidoptera). *Curr Sci* 89:291–298

Li YB, Han LB, Wang HY, Zhang J, Sun ST, Feng DQ, Yang CL, Sun YD, Zhong NQ, Xia GX (2016) The thioredoxin GbNRX1 plays a crucial role in homeostasis of apoplastic reactive oxygen species in response to Verticillium dahliae infection in cotton. *Plant Physiol* 170:2392–2406

Lv SL, Lian LJ, Tao PL, Li ZX, Zhang KW, Zhang JR (2009) Overexpression of Thellungiella halophila H(+)-PPase (TsVP) in cotton enhances drought stress resistance of plants. *Planta* 229:899–910

Malik HJ, Raza A, Amin I, Scheffler JA, Scheffler BE, Brown JK, Mansoor S (2016) RNAi-mediated mortality of the whitefly through transgenic expression of double-stranded RNA homologous to acetylcholinesterase and ecdysone receptor in tobacco plants. *Sci Rep* 6:38469

Ni M, Ma W, Wang X, Gao M, Dai Y, Wei X, Zhang L, Peng Y, Chen S, Ding L, Tian Y, Li J, Wang H, Wang X, Xu G, Guo W, Yang Y, Wu Y, Heuberger S, Tabashnik BE, Zhang T, Zhu Z (2017) Next-generation transgenic cotton: pyramiding RNAi and Bt counters insect resistance. *Plant Biotechnol J* 15:1204–1213

Qaim M (2016) Genetically modified crops and agricultural development, 1st edn. Palgrave Macmillan US

Raza A, Malik HJ, Shafiq M, Amin I, Scheffler JA, Scheffler BE, Mansoor S (2016) RNA interference based approach to down regulate osmoregulators of whitefly (Bemisia tabaci): potential technology for the control of whitefly. *PLoS One* 11:e0153883

Rousseau-Gueutin M, Keller J, Ferreira de Carvalho J, Aïnouche A, Martin G (2018) The intertwined chloroplast and nuclear genome coevolution in plants. In: Ratnadewi D, Hamim (eds) Plant growth and regulation—alterations to sustain unfavorable conditions

Ruhlman T, Verma D, Samson N, Daniell H (2010) The role of heterologous chloroplast sequence elements in transgene integration and expression. *Plant Physiol* 152:2088–2104

Shanahan J, Scheufele D, Lee E (2001) The polls-trends: attitudes about agricultural biotechnology and genetically modified organisms. *Public Opin Q* 65:267–281

Small E (2009) Top 100 food plants. NRC Research Press, Ottawa

Steinrücken HC, Amrhein N (1980) The herbicide glyphosate is a potent inhibitor of 5-enolpyruvylshikimic acid-3-phosphate synthase. *Biochem Biophys Res Commun* 94:1207–1212

Sun L, Zhu L, Xu L, Yuan D, Min L, Zhang X (2014) Cotton cytochrome P450 CYP82D regulates systemic cell death by modulating the octadecanoid pathway. *Nat Commun* 5:5372

Takeyama N, Kiyono H, Yuki Y (2015) Plant-based vaccines for animals and humans: recent advances in technology and clinical trials. *Ther Adv Vaccines* 3:139–154

Tang W, Tu L, Yang X, Tan J, Deng F, Hao J, Guo K, Lindsey K, Zhang X (2014) The calcium sensor GhCaM7 promotes cotton fiber elongation by modulating reactive oxygen species (ROS) production. *New Phytol* 202:509–520

Tiwari S, Verma PC, Singh PK, Tuli R (2009) Plants as bioreactors for the production of vaccine antigens. *Biotechnol Adv* 27:449–467

Twyford AD, Ness RW (2017) Strategies for complete plastid genome sequencing. *Mol Ecol Resour* 17:858–868

Umbeck P, Johnson G, Barton K, Swain W (1987) Genetically transformed cotton (*Gossypium Hirsutum* L.) plants. *Bio/Technology* 5:263–266

Walmsley AM, Arntzen CJ (2000) Plants for delivery of edible vaccines. *Curr Opin Biotechnol* 11:126–129

Wan Q, Guan X, Yang N, Wu H, Pan M, Liu B, Fang L, Yang S, Hu Y, Ye W, Zhang H, Ma P, Chen J, Wang Q, Mei G, Cai C, Yang D, Wang J, Guo W, Zhang W, Chen X, Zhang T (2016) Small interfering RNAs from bidirectional transcripts of GhMML3_A12 regulate cotton fiber development. *New Phytol* 210:1298–1310

Wang C, Lu W, He X, Wang F, Zhou Y, Guo X, Guo X (2016) The cotton mitogen-activated protein kinase kinase 3 functions in drought tolerance by regulating stomatal responses and root growth. *Plant Cell Physiol* 57:1629–1642

Wang L, Wu SM, Zhu Y, Fan Q, Zhang ZN, Hu G, Peng QZ, Wu JH (2017) Functional characterization of a novel jasmonate ZIM-domain interactor (NINJA) from upland cotton (*Gossypium hirsutum*). *Plant Physiol Biochem* 112:152–160

Wedegaertner T, Rathore K (2015) Elimination of gossypol in cottonseed will improve its utilization. *Procedia Environ Sci* 29:124–125

Wolfe KH, Li WH, Sharp PM (1987) Rates of nucleotide substitution vary greatly among plant mitochondrial, chloroplast, and nuclear DNAs. *Proc Natl Acad Sci U S A* 84:9054–9058

Wu Y, Liu F, Yang DG, Li W, Zhou XJ, Pei XY, Liu YG, He KL, Zhang WS, Ren ZY, Zhou KH, Ma XF, Li ZH (2018) Comparative chloroplast genomics of gossypium species: insights into repeat sequence variations and phylogeny. *Front Plant Sci* 9:376

Yang H, Zhang D, Li X, Li H, Zhang D, Lan H, Wood AJ, Wang J (2016) Overexpression of ScALDH21 gene in cotton improves drought tolerance and growth in greenhouse and field conditions. *Mol Breed* 36:34

Zahid KR, Ali F, Shah F, Younas M, Shah T, Shahwar D, Hassan W, Ahmad Z, Qi C, Lu Y, Iqbal A, Wu W (2016) Response and tolerance mechanism of cotton *Gossypium hirsutum* L. to elevated temperature stress: a review. *Front Plant Sci* 7:937

Zaidi SS, Mahfouz MM, Mansoor S (2017) CRISPR-Cpf1: a new tool for plant genome editing. *Trends Plant Sci* 22:550–553

Zaidi SS, Mansoor S, Paterson A (2018a) The rise of cotton genomics. *Trends Plant Sci* 23:953–955

Zaidi SS, Mukhtar MS, Mansoor S (2018b) Genome editing: targeting susceptibility genes for plant disease resistance. *Trends Biotechnol* 36:898–906

Zaidi SS, Vanderschuren H, Qaim M, Mahfouz MM, Kohli A, Mansoor S, Tester M (2019) New plant breeding technologies for food security. *Science* 363:1390–1391

Zehr UB (2010) Cotton. Springer, Berlin, Heidelberg

Chapter 14
Targeted Breeding in Cotton Using CRISPR/Cas9 Genome Editing

Baohong Zhang and Mehboob-ur-Rahman

14.1 Introduction

Cotton is the most important fiber crop; cottonseeds are also an important resource for oil and proteins. Additionally, there are high percentage of certain compounds that have pharmaceutical values, such as vitamin E and gossypol in cottonseeds. As we know, vitamin E is the only vitamin that our human body cannot synthesize and must be taken up from food. Because it contains high oil and cellulosic contents, thus cotton can also be a potential biofuel crop. Breeding high-yield and high-quality cotton cultivars has attracted attentions from both industries and scientists for a long history. Since cotton was domesticated, human has been continuously improving cotton through different ways. At the beginning, natural selection was the dominant method for selecting cotton cultivars with higher yields and better quality. Later, cross breeding was introduced in cotton breeding, which includes simple single cross, multiple crosses, and back crosses. Cross between cultivated cotton species and wild species has also been attempted; however little progress was achieved. All these methods, including natural selection and cross breeding, are called traditional breeding. Traditional breeding technology has made significantly progress on cotton breeding; many cultivars have been bred and widely adopted by cotton farmers around the world. These cultivars have been making significant contribution to human daily life and economic development. However, as timing going, traditional breeding is becoming harder and harder to solve certain problems our human beings

---

B. Zhang (✉)
Department of Biology, East Carolina University, Greenville, NC, USA
e-mail: zhangb@ecu.edu

Mehboob-ur-Rahman
Agricultural Biotechnology Division, National Institute for Biotechnology & Genetic Engineering (NIBGE) College, Faisalabad, Pakistan

Pakistan Institute for Engineering and Applied Science (PIEAS), Islamabad, Pakistan

Mehboob-ur-Rahman et al. (eds.), *Cotton Precision Breeding*,
https://doi.org/10.1007/978-3-030-64504-5_14

are facing, for example, increased labor cost for managing weeds in cotton field. Although herbicides had been developed for controlling weeds, these have the potential to damage cotton plant or only kill certain types of weeds; thus it is nearly impossible to use herbicide to manage all weed species in the cotton field. Another big issue for cotton is insect pest infestation. There are hundreds of insect species which are surviving on cotton. Many insects, such as bollworms, cause significant damage to cotton and impact cotton yield and quality. Although chemical pesticides play an important role for controlling insect pests during the long agriculture history, however, long-term usage of pesticides causes several major issues including environmental pollutions and insect resistance as well as damaging beneficial insects. Chemical pesticides are being sprayed on cotton plants; majority of these are rained off and thus contaminate the surface and groundwater. There are multiple reports indicating the prevalence of pesticide residues in water resources including the rivers and lakes (Barnhoorn and van Dyk 2020; Peters et al. 2013; Terzaghi et al. 2020; Zhang et al. 2008). Long time usage of chemical pesticides also allows insects to develop resistance to these pesticides, particularly at the early of the 1990s. During that time, in China, bollworms were broken out; even the cotton farmers sprayed the pesticides weekly for killing these insects but could not; finally, farmers are supposed to manually pick up these bollworms from the cotton field. These manually picked larvae of insects are often fed by birds which may accumulate pesticides in their bodies. Although cotton breeders and scientists have bred new cultivars which were resistant to insect pests by incorporating traits such as high gossypol content and leaf surface hairs, these new cultivars were not enough to manage the insect pest infestation in the field. In the 1980s, the quick development of transgenic technologies opened a new era for plant breeding, including cotton. Through *Agrobacterium*-mediated gene transformation and gene gun technology, both insect-resistant genes, such as Bt, and herbicide-resistant genes have been inserted into cotton genome and allow cotton to obtain the capability of synthesizing the Bt toxic protein to kill the insects or resistance of herbicides. Transgenic cotton is among the first transgenic plants for commercialization. Currently, transgenic *Bt* cotton and herbicide-resistant cotton have been widely adopted around the world by cotton farmers, and huge economic and social impacts have been achieved (Zhang 2019d).

Although transgenic cotton has brought huge benefits to the world, transgenic technology is based on an individual gene. To develop transgenic cotton, it was the first step to insert desired gene, such as *Bt* gene for insect resistance. However, it is always hard to find such kind of desired genes from other species. During the long evolution history, plants and insects evolved a complicated symbiotic relationship, and there are some cultivars that exhibit tolerance to certain insects. At the moment, there is nearly no cultivar which is completely resistant to a specific insect. Thus, almost all required genes used for commercial purposes have ben excised from bacteria (Zhang 2019a). Another potential issue for current transgenic technology is the random insertion of foreign genes into the plant genome, in which the transgene may alter the expression of certain cotton genes at the site of its integration. These are bottlenecks for current transgenic technology. Because of these, current commercial transgenic cotton is only limited to few traits, majorly insect resistance and herbicide resistance (Zhang 2019d).

Recently developed genome editing technology has opened new dimensions in editing cotton genome precisely, and also the function of genes can be studied. Particularly, as the discovery and application of CRISPR (clustered regularly interspaced short palindromic repeats) and CRISPR-associated (Cas) genes, genome editing has opened a new era for targeted-induced gene mutagenesis and precision breeding in cotton.

14.2 Genome Editing Tools

Genome editing is an assay that has the potential to edit a specific DNA/gene sequence of a genome. During the genome editing process, a molecule scissor, majorly a specific nuclease, cuts the DNA sequence from a specific location and forms a double-stranded break (DSB). Then, the cells employ their own repair mechanisms to repair these breaks through nonhomologous end joining (NHEJ) pathway. During the cut and repair process, some nucleotides are missed or removed from the original sequences (in some cases, it may also insert few nucleotides) which may alter the resultant protein and ultimately the phenotype. In most cases, these changes in gene sequence may also lead to complete silencing of the targeted gene. Of course, during the DNA repair process, if there is a DNA template existing with homology to the sequence flanking the DSB location, DNA repair can also go the second pathway, homology-directed repair (HDR), in which HDR can seal the DSB in an error-free manner. Thus, genome editing not only can knock out/silent an individual gene but also can overexpress the specific individual gene (Zhang and Zhang, 2020; Chen et al. 2019; Doudna and Charpentier 2014; Hsu et al. 2014).

For genome editing, the key is to find a nuclease that can precisely cut the DNA sequence at a desired location. In the past three decades, attempts were made to find such kind of nucleases, and great progress has been made (Doudna and Charpentier 2014; Hsu et al. 2014). According to the type of nucleases, the history of genome editing can be divided into four major stages. The four major families of engineered nucleases in the order of discovery and application in genome editing are meganucleases, zinc finger nuclease (ZFN), transcription activator-like effector nucleases (TALEN), and the clustered regularly interspaced short palindromic repeats (CRISPR) and CRISPR-associated protein (Cas) system. Meganucleases are the first class of nucleases used in genome editing; due to its high specificity, it offers several advantages including high precision, less off-target impact, and much low cell toxicity. However, it is very complicated and time-consuming to design and use meganucleases to edit a specific genome sequence. Thus, as the discovery of ZFN and TALEN, scientists quickly switch the genome editing tool to ZFN and TALEN (Wright et al. 2014). Although it was discovered in 1985, ZFNs were widely modified for the genome editing purpose in the 1990s. ZFNs have two separated domains, DNA-binding domain and DNA-cleavage domains; both are required and necessary for precise genome editing. The DNA-binding domain typically contains three to six individual zinc finger repeats, and each zinc finger recognizes three base pairs (Urnov et al. 2010). Thus, we can synthesize the zinc finger repeats in the DNA-binding domain to bind a specific DNA sequence for cleavage of the DNA-cleavage domain. TALENs are a class of restriction enzymes that can be engineered to cut a specific DNA sequence by fusing a TAL effector DNA-binding domain to a DNA-cleavage domain. Before the CRISPR/Cas9, both ZFNs and TALENs are prominent tools in the field of genome editing. Because both of them edit a DNA sequence through a protein-DNA interaction, as like the meganucleases, it is complicated and time-consuming to construct the ZFN and TALEN system for genome editing that limits the wide and quick application of these two genome editing tools. Since CRISPR/Cas9 was recognized as a genome editing tool (Gasiunas et al. 2012; Jinek et al. 2012), many research laboratories and industries have adopted the CRISPR/Cas9. Currently, CRISPR/Cas9 has become the most popular and robust tool for genome editing.

CRISPR/Cas9 was originally found in prokaryotes, majorly in bacteria and archaea. It is an adoptive immune defense mechanism, in which prokaryotes break down the genetic materials of the invading viruses (Horvath and Barrangou 2010; Makarova et al. 2011). After understanding the molecular mechanisms, scientists quickly adopted this naturally occurred genome editing tool for editing a specific DNA sequence in plants, animals, and microbes. CRISPR/Cas genome editing system is simple and easy, as it only requires designing of one single gRNA for performing CRISPR/Cas9-based genome editing. The principle of CRISPR/Cas9 genome editing is that the single gRNA guides the Cas nuclease to cut the double strands of DNA sequence to generate DSB and then the cell employs its own DNA repair mechanisms to fix the DSB. Based on the DNA repair pathway, NHEJ or HDR, the repair results in different consequences, including gene silencing (knockout because of the frameshift) or knocking in (if there is a DNA template for HDR repair). Cas enzyme activity requires a specific short DNA sequence, called protospacer adjacent motifs (PAM) (Nishimasu et al. 2014). Cas enzymes usually cut DNA sequences around the PAM sequence. Thus, it is very important to find a PAM sequence in the target DNA sequence when using CRISPR/Cas9 edits a specific gene. Cas enzymes are obtained from different bacteria, which recognize different PAM sequences (Shmakov et al. 2017). For example, the well-known, the first identified, and also widely used Cas enzyme is Cas9 obtained from *Streptococcus pyogenes*, named SpCas9. It only recognizes NGG PAM sequence (N can be any nucleotide A, T, G, or C). PAM requirement limits the gene sequences that CRISPR can be accessible; to expand the DNA sequences that CRISPR/Cas9 can edit, scientists have been studying different bacteria and archaea to find new Cas enzymes that can recognize different PAM sequences (Shmakov et al. 2017). Currently, hundreds of Cas enzymes have been identified. However, only a few of them can easily be used for genome editing purpose which include the first Cas, termed Cas9, Cas12, and Cas13. Because the genome editing is started by using Cas9, this technology is called CRISPR/Cas9 genome editing although other Cas proteins, such as Cas12 and Cas13, can be used for genome editing. At the same time, attempt was made to modify currently used Cas enzymes, such as Cas9, to relax PAM requirement. These modified Cas enzymes significantly relax the PAM requirement that will boost the power of CRISPR/Cas9 genome editing technology (Zhang and Zhang 2020). Currently, CRISPR/Cas genome editing has been widely used to target an individual DNA and even RNA sequence for a variety of purposes, including silencing a gene (knockout), overexpression of a gene (knock in), regulating a gene expression (CRISPRi or CRISPRa), regulating epigenome, and mentoring a special sequence (Anzalone et al. 2020).

CRISPR/Cas9 genome editing technology was also quickly adopted and established in cotton for studying the gene function and cotton improvement. Li et al. (2017) reported for the first time that the CRISPR/Cas9 system can be successfully used to knock out an endogenous gene in cotton (Li et al. 2017). Thereafter, CRISPR/Cas9 genome editing tool has been quickly deployed for studying function of several important cotton genes. Currently, more laboratories have established CRISPR/Cas9 genome editing technology for cotton and also used to edit multiple genes in cotton (Janga et al. 2017; Li et al. 2017; Long et al. 2018; Wang et al. 2017, 2018; Zhang et al. 2018; Zhu et al. 2018), including base editing (Qin et al. 2020).

14.3 Protocol for Precisely Editing a Gene in Cotton Using CRISPR/Cas9 Technology

The protocol for performing cotton genome editing is relatively simple. Steps for editing a gene in cotton have been shown in Fig. 14.1, and it was adopted from the first publication in cotton genome editing field (Li et al. 2017). Following is the detailed method for performing CRISPR/Cas9 genome editing in cotton (Li et al. 2017).

14.3.1 gRNA Design and Constructing DNA for Transformation

Designing a good-quality gRNA is a key for genome editing. To design a gRNA, a PAM sequence should be located in the target gene sequence, particularly sit on an exon of this gene. A good thing is that there is multiple online software available which we can adopt to design gRNAs which save lots of time for designing gRNAs.

After designing and synthesizing the gRNAs, the traditional recombination technology is used to insert a gRNA into a specific vector containing the Cas9 protein. Currently, there are lot of vectors commercially available or can be requested from other labs who are performing the similar research. Usually, the DNA constructs are firstly built in *E. coli* and then inserted into T-DNA and Ti plasmid for plant transformation. Finally, Ti plasmids containing the gRNA and Cas9 protein genes are transformed into an agrobacterial strain that are used in cotton transformation, such as LBA4404. Because gRNA is a short sequence, usually U6 promoter is used to

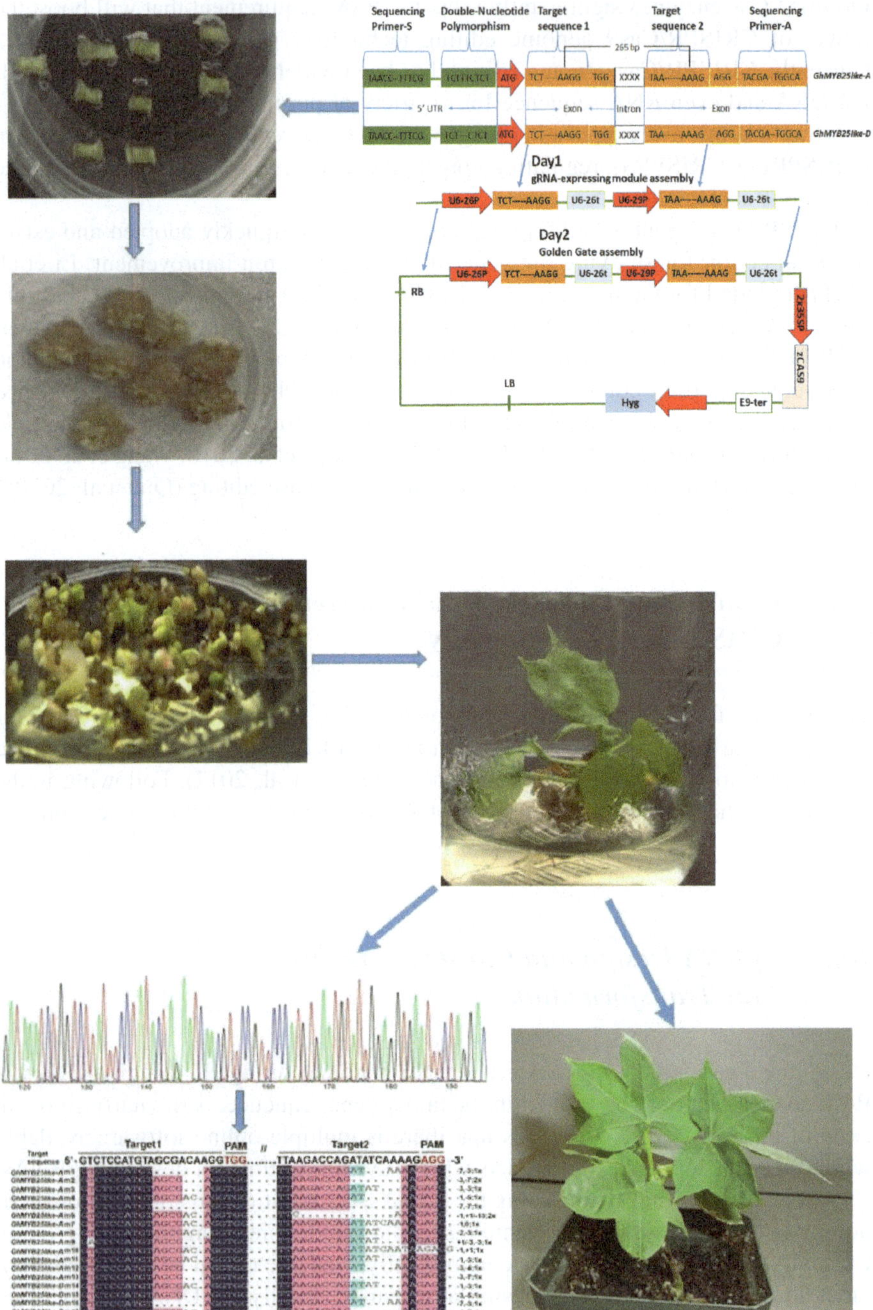

Fig. 14.1 Diagram of CRISPR/Cas9-mediated genome editing in cotton. The protocol is adopted from the first publication in the field of cotton genome editing by CRISPR/Cas9 technology (Li et al. 2017; Li and Zhang 2019)

express gRNA; CaMV35S promotor is commonly used for Cas9 gene. The DNA construction can be ready in a couple of days prior to cotton transformation.

14.3.2 *Transfer CRISPR/Cas9 System into Cotton Plant Cells*

This introduction of CRISPR/Cas9 system is equivalent to the introduction of any other foreign genes in plant, such as *Bt* gene, into cotton. Thus, transformation method used to obtain transgenic plants can be used to edit an individual gene in cotton using CRISPR/Cas technology. These methods can be *Agrobacterium*-mediated gene transformation, gene gun-mediated transformation, or pollen-tube pathway-mediated transformation (Zhang 2019b). For example, *Agrobacterium*-mediated gene transformation is employed to deliver the CRISPR/Cas9 system into plant cell according to a previously reported method (Li et al. 2017). During CRISPR/Cas9 transformation, selection of transgenic cells is important for obtaining genome editing events (Fig. 14.1). Thus, after co-culture of cotton explants with *agrobacteria* containing CRISPR/Cas9 system, selection regents (antibiotics) at callus induction and differentiation stage should be added into the medium to limit the number of non-transgenic/edited cells.

14.3.3 *Select and Confirm CRISPR/Cas9 Genome Editing Events*

There are many methods for determining the genome editing events (Sentmanat et al. 2018). Among them, the T7 endonuclease 1 (T7E1) mismatch detection assay is a widely used method to confirm the genome editing events in plant. PCR, the Tracking of Indels by Decomposition (TIDE), the Indel Detection by Amplicon Analysis (IDAA) assay, and targeted the next-generation sequencing (NGS) are also employed to determine the genome editing events and efficiency (Sentmanat et al. 2018).

14.3.4 *Detect DNA Cleavage and Off-Target Effects*

For genome editing, it is good but not required to determine where and how many nucleotides are deleted or inserted into the targeted DNA sequence. This can be determined by sequencing the targeted region by traditional sequencing technology or deep sequencing assay. Traditional sequencing only tells where the nucleotides are changed and how they are changed. The deep sequencing has the potential to detect off-target effects in the entire genome.

14.3.5 Obtain Homozygous Genome Editing Plants and Even Transgene-Free Genome Editing Events

Because cotton is a heterotetraploid, for many cases, the genome-edited plants are heterozygous which require self-crossing to obtain homozygous. During self-cross process, the inserted CRISPR/Cas9 component may be lost in some plants, but the genome editing events still remain there. In this way, a transgene-free genome editing plant can be obtained.

14.4 CRISPR/Cas Technology Opens a New Era for Precise Breeding in Cotton

Because cotton is allotetraploid, it is relatively difficult to obtain a pure line in a limited time that can be exploited in breeding and agronomic purposes. Although Bt cotton and herbicide-resistant cotton have been adopted by cotton farmers across the world, transgenic technology is still cumbersome than that of the other crops owing to lengthy in vitro regeneration process and random insertion of transgene(s). Since the CRISPR/Cas9 technology was employed to study gene function in plants, it has been catching attention from the cotton scientific community. Currently, many research laboratories and breeders, including the biotechnology companies, have been joining this exciting field to employ CRISPR/Cas genome editing tool for undertaking gene functional study as well as for improving cotton traits. CRISPR/Cas-based genome editing is becoming a new era for transgenic and precise breeding in cotton.

14.4.1 CRISPR/Cas9 for Cotton Fiber Development

Investigating the molecular mechanisms of cotton fiber initiation and development will provide a firm foundation for improving fiber yield and quality and will add in increasing the income of cotton farmers. During the past one decade, swift developments for deep sequencing technology and gene expression analysis have paved the way for the identification of large number of genes potentially associated with cotton fiber initiation and development (Fang et al. 2018; Guan et al. 2014; Ijaz et al. 2019; Wang and Zhang 2015; Wu et al. 2017; Yang et al. 2020), which include both protein-coding genes and small regulatory RNAs, such as microRNAs (miRNAs) (Naoumkina et al. 2016; Xie et al. 2015a, b), one important gene regulator that plays versatile role in plant growth and development as well as response to environmental stresses (Zhang 2015; Zhang et al. 2006). However, due to lake of high-throughput tools for gene function study, the functions of these genes have not yet been elucidated. CRISPR/Cas9 genome editing opened a new way to study the function of genes and its application in cotton fiber improvement.

The first application of CRISPR/Cas9 genome editing platform in cotton was deployed to study the function of MYB-25-like transcription factor in cotton fiber development (Li et al. 2017). Earlier it was shown that *MYB-25-like* gene was dominantly expressed during cotton fiber initiation and early development (Walford et al. 2011). The CRISPR/Cas9-mediated knockouts of MYB-25-like transcription factor also failed to trigger the cotton fiber development (Li et al. 2017; Zhang 2019c). In their study, two gRNAs were designed for targeting two different locations in *MYB25-like* gene which were conserved in both the A and D sub-genomes. It was also indicated that both gRNAs worked very well individually and/or together to generate lots of deletion/insertion mutations with a high efficiency without any off-target effects (Li et al. 2017).

The 14-3-3 proteins belong to the class of conserved regulatory molecules that widely exist in many plant species. There are at least 25 14-3-3 proteins in cotton (Sun et al. 2011). Many of these play a vital role in cotton fiber initiation and elongation (Zhang et al. 2010; Zhou et al. 2015). Transgenic cotton with overexpression of *14-3-3L* enhanced cotton fiber length, while the inhibition of *14-3-3* blocked the fiber initiation and elongation process (Zhou et al. 2015). It was suggested that the *14-3-3* modulated the brassinosteroid signaling pathway (Zhang et al. 2010). Two copies of *14-3-3d* gene were recently knocked out using CRISPR/Cas9 technology (Zhang et al. 2018). ALARP encodes ALARP protein rich in alanine, which is preferentially expressed in cotton fibers. By using CRISPR/Cas9 technology, Zhu et al. (2018) knocked out *ALARP* without detecting any potential off-target effect (Zhu et al. 2018).

14.4.2 CRISPR/Cas9 for Cotton Resistance to Diseases

Cotton production is negatively affected by insect pests and diseases. During the growing season, many diseases can cause problems to cotton growth and development as well as yield and quality. There are several common diseases which infect cotton plant across the cotton-growing regions. Among these, *Verticillium* wilt and *Fusarium* wilt are caused by fungi. Since CRISPR/Cas9 was recognized as a tool for genome editing in plant, it has been quickly used in generating disease-resistant plants. At the moment, multiple genome-edited plants have been obtained with high resistance to different diseases caused by fungus, bacteria, or virus (Arora and Narula 2017; Langner et al. 2018; Zaidi et al. 2018). In cotton, Zhang et al. (2018) successfully knocked out two copies of *14-3-3d* gene by CRISPR/Cas9 genome editing technology. The transgene-free knockout lines showed higher resistance to *Verticillium dahliae* infestation compared to the wild-type plants, which were obtained after extensive selections (Zhang et al. 2018). After inoculating with 10^5 conidia/mL *V. dahliae* for 18 days, the CRISPR/Cas9-edited plants showed significantly reduced disease symptoms, and the disease index was less than that in the wild-type CCR35 plants (Zhang et al. 2018). For instance, the infected plant severity was decreased from ~90% in wild type to ~30% in the CRISPR/Cas-edited plants. The disease index was decreased by more than 50% in the genome-edited plants as compared to the wild type (Zhang et al. 2018).

14.4.3 CRISPR/Cas9 for Cotton Resistance to Abiotic Stress

During the entire growing season, cotton crop is exposed to many environmental stresses, including drought, salinity, and extreme low (chilling) and high (heat) temperatures. These stresses significantly affect cotton growth and development (Zhou et al. 2014). Thus, it is a big challenge for scientists and breeders to breed new cotton cultivars with high tolerance to different environmental stresses. Current CRISPR/Cas9 genome editing technology allows us to quickly achieve resilience to these stresses. A recent study shows that downregulation of *GhHB12* gene, a HD-ZiP transcription factor subfamily I gene, enhanced tolerance to abiotic stresses in cotton, suggesting that knockout of this gene can help developing resilient cotton lines with strong environmental adaptability (He et al. 2020).

14.4.4 CRISPR/Cas9 for Improving Cotton Plant Development

Improving plant morphology also helps in improving plant yield and quality as well as response to environmental stresses. For example, if we develop a cotton variety with deep root system, the cotton plants will uptake more nutrients and water from the deeper layers of soil. Thus, a cotton plant can sustain drought stress without any serious depression to its growth. Stronger and healthy roots also help plant to cope a variety of other abiotic and biotic stresses. Nitric oxide (NO) is an important regulator for root development in plants. Increasing NO concentration makes plant to produce more lateral and adventitious roots. In plant, nitric oxide synthase (NOS) catalyzes the synthesis of NO; however, arginase competes with NOS for arginine (ARG) substrate. Overexpression of *arg* significantly inhibited the NO accumulation in cotton root and then decreased the formation of lateral roots in transgenic cotton (Meng et al. 2015). Knockout of *arg* gene by CRISPR/Cas9 technology significantly enhanced root development in cotton (Wang et al. 2017). The total number of lateral roots and the total root surface area were enhanced in both high and low nitrogen conditions in genome-edited cotton plants as compared to the wild-type cotton (Wang et al. 2017). This suggests that knocking out of *arg* gene not only promotes root development but also enhances tolerance to nitrogen deficiency. This is because that *arg* knockout lines have better root development that will enhance the absorption of water and nutrients of the transgenic cotton plant, and further enhance plant growth and development, and response to various environmental stresses.

The miRNAs control almost all biological process in plant development (Li and Zhang 2016). The miR156 is a well-studied miRNA, which controls seedling development and timing of various developmental stages (vegetative growth to reproductive growth). Recently, the Zhang's lab at East Carolina University employed CRISPR/Cas9 to knock out miR156 in cotton; the miR156 knockout lines of cotton showed restricted growth and are also found sensitive to abiotic stresses, including drought, salinity, and waterlogging.

14.4.5 CRISPR/Cas9 for Improving Other Traits in Cotton

Except the traits mentioned, CRISPR/Cas9 genome editing is also employed to improve other agronomically important traits. Through CRISPR/Cas9 knockout of a transcription factor gene, called *cotton gland formation 3* (*cgf3*), the genome-edited cotton plants displayed glandless phenotype in cotton (Janga et al. 2019). Gao et al. (2020) silenced *CGP1* gene (involves in gland formation) by CRISPR/Cas technology and obtained genome-edited plants which exhibited glandless-like phenotype. However, the CRISPR/Cas9 knockout cotton plants exhibited normal gland structure or density although the gland pigmentation of these plants was affected. It was also revealed that level of gossypols and associated terpenoids compounds were substantially depressed. Such depression in expression of genes involved in biosynthesis of gossypol was observed in the genome-edited plants (Gao et al. 2020). CRISPR/Cas9-mediated knockout of *GhTST2* resulted in significantly decreased Glc content in cotton (Deng et al. 2020).

14.5 Concluding Remarks and Future Perspectives

Although CRISPR/Cas9 genome editing technology has been showing very promising results that can pave the way for initiating precise breeding in cotton (Zhang 2019c), still there is a long way to go for cultivating CRISPR/Cas-based breeding products in the field. Several issues should be addressed to speed the application of CRISPR/Cas9 precise breeding in cotton.

14.5.1 A Highly Efficient Plant Regeneration System Needed for CRISPR/Cas-Based Genome Editing in Cotton

Current genome editing in cotton is still based on the traditional transgenic technology, which requires to transform the CRISPR/Cas9 system into cotton cells and then obtain regenerated plants from the genome editing cells (Li et al. 2017; Li and Zhang 2019). Unfortunately, compared with other crops such as rice, it is hard to obtain plant regeneration through cotton tissue culture. Although great progresses have been made in the past 20 years, the plant regeneration is still limited to a limited number of genotypes in cotton (Juturu et al. 2015; Zhang et al. 2009). Additionally, the majority of cotton cultivars with plant regeneration ability is out of markets with certain undesirable traits; it is hard to obtain regenerated plants from majority of cultivars that are widely adopted by the cotton farmers (Zhang et al. 2009). Thus, more efforts are needed to establish a better plant regeneration system for cotton transformation or discover a new way to make a transgenic cotton (Peng et al., 2021).

14.5.2 Finding Target Genes Is the Key for CRISPR/Cas9-Based Precise Breeding in Cotton

CRISPR/Cas9-based precise breeding is entirely dependent on our comprehensive understanding about the function of cotton genes. Without a targeted gene, it is impossible to improve cotton traits through transgenic or genome editing. In 2012, the draft genome of a diploid cotton *G. raunmondii* was published and allow people for the first time to gain insight into cotton genome stucture, it also opens a new era for cotton whole genome sequencing, gene function study and prcesion breeding (Wang etal., 2012) Since then, multiple reports pertaining to whole genome sequencing of all cultivated cotton species and few others reveal huge information on cotton genome structure and evolution, and thousands of genes have been identified in cotton (Du et al. 2018; Hu et al. 2019; Li et al., 2014; Li et al. 2015; Wang et al. 2019; Zhang et al. 2015). However, key genes involved in cotton fiber development, yield, and quality as well as response to different abiotic and biotic stresses are yet to be identified. Thus, more research should be focused on studying gene function during the next 10 years so that the generated knowledge can be used for improving cotton. However, both CRISPR/Cas9 genome editing and cotton sequencing open a new era for initiating cotton precision breeding, and their role will significantly enhance cotton production (Zhang 2019c).

Acknowledgments The CRISPR/Cas9-related research, including establishment of the CRISPR/Cas9 genome system in cotton, in B.Z. lab is supported in part by Cotton Incorporated and the National Science Foundation (award 1658709).

References

Anzalone AV, Koblan LW, Liu DR (2020) Genome editing with CRISPR-Cas nucleases, base editors, transposases and prime editors. *Nat Biotechnol* 38(7):824–844

Arora L, Narula A (2017) Gene editing and crop improvement using CRISPR-Cas9 system. *Front Plant Sci* 8:1932

Barnhoorn I, van Dyk C (2020) The first report of selected herbicides and fungicides in water and fish from a highly utilized and polluted freshwater urban impoundment. *Environ Sci Pollut Res Int* 27:33393–33398

Chen KL, Wang YP, Zhang R, Zhang HW, Gao CX (2019) CRISPR/Cas genome editing and precision plant breeding in agriculture. In: Merchant SS (ed) Annual review of plant biology, vol 70, pp 667–697

Deng J, Yang X, Sun W, Miao Y, He L, Zhang X (2020) The calcium sensor CBL2 and its interacting kinase CIPK6 are involved in plant sugar homeostasis via interacting with tonoplast sugar transporter TST2(1). *Plant Physiol* 183:236–249

Doudna JA, Charpentier E (2014) The new frontier of genome engineering with CRISPR-Cas9. *Science* 346:1077

Du XM, Huang G, He SP, Yang ZE, Sun GF, Ma XF, Li N, Zhang XY, Sun JL, Liu M, Jia YH, Pan ZE, Gong WF, Liu ZH, Zhu HQ, Ma L, Liu FY, Yang DG, Wang F, Fan W, Gong Q, Peng Z, Wang LR, Wang XY, Xu SJ, Shang HH, Lu CR, Zheng HK, Huang SW, Lin T, Zhu YX, Li FG

(2018) Resequencing of 243 diploid cotton accessions based on an updated A genome identifies the genetic basis of key agronomic traits. *Nat Genet* 50:796

Fang DD, Naoumkina M, Kim HJ (2018) Unraveling cotton fiber development using fiber mutants in the post-genomic era. *Crop Sci* 58:2214–2228

Gao W, Xu F-C, Long L, Li Y, Zhang J-L, Chong L, Botella JR, Song C-P (2020) The gland localized CGP1 controls gland pigmentation and gossypol accumulation in cotton. *Plant Biotechnol J* 18:1573–1584

Gasiunas G, Barrangou R, Horvath P, Siksnys V (2012) Cas9-crRNA ribonucleoprotein complex mediates specific DNA cleavage for adaptive immunity in bacteria. *Proc Natl Acad Sci U S A* 109:E2579–E2586

Guan XY, Song QX, Chen ZJ (2014) Polyploidy and small RNA regulation of cotton fiber development. *Trends Plant Sci* 19:516–528

He X, Luo X, Wang T, Liu S, Zhang X, Zhu L (2020) GhHB12 negatively regulates abiotic stress tolerance in Arabidopsis and cotton. *Environ Exp Bot* 176:104087

Horvath P, Barrangou R (2010) CRISPR/Cas, the immune system of bacteria and archaea. *Science* 327:167–170

Hsu PD, Lander ES, Zhang F (2014) Development and applications of CRISPR-Cas9 for genome engineering. *Cell* 157:1262–1278

Hu Y, Chen JD, Fang L, Zhang ZY, Ma W, Niu YC, Ju LZ, Deng JQ, Zhao T, Lian JM, Baruch K, Fang D, Liu X, Ruan YL, Rahman MU, Han JL, Wang K, Wang Q, Wu HT, Mei GF, Zang YH, Han ZG, Xu CY, Shen WJ, Yang DF, Si ZF, Dai F, Zou LF, Huang F, Bai YL, Zhang YG, Brodt A, Ben-Hamo H, Zhu XF, Zhou BL, Guan XY, Zhu SJ, Chen XY, Zhang TZ (2019) Gossypium barbadense and Gossypium hirsutum genomes provide insights into the origin and evolution of allotetraploid cotton. *Nat Genet* 51:739

Ijaz B, Zhao N, Kong J, Hua JP (2019) Fiber quality improvement in upland cotton (Gossypium hirsutum L.): quantitative trait loci mapping and marker assisted selection application. *Front Plant Sci* 10:1585

Janga MR, Campbell LM, Rathore KS (2017) CRISPR/Cas9-mediated targeted mutagenesis in upland cotton (Gossypium hirsutum L.). *Plant Mol Biol* 94:349–360

Janga MR, Pandeya D, Campbell LM, Konganti K, Villafuerte ST, Puckhaber L, Pepper A, Stipanovic RD, Scheffler JA, Rathore KS (2019) Genes regulating gland development in the cotton plant. *Plant Biotechnol J* 17:1142–1153

Jinek M, Chylinski K, Fonfara I, Hauer M, Doudna JA, Charpentier E (2012) A programmable dual-RNA-guided DNA endonuclease in adaptive bacterial i9mmunity. Science 337:816–821

Juturu VN, Mekala GK, Kirti PB (2015) Current status of tissue culture and genetic transformation research in cotton (Gossypium spp.). *Plant Cell Tissue Org Cult* 120:813–839

Langner T, Kamoun S, Belhaj K (2018) CRISPR crops: plant genome editing toward disease resistance. In: Leach JE, Lindow SE (eds) Annual review of phytopathology, vol 56, pp 479–512

Li F, Fan G, Wang K (2014) Genome sequence of the cultivated cotton Gossypium arboreum. *Nature Genetics*, 46: 567–572

Li C, Zhang BH (2016) MicroRNAs in control of plant development. *J Cell Physiol* 231:303–313

Li C, Zhang B (2019) Genome editing in cotton using CRISPR/Cas9 system. *Methods Mol Biol* (Clifton, NJ) 1902:95–104

Li FG, Fan GY, Lu CR, Xiao GH, Zou CS, Kohel RJ, Ma ZY, Shang HH, Ma XF, Wu JY, Liang XM, Huang G, Percy RG, Liu K, Yang WH, Chen WB, Du XM, Shi CC, Yuan YL, Ye WW, Liu X, Zhang XY, Liu WQ, Wei HL, Wei SJ, Huang GD, Zhang XL, Zhu SJ, Zhang H, Sun FM, Wang XF, Liang J, Wang JH, He Q, Huang LH, Wang J, Cui JJ, Song GL, Wang KB, Xu X, Yu JZ, Zhu YX, Yu SX (2015) Genome sequence of cultivated Upland cotton (Gossypium hirsutum TM-1) provides insights into genome evolution. *Nat Biotechnol* 33:524–U242

Li C, Unver T, Zhang B (2017) A high-efficiency CRISPR/Cas9 system for targeted mutagenesis in Cotton (Gossypium hirsutum L.). *Sci Rep* 7:43902

Long L, Guo D-D, Gao W, Yang W-W, Hou L-P, Ma X-N, Miao Y-C, Botella JR, Song C-P (2018) Optimization of CRISPR/Cas9 genome editing in cotton by improved sgRNA expression. *Plant Methods* 14

Makarova KS, Haft DH, Barrangou R, Brouns SJJ, Charpentier E, Horvath P, Moineau S, Mojica FJM, Wolf YI, Yakunin AF, van der Oost J, Koonin EV (2011) Evolution and classification of the CRISPR-Cas systems. *Nat Rev Microbiol* 9:467–477

Meng ZG, Meng ZH, Zhang R, Liang CZ, Wan JM, Wang YL, Zhai HH, Guo SD (2015) Expression of the rice arginase gene OsARG in cotton influences the morphology and nitrogen transition of seedlings. *PLoS One* 10:e0141530

Naoumkina M, Thyssen GN, Fang DD, Hinchliffe DJ, Florane CB, Jenkins JN (2016) Small RNA sequencing and degradome analysis of developing fibers of short fiber mutants Ligon-lintles-1 (Li-1) and-2 (Li-2) revealed a role for miRNAs and their targets in cotton fiber elongation. *BMC Genomics* 17:360

Nishimasu H, Ran FA, Hsu PD, Konermann S, Shehata SI, Dohmae N, Ishitani R, Zhang F, Nureki O (2014) Crystal structure of Cas9 in complex with guide RNA and target DNA. *Cell* 156:935–949

Peng RH, Jones DC, Liu F, Zhang (2021) From sequencing to genome editing for cotton improvement. *Trends in Biotechnology*, https://doi.org/10.1016/j.tibtech.2020.09.001

Peters K, Bundschuh M, Schäfer RB (2013) Review on the effects of toxicants on freshwater ecosystem functions. *Environ Pollut (Barking, Essex: 1987)* 180:324–329

Qin L, Li J, Wang Q, Xu Z, Sun L, Alariqi M, Manghwar H, Wang G, Li B, Ding X, Rui H, Huang H, Lu T, Lindsey K, Daniell H, Zhang X, Jin S (2020) High-efficient and precise base editing of C center dot G to T center dot A in the allotetraploid cotton (Gossypium hirsutum) genome using a modified CRISPR/Cas9 system. *Plant Biotechnol J* 18:45–56

Sentmanat MF, Peters ST, Florian CP, Connelly JP, Pruett-Miller SM (2018) A survey of validation strategies for CRISPR-Cas9 editing. *Sci Rep* 8:888

Shmakov S, Smargon A, Scott D, Cox D, Pyzocha N, Yan W, Abudayyeh OO, Gootenberg JS, Makarova KS, Wolf YI, Severinov K, Zhang F, Koonin EV (2017) Diversity and evolution of class 2 CRISPR-Cas systems. *Nat Rev Microbiol* 15:169–182

Sun GL, Xie FL, Zhang BH (2011) Transcriptome-wide identification and stress properties of the 14-3-3 gene family in cotton (Gossypium hirsutum L.). *Funct Integr Genomics* 11:627–636

Terzaghi E, Vitale CM, Di Guardo A (2020) Modelling peak exposure of pesticides in terrestrial and aquatic ecosystems: importance of dissolved organic carbon and vertical particle movement in soil. *SAR QSAR Environ Res* 31:19–32

Urnov FD, Rebar EJ, Holmes MC, Zhang HS, Gregory PD (2010) Genome editing with engineered zinc finger nucleases. *Nat Rev Genet* 11:636–646

Walford SA, Wu YR, Llewellyn DJ, Dennis ES (2011) GhMYB25-like: a key factor in early cotton fibre development. *Plant J* 65:785–797

Wang K, Wang Z, Li F, Ye W (2012) The draft genome of a diploid cotton Gossypium raimondii. *Nature Genetics*. 44(10): 1098–1104. https://doi.org/10.1038/ng.2371

Wang QL, Zhang BH (2015) MicroRNAs in cotton: an open world needs more exploration. *Planta* 241:1303–1312

Wang Y, Meng Z, Liang C, Meng Z, Wang Y, Sun G, Zhu T, Cai Y, Guo S, Zhang R, Lin Y (2017) Increased lateral root formation by CRISPR/Cas9-mediated editing of arginase genes in cotton. *Sci China Life Sci* 60:524–527

Wang P, Zhang J, Sun L, Ma Y, Xu J, Liang S, Deng J, Tan J, Zhang Q, Tu L, Daniell H, Jin S, Zhang X (2018) High efficient multisites genome editing in allotetraploid cotton (Gossypium hirsutum) using CRISPR/Cas9 system. *Plant Biotechnol J* 16:137–150

Wang MJ, Tu LL, Yuan DJ, Zhu D, Shen C, Li JY, Liu FY, Pei LL, Wang PC, Zhao GN, Ye ZX, Huang H, Yan FL, Ma YZ, Zhang L, Liu M, You JQ, Yang YC, Liu ZP, Huang F, Li BQ, Qiu P, Zhang QH, Zhu LF, Jin SX, Yang XY, Min L, Li GL, Chen LL, Zheng HK, Lindsey K, Lin ZX, Udall JA, Zhang XL (2019) Reference genome sequences of two cultivated allotetraploid cottons, Gossypium hirsutum and Gossypium barbadense. *Nat Genet* 51:224

Wright DA, Li T, Yang B, Spalding MH (2014) TALEN-mediated genome editing: prospects and perspectives. *Biochem* J 462:15–24

Wu ZG, Yang Y, Huang G, Lin J, Xia YY, Zhu YX (2017) Cotton functional genomics reveals global insight into genome evolution and fiber development. *J Genet Genomics* 44:511–518

Xie FL, Wang QL, Sun RR, Zhang BH (2015a) Deep sequencing reveals important roles of microRNAs in response to drought and salinity stress in cotton. *J Exp Bot* 66:789–804

Xie FL, Wang QL, Zhang BH (2015b) Global microRNA modification in cotton (Gossypium hirsutum L.). *Plant Biotechnol J* 13:492–500

Yang ZR, Qanmber G, Wang Z, Yang ZE, Li FG (2020) Gossypium genomics: trends, scope, and utilization for cotton improvement. *Trends Plant Sci* 25:488–500

Zaidi S, Mukhtar MS, Mansoor S (2018) Genome editing: targeting susceptibility genes for plant disease resistance. *Trends Biotechnol* 36:898–906

Zhang BH (2015) MicroRNA: a new target for improving plant tolerance to abiotic stress. *J Exp Bot* 66:1749–1761

Zhang B (2019a) Agrobacterium-mediated genetic transformation of cotton. *Methods Mol Biol* 1902:19–33

Zhang B (2019b) Transgenic cotton: from biotransformation methods to agricultural application. *Methods Mol Biol* 1902:3–16

Zhang BH (2019c) CRISPR/Cas-based genome editing: a new era for transgenic and precise breeding in cotton. *ICAC Recorder* 37:3–10

Zhang BH (2019d) Transgenic cotton. Humana Press, New York

Zhang D, Zhang B (2020) SpRY: engineered CRISPR/Cas9 harnesses new genome-editing power. *Trends in Genetics*, 36(8):546–548. https://doi.org/10.1016/j.tig.2020.05.001

Zhang BH, Pan XP, Cobb GP, Anderson TA (2006) Plant microRNA: a small regulatory molecule with big impact. *Dev Biol* 289:3–16

Zhang B, Pan X, Venne L, Dunnum S, McMurry ST, Cobb GP, Anderson TA (2008) Development of a method for the determination of 9 currently used cotton pesticides by gas chromatography with electron capture detection. *Talanta* 75:1055–1060

Zhang BH, Wang QL, Liu F, Wang KB, Frazier TP (2009) Highly efficient plant regeneration through somatic embryogenesis in 20 elite commercial cotton (Gossypium hirsutum L.) cultivars. *Plant Omics* 2:259–268

Zhang ZT, Zhou Y, Li Y, Shao SQ, Li BY, Shi HY, Li XB (2010) Interactome analysis of the six cotton 14-3-3s that are preferentially expressed in fibres and involved in cell elongation. *J Exp Bot* 61:3331–3344

Zhang TZ, Hu Y, Jiang WK, Fang L, Guan XY, Chen JD, Zhang JB, Saski CA, Scheffler BE, Stelly DM, Hulse-Kemp AM, Wan Q, Liu BL, Liu CX, Wang S, Pan MQ, Wang YK, Wang DW, Ye WX, Chang LJ, Zhang WP, Song QX, Kirkbride RC, Chen XY, Dennis E, Llewellyn DJ, Peterson DG, Thaxton P, Jones DC, Wang Q, Xu XY, Zhang H, Wu HT, Zhou L, Mei GF, Chen SQ, Tian Y, Xiang D, Li XH, Ding J, Zuo QY, Tao LN, Liu YC, Li J, Lin Y, Hui YY, Cao ZS, Cai CP, Zhu XF, Jiang Z, Zhou BL, Guo WZ, Li RQ, Chen ZJ (2015) Sequencing of allotetraploid cotton (Gossypium hirsutum L. acc. TM-1) provides a resource for fiber improvement. *Nat Biotechnol* 33:531–U252

Zhang Z, Ge X, Luo X, Wang P, Fan Q, Hu G, Xiao J, Li F, Wu J (2018) Simultaneous editing of two copies of Gh14-3-3d confers enhanced transgene-clean plant defense against Verticillium dahliae in allotetraploid upland cotton. *Front Plant Sci* 9:842

Zhou ML, Sun GQ, Sun ZM, Tang YX, Wu YM (2014) Cotton proteomics for deciphering the mechanism of environment stress response and fiber development. *J Proteome* 105:74–84

Zhou Y, Zhang ZT, Li M, Wei XZ, Li XJ, Li BY, Li XB (2015) Cotton (Gossypium hirsutum) 14-3-3 proteins participate in regulation of fibre initiation and elongation by modulating brassinosteroid signalling. *Plant Biotechnol J* 13:269–280

Zhu S, Yu X, Li Y, Sun Y, Zhu Q, Sun J (2018) Highly efficient targeted gene editing in upland cotton using the CRISPR/Cas9 system. *Int J Mol Sci* 19:3000

Chapter 15
Use of CRISPR/Cas System to Create Resistance to Cotton Diseases

Sayyad Ali Raza Bukhari, Muhammad Saeed, and Rob W. Briddon

15.1 Introduction

Plants live in a changing environment surrounded by coevolving pathogens including bacteria, viruses, and fungi. These pathogens impair growth and may result in economic losses to crop plants. To counter pathogens, plants have developed various strategies—the exact mechanism of which varies depending upon the type of pathogen. The first line of defense is recognition of conserved patterns known as pathogen-associated molecular patterns (PAMPs) by transmembrane pattern recognition receptors (PRRs) (Boller and Felix 2009). For viruses PAMPs may include double-stranded RNA (Teixeira et al. 2019). PRRs also perceive internal signals secreted by pathogen-induced cell wall damage termed "danger-associated molecular patterns" (DAMPs). PRR stimulation elicits a response referred to as PRR-triggered immunity (PTI) (Macho and Zipfel 2014). To interfere with PTI, successful nonviral pathogens deliver effectors into the host by the type III secretion system. Effectors weaken the PTI response and allow successful colonization of the pathogen. However, plants have developed a second tier of defense mechanism that works in the cytoplasm and recognizes effectors. Once an effector molecule manages to enter in the cytoplasm, it is detected by products of resistance (R)

S. A. R. Bukhari
Department of Biotechnology, Knowledge Unit of Science, University of Management and Technology, Sialkot, Punjab, Pakistan

M. Saeed (✉)
Department of Botany, Government College University, Faisalabad, Pakistan
e-mail: saeed_pbg@gcuf.edu.pk

R. W. Briddon
National Institute for Biotechnology and Genetic Engineering, Faisalabad, Pakistan

Mehboob-ur-Rahman et al. (eds.), *Cotton Precision Breeding*,
https://doi.org/10.1007/978-3-030-64504-5_15

genes—intracellular nucleotide-binding domain and leucine-rich repeat domains—resulting in a number of cellular responses termed "effector-triggered immunity" (ETI) (Jones and Dangl 2006). In contrast to PAMPs, effector molecules and the ETI response element are coevolving to dominate the pathogens. The struggle to dominate the pathogens is well reflected by their diversity, variability, and dispensability. For rapidly evolving pathogens, ETI is responsible for countering the pathogens. Pathogenic hegemony requires mechanism to evade R-genes and cooperation of host genes to establish a compatible interaction. Host genes that are targeted by effectors and are made to work in favor of a pathogen are called susceptibility (S) genes. These quisling genes can be mutated to obstruct the pathogen. Disrupted S-genes and/or adding multiple R-genes provide resistance to plants against various pathogens as reviewed in van Schie and Takken (2014). The robustness of S-genes is epitomized by the *elf4G* and mildew locus O *(MLO)* genes. In case of mutation, these genes confer resistance against rice tungro disease and powdery mildew disease, respectively (Acevedo-Garcia et al. 2017; Macovei et al. 2018; Langner et al. 2018).

Since the domestication of crop plants, plant breeding has played a significant role in combating diseases by stacking R-genes and/or introducing mutated S-genes (Pavan et al. 2009; St. Clair 2010). Plant breeders have used naturally available genetic resources to select favorable traits. Bottlenecks in the genetic diversity have been overcome by the use of artificial mutagens that substantially increased genetic variation (Siegemund 2000). Various chemicals and radiation have been used to randomly create mutant lines of varied phenotypes (Ahloowalia et al. 2004). Populations of mutant plants are screened for the desirable trait. Such indiscriminate mutations are mostly accompanied by uncertain and often deleterious effects. The procedure is also a time-consuming and labor-intensive process. In view of the high evolution rate of most of the pathogens—viruses, bacteria, and fungi—the product of a long-term breeding plan might not be durable in the field. To combat rapidly evolving pathogens, it is important to develop preemptive defense strategies.

The advent of recombinant DNA technology in the 1970s and subsequent plant transformation procedures in many cases overcame species barriers and paved the way for the transfer of genetic material among distantly related living organisms. However, random incorporation of a transgene can cause unwanted side effects and requires safety studies to comply with biosafety and risk regulations. Despite the promise of genetically modified crops, safety concerns have restricted their use (van Schie and Takken 2014).

Genome editing is an advanced molecular biology technique that allows the addition, deletion, modification, or replacement of DNA at targeted locations in the genome with unprecedented precision and efficiency. In addition to changing base composition of endogenous genetic material, genome editing is able to modify epigenetic marks in a sequence-specific manner (Bortesi and Fischer 2015; Lowder et al. 2015). These capabilities are important for increasing immunity, disrupting susceptibility, and regulating gene expression. One of the major risks associated with advanced crop improvement techniques is indiscriminate insertion of a gene that may result in disruption, upregulation, or downregulation of host genes. Targeted DNA modifications, in contrast, impose lower safety risks (Hartung and Schiemann 2014). With the exception of the gene-editing agent, edited plants are indistinguishable from a natural mutant. These precise mutations can expedite modern plant breeding programs for precise crop improvement (Voytas and Gao 2014).

Homologous recombination was initially exploited in attempts to achieve targeted manipulation of the genome and was called "gene targeting" (GT). GT has been successfully carried out in bacteria and yeast, for which the major process of gene integration is homologous recombination. However, the same is not true for plants. Homologous recombination in higher plants has been demonstrated to be highly inefficient due to random integration—illegitimate recombination—of the transgene even in the presence of homologous ends (Puchta 2002). Several studies have reported homologous recombination-mediated GT in mouse embryonic stem cells (Doetschman et al. 1987; Thomas and Capecchi 1987) and plants (Paszkowski et al. 1988; Offringa et al. 1990), but the efficiency of mutation has been unacceptably low. Despite their unsuitability, the studies have provided significant insight in understanding the mechanism of homologous recombination (Capecchi 1989). Later it was observed that double-strand breaks (DSBs) can enhance homologous recombination-mediated gene targeting (Tovar and Lichtenstein 1992; Chiurazzi et al. 1996).

In higher organisms DSB repair has two competing pathways: homology-directed repair (HDR) and nonhomologous end joining (NHEJ) (Pâques and Haber 1999). HDR is a template-depended faithful repair mechanism that accurately repairs a DSB using information from the template. This is useful for precise editing and GT. NHEJ is non-template-depended repair mechanism that quickly joins the break point with a random insertion or deletion (indel) at the break point. Random integration is attributed to NHEJ, which is the preferred pathway in higher organisms, particularly higher plants, because of the absence of somatic recombination (Puchta 2004; Puchta and Fauser 2013; Knoll et al. 2014).

The advent of nucleases to create DSBs made the prospects of genome editing more positive and has strengthened the capacity and accuracy of genome-editing technologies. There are four tools in genome editing: meganucleases (Epinat et al. 2003), zinc-finger nucleases (ZFNs) (Kim et al. 1996), transcription activator-like effector nucleases (TALENs) (Zhang et al. 2013), and clustered regularly interspaced short palindromic repeats (CRISPR)/CRISPR-associated protein (Cas) as demonstrated in Fig. 15.1. Meganucleases, ZFNs, and TALENs have been around for more than two decades, but a recent boost to genome-editing studies has been due to CRISPR (Silva et al. 2011; Esvelt and Wang 2013; Gaj et al. 2013; Joung and Sander 2013; Sander and Joung 2014). CRISPR provides easy and simple design of constructs and multiplex targeting. All of these technologies have two basic requirements: recognition of specific DNA sequence and an endonuclease to create DSBs. However, all the systems rely on the HDR and NHEJ endogenous repair systems of the target cell (Waterworth et al. 2011).

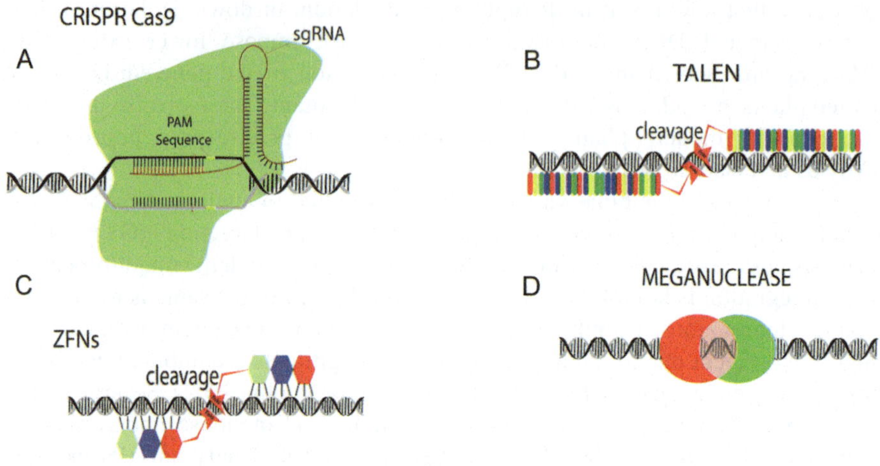

Fig. 15.1 General mechanism for genome editing in plants. (**a**) CRISPR/Cas9: The Cas9, sgRNA, and target DNA complex are shown. The PAM sequence is shown in yellow. Cleavage occurs three nucleotides upstream from the PAM sequence. (**b**) TALEN: Several single nucleotide-binding domains are fused to the Fok1 nuclease domain. (**c**) ZFNs: Several three nucleotide DNA-binding domains are fused to the Fok1 nuclease. (**d**) Meganuclease: Endodeoxyribonuclease domains are shown that recognize a specific DNA sequence

15.2 Meganucleases

Meganucleases, also known as homing nucleases, were first identified in yeast mitochondria (Nicolas et al. 1989; Belfort and Roberts 1997). The nuclease activity of meganucleases has been demonstrated in *Nicotiana plumbaginifolia* protoplasts (Puchta et al. 1993) and transgenic *Nicotiana tabacum* lines (Puchta et al. 1996). Meganucleases have been used for NHEJ mutagenesis in maize (Gao et al. 2010), transgene excision in *Arabidopsis* (Antunes et al. 2012), and the introduction of multiple genes, so-called gene stacking in cotton (D'Halluin et al. 2013). Much of our insight into DSB repair in plants has been obtained by using meganucleases. These are small (~165 amino acids) nucleases suitable for many delivery methods including viral vectors. However, few meganucleases recognize limited sites on a genome. Extensive engineering and high-throughput screening is required to develop new meganucleases with tailored DNA-binding abilities. Redesign is also complicated by the non-modular nature of the binding and cleavage domains. Because of overlap between binding and cleavage domains, engineering one affects the efficiency of the other. These factors limit the use of meganucleases in genome editing (Harrison et al. 2014).

15.3 Zinc-Finger Nucleases

A ZFN is a fusion of specific DNA-binding zinc-finger (ZF) domain with the DNA endonuclease enzyme *Fok*1. Zinc-finger proteins (ZFPs) are a class of eukaryotic transcription factors whose DNA-binding domain—zinc finger—can be modified to bind at a specific DNA sequence. Each zinc-finger domain recognizes three nucleotides. Fusion of an array of zinc-finger domains with *Fok*1 enables the fusion product to generate a nick in a sequence-specific manner. However, to generate DSBs two such fusion products are produced that can be targeted to a single sequence. ZFN-mediated gene targeting and random mutagenesis has been demonstrated in *Drosophila*, *Arabidopsis*, *Glycine max*, and *Zea mays* (Carroll 2011; Puchta and Fauser 2014). Although the engineering of ZFN is easier than meganuclease, the context-dependent functioning of zinc fingers reduces their usefulness as an effective tool in genome engineering. A zinc finger that recognizes AAA may recognize a different triplet when fused with other zinc fingers. The specificity and efficiency of a ZFN cannot be predicted because of the compromising nature of other ZF domains when joined with a second ZF domain. Therefore several ZFNs need to be tested in parallel in an expensive validation process (Ramirez et al. 2008).

15.4 TALENs

A TALEN is a bacterial effector protein that is delivered into host cells during infection. The effector protein has DNA-binding ability and binds to several promoters in the host cell to modulate transcription (Van den Ackerveken et al. 1996; Boch et al. 2009). TALENs have been used extensively in genome engineering in plants (Mahfouz et al. 2011). They have been employed for pathogen resistance in rice by mutating the *OsSWEET14* gene (Li et al. 2012) and herbicide resistance in tobacco by targeting the *ALS* gene (Zhang et al. 2013). They are similar to ZFNs, consisting of an array of TALE domains—recognizing a specific DNA sequence—and a *Fok*I nuclease domain. The modular nature of TALENs overcomes the impediment imposed by ZFNs and is not influenced by nearby TALE array (Cermak et al. 2011). Direct repeats of 33–35 amino acids make a TALE-binding domain. Repeat variable di-residues (RVDs), consisting of two amino acids, recognize a single nucleotide in DNA. The most widely employed RVDs and their nucleotide partner are HD (cytosine), NI (adenine), NG (thymine), and NN (guanine and adenine). This one-to-one correspondence between a single RVD and single DNA base has overcome the design challenges of meganucleases and zinc-finger nucleases. TALENs can recognize any DNA sequence in the genome but are difficult to design and clone. Moreover, it requires costly protein engineering (Holkers et al. 2013).

15.5 CRISPR/Cas System

CRISPR/Cas is a bacterial adaptive immune system that provides defense against bacteriophages and foreign plasmids by sequence-specific cleavage (Barrangou et al. 2007; Bikard et al. 2012; Barrangou 2013; Amitai and Sorek 2016). Although bacteria have other defense mechanisms, such as abortive infection, receptor mutation, and restriction modification, the CRISPR/Cas system is an inheritable adaptive immune response against foreign DNA. This was first reported in 2007 in *Streptococcus thermophiles* (Barrangou et al. 2007).The artificial CRISPR/Cas system has two components: a guide RNA recognizes a specific sequence on the foreign DNA, and Cas, or Cas complex, imparts a DSB. CRISPR/Cas systems can be divided broadly into two classes: Class I uses multiple Cas proteins to cleave foreign DNA, whereas class II employs a single nuclease. By virtue of involving only a single nuclease, class II is most suitable for genome engineering and in recent years has been extensively used for both NHEJ- and HDR-mediated genome engineering in diverse organisms. Class II systems encompass two DNA nucleases, Cas9 and Cas12a (previously Cpf1) as demonstrated in Fig. 15.2. Cas9 creates blunt ends, whereas Cas12a leaves staggered cuts in DNA. Simplicity of use, ease

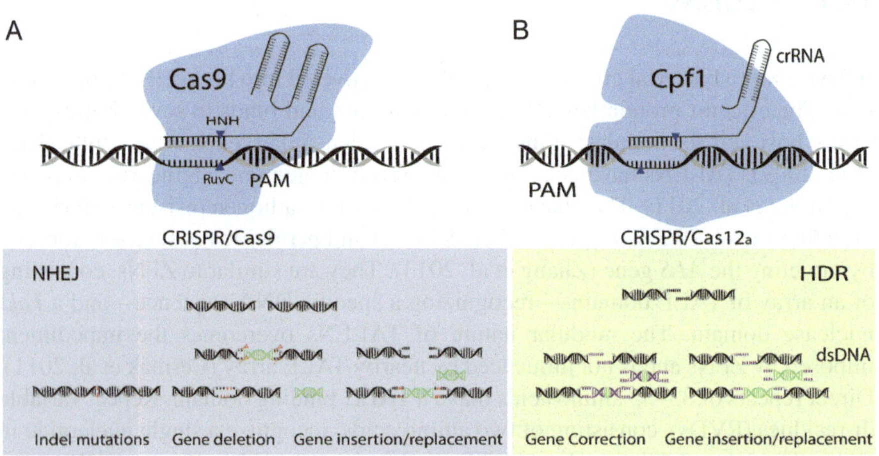

Fig. 15.2 Genome-editing applications of CRISPR/Cas after inducing double breaks in genome. (**a**) Cas9-mediated cleavage results in blunt end breaks which are preferably repaired by NHEJ. NHEJ results in loss-of-function indels. For gene deletion the excised fragment is shown in green. For gene insertion/replacement, the incoming segment is shown in green, and red nucleotides are indels at the site of gene insertion due to NHEJ. (**b**) Cas12a creates staggered cuts which are preferentially repaired by HDR. HDR is suitable for gene correction and gene insertion. For HDR the incoming template has sequences that overlap the target to allow homologous recombination to take place

of cloning, and multiplexing are the characteristics responsible for widespread acceptability of class II CRISPR genome engineering. The RNA-guided class II CRISPR/Cas system requires a protospacer adjacent motif (PAM) for successful cleavage of the target.

In the natural system, Cas9 is guided by CRISPR RNA (crRNA) and trans-activating crRNA (tracrRNA). Initially the crRNA and tracrRNA bind each other, and the crRNA is processed further to make the mature crRNA-tracrRNA complex. This complex guides Cas9 for cleavage at a complementary sequence flanked by the PAM sequence (Mali et al. 2013). However crRNA and tracrRNA can be engineered as a single-guide RNA (sgRNA) which can also direct Cas9 cleavage (Jinek et al. 2012). This was a major milestone toward genome editing in eukaryotes, allowing ease of cloning and multiplexing. Both crRNA-tracrRNA hybrids and sgRNA have been used in genome editing in eukaryotes. Cas12a differs from Cas9 in three main features: Cas12a requires only a crRNA, cleaves at target DNA with a T-rich PAM, and makes staggered cuts. CRISPR/Cas has been widely employed to create so-called knockin (replacement of a sequence) and knockout (disruption of a gene) mutations in plants (Lowder et al. 2015; Chen et al. 2019).

The objective of sequence-specific nucleases is to produce targeted DSBs that initiate the host HDR and NHEJ repair responses. An interesting attribute of foreign DNA integration is its proclivity to insert into a preexisting DSB (Puchta et al. 1996). If foreign DNA sequence has homology with DNA flanking a break site, the break will be repaired by HDR; otherwise the break is repaired by NHEJ. Staggered, rather than blunt, breaks favor repair by HDR (Moreno-Mateos et al. 2017).

15.6 Genome Editing by NHEJ

In plants, NHEJ is the preferred pathway for the repair of DSBs. However, NHEJ is an error-prone repair system that generates a small number of indels at the site of the break. Indels in protein-coding regions result in loss-of-function mutations as demonstrated in Fig. 15.2. Indels in promoter sequences can render transcription factors unable to bind the promoter and preclude transcription. In disease resistance, this strategy has been used in transcriptional control or disruption of S-genes. Loss-of-function mutations can be achieved by stable integration of nuclease by *Agrobacterium*-mediated transformation driven by cell-specific, constitutive, or inducible promoters. A nuclease-guiding sgRNA can be integrated with the nuclease or agro-infiltrated into Cas9-expressing plants. Indels can be detected using T7E1 or Surveyor nuclease assays to detect mismatches in heteroduplexes, loss of restriction or primer binding sites, and deep sequencing (Zischewski et al. 2017). The use of two sgRNAs targeting two sites across a gene can lead to gene deletion (see Fig. 15.2).

15.7 Genome Editing by HDR

Although NHEJ is very efficient in plants for production of knockout mutations on a large scale, it is devoid of the precision necessary for more sophisticated genome editing. Site-specific DSB repair, by adding a homologous sequence, can result in precise genome editing and the precise addition of foreign genes by HDR. The technique can be used to generate allelic variants and gene targeting. HDR-mediated repair has been widely used in many organisms. However, it has been found to be quite challenging in plants due to the less efficient HDR mechanism and hurdles in delivering the donor template. HDR predominates during the S and G2 phases of the cell cycle, with illegitimate recombination hindering HDR during the other phases. Several strategies have been investigated to increase HDR-mediated editing. Increasing the concentration of donor template can increase the efficiency of HDR-mediated repair, achieved using biolistic rather than *Agrobacterium*-mediated delivery (Sun et al. 2016). Use of a virus replicon to deliver the donor template has also been shown to enhance the efficiency of HDR (Baltes and Voytas 2015). Since the two repair pathways compete for DSB repair, the evidence supports the idea of suppressing NHEJ and stimulating HDR (Puchta et al. 1996; Qi et al. 2013). The expression of the bacterial *RecA* has been used to improve the efficiency of HDR in tobacco (Reiss et al. 2000). Staggered, rather than blunt, breaks have also been found to enhance HDR-mediated repair. Staggered cuts can be achieved using nickase Cas9 (nCas9) or Cas12a.

15.8 CRISPR/Cas Mutation for Resistance to Viruses

Phytopathogenic viruses cause significant economic losses to crops. Viruses are categorized as either having DNA or RNA genomes. Cotton is affected by both DNA and RNA viruses. The DNA-containing viruses belong to the family *Geminiviridae* and cause cotton leaf curl disease in the Old World. The RNA-containing viruses belong to the genus *Polerovirus* (family *Luteoviridae*), which cause cotton bunchy top disease (Reddall et al. 2004), and the genus *Ilarvirus* (family *Bromoviridae*) (Vinodkumar et al. 2017). In South Asia, viruses with single-stranded DNA genomes of the genus *Begomovirus* (family *Geminiviridae*) cause the most significant losses to cotton production (Brown et al. 2015).

The virus family *Geminiviridae* is classified into seven genera, and the genus *Begomovirus* encompasses the majority of the known geminiviruses that infect only dicotyledonous plants (Zerbini et al. 2017). Begomoviruses are transmitted plant-to-plant by whiteflies of the *Bemisia tabaci* species complex (Brown et al. 2000; Perring 2001). The *B. tabaci* species complex is a group of reproductively isolated and morphologically indistinguishable populations (cryptic species), which each have differing specificities for transmitting begomoviruses (Perring 2001; Polston et al. 2014).

Studies have sought to generate virus-resistant plants using CRISPR/Cas9 in two ways, by directly targeting conserved regions of the virus genome and by mutating relevant S-genes. Studies have focused on generating resistance against *Beet severe curly top virus* (BSCTV; genus *Curtovirus*, family *Geminiviridae*) and *Bean yellow dwarf virus* (BeYDV; genus *Mastrevirus*, family *Geminiviridae*) in *N. benthamiana* and *Arabidopsis*, respectively (Baltes et al. 2015; Ji et al. 2015). Both studies targeted coding as well as noncoding regulatory elements and found reduced virus titer at varying levels depending upon target and expression level of Cas9 and sgRNA—the higher the expression the better the resistance (Ji et al. 2015). *Tobacco rattle virus* (family *Virgaviridae*, genus *Tobravirus*) has been used as a vector for the delivery of sgRNAs into Cas9-expressing transgenic plants to target the replication-associated protein, coat protein (CP), and intergenic region (IR) of the begomoviruses *Tomato yellow leaf curl Sardinia virus* (TYLCSV) and C*otton leaf curl Khokhran virus* (CLCuKoV) (Ali et al. 2015a, b). All tested sgRNAs reduced the accumulation of virus in comparison to control plants but targeting the origin of viral virion-strand DNA replication in the IR region proved to be the most successful at interfering with virus replication. However, two studies also showed that the use of CRISPR/Cas-generated virus resistance can select for editing-resistant (resistance breaking) variants of the viruses (Ali et al. 2016; Mehta et al. 2019).

It has been observed that cleavage domains of Cas9 are not essential for suppression of virus in plants. Deactivating mutational ability of Cas9 results in mutated Cas9 (dCas9) that is still able to suppress replication and systemic movement of geminiviruses in the presence of IR-targeted sgRNA (Khan et al. 2019). Interestingly, the study suggests that mere binding of the Cas9 is enough to suppress virus accumulation and cleavage of the virus genome is not necessary.

Conventional Cas9 from *Streptococcus pyogenes* targets only DNA. However, variants of Cas9 from *Francisella novicida* (FnCas9) and Cas13a—another class II Cas nuclease—from *Leptotrichia wadei* (LwaCas13a) have been shown to recognize and cleave RNAs (Abudayyeh et al. 2017). FnCas9 has been shown to mediate RNA-guided RNA cleavage in *N. benthamiana* and *Arabidopsis*, suppressing the accumulation of the RNA viruses *Cucumber mosaic virus* and *Tobacco mosaic virus* by 40–80% in comparison to control plants (Zhang et al. 2018). Furthermore, it was noted that the catalytic activity of FnCas9 was not essential to attenuate virus infection, suggesting an interference activity due to the sgRNA-guided binding of Fncas9 to viral RNA. LwaCas12a was also employed for RNA-guided cleavage of *Turnip mosaic virus* (TuMV) in *N. benthamiana.* The study generated a recombinant virus by fusing TuMV with green fluorescent protein (GFP) to create a system for easily monitoring replication, infection, and spread of the virus (Aman et al. 2018). Better virus resistance was obtained with a crRNA targeting the helper-component proteinase-coding sequences than a crRNA targeting the CP-encoding sequences.

About 100 S-genes have been identified that support important steps in the virus cycle and confer resistance upon inactivation (Nicaise 2014; van Schie and Takken 2014). Most S-genes are eukaryotic translation initiation factors (eIF) 4E/4G or their isoforms (Ruffel et al. 2006; Abdul-Razzak et al. 2009). The eIF4Es were observed to confer loss of susceptibility in mutant *A. thaliana* against *Tobacco etch virus* (TEV), a *potyvirus* (Lellis et al. 2002), and crops such as tomato, pepper, and lettuce (Ruffel et al. 2002, 2005). RNA-dependent RNA polymerase has two allelic forms, Ty1 and Ty3. Their silencing using RNAi was shown to cause loss of susceptibility to the begomovirus *Tomato yellow leaf curl virus* (Verlaan et al. 2013). Synaptotagmins are well-characterized calcium sensors in animals. They were first thought to be exclusively present in animals. However, they have been identified in plants and found to have role in viral diseases. *Arabidopsis* synaptotagmin (SYTA) regulates endosome recycling. In case of virus infection, it interacts with the movement protein (MP) helping regulation of begomoviruses in *Cabbage leaf curl virus* (CaLCuV) and TMV. Inhibition in the systemic infection of CaLCuV and hindrance in the cell-to-cell movement of the movement protein of TMV and CaLCuV has been observed in SYTA knockdown lines (Lewis and Lazarowitz 2010).

15.9 CRISPR/Cas Mutation for Resistance to Fungi

Fungal pathogens are responsible for several diseases of cotton including rust, canker, rot, wilt, spot, anthracnose, and mildew (Wang et al. 2004; Lutfunnessa and Shamsi 2011). Along with causing yield losses, fungi may reduce the quality of the harvested products. Additionally, foods contaminated with fungal mycotoxins are a serious concern. Analyses of the interaction between plants and fungi uncovered several candidate genes for exploitation by CRISPR/Cas system to create resistance.

A classic example of an S-gene is the *MLO* gene. It confers susceptibility to powdery mildew in barley (Jorgensen 1992). Its role in susceptibility was confirmed in both dicotyledons (*Arabidopsis*, grape, tomato, and pea) and monocotyledons (rice and wheat) (Consonni et al. 2006; Bai et al. 2007). Loss-of-function mutant *mlo* confers broad-spectrum resistance to multiple strains of powdery mildew species *Golovinomyces orontii* and *G. chicoracerarum* (Consonni et al. 2006) and the bacterium *Xanthomonas campestris* in *A. thaliana* (Kim and Hwang 2012). Recently all three homeoalleles were targeted by CRISPR/Cas9 in wheat (Wang et al. 2014). Transgene-free *mlo* mutant grape was created by protoplast transformation of in vitro translated Cas9/sgRNA complex (Malnoy et al. 2016). Although mutations were detected, plants were not regenerated and a resistance assay was not performed. However, a similar study reinforced the idea by demonstrating decreased susceptibility in RNAi-mediated silenced *MLO* against *Erysiphe necator* in grape (Pessina et al. 2016). Similarly, non-transgenic *MLO*-mutated tomato plants were produced by using two sgRNAs spaced 42 bp apart to delete a segment of the Mlo gene followed by segregation and selection of T-DNA-free progeny. Furthermore, no off-target effects were noted. The new non-transgenic variety was named "Tomelo" and is resistant to *Oidium neolycopersici* (Nekrasov et al. 2017).

Verticillium wilt is a cotton vascular fungal disease caused by *Verticillium dahliae*, limiting cotton yield and quality. To understand the disease, a transcriptomics study was conducted that revealed downregulation of HDTF1 in cotton upon exposure to the pathogen, suggesting a role in disease susceptibility. HDTF1 encodes a transcription factor that putatively downregulates the jasmonic acid (JA) pathway. Partial suppression of HDTF1 by virus-induced gene silencing resulted in resistance against *V. dahlia* and *Botrytis cinerea*. Furthermore, it was revealed that JA accumulated in HDTF1 suppressed plants (Gao et al. 2016).

In another study grape (*Vitis vinifera*) resistance was obtained by mutating a WRKY transcription factor. Four sgRNAs were used and both monoallelic and biallelic mutants were regenerated. Regenerants were tested against *Botrytis cinerea*. All mutants expressed resistance, but the biallelic mutants showed stronger resistance in comparison to monoallelic mutants (Wang et al. 2018).

Resistance to rice blast, caused by *Magnaporthe oryzae*, was developed by CRISPR/Cas9 loss-of-function disruption of the OsERF922 and OsSEC3A genes in rice (Wang, Wang et al. 2016; Ma et al. 2018a). OsERF922 is a transcription factor implicated in multiple stress responses, whereas OsSEC3A is a part of the exocyst complex involved in morphogenesis. Mutation of OsERF922 resulted in resistance to blast without negative pleiotropic effects on plant growth. However, Ossec3a mutant plants showed pleiotropic effects along with resistance to *M. oryzae*. Mutant plants showed higher levels of salicylic acid (SA) and upregulation of pathogenesis- and SA-related genes but also a dwarf stature (Ma et al. 2018b). Altogether, CRISPR/Cas-mediated genome editing is a useful approach to expedite the generation of disease resistance and allows unprecedented control over previously random gene transfer methods.

Methyl jasmonate (MeJA) plays a crucial role in plant defense against pathogenic microorganisms. Downstream molecular mechanism of this defense response is better characterized than the upstream molecular signaling. MYC2 was found to be critical in jasmonic acid (JA) signaling in case of saprotrophic infection. The knockdown mutants of simyc2 were observed to reduce disease resistance against *Botrytis cinerea* in tomato plants. The finding suggests a potential role of MYC2 as positive regulator of resistance against fungal pathogens. Overexpression or stacking of MYC2 may provide enhanced resistance in crop plants against saprotrophic pathogens (Jiang et al. 2015; Wang et al. 2015; Glowacz et al. 2017; Shu et al. 2020).

Several *Arabidopsis* mutants (powdery mildew resistant 1–powdery mildew resistant 4) have shown resistance against a variety of fungal pathogens that include *Golovinomyces cichoracearum*, downy mildew, and *G. orontii*. PMR4 homologue silencing in tomato resulted in enhanced resistance against powdery mildew as compared to control plants. Similarly, CRISPR/Cas9-mediated gene knockout of PMR4 resulted in enhanced resistance against powdery mildew disease (Santillán Martínez et al. 2020).

15.10 CRISPR/Cas Mutation for Resistance to Bacteria

Several studies have shown the use of CRISPR/Cas9 system for control of bacterial diseases. Bacterial blight is a major threat to many economically important crops. In a series of studies, it was determined that SWEET (Sugars will eventually be exported transporters) genes act as susceptibility genes in plants. SWEET genes encode sucrose transporters and are upregulated by the bacterial effector protein PthXo2 of *Xanthomonas oryzae* pv. *oryzae*, contributing to pathogen virulence (Zhou et al. 2015). The TALEN approach was applied to mutate the promoter of OsSWEET14, and the effector was found to be unproductive in upregulating the mutated promoter resulting in resistance (Li et al. 2012). Similarly knockout mutations of OsSWEET13 in rice produced using CRISPR/Cas9 resulted in resistance against bacterial blight. The resistance was observed without any yield penalty as compared to nonmutant control plants (Zhou et al. 2015; Zeng et al. 2020).

Citrus canker, caused by *Xanthomonas citri* subsp. *citri*, is a serious problem in the citrus industry. Similar to bacterial blight, the effector PthA4 is produced by the *Xanthomonas citri* subsp. *citri* and binds the promoter region of the susceptibility gene *Lateral organ boundaries* (CsLOB1) gene leading to upregulation. CRISPR/Cas9-mediated mutation of the promoter region of the CsLOB1 gene in grapefruit rendered the plant resistant to canker by preventing the binding of the effector protein (Peng et al. 2017). The concept was further bolstered by deleting a portion of the effector-binding elements from both alleles of CsLOB1 of grapefruit resulting in plants with resistance to canker (Peng et al. 2017).

15.11 CRISPR-Mediated Resistance: S-Genes vs R-Genes

Genome-editing technology for disease resistance has focused on mutating host S-genes rather than stacking R-genes. Several transgenic approaches have stacked multiple R-genes to confer pathogen resistance (Zhu et al. 2012; Zhang et al. 2014). Before concluding which one is better, it is important to understand the mechanism of resistance conferred and ease of methodology to achieve the desired results. Whereas R-gene-mediated resistance is inherited in a dominant manner and triggers resistance upon recognition of a pathogen-derived avirulence factor, S-gene-mediated resistance is inherited in recessive manner. When in recessive genetic makeup, S-genes halt a fundamental step in the life cycle of the pathogen as demonstrated in Fig. 15.3. Simple point mutations in R-genes can incapacitate R-gene recognition of the avirulence factor, enabling the pathogen to evade the resistance and establish disease. S-genes, in contrast, are already nonfunctional. Similarly, the high mutation rate of pathogens can modify the avirulence factor, allowing the pathogen to evade R-gene-mediated recognition much more easily than S-gene-mediated resistance. Pathogens often release multiple effectors which are mostly encoded by rapidly mutating genomic locations (Rep and Kistler 2010; Ravensdale et al. 2011). The durability of R-gene resistance depends upon the fitness costs of losing an effector. Conserved effectors recognizing R-genes provide more durable resistance because of a high fitness penalty (Leach et al. 2001; McDonald and Linde 2002). To circumvent S-gene-mediated resistance, pathogens have to overcome dependence on host factors that can be achieved by acquiring a new function or developing several alternate strategies, which is difficult to achieve. It is likely that

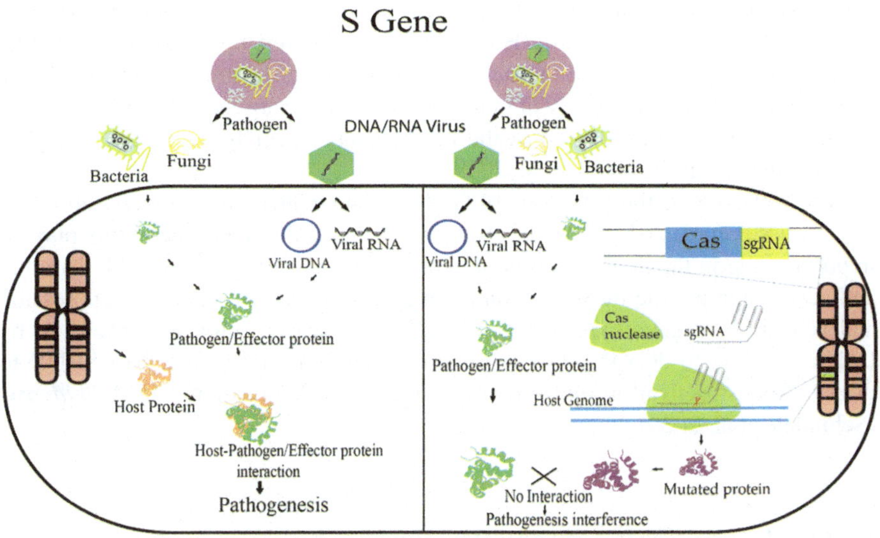

Fig. 15.3 S-gene loss-of-function mutation-mediated resistance against biotic factors. On the left, a host factor interacts with a pathogen protein to establish the disease. On the right, mutation of the corresponding host factor prevents the interaction and causes disease interference

S-gene-mediated resistance will provide more durable resistance than R-gene-based resistance. However, the fitness costs associated with S-gene mutations may limit their use. A combination of both R- and S-genes will be a promising avenue to investigate.

15.12 Beyond Gene Editing

The capabilities of CRISPR/Cas9 go beyond GT and knockout and typically involve either nCas9 or deactivated Cas9 (dCas9). nCas9, a mutant Cas9 with ability to cleave single-stranded DNA (ssDNA), is used for base editing in plant breeding, while dCas9, a mutant Cas9 that is unable to cleave ssDNA, is used for reprogramming gene expression. The versatility of nCas9 and dCas9 is reflected in the fact that both can act as a scaffold for sgRNA-mediated targeting of other proteins or enzymes (Bortesi and Fischer 2015).

Cytidine deaminase or adenine deaminase fused to nCas9 causes transitional mutation of cytosine to thymine and guanine to adenine, respectively. The base-editing ability of nCas9 has been observed in many crop plants such as rice, wheat, and maize. The transition was observed within 3–9 nucleotide bases of the protospacer in both protoplasts and transgenic plants. The point mutation is attributed to many physiologically important traits as demonstrated in many genome-wide association studies (Henikoff and Comai 2003; Zhao et al. 2011). In rice, the simultaneous conversion of C/G to T/A was observed by the introduction of both cytidine and adenine deaminase (Hua et al. 2019). Interestingly, genome-wide off-target mutations are revealed in rice. The majority of these mutations were C to T conversion. The high frequency of mutations, even in the absence of sgRNA, reveals a need for further optimization of the mechanism, especially for cytidine deaminase-mediated transition (Jin et al. 2019).

A dCas9 expands the versatility of Cas9 to gene regulation or fluorescent imaging (Dreissig et al. 2017; Lowder et al. 2015). Several epigenome marks play an important role in the determination of expression status of chromatin. Methylation and acetylation are the most important epigenetic marks and result in suppression and activation of gene expression, respectively. The activation is carried out by fusion of VP64 to dCas9, while repression is carried out by fusion of SRDX to dCas9. Robust activation and repression has been demonstrated in *Arabidopsis* (Schindele et al. 2020).

15.13 Large-Scale Genome Restructuring

Large-scale deletion, inversion, translocation, and recombination provide an additional source of variability and biodiversity for plant breeders. Several studies have shown that site-directed double-strand breaks can induce large-scale deletions, inversions, and translocations (Siebert and Puchta 2002; Pacher et al. 2007). Recently, CRISPR/Cas9 has been employed for induction of large-scale deletions, ranging from 13 to 120 kb using tissue-specific expression of Cas9 in *Arabidopsis*, and small deletions of less than 100 bps in *N. benthamiana* (Ordon et al. 2017; Durr et al. 2018). Inversions are particularly important as it leaves the homologous chromosomes inaccessible for homologous recombination. The obstacle in homologous recombination can establish trait linkage in plants.

15.14 Multiplexing: A Tool for Accelerating Genetic Diversity

One of the major potential advantages of CRISPR/Cas9 in modern breeding programs is its multiplexing capability. Multiplexing can massively accelerate the breeding process, can enhance the biodiversity, and can produce outcomes difficult to attain using classical breeding approaches.

CRISPR/Cas9 offers two alternative approaches to enhance genetic diversity: (1) targeting of different homoalleles using sgRNA and (2) using multiple sgRNAs for targeting unrelated sequences within a genome. The reports have demonstrated mutations at multiple target sites in both model plants and crops plants (Mercx et al. 2017; Li et al. 2018b).

Three genes involved in anthocyanin metabolism were targeted in purple leaf rice lines to generate white leaf rice mutants. Similarly, four mutations were observed in the GABA metabolic pathway genes targeting five genes (Ma et al. 2015; Li et al. 2018b). One gene was left un-mutated perhaps due to high GC content for the gene. The same group targeted multiple genes in the carotenoid metabolic pathway of tomato to simultaneously mutate multiple genes using multiplexed CRISPR/Cas9 (Li et al. 2018a). Recently, up to eight physiologically important rice genes have been simultaneously mutated, generating homozygous as well as heterozygous mutated lines (Shen et al. 2017). The recent success in multiplexing multiple traits in polyploidy plants offers unprecedented opportunities for trait improvement and enhancing genetic diversity for use in modern breeding programs.

15.15 Conclusion

Several genome-editing tools have been developed for genetic manipulation. The advent of CRISPR/Cas9 technology has expedited the approach to explore the capacities of genome-editing tools. The major applications of genome-editing tools in plant-pathogen interaction are site-specific gene targeting and site-specific random mutation. The first is suitable for R-gene stacking by HDR, whereas the second paves the way to the knockout of S-genes by NHEJ. The inactivation of S-genes is preferred over the addition of R-genes for two reasons:

- HDR-mediated site-specific gene targeting requires expertise and is less efficient in plants in comparison to NHEJ-mediated gene knockdown.
- Due to the high mutation rate of pathogens, it is easy for them to circumvent R-gene-mediated resistance. To overcome an S-gene-mediated resistance requires pathogen to acquire a new function that is comparatively difficult.

Many S-gene mutations will suffer fitness costs, and screening of new S-genes with minimal fitness costs for a particular pathogen can be a tedious task. However, a similar challenge is also faced in exploring effective R-genes in a specific crop. In view of the technical demands and future benefits, S-genes offer a way forward. However, reinforcing plant immunity by the addition of an effective R-gene can enhance resistance in plants.

References

Abdul-Razzak A, Guiraud T, Peypelut M, Walter J, Houvenaghel MC, Candresse T, Gall OLE, German-Retana S (2009) Involvement of the cylindrical inclusion (CI) protein in the overcoming of an eIF4E-mediated resistance against Lettuce mosaic potyvirus. *Mol Plant Pathol* 10(1):109–113

Abudayyeh OO, Gootenberg JS, Essletzbichler P, Han S, Joung J, Belanto JJ, Verdine V, Cox DBT, Kellner MJ, Regev A, Lander ES, Voytas DF, Ting AY, Zhang F (2017) RNA targeting with CRISPR-Cas13. *Nature* 550(7675):280–284

Acevedo-Garcia J, Spencer D, Thieron H, Reinstädler A, Hammond-Kosack K, Phillips AL, Panstruga R (2017) mlo-based powdery mildew resistance in hexaploid bread wheat generated by a non-transgenic TILLING approach. *Plant Biotechnol J* 15(3):367–378

Ahloowalia BS, Maluszynski M, Nichterlein K (2004) Global impact of mutation-derived varieties. *Euphytica* 135(2):187–204

Ali Z, Abul-faraj A, Piatek M, Mahfouz MM (2015a) Activity and specificity of TRV-mediated gene editing in plants. *Plant Signal Behav* 10(10):e1044191

Ali Z, Abulfaraj A, Idris A, Ali S, Tashkandi M, Mahfouz M (2015b) CRISPR/Cas9-mediated viral interference in plants. *Genome Biol* 16(1):238

Ali Z, Ali S, Tashkandi M, Zaidi S S-e-A, Mahfouz MM (2016) CRISPR/Cas9-mediated immunity to geminiviruses: differential interference and evasion. *Sci Rep* 6:26912

Aman R, Ali Z, Butt H, Mahas A, Aljedaani F, Khan MZ, Ding S, Mahfouz M (2018) RNA virus interference via CRISPR/Cas13a system in plants. *Genome Biol* 19(1):1

Amitai G, Sorek R (2016) CRISPR-Cas adaptation: insights into the mechanism of action. *Nat Rev Microbiol* 14(2):67–76

Antunes MS, Smith JJ, Jantz D, Medford JI (2012) Targeted DNA excision in Arabidopsis by a re-engineered homing endonuclease. *BMC Biotechnol* 12(1):86

Bai Y, Pavan S, Zheng Z, Zappel NF, Reinstädler A, Lotti C, De Giovanni C, Ricciardi L, Lindhout P, Visser R, Theres K, Panstruga R (2007) Naturally occurring broad-spectrum powdery mildew resistance in a central american tomato accession is caused by loss of Mlo function. *Mol Plant-Microbe Interact* 21(1):30–39

Baltes NJ, Voytas DF (2015) Enabling plant synthetic biology through genome engineering. *Trends Biotechnol* 33(2):120–131

Baltes NJ, Hummel AW, Konecna E, Cegan R, Bruns AN, Bisaro DM, Voytas DF (2015) Conferring resistance to geminiviruses with the CRISPR-Cas prokaryotic immune system. *Nat Plants* 1:15145

Barrangou R (2013) CRISPR-Cas systems and RNA-guided interference. *Wiley Interdiscip Rev RNA* 4(3):267–278

Barrangou R, Fremaux C, Deveau H l n, Richards M, Boyaval P, Moineau S, Romero DA, Horvath P (2007) CRISPR provides acquired resistance against viruses in prokaryotes. *Science* 315(5819):1709–1712

Belfort M, Roberts RJ (1997) Homing endonucleases: keeping the house in order. *Nucleic Acids Res* 25(17):3379–3388

Bikard D, Hatoum-Aslan A, Mucida D, Marraffini LA (2012) CRISPR interference can prevent natural transformation and virulence acquisition during in vivo bacterial infection. *Cell Host Microbe* 12(2):177–186

Boch J, Scholze H, Schornack S, Landgraf A, Hahn S, Kay S, Lahaye T, Nickstadt A, Bonas U (2009) Breaking the code of DNA binding specificity of TAL-Type III effectors. *Science* 326(5959):1509

Boller T, Felix G (2009) A renaissance of elicitors: perception of microbe-associated molecular patterns and danger signals by pattern-recognition receptors. *Annu Rev Plant Biol* 60(1):379–406

Bortesi L, Fischer R (2015) The CRISPR/Cas9 system for plant genome editing and beyond. *Biotechnol Adv* 33(1):41–52

Brown JK, Perring TM, Cooper AD, Bedford ID, Markham PG (2000) Genetic analysis of Bemisia (Hemiptera: Aleyrodidae) populations by isoelectric focusing electrophoresis. *Biochem Genet* 38(1):13–25

Brown JK, Zerbini FM, Navas-Castillo J, Moriones E, Ramos-Sobrinho R, Silva JCF, Fiallo-Olivé E, Briddon RW, Hernández-Zepeda C, Idris A, Malathi VG, Martin DP, Rivera-Bustamante R, Ueda S, Varsani A (2015) Revision of Begomovirus taxonomy based on pairwise sequence comparisons. *Arch Virol* 160(6):1593–1619

Capecchi MR (1989) Altering the genome by homologous recombination. *Science* 244(4910):1288

Carroll D (2011) Genome engineering with zinc-finger nucleases. *Genetics* 188(4):773–782

Cermak T, Doyle EL, Christian M, Wang L, Zhang Y, Schmidt C, Baller JA, Somia NV, Bogdanove AJ, Voytas DF (2011) Efficient design and assembly of custom TALEN and other TAL effector-based constructs for DNA targeting. *Nucleic Acids Res* 39(12):e82–e82

Chen K, Wang Y, Zhang R, Zhang H, Gao C (2019) CRISPR/Cas genome editing and precision plant breeding in agriculture. *Annu Rev Plant Biol* 70(1):667–697

Chiurazzi M, Ray A, Viret JF, Perera R, Wang XH, Lloyd AM, Signer ER (1996) Enhancement of somatic intrachromosomal homologous recombination in Arabidopsis by the HO endonuclease. *Plant Cell* 8(11):2057

Consonni C, Humphry ME, Hartmann HA, Livaja M, Durner J, Westphal L, Vogel J, Lipka V, Kemmerling B, Schulze-Lefert P, Somerville SC, Panstruga R (2006) Conserved requirement for a plant host cell protein in powdery mildew pathogenesis. *Nat Genet* 38:716

D'Halluin K, Vanderstraeten C, Van Hulle J, Rosolowska J, Van Den Brande I, Pennewaert A, D'Hont K, Bossut M, Jantz D, Ruiter R, Broadhvest J (2013) Targeted molecular trait stacking in cotton through targeted double-strand break induction. *Plant Biotechnol J* 11(8):933–941

Doetschman T, Gregg RG, Maeda N, Hooper ML, Melton DW, Thompson S, Smithies O (1987) Targeted correction of a mutant HPRT gene in mouse embryonic stem cells. *Nature* 330(6148):576–578

Dreissig S, Schiml S, Schindele P, Weiss O, Rutten T, Schubert V, Gladilin E, Mette MF, Puchta H, Houben A (2017) Live-cell CRISPR imaging in plants reveals dynamic telomere movements. *Plant J* 91(4):565–573

Durr J, Papareddy R, Nakajima K, Gutierrez-Marcos J (2018) Highly efficient heritable targeted deletions of gene clusters and non-coding regulatory regions in Arabidopsis using CRISPR/Cas9. *Sci Rep* 8(1):1–11

Epinat J-C, Arnould S, Chames P, Rochaix P, Desfontaines D, Puzin C, Patin A, Zanghellini A, Paques F, Lacroix E (2003) A novel engineered meganuclease induces homologous recombination in yeast and mammalian cells. *Nucleic Acids Res* 31(11):2952–2962

Esvelt KM, Wang HH (2013) Genome-scale engineering for systems and synthetic biology. *Mol Syst Biol* 9(1):641

Gaj T, Gersbach CA, Barbas Iii CF (2013) ZFN, TALEN, and CRISPR/Cas-based methods for genome engineering. *Trends Biotechnol* 31(7):397–405

Gao H, Smith J, Yang M, Jones S, Djukanovic V, Nicholson MG, West A, Bidney D, Falco SC, Jantz D, Lyznik LA (2010) Heritable targeted mutagenesis in maize using a designed endonuclease. *Plant J* 61(1):176–187

Gao W, Long L, Xu L, Lindsey K, Zhang X, Zhu L (2016) Suppression of the homeobox gene HDTF1 enhances resistance to Verticillium dahliae and Botrytis cinerea in cotton. *J Integr Plant Biol* 58(5):503–513

Glowacz M, Roets N, Sivakumar D (2017) Control of anthracnose disease via increased activity of defence related enzymes in 'Hass' avocado fruit treated with methyl jasmonate and methyl salicylate. *Food Chem* 234:163–167

Harrison MM, Jenkins BV, O'Connor-Giles KM, Wildonger J (2014) A CRISPR view of development. *Genes Dev* 28(17):1859–1872

Hartung F, Schiemann J (2014) Precise plant breeding using new genome editing techniques: opportunities, safety and regulation in the EU. *Plant J* 78(5):742–752

Henikoff S, Comai L (2003) Single-nucleotide mutations for plant functional genomics. *Annu Rev Plant Biol* 54(1):375–401

Holkers M, Maggio I, Liu J, Janssen JM, Miselli F, Mussolino C, Recchia A, Cathomen T, Goncalves MAFV (2013) Differential integrity of TALE nuclease genes following adenoviral and lentiviral vector gene transfer into human cells. *Nucleic Acids Res* 41(5):e63–e63

Hua K, Tao X, Zhu J-K (2019) Expanding the base editing scope in rice by using Cas9 variants. *Plant Biotechnol J* 17(2):499–504

Ji X, Zhang H, Zhang Y, Wang Y, Gao C (2015) Establishing a CRISPR-Cas-like immune system conferring DNA virus resistance in plants. *Nat Plants* 1:15144

Jiang L, Jin P, Wang L, Yu X, Wang H, Zheng Y (2015) Methyl jasmonate primes defense responses against Botrytis cinerea and reduces disease development in harvested table grapes. *Sci Hortic* 192:218–223

Jin S, Zong Y, Gao Q, Zhu Z, Wang Y, Qin P, Liang C, Wang D, Qiu J-L, Zhang F, Gao C (2019) Cytosine, but not adenine, base editors induce genome-wide off-target mutations in rice. *Science* 364(6437):292

Jinek M, Chylinski K, Fonfara I, Hauer M, Doudna JA, Charpentier E (2012) A programmable dual-RNA-guided DNA endonuclease in adaptive bacterial immunity. *Science (New York, NY)* 337(6096):816–821

Jones JD, Dangl JL (2006) The plant immune system. *Nature* 444(7117):323–329

Jorgensen IH (1992) Discovery, characterization and exploitation of Mlo powdery mildew resistance in barley. *Euphytica* 63(1):141–152

Joung JK, Sander JD (2013) TALENs: a widely applicable technology for targeted genome editing. *Nat Rev Mol Cell Biol* 14(1):49–55

Khan Z, Khan SH, Ahmad A, Aslam S, Mubarik MS, Khan S (2019) CRISPR/dCas9-mediated inhibition of replication of begomoviruses. *Int J Agric Biol* 21:711–718

Kim DS, Hwang BK (2012) The pepper MLO gene, CaMLO2, is involved in the susceptibility cell-death response and bacterial and oomycete proliferation. *Plant J* 72(5):843–855

Kim YG, Cha J, Chandrasegaran S (1996) Hybrid restriction enzymes: zinc finger fusions to Fok I cleavage domain. *Proc Natl Acad Sci U S A* 93(3):1156–1160

Knoll A, Fauser F, Puchta H (2014) DNA recombination in somatic plant cells: mechanisms and evolutionary consequences. *Chromosom Res* 22(2):191–201

Langner T, Kamoun S, Belhaj K (2018) CRISPR crops: plant genome editing toward disease resistance. *Annu Rev Phytopathol* 56(1):479–512

Leach JE, Vera Cruz CM, Bai J, Leung H (2001) Pathogen fitness penalty as a predictor of durability of disease resistance genes. *Annu Rev Phytopathol* 39(1):187–224

Li T, Liu B, Spalding MH, Weeks DP, Yang B (2012) High-efficiency TALEN-based gene editing produces disease-resistant rice. *Nat Biotechnol* 30:390

Li X, Wang Y, Chen S, Tian H, Fu D, Zhu B, Luo Y, Zhu H (2018a) Lycopene is enriched in tomato fruit by CRISPR/Cas9-mediated multiplex genome editing. *Front Plant Sci* 9:559

Li R, Li R, Li X, Fu D, Zhu B, Tian H, Luo Y, Zhu H (2018b) Multiplexed CRISPR/Cas9-mediated metabolic engineering of γ-aminobutyric acid levels in Solanum lycopersicum. *Plant Biotechnol J* 16(2):415–427

Lellis AD, Kasschau KD, Whitham SA, Carrington JC (2002) Loss-of-susceptibility mutants of *Arabidopsis thaliana* reveal an essential role for eIF(iso)4E during potyvirus infection. *Curr Biol.* 12(12):1046–1051

Lewis JD, Lazarowitz SG (2010) *Arabidopsis synaptotagmin* SYTA regulates endocytosis and virus movement protein cell-to-cell transport. *Proc Natl Acad Sci U S A* 107(6):2491–2496

Lowder LG, Zhang D, Baltes NJ, Paul JW, Tang X, Zheng X, Voytas DF, Hsieh T-F, Zhang Y, Qi Y (2015) A CRISPR/Cas9 toolbox for multiplexed plant genome editing and transcriptional regulation. *Plant Physiol* 169(2):971–985

Lutfunnessa RJF, Shamsi S (2011) Fungal diseases of cotton plant Gossypium hirsutum L. in Bangladesh. *Dhaka Univ J Biol Sci* 20(2):139–146

Ma X, Zhang Q, Zhu Q, Liu W, Chen Y, Qiu R, Wang B, Yang Z, Li H, Lin Y, Xie Y, Shen R, Chen S, Wang Z, Chen Y, Guo J, Chen L, Zhao X, Dong Z, Liu Y-G (2015) A Robust CRISPR/Cas9 system for convenient, high-efficiency multiplex genome editing in monocot and dicot plants. *Mol Plant* 8(8):1274–1284

Ma J, Sodmergen CL, Chen J, Wang M, Wang S, Ren Y, Cheng Z (2018a) Corrigendum: disruption of OsSEC3A increases the content of salicylic acid and induces plant defense responses in rice. *J Exp Bot* 69(7):1817–1817

Ma J, Chen J, Wang M, Ren Y, Wang S, Lei C, Cheng Z, Sodmergen (2018b) Disruption of OsSEC3A increases the content of salicylic acid and induces plant defense responses in rice. *J Exp Bot* 69(5):1051–1064

Macho AP, Zipfel C (2014) Plant PRRs and the activation of innate immune signaling. *Mol Cell* 54(2):263–272

Macovei A, Sevilla NR, Cantos C, Jonson GB, Slamet-Loedin I, Čermák T, Voytas DF, Choi I-R, Chadha-Mohanty P (2018) Novel alleles of rice eIF4G generated by CRISPR/Cas9-targeted mutagenesis confer resistance to Rice tungro spherical virus. *Plant Biotechnol J* 16(11):1918–1927

Mahfouz MM, Li L, Shamimuzzaman M, Wibowo A, Fang X, Zhu J-K (2011) De novo-engineered transcription activator-like effector (TALE) hybrid nuclease with novel DNA binding specificity creates double-strand breaks. *Proc Natl Acad Sci U S A* 108(6):2623–2628

Mali P, Esvelt KM, Church GM (2013) Cas9 as a versatile tool for engineering biology. *Nat Methods* 10(10):957–963

Malnoy M, Viola R, Jung M-H, Koo O-J, Kim S, Kim J-S, Velasco R, Nagamangala Kanchiswamy C (2016) DNA-free genetically edited grapevine and apple protoplast using CRISPR/Cas9 ribonucleoproteins. *Front Plant Sci* 7:1904–1904

McDonald BA, Linde C (2002) Pathogen population genetics, evolutionary potential, and durable resistance. *Annu Rev Phytopathol* 40(1):349–379

Mehta D, Stürchler A, Anjanappa RB, Zaidi SS-EA, Hirsch-Hoffmann M, Gruissem W, Vanderschuren H (2019) Linking CRISPR-Cas9 interference in cassava to the evolution of editing-resistant geminiviruses. *Genome Biol* 20(1):80–80

Mercx S, Smargiasso N, Chaumont F, De Pauw E, Boutry M, Navarre C (2017) Inactivation of the β (1, 2)-xylosyltransferase and the α (1, 3)-fucosyltransferase genes in Nicotiana tabacum BY-2 cells by a multiplex CRISPR/Cas9 strategy results in glycoproteins without plant-specific glycans. *Front Plant Sci* 8:403

Moreno-Mateos MA, Fernandez JP, Rouet R, Vejnar CE, Lane MA, Mis E, Khokha MK, Doudna JA, Giraldez AJ (2017) CRISPR-Cpf1 mediates efficient homology-directed repair and temperature-controlled genome editing. *Nat Commun* 8(1):2024

Nekrasov V, Wang C, Win J, Lanz C, Weigel D, Kamoun S (2017) Rapid generation of a transgene-free powdery mildew resistant tomato by genome deletion. *Sci Rep* 7(1):482

Nicaise V (2014) Crop immunity against viruses: outcomes and future challenges. *Front Plant Sci* 5:660

Nicolas A, Treco D, Schultes NP, Szostak JW (1989) An initiation site for meiotic gene conversion in the yeast Saccharomyces cerevisiae. *Nature* 338(6210):35–39

Offringa R, de Groot MJ, Haagsman HJ, Does MP, van den Elzen PJ, Hooykaas PJ (1990) Extrachromosomal homologous recombination and gene targeting in plant cells after Agrobacterium mediated transformation. *EMBO J* 9(10):3077–3084

Ordon J, Gantner J, Kemna J, Schwalgun L, Reschke M, Streubel J, Boch J, Stuttmann J (2017) Generation of chromosomal deletions in dicotyledonous plants employing a user-friendly genome editing toolkit. *Plant J* 89(1):155–168

Pacher M, Schmidt-Puchta W, Puchta H (2007) Two unlinked double-strand breaks can induce reciprocal exchanges in plant genomes via homologous recombination and nonhomologous end joining. *Genetics* 175(1):21

Pâques F, Haber JE (1999) Multiple pathways of recombination induced by double-strand breaks in Saccharomyces cerevisiae. *Microbiol Mol Biol Rev* 63(2):349–404

Paszkowski J, Baur M, Bogucki A, Potrykus I (1988) Gene targeting in plants. *EMBO J* 7(13):4021–4026

Pavan S, Jacobsen E, Visser RGF, Bai Y (2009) Loss of susceptibility as a novel breeding strategy for durable and broad-spectrum resistance. *Mol Breed* 25(1):1

Peng A, Chen S, Lei T, Xu L, He Y, Wu L, Yao L, Zou X (2017) Engineering canker-resistant plants through CRISPR/Cas9-targeted editing of the susceptibility gene CsLOB1 promoter in citrus. *Plant Biotechnol J* 15(12):1509–1519

Perring TM (2001) The Bemisia tabaci species complex. *Crop Prot* 20(9):725–737

Pessina S, Lenzi L, Perazzolli M, Campa M, Dalla Costa L, Urso S, Valè G, Salamini F, Velasco R, Malnoy M (2016) Knockdown of MLO genes reduces susceptibility to powdery mildew in grapevine. *Hortic Res* 3:16016

Polston JE, De Barro P, Boykin LM (2014) Transmission specificities of plant viruses with the newly identified species of the Bemisia tabaci species complex. *Pest Manag Sci* 70(10):1547–1552

Puchta H (2002) Gene replacement by homologous recombination in plants. *Plant Mol Biol* 48(1):173–182

Puchta H (2004) The repair of double-strand breaks in plants: mechanisms and consequences for genome evolution. *J Exp Bot* 56(409):1–14

Puchta H, Fauser F (2013) Gene targeting in plants: 25 years later. *Int J Dev Biol* 57:629–637

Puchta H, Fauser F (2014) Synthetic nucleases for genome engineering in plants: prospects for a bright future. *Plant J* 78(5):727–741

Puchta H, Dujon B, Hohn B (1993) Homologous recombination in plant cells is enhanced by in vivo induction of double strand breaks into DNA by a site-specific endonuclease. *Nucleic Acids Res* 21(22):5034–5040

Puchta H, Dujon B, Hohn B (1996) Two different but related mechanisms are used in plants for the repair of genomic double-strand breaks by homologous recombination. *Proc Natl Acad Sci U S A* 93(10):5055–5060

Qi Y, Zhang Y, Zhang F, Baller JA, Cleland SC, Ryu Y, Starker CG, Voytas DF (2013) Increasing frequencies of site-specific mutagenesis and gene targeting in Arabidopsis by manipulating DNA repair pathways. *Genome Res* 23(3):547–554

Ramirez CL, Foley JE, Wright DA, Muller-Lerch F, Rahman SH, Cornu TI, Winfrey RJ, Sander JD, Fu F, Townsend JA, Cathomen T, Voytas DF, Joung JK (2008) Unexpected failure rates for modular assembly of engineered zinc fingers. *Nat Methods* 5(5):374–375

Ravensdale M, Nemri A, Thrall PH, Ellis JG, Dodds PN (2011) Co-evolutionary interactions between host resistance and pathogen effector genes in flax rust disease. *Mol Plant Pathol* 12(1):93–102

Reddall A, Ali A, Able JA, Stonor J, Tesoriero L, Wright PR, Rezaian MA, Wilson LJ (2004) Cotton bunchy top: an aphid and graft transmitted cotton disease. *Australas Plant Pathol* 33(2):197–202

Ruffel S, Dussault MH Palloix A, Moury B, Bendahmane A, Robaglia C, Caranta C (2002) A natural recessive resistance gene against potato virus Y in pepper corresponds to the eukaryotic initiation factor 4E (eIF4E). *Plant J* 32(6):1067–1075

Ruffel S., Gallois JL, Lesage ML, Caranta C (2005) The recessive potyvirus resistance gene pot-1 is the tomato orthologue of the pepper pvr2-eIF4E gene. *Mol Genet Genomics*: 274(4):346–353

Ruffel S, Gallois JL, Moury B, Robaglia C, Palloix A, Caranta C (2006) Simultaneous mutations in translation initiation factors eIF4E and eIF(iso)4E are required to prevent pepper veinal mottle virus infection of pepper. *J Gen Virol* 87(Pt 7):2089–2098

Reiss B, Schubert I, Köpchen K, Wendeler E, Schell J, Puchta H (2000) RecA stimulates sister chromatid exchange and the fidelity of double-strand break repair, but not gene targeting, in plants transformed by Agrobacterium. *Proc Natl Acad Sci U S A* 97(7):3358–3363

Rep M, Kistler HC (2010) The genomic organization of plant pathogenicity in Fusarium species. *Curr Opin Plant Biol* 13(4):420–426

Sander JD, Joung JK (2014) CRISPR-Cas systems for editing, regulating and targeting genomes. *Nat Biotechnol* 32(4):347–355

Santillán Martínez MI, Bracuto V, Koseoglou E, Appiano M, Jacobsen E, Visser RGF, Wolters A-MA, Bai Y (2020) CRISPR/Cas9-targeted mutagenesis of the tomato susceptibility gene PMR4 for resistance against powdery mildew. *BMC Plant Biol* 20(1):284–284

Schindele A, Dorn A, Puchta H (2020) CRISPR/Cas brings plant biology and breeding into the fast lane. *Curr Opin Biotechnol* 61:7–14

Shen L, Hua Y, Fu Y, Li J, Liu Q, Jiao X, Xin G, Wang J, Wang X, Yan C, Wang K (2017) Rapid generation of genetic diversity by multiplex CRISPR/Cas9 genome editing in rice. *Sci China Life Sci* 60(5):506–515

Shu P, Li Z, Min D, Zhang X, Ai W, Li J, Zhou J, Li Z, Li F, Li X (2020) CRISPR/Cas9-mediated SlMYC2 mutagenesis adverse to tomato plant growth and MeJA-induced fruit resistance to Botrytis cinerea. *J Agric Food Chem* 68(20):5529–5538

Siebert R, Puchta H (2002) Efficient repair of genomic double-strand breaks by homologous recombination between directly repeated sequences in the plant genome. *Plant Cell* 14(5):1121

Siegemund F (2000) Mutation breeding. Theory and practical applications. Cambridge: Cambridge University Press 353 pages, 18 figures, 5 tables; $ 120.00 ISBN 0-85404-750-6. *Acta Biotechnol* 20(2):97–98

Silva G, Poirot L, Galetto R, Smith J, Montoya G, Duchateau P, Paques F (2011) Meganucleases and other tools for targeted genome engineering: perspectives and challenges for gene therapy. *Curr Gene Ther* 11(1):11–27

St. Clair DA (2010) Quantitative disease resistance and quantitative resistance loci in breeding. *Annu Rev Phytopathol* 48(1):247–268

Sun Y, Zhang X, Wu C, He Y, Ma Y, Hou H, Guo X, Du W, Zhao Y, Xia L (2016) Engineering herbicide-resistant rice plants through CRISPR/Cas9-mediated homologous recombination of acetolactate synthase. *Mol Plant* 9(4):628–631

Teixeira RM, Ferreira MA, Raimundo GAS, Loriato VAP, Reis PAB, Fontes EPB (2019) Virus perception at the cell surface: revisiting the roles of receptor-like kinases as viral pattern recognition receptors. *Mol Plant Pathol* 20(9):1196–1202

Thomas KR, Capecchi MR (1987) Site-directed mutagenesis by gene targeting in mouse embryo-derived stem cells. *Cell* 51(3):503–512

Tovar J, Lichtenstein C (1992) Somatic and meiotic chromosomal recombination between inverted duplications in transgenic tobacco plants. *Plant Cell* 4(3):319

Van den Ackerveken G, Marois E, Bonas U (1996) Recognition of the bacterial avirulence protein AvrBs3 occurs inside the host plant cell. *Cell* 87(7):1307–1316

van Schie CCN, Takken FLW (2014) Susceptibility genes 101: how to be a good host. *Annu Rev Phytopathol* 52(1):551–581

Verlaan MG, Hutton SF, Ibrahem RM, Kormelink R, Visser RGF, Scott JW, Edwards JD, Bai Y (2013) The Tomato Yellow Leaf Curl Virus resistance genes Ty-1 and Ty-3 are allelic and code for DFDGD-class RNA-dependent RNA polymerases. *PLoS Genet* 9(3):e1003399

Vinodkumar S, Nakkeeran S, Malathi VG, Karthikeyan G, Amala Balu P, Mohankumar S, Renukadevi P (2017) Tobacco streak virus: an emerging threat to cotton cultivation in India. *Phytoparasitica* 45(5):729–743

Voytas DF, Gao C (2014) Precision genome engineering and agriculture: opportunities and regulatory challenges. *PLoS Biol* 12(6):e1001877–e1001877

Wang YQ, Chen DJ, Wang DM, Huang QS, Yao ZP, Liu FJ, Wei XW, Li RJ, Zhang ZN, Sun YR (2004) Over-expression of Gastrodia anti-fungal protein enhances Verticillium wilt resistance in coloured cotton. *Plant Breed* 123(5):454–459

Wang Y, Cheng X, Shan Q, Zhang Y, Liu J, Gao C, Qiu J-L (2014) Simultaneous editing of three homoeoalleles in hexaploid bread wheat confers heritable resistance to powdery mildew. *Nat Biotechnol* 32:947

Wang L, Jin P, Wang J, Jiang L, Shan T, Zheng Y (2015) Methyl jasmonate primed defense responses against Penicillium expansum in sweet cherry fruit. *Plant Mol Biol Report* 33(5):1464–1471

Wang F, Wang C, Liu P, Lei C, Hao W, Gao Y, Liu Y-G, Zhao K (2016) Enhanced rice blast resistance by CRISPR/Cas9-targeted mutagenesis of the erf transcription factor gene OsERF922. *PLoS One* 11(4):e0154027

Wang X, Tu M, Wang D, Liu J, Li Y, Li Z, Wang Y, Wang X (2018) CRISPR/Cas9-mediated efficient targeted mutagenesis in grape in the first generation. *Plant Biotechnol J* 16(4):844–855

Waterworth WM, Drury GE, Bray CM, West CE (2011) Repairing breaks in the plant genome: the importance of keeping it together. *New Phytol* 192(4):805–822

Zeng X, Luo Y, Vu NTQ, Shen S, Xia K, Zhang M (2020) CRISPR/Cas9-mediated mutation of OsSWEET14 in rice cv. Zhonghua11 confers resistance to Xanthomonas oryzae pv. oryzae without yield penalty. *BMC Plant Biol* 20(1):313–313

Zerbini FM, Briddon RW, Idris A, Martin DP, Moriones E, Navas-Castillo J, Rivera-Bustamante R, Roumagnac P, Varsani A, Ictv Report Consortium (2017) ICTV virus taxonomy profile: geminiviridae. *J Gen Virol* 98(2):131–133

Zhang Y, Zhang F, Li X, Baller JA, Qi Y, Starker CG, Bogdanove AJ, Voytas DF (2013) Transcription activator-like effector nucleases enable efficient plant genome engineering. *Plant Physiol* 161(1):20–27

Zhang C, Liu L, Wang X, Vossen J, Li G, Li T, Zheng Z, Gao J, Guo Y, Visser RGF, Li J, Bai Y, Du Y (2014) The Ph-3 gene from Solanum pimpinellifolium encodes CC-NBS-LRR protein conferring resistance to Phytophthora infestans. *Theor Appl Genet* 127(6):1353–1364

Zhang T, Zheng Q, Yi X, An H, Zhao Y, Ma S, Zhou G (2018) Establishing RNA virus resistance in plants by harnessing CRISPR immune system. *Plant Biotechnol J* 16(8):1415–1423

Zhao K, Tung C-W, Eizenga GC, Wright MH, Ali ML, Price AH, Norton GJ, Islam MR, Reynolds A, Mezey J (2011) Genome-wide association mapping reveals a rich genetic architecture of complex traits in Oryza sativa. *Nat Commun* 2(1):1–10

Zhou J, Peng Z, Long J, Sosso D, Liu B, Eom J-S, Huang S, Liu S, Vera Cruz C, Frommer WB, White FF, Yang B (2015) Gene targeting by the TAL effector PthXo2 reveals cryptic resistance gene for bacterial blight of rice. *Plant J* 82(4):632–643

Zhu S, Li Y, Vossen JH, Visser RGF, Jacobsen E (2012) Functional stacking of three resistance genes against Phytophthora infestans in potato. *Transgenic Res* 21(1):89–99

Zischewski J, Fischer R, Bortesi L (2017) Detection of on-target and off-target mutations generated by CRISPR/Cas9 and other sequence-specific nucleases. *Biotechnol Adv* 35(1):95–104

Chapter 16
Critical Evaluation of GM Cotton

G. Balasubramani, K. P. Raghavendra, Joy Das, Rakesh Kumar, H. B. Santosh, J. Amudha, Sandhya Kranthi, and K. R. Kranthi

16.1 Introduction

Genetically modified (GM) crops mainly aimed to increase crop protection through the introduction of resistance against insects and diseases and tolerance to herbicides. Since their approval for commercial cultivation, they have been adopted in no less than 70 countries across the globe during the last 2 decades. During the span of 23 years (1996–2018), global area coverage of GM crops with now 2.5 billion hectares amounted to an upsurge of ~113-fold since 1996 (ISAAA 2018). Till 2018, GM cotton held a global area of 24.9 million hectares (Mha), making it the third most adopted GM crops, ranked right after soybean and maize (Burkitbayeva et al. 2016; ISAAA 2018). As per 2017 FAO global crop area coverage report, GM cotton shared 76% of the global cotton area in 2018 (ISAAA 2018). Notably, among all GM crops, Bt cotton has gained immense popularity and acceptability among smallholder farmers across the globe (Burkitbayeva et al. 2016). No wonder, Bt cotton has been the sole adopted and cultivated GM crops among several countries, particularly in developing nations like India (11.6 mha), Pakistan (2.8 mha), Myanmar (0.3 mha), Sudan (0.2 mha), Mexico (0.2 mha), and Eswatini (<0.1 mha) (ISAAA 2018). As a matter of concern, most of such developing nations are still lacking a sound regulatory milieu for strict monitoring of illegal and unauthorized dissemination of GM seeds or planting materials among the local cotton growers. As for instance, 6 out of 11 top GM crop-growing countries around the globe had recorded illegal cross-border permeation and released unauthorized GM seeds for sale preceding any approval from concerned regulatory authorities. Interestingly, all those

G. Balasubramani · K. P. Raghavendra · J. Das · R. Kumar · H. B. Santosh · J. Amudha
Central Institute for Cotton Research, Nagpur, India

S. Kranthi · K. R. Kranthi (✉)
International Cotton Advisory Committee, Washington, DC, USA
e-mail: keshav@icac.org

Mehboob-ur-Rahman et al. (eds.), *Cotton Precision Breeding*,
https://doi.org/10.1007/978-3-030-64504-5_16

six aforementioned countries happened to be developing nations, namely, China, India, Pakistan, Brazil, Bolivia, and Paraguay (Ramaswami et al. 2012; Sinebo and Maredia 2016). On the contrary, such unauthorized and reckless adoption of illegal GM seeds by allured farmers at times resulted in the cultivation of substandard GM crops with below par yield and quality traits. This often turns out to be disastrous, particularly for resource-poor farmers, who are easy targets by opportunistic local vendors who persuade and lure farmers to pay hefty amounts to purchase illegitimate spurious GM seeds. As a result, the poor farmers are often left marginalized and empty-handed with meager returns from their crops at the end. Moreover, this problem turns more troublesome if there is an accidental admixture of illegal GM seeds among non-GM or authorized GM seeds. GM crops containment is also necessary to restrict unwanted "transgene" (a gene which is artificially introduced into the other organism) gene flow across cross-compatible species, including wild or weed relatives (Ryffel 2014). Cotton being primarily a self-pollinated crop, most transgene flow in cotton is linked to uncontrolled GM seeds dissemination via livestocks and secondary cross-pollinations (Ryffel 2014), besides illicit man-made ventures. Hence, it is extremely crucial to trace illegitimate GM crops including largely popular Bt cotton, with robust precision to facilitate firm regulation over GM entities.

16.2 Development of Transgenic Plant

GM plants are also called transgenic plants whose DNA is modified with new useful traits using genetic engineering techniques which include resistance to insects, diseases, tolerance to adverse environmental conditions, high yield and nutritional quality, production of edible vaccine, etc. The process of transgenic development is to assemble a gene or combination of genes isolated from various sources to develop an improved crop plant. The very first transgenic plant was tobacco expressing antibiotic resistance gene in 1982 (Herrera-Estrella 1983). The first field trials of genetically engineered plants were for herbicide-resistant tobacco plants cultivated in France and the USA in 1986 (Clive 1996). The first genetically engineered insect-resistant plant was developed by incorporating genes from *Bacillus thuringiensis)* (Bt) that produced insecticidal proteins into tobacco (Vaeck et al. 1987). The first genetically modified food crop tomato (Flavr Savr™) (Kramer and Redenbaugh 1994) for consumption was developed for delayed ripening by Calgene's (USA) in 1994. Transgenic cotton resistant to lepidopteron insect with Bt gene *(cry1Ac)* is the first transgenic plants globally commercialized in 1995 (USA). In 1995, Bt cotton (Monsanto), bromoxynil-resistant cotton (Calgene), glyphosate-resistant soybeans (Monsanto), Bt maize (Ciba-Geigy), virus-resistant squash (Asgrow), and additional delayed ripening tomatoes (DNAP, Zeneca/Peto, and Monsanto) were developed. The principles involved in the generation of transgenic plant are discussed below. The process of genetic engineering and development of transgenic plants requires the following steps:

(a) Isolation of genes/gene constructs
(b) Genetic transformation methods
(c) Regeneration of plants through somatic embryogenesis

16.2.1 Isolation of Genes

Isolation of genes for agronomically important traits like higher yield, improved quality, pest and disease resistance, herbicide resistance, and tolerance to heat, cold, and drought is possible to produce millions of copies and determine their nucleotide sequence. Smith and Welcox (1970) discovered restriction enzymes which cut DNA at specific places, enabling to isolate genes from an organism (Roberts 2005). The identified gene of interest should be characterized for its regulation, effect on the plant, and interaction with other genes in the same biochemical pathway. Once a gene has been isolated, it is cloned in a bacterial vector with requisite modifications before transferring into a plant. Genetic engineering has broken down the species boundary, as the entire organisms have DNA as the basic material. The genes for transformation can be obtained from a wide range of sources, like the primitive organisms and virus to multicellular organism including human, and also can be synthesized de novo from the complete genome sequences of the organism. Much effort in recent years has been devoted to identifying potential target genes for use in genetic engineering for economically important traits especially biotic and abiotic stress resistance and improvement of quality traits. The process has been accelerated by reference to the rapidly expanding bioinformatics databases, by progress in elucidating the plant and bacterial genomes. There is no doubt that the use of insect resistant and herbicide tolerant (Howe and Jander 2008), singly and in combination, have been successful in practice, aside from social and environmental concerns. Gene pyramiding or stacking appears to confer relatively greater benefit as reported in case of increased expression of biotic stress regulatory gene (Claire 2005). The gene of interest should be free from technical complexity, issues of food/feed safety, and consumer health risk.

16.2.2 Gene Construct

Gene construct can be defined as engineered DNA fragment to be transferred, integrated, and expressed in the genome of the target plant. Apart from the gene of interest, promoter (“starter”) and a terminator (“stop signal”) are required for expression. In most cases, additional sequences are included, e.g., marker genes, which are essential for the selection of transformants and this gene also accompanied by a promoter and a terminator. In case of *Agrobacterium* gene construct, the left and right border sequences are essential units flanking the abovementioned genes, and these are collectively called as gene construct. The promoter region is typically located at the 5′ upstream of a gene. Promoters are known for their function in governing gene expression, to an on/off switch. The promoters can be categorized into three main groups: constitutive promoters, tissue-specific promoters, and inducible promoters (Hernandez-Garcia and Finer 2014). The upstream of each gene contains regulatory information about how and when the gene is to be expressed. The area binds to proteins (RNA polymerase) that are needed for gene expression (transcription). All genes must have a promoter in order to be expressed. Genes transferred by genetic engineering must be accompanied by a promoter. Some promoters are active in all cells at all times called constitutive promoter (e.g., 35S CAMV promoter), while others are specific to different organisms or tissue types (e.g., seed specific). Others are sensitive to external signals such as temperature or the presence of a certain chemical. Such promoters can be used as controllable on/off switches for genes. The 35S CaMV promoter is very strong well-known constitutive promoter and widely for the plant transformation. It was discovered at the beginning of the 1980s, by Chua and collaborators at the Rockefeller University. The antibiotics kanamycin (*nptII*, encoding neomycin phosphotransferase) and hygromycin (*hptIV*, encoding hygromycin phosphotransferase, isolated from *E. coli*) are mostly used and in herbicides glyphosate (EPSPS, 5-enolpyruvate shikimate-3-phosphate synthase). PMI marker gene that allows metabolic selection for transgenic plants was used in golden rice (Hoa et al. 2003). Marker-free transgenics were developed with site-specific recombinases that cleave a marker gene within two specific sites (Hare and Chua 2002). The common reporter genes used to monitor plant transgene expression include *gus* (beta-glucuronidase), beta-galactosidase (*LacZ*), green fluorescent protein (GFP), luciferase (*Luc*), and *Chloramphenicol acetyltransferase* (CAT) (Jefferson et al. 1987), and reporter gene is most commonly used in the gene construct.

16.2.3 Genetic Transformation

16.2.3.1 Vector-Mediated Gene Transfer

Agrobacterium tumefaciens was discovered in the early 1970s, and it naturally transfers DNA (T-DNA) with "*onc* genes" in the Ti plasmid into the plant cell and produces crown gall (Christie and Gordon 2014). *A. tumefaciens* used for plant transformation are disarmed by removing the tumor-promoting and opine-synthesis genes and replaced with the desired foreign gene or selective markers, enabling the incorporation of foreign genes into plant's genome, transiently or stably (Van Montagu and Zambryski 2013). Vector is a DNA molecule which consists of insert (transgene), origin of replication, and a larger sequence that serves as the "backbone" of the vector. There are a number of vectors available; however, for genetic transformation in plant system, plasmid vectors are commonly used. Many plasmids are commercially available for such uses. The gene of interest is cloned along

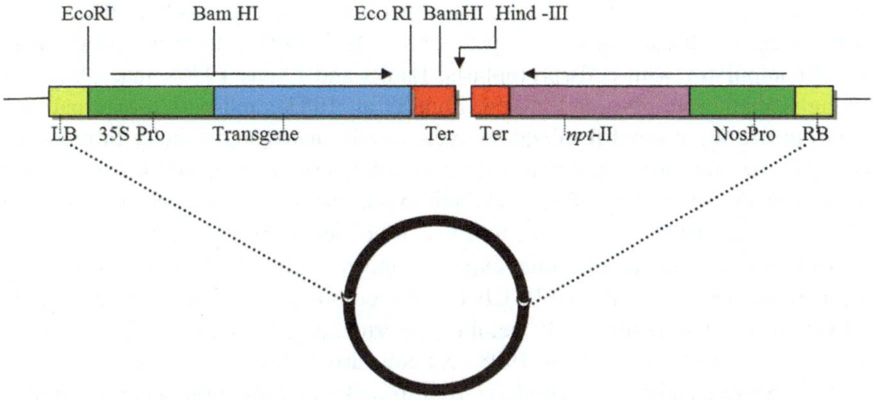

Fig. 16.1 A typical plant transformation binary vector map. *LB* left border of T-DNA, *35S Pro* CaMV promoter (constitutive), *Ter* terminator, *NosPro* nopaline promoter of *Agrobacterium*, *npt-II* neomycin phosphotransferase II gene, *RB* right border of T-DNA

with other selection marker genes (antibiotics) at multiple cloning sites (MCS, or polylinker), which has restriction site in a short region for the insertion of DNA fragments. T-DNA has left and right border repeat sequences which are essential genetic element (Fig. 16.1). T-DNA border repeat sequences (T-DNA borders) contain 25 bp that are highly conserved in all Ti and Ri plasmids (Barker et al. 1983). The binary vector system has cassettes with T-DNA (l0–15 kb) which are subsequently transferred into the plant genome with the *vir* genes residing on a separate helper plasmid that produces stable transgenics. A co-integration system developed through homologous recombination between disarmed Ti plasmid and an intermediate vector could be used to transfer many genes (about 150 kb) (De Framond et al. 1983). Among all the methods of transformation, *Agrobacterium* mediation is a good system to obtain transgenic plants with lower copy number, intact transgene, appropriate segregation, and transgene expression (Dai et al. 2001).

16.2.3.2 Direct Gene Transfer (Physical Methods): Particle Bombardment/Microprojectile

Particle bombardment was first described as a method for the production of transgenic plants (Sanford et al. 1987). In this method using a "gene gun" (Helios® Bio-Rad), the naked DNA/plasmid carrying the gene were coated with tungsten/gold particles and shot into the target tissue/plant cell under high pressure of helium gas. The fast-moving particles penetrate through the plant cell wall, directing the coated DNA into the nucleus. The efficiency of transformation is highest with gold particles in the range of 0.7–1.0 μm mean diameter (Southgate et al. 1995) especially for the recalcitrant cereals and major agronomic crops (McCabe and Christou 1993) as well as in the reduction of the amount of coated DNA on the microcarriers via biolistic gun (Sivamani et al. 2009). The other direct methods such as electroporation (using electrical pulse with protoplasts, Hui 1995), microinjection (using micromanipulator with cells/protoplasts, Banks and Evans 1976), macroinjection (using immature embryo and pollen, Zhou et al. 1983), and pollen-tube pathway (PTP) utilize the normal fertilization cycle to eliminate the difficulty in regeneration. PTP-based transformation is an injection/delivery of naked DNA/drop of DNA solution to the stigma/top of the style into ovaries to produce transformed progeny (Touraev et al. 1997). This procedure was tried first in rice, wheat, and soybean. This method minimizes the time, expense, and recalcitrant plant cell culture and regeneration (Wang et al. 2013). Chemical methods such as polyethylene glycol (PEG) are used to disrupt cell membrane permeating the entry of foreign DNA (Lazzeri et al. 1991; Kofer et al. 1998). Silicon carbide fibers were used for wounding to improve frequency of *Agrobacterium*-based transformation in cotton (Arshad et al. 2013). However, *Agrobacterium tumefaciens*-mediated transformation is the best method over other transformation methods since, reduction in transgene copy number and intact gene sequence integration and segregation (Jones et al. 2005).

16.2.4 Regeneration of Plants Through Somatic Embryogenesis

The development of transgenic plants requires an efficient regeneration system. Regeneration through somatic embryogenesis is ideal over organogenesis because the entire plant is regenerated through a single somatic cell (Merkle et al. 1995). Shoemaker et al. (1986) induced somatic embryos in *G. hirsutum* cultivars Coker 201 and Coker 315 by manipulating culture media. The somatic embryos were derived from isodiametric, densely cytoplasmic cells and regenerated embryos from the hormone-free medium. Later a number of groups have regenerated Coker lines by somatic embryogenesis (Trolinder and Goodin 1987; Finer 1988; Firoozabady and DeBoer 1993) and other lines Sicala, Siokara (Cousins et al. 1991; Rangan and Rajasekaran 1996), Simian (Zhang et al. 2001), and Acala (Rangan 1993; Rangan and Rajasekaran 1996). Perlak et al. (1990) introduced *cry1Ab* and *cry1Ac* genes into cotton (*G. hirsutum*) plants, and transformed plants showed a high level of resistance to *Helicoverpa*. Bt gene inserted Coker-312 plants were used as mother plant to transfer Bt trait to other cotton cultivars by back crossing method.

The development of transgenic plants is severely constrained by the poor regenerative capacity of cotton plants (Zhang et al. 2011). Plant regeneration through somatic embryo is a long and complicated process. Initially, the transformed cells dedifferentiate into calli on the culture medium; then after several weeks, embryogenic callus undergoes into four stages like globular, heart-shaped, torpedo, and cotyledonary stages, which ultimately grows into a complete plant. The process of converting a non-embryogenic callus into embryogenic callus is the key bottleneck step (Zhang et al. 2011; Shang et al. 2009; Rajeswari et al. 2010). Somatic embryos formation is a complex dynamic process that involves many intracellular metabolic changes and is influenced by various external environmental factors. The underlying temporal-specific expression of genes plays an important role in this process. Sun et al. (2018) studied the molecular mechanism of conversion from non-embryogenic callus into embryogenic callus in cotton which allows for the identification of novel genes involved. They compared transcriptome changes in the transformation from non-embryogenic callus into embryogenic callus and identified 46 transcripts that may contribute to initiating embryogenic shift. Analyzing the transcriptional activity of genes during the transition from non-embryogenic callus to embryogenic callus may help to reveal the molecular mechanisms involved in the acquirement of embryogenic potential, which, in turn, may provide ideas to facilitate the induction of embryogenic callus to further promote regeneration of a wider range of cotton cultivars for genetic modification.

16.2.5 Transgene Integration and Inheritance

The transgene integration into plant cells may be stable or transient. If the transgene integrates into the nucleus of the plant cell, then it is a stable transformation. The transgene gets integrated into the chromosome of the genome, and the copy number replicates along with the chromosome, and they are inherited in the next generation of the transgenic plant. In the transient expression, the transgene is not integrated into the nucleus, and they remain for a limited period and are not replicated and lost through cell division. Stable integration and expression depend on the gene construct and transformation method (Gelvin 2005). The transgene of a stable transgenic plant inherit in a Mendelian fashion. However, transgene inheritance depends on the location of integration and the copy number (Tizaoui and Kchouk 2012).

16.3 GM Cotton Detection Methods

Detection methods for genetically modified crops (GM crops) are primarily divided into gene-based and protein-based approaches (Holst-Jensen et al. 2012; Yu-jia et al. 2020). In addition, phenotype-mediated detection of GM crops is also considered as a cheap and effective tool (Holst-Jensen et al. 2012; Kamle et al. 2017) (Fig. 16.2). Traits like herbicide tolerance (HT) and insect resistance (IR) are often monitored through phenotyping via specific bioassays which are deemed to be accurate and user-friendly approach for quality assurance of GM crops (Kamle et al. 2017). For instance, phenotype-based herbicide bioassay for detection of glyphosate-tolerant Roundup Ready cotton is commercially available (Lübeck 2002). However, phenotype-based detection of GM crops has a narrow spectrum of application. As a matter of fact, for precise phenotype-mediated detection, the trait pertaining to the GM crops should ideally be phenotypically distinctive. Moreover,

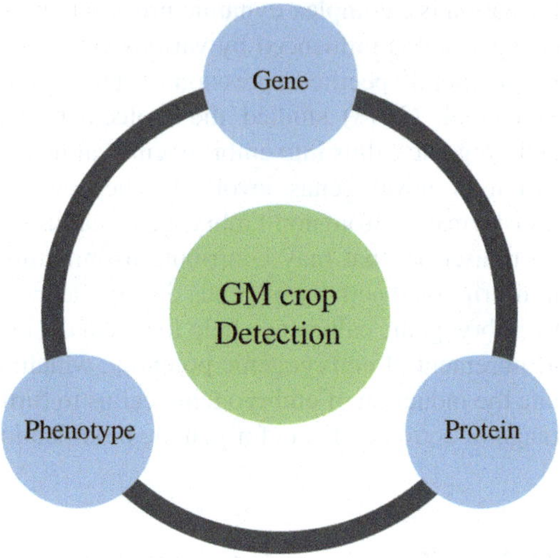

Fig. 16.2 The three basic biological raw materials for detection of GM crops

a GM crop must be grown for phenotype-based trait assessment which can be time-consuming as well. Hence, phenotype-based approaches for GM detection are not always apropos.

16.3.1 Gene-Based Approaches

Gene-/DNA-based approaches for GM detection encompass several biotechnological tools which include PCR-based tools, LAMP-PCR, microarray tools, etc. (Table 16.1). Moreover, the classical genomics tool, namely, southern blot analysis, can precisely reveal exact copy number of transgene integration in host crops (Kamle et al. 2017). Nevertheless, most of these gene-/DNA-based detection tools for GM crops are based upon four analytical modules, viz., species-/taxon-specific, element-specific, construct-specific, and event-specific (reviewed in detail by Holst-Jensen et al. 2012). Taxon-/species-specific modules are used to trace any specific DNA/gene sequence exclusive to a particular species. For instance, cotton-specific stearoyl-acyl carrier protein desaturase (*Sad1*) gene is often targeted as an endogenous reference for detection of GM cotton (Yang et al. 2005a, b). Element-specific modules can trace discrete inserted DNA motifs like promoter or terminator or intron/exon, etc., while construct-specific modules target the region bridging two distinct adjoining DNA elements, viz., promoter-transgene junction representing man-made transgene constructs (Randhawa et al. 2016). Event-specific modules target a DNA sequence motif spanning the region of junction between the transgene construct and the host's genomic DNA. Event-specific target sequence motif represents a unique signature tag pertaining to particular GM crops and, hence, provides explicit detection of a specific GM event with an irrefutable edge over the other

Table 16.1 Tools for gene-/DNA-based approach for GM detection, their applications, and key advantages

DNA-based GM crops detection tools	Application	Key advantages
Quantitative real-time PCR (Fraiture et al. 2015)	Can monitor PCR products with respect to time and quantify those amplicons by measuring a specific signal, often emitted through fluorescence signal	Sensitive, precise, and ease of quantification pertaining to GM crops detection
Multiplex PCR (Kamle et al. 2017)	Can concurrently target and amplify several DNA sequences in a single (PCR) reaction	Time-saving and cost-effective, yet sensitive, tool for GM detection
LAMP-PCR (Singh et al. 2020)	An isothermal nucleic acid amplification technique where target DNA sequence(s) can be amplified at constant temperature by the action of *Bst* polymerase (extracted from *Bacillus stearothermophilus*). The results can be obtained rapidly and visually observed by simple colorimetric approach	An excellent alternative to conventional PCR; does not require thermocycler and can be performed at a constant temperature. Simple, rapid, and practically adaptable tool, amenable for on-site application
Real-time LAMP-PCR (Randhawa et al. 2015)	An advanced LAMP-PCR where targets can be monitored with respect to time measured using a fluorescent dye-based detection of LAMP-PCR amplicons	Quick and sensitive, found to detect up to two target copies within 35 min, amenable for on-site application for detection of GM crops
Genome walking/ Long Template Rapid Amplification of gDNA Ends (LT-RADE) (Spalinskas et al. 2013)	Restriction-independent genome walking technique to detect unknown DNA regions adjoining a known DNA segment of transgene elements	User-friendly, can trace flanking regions of transgenic insert
Padlock probe ligation—microarray (Prins et al. 2008, 2010)	Padlock probes are detection probes which recognize and hybridize target sites within DNA sequence. Upon hybridization, the probe gets linearized and ligated to form a circular probe molecule, which can be amplified by using labeled primer, thus facilitating detection of target site. This can be combined with microarray detection module for high-throughput detection	Suitable for simultaneous detection of multiple unauthorized GM crops samples

(continued)

Table 16.1 (continued)

DNA-based GM crops detection tools	Application	Key advantages
Digital PCR (dPCR) (Vogelstein and Kinzler 1999; Yu-jia et al. 2020)	dPCR sample mix containing similar components as that in qPCR is distributed and partitioned into large number of individual wells before amplification, with an assumption that each well would contain either 1 or 0 targets following Poisson distribution. Following this, PCR amplification is carried out, to determine and quantify the number of positive vs negative reactions measured via fluorescence signal	Advanced form of qPCR, but independent of internal control and endogenous reference genes; absolute quantification of target gene(s) in GM crops can be calculated using Poisson's statistics
PCR capillary gel electrophoresis technology (Fraiture et al. 2015)	Fluorescently labeled primers are used for multiplexed PCR which facilitate distinguishing between varied amplicons of the same size	Resolution power is much higher as compared to conventional gel electrophoresis while detecting PCR products from multiplexed assay and, hence, suitable for GM crops detection

analysis modules (Holst-Jensen et al. 2012; Randhawa et al. 2016). Several reports for module-specific GM detection in cotton are cited in Table 16.2.

16.3.2 Protein-Based Approaches

Protein-based detection of GM crops is majorly restricted to fresh or frozen samples, as proteins are usually prone to denaturation due to heat or rigorous processing of plant samples. However, protein-based detection tools are remarkably reliable, as those can trace the actual product of the transgene from GM crops. Tools like enzyme-linked immunosorbent assay (ELISA), immunoassay strips, lateral flow strips, etc. are some of the popular and unswerving GM crops detection techniques that are often utilized as working principles behind several commercially available user-friendly kits. For instance, a series of ELISA and immunoassay-based detection kits, namely, "Bt-quant," "Bt-GUS," "Bt-Zygosity," and "Bt-Express," had been developed by ICAR-Central Institute of Cotton Research (India) for rapid detection and quantification of *cry* gene/*cry* toxin from Bt cotton transgenic plants (*http://www.cicr.org.in/tech_bank/Bt_kit.pdf; http://www.cicr.org.in/tech_bank/bt_express.pdf*). Especially, the "Bt-Express" kit became extremely popular and admired among cotton growers across India. It was largely because the kit was designed to be so simple and user-friendly that even illiterate farmers could detect presence or absence of Bt toxin from plant samples with ease, and hence, this detection kit deserves special mention. Apart from those, conventional proteomics techniques like western blot analysis (WB), mass spectrometry (MS), and two-

Table 16.2 Gene/DNA and protein-based tools for GM cotton detection

Detection approach	Basis of detection	GM detection tool	Detection modules	Application in GM cotton detection	References
Gene level (DNA)	PCR-based	qPCR	Event-specific	Cotton event DAS-81910-7	JCR GMOMETHODS, *http://gmo-crl.jrc.ec.europa.eu/gmomethods/*
		rtPCR	Element-specific	*cry2Ab2* detection	Dinon et al. (2011)
		qPCR	Taxon-specific	Fiber-specific acyl carrier protein	JCR GMOMETHODS, *http://gmo-crl.jrc.ec.europa.eu/gmomethods/*
		Multiplex real-time PCR	Taxon-specific	Stearoyl-acyl carrier protein desaturase gene	Cottenet et al. (2013)
		Multiplex PCR CGE	Taxon-specific	Acyl carrier protein 1 gene from cotton	Nadal et al. (2006)
		Multiplex PCR microarray	Taxon-specific	Stearoyl-acyl carrier protein desaturase gene	Prins et al. (2008)
		dPCR	Taxon-specific	*adhC* (alcohol dehydrogenase C gene from cotton)	Brod et al. (2014)
	DNA walking	LT-RADE (genome walking)	Construct-specific	LLCOTTON25 detection	Spalinskas et al. (2013)
	LAMP-PCR	LAMP-PCR	Construct-specific	*cry1Ac*, *cry2Ab2*, and *cp4-epsps* genes	Singh et al. (2020)
		Real-time LAMP	Event-specific	MON531 and MON15985	Randhawa et al. (2015)
	Microarray-based	PPLMD	Event-specific	*bar* MON1445	Prins et al. (2008)
	NGS		–	*cry1Ac*	Debode et al. (2019)

(continued)

Table 16.2 (continued)

Detection approach	Basis of detection	GM detection tool	Detection modules	Application in GM cotton detection	References
Protein level	ELISA	Monoclonal antibody-based sandwich ELISA	–	*vip3Aa*	Liu et al. (2020)
	Western blot	Western blot immunodetection	–	*cry*10Aa	Ribeiro et al. (2017)
	Immunostrips/ dipsticks	Immunostrip assay	–	*cry1Ac* and *cry2Ab*	Siddiqui et al. (2019)

NB: *PCR* polymeric chain reaction, *rtPCR* real-time PCR, *CGE* capillary gel electrophoresis, *qPCR* quantitative PCR, *dPCR* digital PCR, *LT-RADE* Long Template Rapid Amplification of gDNA Ends, *LAMP* loop-mediated isothermal amplification, *PPLMD* padlock probe ligation—microarray, *bar* phosphinothricin-*N*-acetyltransferases gene of *Streptomyces hygroscopicus*, *vip* vegetative insecticidal protein, *pat* phosphinothricin-*N*-acetyltransferase, *NGS* next-generation sequencing dimensional SDS gel electrophoresis (2-DE) could be effectively employed to detect targeted proteins from GM crops (Ruebelt et al. 2006) (Table 16.2). In addition, using those techniques, the after-effects of transgene integration into GM crops with respect to their wild-type counterparts can be vividly studied. For instance, a comparative study encompassing Bt cotton and its non-transgenic counterpart was conducted to trace any transgene-induced unintended effects on Bt cotton. For this purpose, a combination of WB, 2-DE, and MS tools was used, which revealed that exogenous DNA could influence the growth and photosynthesis in GM cotton (Wang et al. 2015).

16.3.3 Advanced GM Detection Techniques

Apart from typical genomics and proteomics tools, several biophysical technique-based approaches are gaining popularity for their applications in detection of GM crops. Techniques like near-infrared (NIR) spectroscopy, surface plasmon resonance (SPR), surface-enhanced Raman scattering (SERS) spectroscopy, and biosensor-based detection are some of the biophysical tools which may be used in detection of GM crops (reviewed by Kamle et al. 2017), including GM cotton. In fact, SPR technology which is commonly used to study protein-ligand interactions could detect transgenic *cry1Ac* cotton with high accuracy, sensitivity, and rapidity (Zhao et al. 2013).

GM cotton is widely adopted and cultivated across the globe. The cutting-edge technology of BT cotton has been widely popular among cotton growers, including resource-poor farmers across the globe. Several countries, however, maintain strict biosafety regulatory monitoring following series of norms and standards before releasing and commercializing GM crops, including GM cotton. At times, to evade such strict biosafety regimes, local farmers and avaricious vendors often adopt illegitimate shortcuts, in a quest to maximize profits and, henceforward, promote spurious and illegal spread of unauthorized GM seeds and planting materials, overlooking the detrimental effects in the long run. Hence, to circumvent this, a set of robust yet simple and handy GM crops detection regimes is certainly the need of the hour. This will not only safeguard the authenticity of germplasm but also aid in delivering better quality of GM products, while serving the interest of farmers.

16.3.4 Characterization Under Contained Trials

After completion of molecular characterization, the transgenic plants and events should be subjected to contained (lab/growth chamber/greenhouse) trial for further study. Contained use is defined as any activity on GM crops should be quarantined/restricted for the safety of humans and the environment. During contained study, biosafety regulatory authority's clearance is required. Primarily the type of information requested by the regulatory authority would be related to the details of plant species, biological document, category of genetic manipulation, detailed molecular characterization of the inserted DNA, as well as host genomic flanking sequences, e.g., the vector used and its resource, details of all functional nucleotides, primers used to amplify specific sequences of DNA, details of descriptions and functions, insertion site(s) and copy numbers of all inserted DNA (transgenes, regulatory sequences, vector backbone), etc.

16.3.5 Event-Specific Flanking Sequence Identification

The point of foreign gene/transgene integration and successful expression in a new genetic background made the transgene is a unique event. Event means the transgene integration at a particular host plant DNA of a single cell, which is transformed and regenerated into a complete transgenic plant. Each event will have different points of transgene integration and its positional effects in the host genome, and the introgressed position in the chromosome of host cell remains unchanged during segregation. Contained trials should be conducted to screen and identify the best event. A number of events are generated during transformation. Forwarding all the transgenic events is difficult and laborious and requires large investments. Thus, one should select the best event satisfying all the set parameter during contained event selection trial and forwarded to biosafety research trials.

The molecular details of transgene integration in the host plant and flanking sequences of the transgene are important for biosafety studies and tracing the transgenic event (Yang et al. 2013). Formerly, PCR-based methods like TAIL-PCR, genome walking, and DNA sequencing have been used to determine the point of transgene integration flanked with host DNA (Nakayama et al. 2001). However, these methods are time-consuming, complex, or cumbersome and may not work if the deletion, modification, or rearrangements occurred in transgene sequence during insertion (Wang et al. 2010). With the advent of high-throughput next-generation sequencing (NGS) technology, whole genome sequences can be obtained precisely within short period of time at low cost. NGS has been widely used in many crops to identify flanking sequences of transgene integration and its location in the chromosomes (Inagaki et al. 2015; Pauwels et al. 2015). Whole genome sequencing (WGS) with targeted bioinformatics analysis is a more sensitive and toil-effective method for characterization of GM plants. The WGS technology can divulge nucleotide sequence variations including single nucleotide polymorphism and InDels, which could detect even small sequence modifications (Pauwels et al. 2015). The WGS information could be used in evaluation of the potential toxicity, allergenicity of GM plants by verification of potential similarities in databases of toxins, targets of toxins, allergenic proteins, and anti-nutritional factors (Guo et al. 2016).

16.4 Confined Field Trials

The development of a genetically engineered (GE) crop plant follows a progression from experimentation in laboratory and other contained facilities to field studies and eventually to cultivation after pre-market environmental risk and food/feed safety assessments have been conducted by the appropriate regulatory authorities (Garcia-Alonso et al. 2014; Rüdelsheim 2015). The most early-phase activities of research and development are performed in laboratories, growth rooms, net house, and glass-houses known as contained trials. Under these conditions there is a physical barrier(s) that contains the material to avoid its direct contact with the environment. The Cartagena Protocol on Biosafety to the Convention on Biological Diversity defines "contained" use as "any operation, undertaken within a facility, installation or other physical structure, which involves living modified organisms that are controlled by specific measures that effectively limit their contact with, and their impact on, the external environment." The contained studies are followed by small-scale, proof-of-concept field trials and then by larger trials to further characterize and multiply material (principally seed) of the transformation events. These regulated field trials are known as confined field trials (CFTs) and are conducted with the permission from the appropriate competent authorities. Confined field trials (CFTs) are field experiments of growing a regulated, GE plant in the environment under specific terms and conditions that are intended to mitigate the establishment and spread of the plant. These are experimental activities conducted on a limited scale to collect data, including the data on potential biosafety impacts under the conditions of reproductive isolation known to mitigate dissemination of experimental plant, its persistence in the environment, and its introduction into food chain (GEAC 2015a, b). These represent greater environmental exposure than the contained studies and smaller degree of exposure than the commercial cultivation. These trials are meant to balance safety and exposure to environment and are considered as an essential component of GM research and development throughout the world.

The Cartagena Protocol on Biosafety (*http://bch.cbd.int/protocol/background/*) governs the movements of living modified organisms (LMOs) from one country to another and, therefore, does not govern the CFTs as CFTs are conducted basically to check the introduction of transgene or the transgenic plant into the environment. Therefore, many countries have developed or are developing regulatory frameworks for safe handling of GM crops. Though the guidelines and standard operating procedures for conduct and monitoring of the different types of CFTs vary with the country, the basic objectives of CFTs remain the same. CFTs are conducted to evaluate agronomic performance; to collect data on potential ecological and biosafety impacts; to understand the weedy characteristics of GE crop; to study the environmental fate of novel plant-expressed proteins; to understand the impact on beneficial, endangered, or other organisms; and to generate plant tissue for nutritional analyses, novel protein expression studies, feeding studies, and other studies (GEAC 2015b).

Conduct of CFTs involves the assignment of responsibilities and obtaining permits. Permitted CFTs are performed under a regime of management practices designed to confine the trials so as to prevent the accidental release of plant material from the trial site, trait introgression into populations of sexually compatible species, or establishment of populations of the experimental GE plant in the environment (Garcia-Alonso et al. 2014). Since the CFTs are designed to understand the potential environmental impacts of GE crop, a regulatory oversight is required for conduct of CFTs, their monitoring, and risk management. It is the responsibility of Trail-in-charge to conduct the CFTs adhering to the norms and guidelines of the regulatory authorities. Safe and successful conduct of CFTs can only be accomplished through a combination of robust regulatory framework, science-based risk mitigation measures, trained personnel dedicated to abiding the terms and conditions of trial authorization, and a qualified monitoring staff (GEAC 2015b). Any weakness in any of these components puts the trust of public on regulatory system into a great risk. Public opinion and perception are considered crucial for acceptance and proliferation of GM technology. Potential risk mitigation procedures are to be in place that can prevent potential negative impacts of the possible known and unknown hazards.

The CFTs are conducted typically and in accordance with internationally accepted approaches to environmental risk assessment (ERA) of GE plants (OECD 1992; SCBD 2000), wherein a comparative assessment is followed where the GE plant is compared to its conventional counterpart, usually the isogenic or a near-isogenic line, which is included in the CFT as a control. Trial endpoints vary depending on the risk hypothesis being tested, but most CFTs aim at identifying any differences between the GE event and its non-GE comparator resulting from intended or unintended consequences of the genetic modification across a range of agroecosystems (OECD 1992; SCBD 2000). Design of CFTs is optimized to obtain data relevant to risk hypotheses while minimizing confounding factors that may interfere with the comparison (Garcia-Alonso et al. 2014).

16.4.1 Types of CFTs

The types of CFTs vary with the country as per the objective of study. Different types of CFTs permitted in India (GEAC 2015a, b) for evaluation of transgenic cotton are provided as follows:

1. *Event selection trials:* Many transgenic events with same gene(s) within the same crop are developed through the process of genetic engineering. These events have varying potential and utility owing to their site of integration in the plant genome, copy number, effect of background genome on its expression, nontargeted effects, and spatiotemporal stability of gene expression across generations. Therefore, it is desirable to evaluate these events under confined conditions to select the most promising event for further evaluation and commercialization. These trials are usually conducted in pots for preliminary evaluation of phenotypic expression.

2. *Biosafety Research Level-I (BRL-I) trials:* Review Committee on Genetic Manipulation (RCGM), Department of Biotechnology (DBT), Ministry of Science and Technology (MoS&T), Government of India, is the regulatory authority for BRL-I trials. These trials are limited in size to no more than 1 acre (0.4 ha) per trial site location and a maximum cumulative total of 20 acres (8.1 ha) for all locations for each plant species/construct combination, per applicant, per crop season.
3. *Biosafety Research Level-II (BRL-II) trials:* Genetic Engineering Appraisal Committee (GEAC) which functions in the Ministry of Environment Forest and Climate Change (MoEF&CC), Government of India, is the regulatory authority for BRL-II trials. These trials are limited in size to no more than 2.5 acre (1 ha) per trial site location, and a number of locations are decided on case-by-case basis for each plant species/construct combination, per applicant, per crop season.
4. *Experimental seed production:* With permission of RCGM and GEAC, the seeds of the selected events can be produced in confined conditions by the applicant for the next phase of trials.
5. *Production of planting material for food and feed studies:* Toxicity and feeding studies for assessment of food and feed safety demand plant material which can be generated under confined field conditions with prior permission of RCGM and GEAC.
6. *Other environmental safety studies:* In addition to BRL-I and BRL-II trials, developer may have to undertake some studies to generate specific information using specific experimental designs. With prior permission of RCGM and GEAC, such trials can be conducted as per guidelines and operating procedures of competent authorities.

16.4.2 Conduct and Monitoring of CFTs

The process of conduct of CFTs starts with the application by the developer which contains key information about description of genetically engineered plant, unmodified counterpart plant, site information with detailed map of the location, experimental protocols, trial management and risk mitigation procedures, and other details required by the concerned competent authority. Many countries have developed detailed guidelines and standard operating procedure for conduct and monitoring of CFTs (Adair and Irwin 2008; CFIA 2000, 2007; EU 2009; OECD 1986; OGTR 2001, 2013; EC 2001; Rüdelsheim 2015; USDA APHIS 2011; DBT 2008; GEAC 2015a, b). The transport, storage, evaluation, harvest, and postharvest operations were regulated through monitoring committees having experts and qualified staff. Lot of documentation and record keeping has to be attended by the concerned personnel at every stage (transport, storage, evaluation, harvest, and postharvest operations) which should be made available for review and scrutiny when demanded by the monitoring committee and/or associated competent authorities.

While evaluating the performance of regulated GE plant of upland cotton (*G. hirsutum*) in CFTs, it must be ensured that the pollen dispersal to other sexually compatible species like Egyptian cotton (*G. barbadense*) and other wild species is prevented. This can be achieved through following combination of different reproductive isolation methods like spatial isolation, temporal isolation, and early termination in line with the guidelines of the regulatory authorities. Spatial isolation is maintaining isolation distance between field trials of the regulated GE plant from the other plants of the same species or from sexually compatible relatives. The isolation distance of 50 m is to be maintained for cotton CFTs, and this isolation distance should be kept free from any other plants of the same or related species. Isolation areas surrounding the CFTs are to be regularly monitored, and prohibited plants found, if any, should be removed before flowering or seed set and should be rendered nonviable using appropriate methods at the trial site. Maintaining reproductive isolation through temporal isolation alone may not be suggestive as cotton is a long-duration crop with longer flowering window. Nevertheless, temporal isolation can be combined with spatial isolation for better risk mitigation in CFTs. Wherever it is compatible with the experimental objectives, early termination can be thought out wherein trial plants are destroyed before anthesis. Depending on the kind of material under trial and associated factors, devitalization of material under CFTs could be achieved through high temperature (autoclave sterilization), low temperature (freezing), chemical treatment (methyl bromide, chloropicrin, and chemosterilants), disinfectants, composting, and desiccation (Rüdelsheim 2015). Monitoring plays an important role in achieving the containment and safety. Designated teams or committees can undertake the monitoring of CFTs which include regular inspection visits to trial site before, during, and after conduct of trial; verification and inspection of different documents, reports, inventories, maps, etc.; inspection of storage facility; preparation of monitoring reports; and suggesting corrective action in case of noncompliance of guidelines and regulatory norms.

16.4.3 Global Status of CFTs: Crop and Trait-Wise Trends

CFTs were conducted for the first time in Canada and the USA in 1987. Since then thousands of CFTs are conducted and are regarded as an essential activity in the development of GM crops intended for commercial cultivation. Although information on the actual performance of CFTs is not systematically available, some information about regulatory applications for CFTs is publicly available in most countries with genetically modified organism (GMO) legislation. One can gain insights about traits and crops which are expected for commercial release in coming years through analysis of annual number of CFTs and the species and traits of the GM plants involved. Smets and Rüdelsheim (2018) conducted the comparative analysis CFTs during 2014–2017 using the information available in the public domain and after amicably addressing the associated challenges like inconsistencies in definition of GMO, basic unit of CFTs, form and format of available data, and data gaps across countries. The comparative analysis revealed that the annual number of CFTs worldwide declined from 14,307 in 2014 to 6346 in 2017 although regional differences were observed. The highest number of CFTs was noted for maize (*Zea mays*; 21,846 CFTs) followed by soya bean (*Glycine max*; 10,896 CFTs), cotton (*Gossypium hirsutum*; 3045 CFTs), and oilseed rape (*Brassica napus*; 1425 CFTs). Among the traits, herbicide tolerance dominated the other traits followed by insect resistance and abiotic stress tolerance. Research institutes (not-for-profit research organizations) accounted for only 4.2% of all CFTs, while the rest 95.5% is accounted the industry. Industry comprised of both multinational players and smaller enterprises performed CFTs in one or more countries, in main crops, while public research institutes usually acted locally, focusing their efforts on lesser crops.

16.4.4 Transportability of CFTs Data: Need for Harmonization of Protocols Across Countries

In-country confined field trials are mandatory for unrestricted release of GM crops for commercial cultivation. It is important that the environment risk assessment for GM crops is done as efficiently and effectively as possible to avoid needless duplication of studies and to reduce unnecessary regulation in light of accumulated evidence and experience (Fedoroff et al. 2010; Raybould 2007). In a situation where GM crop is cultivated in a country where it is approved and crop produce or its products are targeted for import in other countries, transportability of confined field trial data of GM crops is advocated (Nakai et al. 2015) as it is considered particularly beneficial to public sector product developers and small enterprises who cannot afford to replicate redundant confined field trials (Garcia-Alonso et al. 2014). There is growing experimental evidence to consistently show that the differences between locations, years, genetic backgrounds, and agronomic practices contribute more to endpoint variation than the process of transgenesis (Harrigan et al. 2010; Ricroch 2012). The differences in endpoint measurements are often detected between different varieties of the same crop planted under very different conditions, but not between the GE and its non-GE counterpart grown under similar conditions (Garcia-Alonso et al. 2014). Conduct of CFTs at multiple sites and in multiple countries is both a logistical and financial challenge as they are highly regulated and resource intensive. It is more compounding to conduct the CFTs when there is a sufficient data already available from the earlier CFTs and peer-reviewed literature as well as practical experience crop breeding and cultivation to support the transportability of CFTs data for consideration by the competent authority to commercial release of GM crop or product in question. Since the CFTs are designed and conducted for comparative assessment of GM and its non-GM counterpart under controlled conditions, they are considered amenable to transportability. Efficient transportability of CFTs data could be achieved through harmonization of protocols at international level.

16.5 Food and Feed Safety Assessment

Successful applications of recombinant DNA technologies were reported in the early 1970s. Asilomar Conference 1975 by the scientific community was the first step toward fixing of guidelines for biotechnology and biosafety regulations. First formal guidelines for regulation of rDNA work in the USA were published in 1976 by the US National Institutes of Health (NIH). Gaining of momentum and rolling out of the biotech product in health and agriculture sector in the early 1980s resulted in the development of first international biosafety guidelines for use of GMOs in industry, agriculture, and environment based on the report on "Recombinant DNA safety Considerations 1986 from Organization of Economic Co-operation and Development (OECD)." Agreement on Cartagena Protocol on Biosafety resulted in the addition of more number of countries with biosafety regulatory framework. The Cartagena Protocol on Biosafety to the Convention on Biological Diversity was finalized and adopted on 29 January 2000 in Montreal, Canada, by more than 130 countries.

16.5.1 Food and Feed Safety: India

Food Safety and Standards Authority of India, Ministry of Health and Family Welfare (MoHFW), is a nodal organization for implementation of Food Standards and Safety Act, 2006, which includes genetically modified foods within the definition of food. Guidelines for the safety assessment of foods derived from genetically engineered (GE) plants, 2008, developed by Indian Council of Medical Research (ICMR), New Delhi, based on guidelines and principles of Codex Alimentarius Commission, 2003.

Framework safety assessment consists of well-structured questions that facilitate the realistic assessment of the safety of food and feed. The guidelines include dossier preparation checklists as an appendix seeking the following information from the applicant for the GE plants:

- Description of genetic-engineered plants includes pedigree, transformation events, and type and purpose of modification.
- Description of the non-transgenic host plant and its use as food includes botany, center of origin, and traits of plant harmful to human health, genotype and phenotype of host plant with relevance to safety, and history of safe use as food.
- Description of the donor organisms covering taxonomic classification, production of toxins, anti-allergens, anti-nutrients which concern to human health if any, pathogenicity nature of organisms, organisms presence in the food chain.
- Description of the genetic modification(s) comprising method of transformation of genetic material, description of all genetic material including their source and function, and description of modifications that affect expression of protein(s).

- Characterization of the genetic modification(s) covering sequence and structural details of target genetic materials inserted in the genome, copy number, gene rearrangement if any, event characterization data including flanking sequence of transgene cassette and details on generation of new fusion protein if any, additional information on the details of gene products of the inserted fragment, quantity of expression, tissue specificity, inheritance of gene, and position effect.
- Compositional analyses of key components: key nutrients, anti-nutrients, and major (fats, proteins, carbohydrates) and minor compounds (minerals, vitamins) of GE plants will be analyzed and compared with equivalent analysis with non-GM counterpart at the same point of time.
- Assessment of possible toxicity and assessment of possible allergenicity (proteins).

Protocols for food and feed safety assessment of GE crops in 2008 have been made available by the Department of Biotechnology (DBT), India, based on guidance and peer-reviewed publications available from the Codex Alimentarius Commission, FAO, WHO, OECD, and International Life Science Institute. It mainly contains acute oral safety limit study in rats and mice, sub-chronic feeding study in rodents, protein thermal stability, pepsin digestibility assay, and livestock feeding study (Codex Alimentarius 2003; DBT 1998; FAO/WHO 2001; OECD 1998, 2000a, b, 2002).

16.5.2 Acute Oral Safety Limit Study in Rats or Mice

Target gene products such as proteins are the test material for assessment of toxicity. Acute toxicity tests with proteins are preferred method for toxicity assessment worldwide. It has been observed that the proteins with toxic nature are effective even at low concentration and shorter time; hence acute oral safety studies are being suggested in protocol and guidelines to assess potential toxicity (Jones and Maryanski 1991; EPA 2000; NRC 2000). Acute oral toxicity test with purified protein and sub-chronic 90-day toxicity with whole plant material should be undertaken along with daily intake food/feed. Mortality, morbidity, or evident toxicity is considered for interpreting the toxicity potential of test substance.

16.5.3 Sub-chronic Feeding Study in Rodents

Ninety-day feeding studies are recommended with transgenic crops in rodents to evaluate the safety of edible parts of the plant as a food. The 90 days' study is advised to know the possible health hazards arise due to repeated exposure of the food containing the test substance.

16.5.4 Protein Thermal Stability

Thermal denaturation of protein molecule leads to loss of structure and function. It is necessary to study the possible allergenicity of newly expressed introduced protein. There is strong correlation between heat stability and allergenic potential. Hence, the protein thermal stability protocol is done to measure the thermolability of recombinant protein when exposed to heat. The purified protein is incubated at different temperature range from 25 to 95 °C for up to 30 min and cooled rapidly. Biological activity of the proteins samples will be used to assess the thermal stability. If the tested protein showed biological activity at higher temperature, it demands further tests to rule out likelihood of the test protein being allergenic.

16.5.5 Pepsin Digestibility Assay

Safe dietary proteins are characterized by their natural easy digestibility and act as dietary source of amino acids. For several food allergens, a correlation is observed between resistance to digestion by pepsin and their allergenic potential (Astwood et al. 1996). Pepsin digestibility is an assessment method to evaluate the potential digestibility of target test proteins in vitro using simulated gastric and intestinal fluids.

16.5.6 Livestock Feeding Study

The GM crop product should be mixed with animal food and feed to livestock such as milking cows or buffalos, goat, broiler chicken, fish, and rats and assess the safety level, and test data generated must be submitted to regulatory authority for approval.

16.6 Report on Environmental Assessment and Impact

Genetically modified (GM) crops have great potential to solve many of perceived problems in agriculture and feed the world. However, it attracts many safety issues, unknown fear, and strong opposition from few scientific and nonscientific groups. In the present contest, GM technology would be the best tool to improve the crop yield and reduce production challenges (Carzoli et al. 2018). Still many countries are hesitant to move forward with establishing biosafety laws and commercializing GM crops, primarily due to risk sensitivity and fear spread by anti-biotech groups. In fact, GM crops are well-studied and evaluated technology for safety to humans and the environment before its introduction into commercial cultivation.

It undergoes stringent biosafety tests formulated by international environment experts and qualifies for environment release. Most of the countries follow these standard operating protocols to assess the potential risk and safety of genetically modified organisms/crops to the environment and living organisms. These include unintentional effects such as impact on nontarget organisms, persistence of modified plants or its residue in the environment, possibility of invading into the new habitats, transfer of genes from GM to other species, etc. Insect resistance Bt cotton has been a rapidly adopted technology since its introduction in 1996. Farmers are using Bt cotton technology globally and benefited through increased productivity, reduced pesticide spray, and minimized environment contamination (Purcell and Perlak 2004). However, Bt cotton haggard on yield advantage, environment risk and resistance development by the insects.

16.6.1 Impact on Pesticide Usage

Cotton is one of the highest pesticide-consuming crops especially to control bollworms, which upset the environment through polluting land and water and poisoning humans and animals. Bt cotton was developed to control American bollworm *Helicoverpa armigera* (*Hubner*) families (Cunningham and Zalucki 2014). It was a minor pest of cotton, in India prior to 1980, but became a major pest due to indiscriminate use of synthetic pyrethroids and increased area under long-duration American cotton *Gossypium hirsutum* hybrid (Kranthi 2016). Bt cotton was introduced in India in 2002 mainly to control bollworm complexes. Evidently the Bt technology had significantly reduced chemical pesticide use and helped farmers not to depend on high-priced pesticides like Spinosad (broad-spectrum insecticide) and Indocarp (bollworm) which were used more frequently in conventional cotton (Veettil et al. 2014). Bt crops may favor biocontrol services and enhance economic benefit, and it was established with field studies indicating that Bt crops protected natural enemies in comparison with non-Bt crops which rely on conventional insecticide (Romeis et al. 2008). A number of studies have been reported that Bt cotton has led to a notable decline in acute pesticides poisoning cases among cotton growers in India and China (Pray et al. 2002; Huang et al. 2005; Kousar and Qaim 2011). Lu et al. (2012) reported a significant level of boost to predators such as ladybirds, lacewing, and spiders in China by adopting Bt cotton. In South Africa 90% of the smallholder Bt cotton producers achieved significant reduction in pesticide use (Ismael et al. 2001). In China, insecticide applications were reduced by an average of 67% and the kilogram of active ingredient by 80% (Huang et al. 2002a, b), and growers in the USA reduced insecticide use by 18.70 lakh pounds of active ingredient per year in 2001 (Gianessi et al. 2002). The reduction in Environmental Impact Quotient (EIQ) through Bt technology adoption has increased from 39% during 2002–2004 to 68% during 2006–2008. Bt adoption has contributed to higher environmental efficiency (Veettil et al. 2016). Krishna and Qaim (2012) analyzed the advantages of transgenic Bt cotton over time, using a panel survey of farmers covering a decade. They claimed that Bt cotton pesticide applications increased yield gains in India. In the 2006–2008 periods, the Bt-induced net reductions in pesticide quantity were 52%. Bt cotton has also reduced pesticide application among the few remaining non-Bt farmers, because widespread adoption has contributed through area-wide suppression of bollworm populations. However, the secondary pests have increased. The recent report by Kranthi and Stone (2020) compared 20 years of Bt cotton data generated in India and showed that Bt adoption area was steadily increasing, but yield was not significant. However, a strong pesticide reduction was recorded initially. The transgenic cotton containing *cry1Ac* found still very efficient to control *H. armigera*, whereas the nontarget pest especially sucking pest, pink bollworm menace cotton cultivation in India and farmers are forced to spray more pesticides.

16.6.2 Pollen Flow

Outcrossing is the unintentional breeding of domestic crops with its related plants. Over the course of evolution, crop species like wheat, potatoes, corn, canola, and numerous others were modified from their original form because of hybridization with related species or weeds or cultivated strains growing nearby. Through this long-established mechanism for gene transfer, any gene in a cultivated crop can be transferred to its wild and semidomesticated relatives. There is potential risk that genes introduced in GM may "escape" (via pollen) to wild or weedy related species growing nearby which is often cited as one of the major risks (Daniell 2002). The degree of gene flow through pollen depends on a number of factors such as sexually compatible species, inheritance of the traits, size of pollen, viability of pollen over time, and distance. Transgene flow in the field between compatible plants can occur when they are close enough for pollen to reach a receptive stigma, the plants have synchronous flowering, and there are no reproductive barriers. Case-by-case study is required for a complete risk analysis. Combination of factors such as difference in experimental design, environmental conditions, and changing pollinator populations affects the results. Now more than two decades over, so far no outcrossed Bt cotton has emerged as a product of gene flow or Bt weeds or more sexually compatible Bt-wild species with transgenes. Experiments have confirmed that most cross-pollination in cotton occurs over distances of less than 50 m (Xanthopoulos and Kechagia 2000: Zhang et al. 2005). But honey bees can travel two miles or more which suggests that the 750 m radius at which pollen from Bt cotton is capable of outcrossing of non-Bt cotton resulted from movement of honey bees from Bt to non-Bt cotton fields (Beekman and Ratnieks 2000). The greatest risk is from herbicide-resistant crops to related weed species. Unwanted gene flow could be prevented or reduced using different barriers such as isolation distance, avoid synchronizing the flowering time, protective vegetation barriers, male sterility, etc. (Ellstrand 2003). Although gene flow has occurred, no examples have demonstrated an adverse environmental effect of gene flow from a transgenic crop to a wild, related plant species.

Heuberger et al. (2010) investigated seed-mediated and pollen-mediated gene flow in the seed production field. Seed-mediated gene flow yields adventitious Bt cotton plant from seed bags, and human error comprises over 15%; in contrast pollen-mediated gene flow affected less than 1% from field edges. Variation in outcrossing was better explained by the area of Bt cotton fields within 750 m of seed production fields than by area of Bt cotton within larger or smaller spatial scales. Variation in outcrossing was also positively associated with the abundance of honey bees (Heuberger et al. 2010).

16.6.3 Impact on Soil Microbial Activity

An imperative aspect of the biosafety assessment of genetically modified plants is to study its impact on soil ecosystem including changes in plant-associated microflora. Microorganisms present in the rhizosphere are affected by root exudates and play an important role in the growth and ecological fitness of their plant host. Rhizosphere microorganisms are considered to be an important component of soil ecological system (Li et al. 2018a, b). The report of the FAO on environmental effect of GM crops recommended that the environmental effect of Bt crops or any transgene protein should be assessed on a case-by-case basis including their potential impact on local soil microflora and biodiversity. Microbes are in close contact with all three soil phases (solid, water, and air), and they can sensitively and rapidly probe responses to soil perturbations. Bt cotton expressing *cry* proteins has no adverse effect on microbial population and enzymatic activity of rhizosphere soil (Zaman et al. 2015). Soil amended with purified *cry* protein and Bt-cotton tissue proteins decrease rapidly with a half-life of approximately 4 and 7 days, respectively (Palm et al. 1996). The difference in the bacterial population of Bt and non-Bt cotton soil might be attributed to variation in root exudates, quantity, composition, and root characteristics (Yasin et al. 2016). Microbes' dependent phosphatase activity significantly increased in the rhizosphere of Bt cotton as compared to non-Bt cotton fields. The higher soil enzyme activities might be due to more organic matter content, microbial activity, and available nutrients compared to non-Bt (Yasin et al. 2016; Singh et al. 2013). Bt cotton plant material had positive effect on acid and alkaline phosphatase activities, and alkaline activity was much higher than acid phosphatase activity (Sun et al. 2007), because alkaline phosphatase is associated with microorganisms, while the acid phosphatase is predominantly due to plants. Shen et al. (2006) recorded no significant differences in soil enzymes such as urease, phosphatase, DHA, PO, proteases, invertases, cellulases, and arylsulfatases in Bt and non-Bt cotton soil. More dehydrogenase activity was recorded in Bt cotton rhizosphere in contrast to non-Bt rhizosphere and could be due to presence of higher bacterial biomass. Dehydrogenase activity is also often used as an alternative to substrate-induced respiration and has been found to be correlated with microbial activity (Chaperon and Sauve 2007). In Bt rhizosphere the available phosphorus, Zn, and Fe contents were observed higher as compared to non-Bt field, and this might be due to high root biomass-mediated exudates (Beura and Rakshit 2011). Cation exchange capacity, total nitrogen, extractable phosphorus, extractable potassium, active carbon, and Fe and Zn contents were higher in the rhizosphere of Bt cotton genotypes compared with non-Bt cotton genotypes (Yasin et al. 2016). Sarkar et al. (2009) demonstrated that the growth of Bt cotton had positive impact on most of the microbial and biochemical indicators, as microbial biomass carbon, microbial biomass nitrogen, microbial biomass phosphorus, and a range of soil enzyme activities and cultivation of Bt cotton appear to be no risk to soil ecosystem functions. Bt toxin released from root exudates and biomass of Bt corn has no apparent effect on earthworms, nematodes, protozoa, bacteria, and fungi in soil (Saxena and Stotzky 2001). Velmourougane and Sahu (2013) studied the effect of Bt protein on soil biological, microbiological, and diversity attributes at 0–30 cm soil depth and found no adverse effect of Bt cotton on soil microbial activity. Thus, the cultivation of Bt cotton expressing protein had apparently no negative effect on metabolic, microbiological activities and nutrient dynamics of soils.

16.6.4 Impact on Biodiversity

Human activities keep on dominating the planet causing rapid ecosystem changes and massive loss of biodiversity (biological diversity refers to the variety of life on earth) across the planet. Major direct threats to biodiversity include global industrialization, urbanization, and now development and use of genetically modified plants or organisms (Verma 2013). Still there is difference of opinion about the potential for "transgenes" to spread and establish in natural ecosystems. For example, transgenes are expected to have neutral or deleterious effects on antibiotic resistance, reproductive sterility, development of superweeds, etc. Further, people have been worried that the biodiversity will be threatened due to encouraging use of GM crops, which would favor monoculture and affect ecological stability (James et al. 1998; Shiva 1993; Sweet and Shepperson 1997). Although biotech new technology has many advantages to complement conservation of biodiversity, genetic pollution or gene contamination is among the environmental concerns by GM crops which need to be cleared or ascertained before releasing into the commercial use, because it can pollute the biodiversity and the natural or wild genetic pool which may cause irreversible damage (Yohannes and Woldesemayate 2017).

The rhizosphere microbial community structure is a rapid and sensitive indicator of anthropogenic (man-made) effects on soil ecology. Microorganisms present in the rhizosphere are affected by root exudates and play an important role in the growth and ecological fitness of their plant host. Kapur et al. (2010) assessed the culturable and non-culturable microbial diversities in Bt cotton and non-Bt cotton soil and found that cropping of Bt cotton did not affect adversely the diversity of the microbial communities. Li et al. (2018a, b) analyzed the diversity and dynamics of rhizosphere fungal community on lateral and taproots of Bt cotton using qPCR and 18S rRNA gene sequencing and found no significant differences in population sizes of Bt and conventional cotton varieties root zones. Further, they suggested that the dominant and rare fungal taxa differentially contribute to community dynamics in different root microhabitats of both Bt and conventional cotton variety. Most studies have been carried out to assess the accumulation and persistence of Bt proteins in soil in which Bt crops have been continuously cultivated for several years (Head et al. 2002; Dubelman et al. 2005; Icoz et al. 2008; Zhang et al. 2019). The association between functional groups microorganisms involved in C, N, and P recycling and their influence on plant growth are potential indicators of the impacts of disturbance on the soil environment (Ferreira et al. 2003). Zhang et al. (2019) recorded significantly higher soil microbial communities in the transgenic Bt cotton field than the non-transgenic cotton field. Further, carbon sources including amino acids, amines, and carbohydrates were utilized significantly by the soil microbial communities. However, *cry1Ac* protein did not accumulate in the fields for the next crop season, but the functional diversity of soil microbial communities was affected continuously.

Herbicide-resistant crops are great concern for the loss of weed diversity, which would come out due to gene flow from herbicide-resistant crops to weeds. The currently available herbicide-resistant GM crops confer broad-spectrum herbicides like glufosinate and glyphosate. Since the introduction of glyphosate-resistant GM crops, about 38 weed species worldwide have been identified that have developed resistance to glyphosate, distributed across 37 countries and in 34 different crops and 6 non-crop situation (Heap and Duke 2018). Almost 50% surveyed farms are infested with glyphosate-resistant "superweeds," and these weeds are spreading very fast (www.ucsusa.org/superweeds). Thus, continuously spraying chemical toxic herbicide may alter the diversity of weeds field habitats. The diversity of weeds edible green leaf weeds get completely devastated in the agricultural landscape, which also affects reduction in the diversity of beneficial insects (Tappeser et al. 2014). Continuous use of herbicide has led to modification in the foraging behavior of insects. The best example is the reduced emigration and excessive feeding on crickets by wolf spiders in response to glyphosate application in the Western USA (Wrinn et al. 2012; Marchetti 2014). Another well-known example is the reduction of monarch butterfly populations in the USA and Mexico, which is due to nonavailability of milkweeds by continuous application of herbicide spray. Milkweeds are the main host plant for the monarch larvae (Brower et al. 2012). Increased application of glyphosate results in massive mortality of aquatic life on farmlands (Isenring 2010). Significant reduction in genetic diversity and variable population frequencies of many insects and weeds have been observed as a consequence of gene flow (NAS 2016). In the future glyphosate-resistant weeds are going to be a great threat to sustained weed control in major agronomic crops.

16.7 Insect Resistance to Bt Crops

After the first commercial release in 1995 and subsequent widespread adoption of insecticidal *cry* toxin (crystal protein from *Bacillus thuringiensis* bacterium)-expressing Bt crops, the evolution of resistance was anticipated in the target pest populations. However, despite the remarkable ability of pest populations to quickly adapt to a myriad of pest control strategies, the cases of field-evolved resistance in important target pests were not documented till 2003 (Tabashnik et al. 2003). Usually, the field populations of key target pests surviving on both Bt and nearby non-Bt host plants were continuously monitored for any sign of field-evolved resistance, which is defined as genetically based decrease in susceptibility of one or more field populations to a toxin in the field (Tabashnik and Carrière 2017). In 2008, based upon the extensive field monitoring datasets spanning before and after Bt commercialization (1992–2006), the first case of field-evolved resistance in some field populations *Helicoverpa zea* against *cry1Ac*-expressing Bt cotton was reported in the USA (Tabashnik et al. 2008). Thereafter, the total number of cases of field-evolved resistance with practical significance has gradually increased from 3 in 2005 to 22 cases in 2018 (Smith et al. 2019; Tabashnik et al. 2020), comprising of 10 insecticidal toxins (9 *cry* and 1 *vip3a* toxin) targeted against some 8 pest species (6 lepidopteron and 2 coleopteron) in 6 Bt crop-growing countries including 12 cases in the USA, 3 in Argentina, and 2 each in Brazil, India, Canada, and South Africa (Fig. 16.3).

In general, pest responses to Bt crops have been classified into three main categories, viz., (1) practical resistance, (2) early warning resistance, and (3) no decreases in susceptibility (Tabashnik et al. 2013; Tabashnik and Carrière 2019). Practical resistance and early warning resistance are field-evolved resistance characterized by a genetically based decrease in susceptibility in field-selected population upon exposure to a toxin in the field. In addition, the likelihood of evolution of cross-resistance against two or more toxins in field is also included. The practical resis-

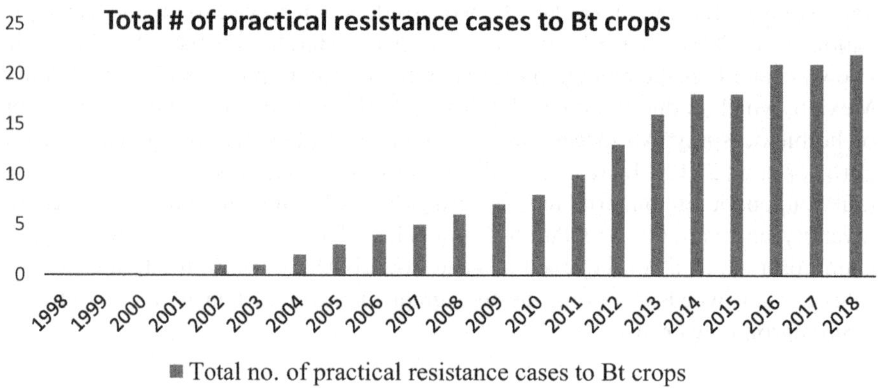

Fig. 16.3 Total number of practical resistance cases to Bt crops

tance can be defined as cases of a field-evolved resistance where more than 50% individuals have been found to be resistant to a toxin with practical field control problems evidenced with a significant reduction of Bt crop efficacy (Tabashnik et al. 2013; Tabashnik and Carrière 2017), while early warning resistance includes all cases of field-evolved resistance with a statistically significant decline in genetically based susceptibility without any evidence of reduced efficacy of Bt crop. The category 3 consists of all cases of pest responses to Bt crops where field resistance monitoring data suggests no sign of a statistically significant decrease in susceptibility in field pest population to the expressed toxin in Bt crops.

16.7.1 Insect Responses to Bt Cotton

A total of 19 pest responses to Bt cotton have been documented with 4 cases each in the category of practical resistance and early warning resistance, while 11 cases fit into the category 3 characterized with no significant decreases in susceptibility (Tables 16.3 and 16.4). In Bt cotton, cases of field-evolved resistance with practical implications are mainly associated with two major lepidopteron cotton pests, viz., *Helicoverpa zea* and *Pectinophora gossypiella* (pink bollworm) in the USA and India, respectively.

16.7.2 Reports of Insect Resistance to Bt Cotton in the USA

The analysis of field monitoring data (1992–2006) for some populations of *Helicoverpa zea* surviving on *cry1Ac*-expressing Bt cotton from southeastern states (Arkansas and Mississippi) of the USA reveals the first incidence of field-evolved resistance (Luttrell et al. 1999; Ali and Luttrell 2007; Ali et al. 2007; Tabashnik et al. 2008, 2009). In general, resistance ratio (RR) value greater than 10 often implies a significant decline in the heritable susceptibility in tested pest population (Tabashnik 1994). The laboratory bioassays with field-selected strains of *Helicoverpa zea* from Arkansas and Mississippi collected during 2003–2006 identified several strains with significantly increased resistance ratio from 1.2 to >(50–1000) compared to the population sampled during 1992–1993 (pre-Bt commercialization) (Ali et al. 2007; Tabashnik et al. 2008, 2009). The results of laboratory bioassays using diet-incorporated *cry1Ac* toxin and percent survival of field-resistant *Helicoverpa zea* larvae have suggested for a substantial decrease in the genetic-based larval susceptibility to *cry1Ac*. In December 2002, the second-generation pyramided Bt cotton called Bollgard II (*cry1Ac* + *cry2Ab*) has been commercially released and widely adopted. Field monitoring data from Bollgard II during 2002–2005 revealed a significant gain in the percentage (from 0 to 50%) of individuals of *Helicoverpa zea* populations with RR >10 and *cry1Ac* LC_{50} > 150 µg per mL of diet and indicated a positive correlation between percent survival

Table 16.3 Global status of responses of field populations of target pests against insecticidal toxins expressed in Bt cotton[a]

Category 1 (practical resistance: >50% resistant individuals with reduced field efficacy of Bt reported)							
Pest species	Country	Toxin	Year[b]	Years[c]	High toxin dose	Low initial resistance allele frequency	References
Helicoverpa zea	USA	*cry1Ac*	1996	6	No	No	Luttrell et al. (1999); Ali et al. (2007); Tabashnik et al. (2008)
	USA	*cry2Abcry2Ab*	2003	2	–	No	
Pectinophora gossypiella	India	*cry1Ac*	2002	6	No	–	Dhurua and Gujar (2011); Naik et al. (2018)
	India	*cry2Abcry2Ab*	2006	8	–	–	Naik et al. (2018)

– Data not available
[a]Data adopted from Tabashnik et al. (2013), Tabashnik and Carrière (2017, 2019)
[b]First year of Bt cotton introduction in the region surveyed for field pest population
[c]Total years from the initial year of Bt cotton introduction to the first evidence of field-evolved resistance in the region surveyed for pest resistance

and *cry2Ab* resistance (Ali and Luttrell 2007; Tabashnik et al. 2009). Similarly, LC_{50} values of both *cry1Ac* and *cry2Ab* in the 61 strains of *Helicoverpa zea* sampled during 2004–2006 were also found in positive correlation, $r = 0.32$. In comparison to susceptible strains, compelling data with five- to sevenfold high LC_{50} for *cry1Ac* and four- to sixfold increased survival on Bollgard II cotton leaves suggested that field-evolved resistance for *cry2Ab* in *Helicoverpa zea* was positively associated with *cry1Ac* resistance. The typically weak but statistically significant cross-resistance to *cry2Ab* in some field populations of *Helicoverpa zea* has unusually accelerated the *cry2Ab* resistance (Tabashnik et al. 2009; Tabashnik and Carrière 2013).

16.7.3 Reports of Pest Resistance to Bt in India

So far, Bt cotton is the only transgenic crop approved for commercial planting in India. The Bt cotton technology is known as Bollgard® (expressing single *cry1Ac* toxin) which was commercialized for the first time in India in 2002 and planted in 50,000 ha (Choudhary and Gaur 2010). Later on, in 2006, Bollgard II was also released for commercial planting with an idea of delaying the evolution of pest resistance and for continued benefits from Bt technology (Choudhary and Gaur 2010; Naik et al. 2018). The cotton cultivation in India is primarily dominated by Bt

Table 16.4 Global status of responses of field populations of target pests against insecticidal toxins expressed in Bt cotton[a]

Pest species	Country	Toxin	Year[b]	Years[c]	References
Category 2: early warning resistance cases (field-evolved resistance with statistically significant decline in genetically based susceptibility without any evidence of reduced efficacy of Bt crop)					
Helicoverpa zea	India	*cry1Ac*	2002	12	Kukanur et al. (2018)
	Pakistan	*cry1Ac*	2010	3	Rashid et al. (2008)[d]
	USA	vip3Aa	2010	8	Yang et al. (2019)
	China	*cry1Ac*	1997	20	Dandan et al. (2019)
Category 3: no significant reduction in genetically based susceptibility in field pest population					
Earias biplaga	South Africa	*cry 1Ac*	1998	15	Fourie et al. (2017)
Helicoverpa armigera	Australia	*cry 1Ac*	1996	16	Bird (2015)
	Australia	*cry2Abcry2Ab*	2004	11	
Helicoverpa punctigera	Australia	*cry 1Ac*	1996	19	Walsh et al. (2018)
	Australia	*cry2Abcry2Ab*	2004	11	Bird (2015)
Helicoverpa virescens	Mexico	*cry1Ac*	1996	11	Blanco et al. (2009)
	USA	*cry 1Ac*	1996	11	
	USA	*cry2Abcry2Ab*	2003	2	Ali et al. (2007)
Pectinophora gossypiella	China	*cry1Ac*	2000	15	Wan et al. (2017)
	USA	*cry 1Ac*	1996	12	Tabashnik and Carrière (2019)
	USA	*cry2Abcry2Ab*	2003	5	

[a]Data adopted from Tabashnik et al. (2013), Tabashnik and Carrière (2017, 2019)
[b]First year of Bt cotton introduction in the region surveyed for field pest population
[c]Total years from the initial year of Bt cotton introduction to the recent year of collection of field monitoring data
[d]Although illegal Bt cotton cultivation has been reported since 2004, commercial planting got official approval only in 2010 onward

cotton hybrids developed through the breeding of Bt variety expressing *cry* toxin(s) with non-Bt cultivar (Choudhary and Gaur 2010). In 2018, the total area under Bt cotton hybrids increased to 50,000 ha in 2002 to 11.6 million ha in 2018, benefiting more than 7.5 million farm families (ISAAA 2018). During the initial years of its introduction, the Bt cotton technology has been very effective in controlling major devastating pest of cotton including *Pectinophora gossypiella*, *Helicoverpa armigera*, *Earias vittella* (spotted bollworm), and *Earias insulana* (spiny bollworm) (Dhurua and Gujar 2011; Naik et al. 2018). In 2009, pink bollworm larvae surviving on *cry1Ac*-expressing Bt cotton plants from fields of four districts of Gujarat state

(India), viz., Rajkot, Amreli, Bhavnagar, and Junagarh, have confirmed the evolution of *cry1Ac* resistance in pink bollworm in laboratory bioassays (Monsanto Cotton India *http://www.monsantoindia.com/monsanto/layout/pressreleases/mmb*). However, the first case of field-evolved resistance with practical field control problem was reported from Amreli district of Gujarat state (Dhurua and Gujar 2011). In the 5-day-old offsprings of field-resistant *Pectinophora gossypiella* collected from Amreli district of Gujarat in 2008, a significantly higher resistance ratio (RR 44) was found to be associated with increased *cry1Ac* LC_{50} concentration (mean value of 1.64 μg per mL of diet). At the same time, the most susceptible population was recorded with a mean LC_{50} value of 0.050 μg per mL. In addition, a significant reduction in mortality (24–31 v/s 97%) in resistant insects was also recorded in bioassays using 1 μg per mL *cry1Ac*. In another study, the field-evolved resistance to *cry1Ac* in field-resistant pink bollworm progenies was also confirmed (Mohan et al. 2016). The study has found that offsprings of field-collected pink bollworm (Amreli, Gujarat) showed a sizable tolerance to *cry1Ac* concentrations of 1.0–10 μg per mL of diet. In a recent report, the pink bollworm response to Bollgard® (single *cry1Ac* toxin) and Bollgard II (*cry1Ac* + *cry2Ab*) was evaluated in the resistance monitoring data collected during 2010–2017 (Naik et al. 2018). A significantly higher resistance ratio to *cry1Ac* (26–262) and *cry2Ab* (1–108) with a substantially high percentage of average pest survival (28.85–72.49%) on Bt-II cotton (expressing *cry1Ac* + *cry2Ab*) was reported. The mean LC_{50} for *cry1Ac* has increased from 0.330 to 6.938 μg/mL from 2013 to 2017, while for *cry2Ab* increased from 0.014 to 12.51 μg/mL during 2013–2017. The study thus confirms the evidence of field-evolved resistance against *cry1Ac* + *cry2Ab* 2 expressing Bollgard II in India, particularly in central and south cotton-growing states.

However, with the introduction of Bt crops, refuge strategy (growing non-Bt hosts along with Bt crops) has also been extensively adopted to delay the evolution of resistance (Tabashnik 2008; Hutchison et al. 2010). The predictions from population genetic models suggested that the evolution of resistance can be postponed to more than 20 years with $\geq$ 5% refuge cover in a condition that the estimated initial resistance allele frequency should be 0.001 and the resistance preferably be governed by two alleles with single locus genetic architecture with completely recessive inheritance (h = 0) (Tabashnik et al. 2008). However, in most cases of field-evolved resistance reported globally, the non-recessive inheritance of resistance alleles along with low refuge abundance, not meeting high toxin dose standards, was considered among the major factors contributed to rapid evolution of field-evolved resistance (Tabashnik et al. 2008; Tabashnik and Carrière 2019).

16.8 Reports on Weed Resistance to Herbicides

Among the various tools, herbicides play essential roles in weed management of almost all the agricultural systems. However, frequent and indiscriminate use of few selected herbicides sharing a similar mode of action has resulted in the increasing cases of the evolution of herbicide resistance in weeds (Beckie 2006). WSSA (the Weed Science Society of America) is a nonprofit professional society; aims to promote research, education, and awareness of weeds in managed and natural ecosystems; and defines herbicide resistance as "the inherited ability of a plant to survive and reproduce following exposure to a dose of herbicide normally lethal to the wild type" (http://wssa.net/). Globally, a total of 514 unique cases of herbicide resistance involving 262 weed species (152 dicots and 110 monocots) with confirmed herbicide resistance against 167 different herbicides targeting 23 of the 26 known herbicide sites of action have been documented in a regularly updating "International Herbicide-Resistant Weed Database" (http://www.weedscience.org/home.aspx) as on data updated in July 2020 (Table 16.5). In cotton, a total of 81 cases of herbicide resistance in 18 different weed species have been documented from 8 countries (Fig. 16.4). Palmer amaranth (*Amaranthus palmeri*) is a leading weed documented in 24 cases of multiple resistances with 5 sites of herbicide actions including EPSP synthase inhibitors (G/9), ALS inhibitors (B/2), PPO inhibitors (E/14), microtubule inhibitors (K1/3), and long-chain fatty acid inhibitors (K3/15).

16.9 GM Cotton Products Commercialized

Even though GM products have been developed for many economic traits, only GM products for insect and herbicide resistance have been commercialized successfully and occupied more than 90% of the total cultivated area. Globally GM cotton is available in the form of insect and herbicide tolerance either as single or stacked traits. In addition recently GM cotton for low gossypol developed by Texas A&M University was commercialized in the USA. Details of gene and transgenic events of insect and herbicide tolerance commercialized (Source: ISAAA's GM Approval Database. *http://www.isaaa.org/gmapprovaldatabase/*) are enlisted in Tables 16.6 and 16.7.

16.10 Global Status of Adoption of GM Cotton

GM cotton is one of the major crops that were first granted permission for commercial cultivation, and today it occupies 13% of the global area (Paul and Hennig 2019). Thirty-eight countries contribute to 98% of global cotton production. Genetically modified cotton is cultivated in 23 countries. Ninety percent of global cotton production comes from just ten countries—Australia, Brazil, Burkina Faso, China, India, Pakistan, Turkey, Turkmenistan, Uzbekistan, and the USA—and of these top cotton-producing countries, Turkey is the only exception as it does not cultivate GM cotton.

Turkey, Greece, and Spain serve a classic example for obtaining high yields without GM technologies. High yields are obtained in Turkey, without the adoption

Table 16.5 Global status of total number of confirmed cases of weed resistance to herbicides

Sr no	Weed species	Common name	Cases no.	Country	Herbicide site of action (WSSA code)
1	*Amaranthus palmeri*	*Palmer amaranth*	24	Brazil, Israel, Mexico, and USA	EPSP synthase inhibitors (G/9), ALS inhibitors (B/2), PPO inhibitors (E/14), microtubule inhibitors (K1/3), long-chain fatty acid inhibitors (K3/15)
2	*Conyza canadensis*	*Horseweed*	9	USA	EPSP synthase inhibitors (G/9), ALS inhibitors (B/2)
3	*Eleusine indica*	*Goosegrass*	9	Brazil and USA	ACCase inhibitors (A/1), EPSP synthase inhibitors (G/9), microtubule inhibitors (K1/3)
4	*Sorghum halepense*	*Johnsongrass*	7	Greece, Israel, and USA	ACCase inhibitors (A/1), ALS inhibitors (B/2), microtubule inhibitors (K1/3)
5	*Xanthium strumarium*	*Common cocklebur*	7	USA	Nucleic acid inhibitors (Z/17)
6	*Amaranthus tuberculatus (=A. rudis)*	*Tall waterhemp*	6	USA	EPSP synthase inhibitors (G/9), ALS inhibitors (B/2)
7	*Amaranthus retroflexus*	*Redroot pigweed*	3	Brazil	ALS inhibitors (B/2), photosystem II inhibitors (C1/5), PPO inhibitors (E/14)
8	*Lolium perenne* ssp. *multiflorum*	*Italian ryegrass*	3	USA	EPSP synthase inhibitors (G/9)
9	*Echinochloa colona*	*Junglerice*	2	Australia and USA	EPSP synthase inhibitors (G/9)
10	*Ambrosia artemisiifolia*	*Common ragweed*	2	USA	EPSP synthase inhibitors (G/9)
11	*Amaranthus spinosus*	*Spiny amaranth*	2	USA	EPSP synthase inhibitors (G/9), ALS inhibitors (B/2)
12	*Sonchus oleraceus*	*Annual sowthistle*	1	Australia	EPSP synthase inhibitors (G/9)
13	*Amaranthus viridis*	*Slender amaranth*	1	Brazil	ALS inhibitors (B/2), photosystem II inhibitors (C1/5)
14	*Ageratum conyzoides*	*Tropical whiteweed*	1	Brazil	ALS inhibitors (B/2)
15	*Digitaria sanguinalis*	*Large crabgrass*	1	China	ACCase inhibitors (A/1)
16	*Digitaria insularis*	*Sourgrass*	1	Paraguay	EPSP synthase inhibitors (G/9)
17	*Ambrosia trifida*	*Giant ragweed*	1	USA	EPSP synthase inhibitors (G/9)
18	*Kochia scoparia*	*Kochia*	1	USA	EPSP synthase inhibitors (G/9)

Source: INTERNATIONAL HERBICIDE-RESISTANT WEED DATABASE (*http://www.weed-science.org/Pages/crop.aspx*)

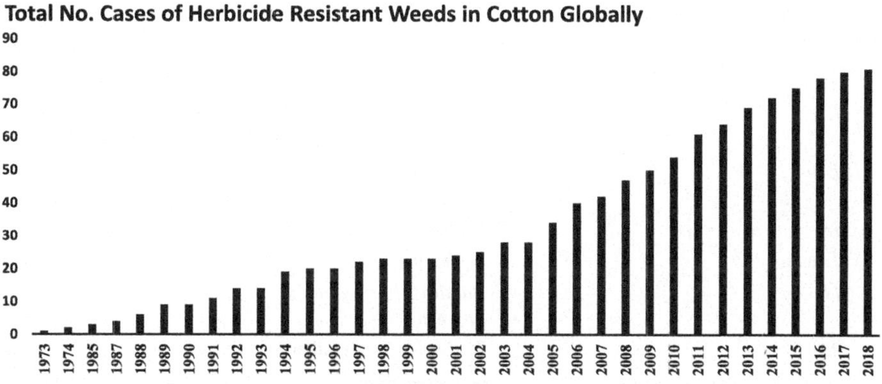

Fig. 16.4 Year-wise distribution of the total number of cases of herbicide-resistant weed species in cotton cropping system globally. Palmer amaranth (*Amaranthus palmeri*) is a leading weed documented in 24 cases of multiple resistances with 5 sites of herbicide actions including EPSP synthase inhibitors (G/9), ALS inhibitors (B/2), PPO inhibitors (E/14), microtubule inhibitors (K1/3), and long-chain fatty acid inhibitors (K3/15)

of Bt cotton. Turkey is an important producer and consumer of cotton. Turkey produces 10,000 tonnes of organic cotton which was expected to increase to 15,000 tonnes in the current 2020–2021 season. The private sector today provides almost all the hybrid cotton seeds in Turkey. Cotton is cultivated in three regions (GAP region, Cukurova region, and Aegean region), and the most dominant region is the GAP region that accounts for about 60.0% of the cotton acreage. For some time now, several steps were taken by the government to benefit cotton production in Turkey. In the GAP region, dams and irrigation channels were constructed that were expected to facilitate an irrigated area of 650,000 ha of land. Open canal system of irrigation was replaced with closed systems. Financial assistance and technical guidance for drip irrigation was provided by the government. Government incentivized cotton production by giving a bonus of 0.8 lira (US 12 cents) for every kilogram of cotton produced. Licensed storage facilities were set up in GAP and Izmir for 15,000 tonnes and 10,000 tonnes, respectively. Turkey spends US$ 77, US$ 400, US$ 546, and US$ 26 per hectare on seeds, fertilizers, pesticides, and manpower, respectively. The cost of cultivation is US$ 413 per hectare, and the production cost is US$ 1.55 per kg of lint (including seed value) and US$ 0.59 per kg of seed cotton (ICAC Cotton Data Book 2020).

Table 16.6 Details of genes for insect resistance and herbicide tolerance in commercialized GM cotton

Genes for insect resistance	Source of genes	Target trait
cry1Ac	*Bacillus thuringiensis* subsp. *kurstaki* strain HD73	*Resistance to lepidopteran insects*
cry1Ab	*B. thuringiensis* subsp. kurstaki	
cry1A	*B. thuringiensis*	
cry1Ab-Ac	*B. thuringiensis*	
cry1F	*Bacillus thuringiensis* var. *aizawai*	
cry2Ab2	*B. thuringiensis* subsp. *kumamotoensis*	
cry2Ae	*B. thuringiensis* subsp. dakota	
vip3A(a) codes for vegetative insecticidal protein	*Bacillus thuringiensis* strain AB88	
cpTI-trypsin inhibitor	*Vigna unguiculata*	*Broad spectrum*
mCry51Aa2	*B. thuringiensis*	*Hemipteran insects Lygus hesperus and L. lineolaris*
Herbicide tolerance in cotton		
cp4 epsps (aroA:CP4)	*Agrobacterium tumefaciens* strain CP4	*Glyphosate tolerance*
2mepsps	*Zea mays*	
pat gene coding for phosphinothricin-*N*-acetyltransferase (PAT)	*Streptomyces viridochromogenes*	*Glufosinate tolerance*
Bar phosphinothricin-*N*-acetyltransferase (PAT) enzyme	*Streptomyces hygroscopicus*	
Bxn produces nitrilase enzyme	*Klebsiella pneumoniae* subsp. *ozaenae*	*Bromoxynil tolerance*
S4-HrA herbicide-tolerant acetolactate synthase (ALS)	*Nicotiana tabacum* cv. Xanthi	Sulfonylurea tolerance
Dmo dicamba monooxygenase enzyme	*Stenotrophomonas maltophilia strain DI-6*	*Dicamba tolerance*
aad-12 aryloxyalkanoate di-oxygenase 12 (AAD-12) protein	*Delftia acidovorans*	*2,4,-D tolerance*
Modified p-hydroxyphenylpyruvate dioxygenase (hppd) enzyme (hppdPF W336)	*Pseudomonas fluorescens strain A32*	*Isoxaflutole tolerance*
Low gossypol seed		
dCS D*-cadinene synthase gene*	*Gossypium hirsutum*	*Low gossypol seed*

Source: ISAAA's GM Approval Database. *http://www.isaaa.org/gmapprovaldatabase/*

Table 16.7 Details of commercialized GM cotton transgenic events for insect resistance and herbicide tolerance

Event name	Gene	Trait	Country and year of approval	Trade name/ developer
GM cotton for insect resistance				
MON531	*cry1Ac*	Resistance to lepidopteran insects	USA 1995, Mexico 1996, South Africa 1997, Argentina 1998, India 2002, Australia 2003, Colombia 2003, Brazil 2005, Paraguay 2007, Costa Rica 2008, Sudan 2012	Bollgard™ cotton, Ingard™
MON757	*cry1Ac*	Resistance to lepidopteran insects	USA 1995, South Africa 1997	Bollgard™ cotton
MON1076	*cry1Ac*	Resistance to lepidopteran insects	USA 1995, South Africa 1997	Bollgard™ cotton
GK12	*cry1Ab-Ac*	Resistance to lepidopteran insects	China 1997	Chinese Academy of Agricultural Sciences
281-24-236	*cry1F*	Resistance to lepidopteran insects	Mexico 2004, USA 2004	Dow AgroSciences LLC
3006-210-23	*cry1Ac*	Resistance to lepidopteran insects	Mexico 2004, USA 2004	Dow AgroSciences LLC
COT102 (IR102)	*vip3A(a)*	Resistance to lepidopteran insects	USA 2005, Australia 2018, Costa Rica (seed production only) 2017	VIPCOT™ cotton
COT102 (IR102)	*vip3A(a)*	Resistance to lepidopteran insects	USA 2005, Australia 2018	VIPCOT™ cotton
Event 1	*cry1Ac*	Resistance to lepidopteran insects	India 2006, Eswatini 2018, Ethiopia 2018	JK 1
GFM *cry1A*	*cry1Ab-Ac*	Resistance to lepidopteran insects	India 2006	Nath Seeds/Global Transgenes Ltd (India)
MLS 9124	*cry1C*	Resistance to lepidopteran insects	India 2009	Metahelix Life Sciences Pvt. Ltd (India)
GHB119	*cry2Ae*	Resistance to lepidopteran insects	USA 2011	BASF

(continued)

Table 16.7 (continued)

Event name	Gene	Trait	Country and year of approval	Trade name/ developer
COT67B	*cry1Ab*	Resistance to lepidopteran insects	USA 2011, Costa Rica (seed production only) 2017	Syngenta
MON15985	*cry1Ac* *cry2Ab2*	Resistance to lepidopteran insects	USA 2002, Australia 2002, South Africa 2003, India 2006, Brazil, Burkina Faso, Costa Rica 2009	Bollgard II™ cotton
281-24-236 × 3006-210-23 (MXB-13)	*cry1Ac* *cry1F*	Resistance to lepidopteran insects	Mexico 2004, Australia 2009, Brazil 2009, Costa Rica 2009	WideStrike™ cotton
COT102 × COT67B	*vip3A*(a) *cry1Ab*	Resistance to lepidopteran insects	Costa Rica 2009	VIPCOT™ cotton
COT102 × MON15985	*vip3A(a)* *cry1Ac,* *cry2Ab2*	Resistance to lepidopteran insects	Australia 2014	Bollgard® III
281-24-236 × 3006-210-23 × COT102	*cry1Ac* *cry1F,* *vip3A(a)*	Resistance to lepidopteran insects	Brazil 2018	Dow AgroSciences LLC
MON88702	*mcry51Aa2*	Resistance to hemipteran insects	Australia, Canada, New Zealand (food), Japan, USA 2018 (food and feed purpose)	Monsanto Company (including fully and partly owned companies)
GM cotton for herbicide tolerance				
BXN10211/ BXN10215 (10215)/ BXN10222 (10222)/ BXN10224 (10224)	*bxn*	*Bromoxynil tolerance*	USA 1994	BXN™ Plus Bollgard™ cotton
MON1445	*cp4 epsps (aroA:CP4)*	*Glyphosate tolerance*	USA 1995, Mexico 2000, South Africa 2000, Argentina 2001, Australia 2003, Japan 2004, Columbia 2004, Brazil 2008, Costa Rica 2008, Paraguay 2013	Roundup Ready™ cotton
MON1698	*cp4 epsps (aroA:CP4)*	*Glyphosate tolerance*	USA 1995, Mexico 2000, South Africa 2000	Roundup Ready™ cotton
19-51a	*S4-HrA*	Sulfonylurea tolerance	USA 1996	DuPont (Pioneer Hi-Bred Int. Inc.)

(continued)

Table 16.7 (continued)

Event name	Gene	Trait	Country and year of approval	Trade name/ developer
LLCotton25	*bar*	*Glufosinate tolerance*	USA 2003, Brazil 2008, Costa Rica 2009, Colombia 2010	Fibermax™ LibertyLink™
MON88913	*cp4 epsps (aroA:CP4)*	*Glyphosate tolerance*	USA 2004, Mexico 2006, Australia 2006, Japan 2006, South Africa 2007, Costa Rica 2009, Brazil 2011	Roundup Ready™ Flex™ cotton
GHB614	*2mepsps*	*Glyphosate tolerance*	USA 2009, Costa Rica 2009, Brazil 2010, Argentina 2012, Australia 2016	GlyTol™
GHB614 × LLCotton25	*2mepsps, bar*	*Glyphosate tolerance Glufosinate tolerance*	Brazil 2012, Colombia 2013, Argentina 2015	GlyTol™ LibertyLink™
MON88701	*Dmo bar*	*Dicamba tolerance Glufosinate tolerance*	USA 2014, Brazil 2017, Costa Rica (seed production only) 2016	Monsanto Company (including fully and partly owned companies)
81910	*pat*	*Glufosinate tolerance*	USA 2015, Brazil 2018, Costa Rica (seed production only) 2016	Dow AgroSciences LLC
MON88701 × MON88913	*dmo, bar, cp4 epsps (aroA:CP4)*	*Dicamba tolerance Glufosinate tolerance Glyphosate tolerance*	Australia 2016 Brazil 2018	Monsanto Company (including fully and partly owned companies)
GHB811	*hppdPF W336, 2mepsps*	*Isoxaflutole tolerance Glyphosate tolerance*	USA 2018, Brazil 2019	BASF
Insect resistance and herbicide tolerance				
31807	*Bxn cry1Ac*	Oxynil tolerance Resistance to lepidopteran insects	USA (1997)	BXN™ Plus Bollgard™ cotton
31808	*Bxn cry1Ac*	Oxynil tolerance Resistance to lepidopteran insects	USA (1997)	BXN™ Plus Bollgard™ cotton

(continued)

Table 16.7 (continued)

Event name	Gene	Trait	Country and year of approval	Trade name/ developer
MON531 × ON1445	*cry1Ac* *cp4 epsps* *(aroA:CP4)*	Resistance to lepidopteran insects Glyphosate tolerance	Mexico 2002, Australia 2003, Japan 2004, South Africa 2005, Columbia 2007, Argentina 2009, Brazil 2009, Costa Rica 2009, Paraguay 2013	Roundup Ready™ Bollgard™ cotton
GHB119	*cry2Ae* *bar*	Resistance to lepidopteran insects Glufosinate tolerance	USA 2011	Bayer Crop Science
T303-3	*cry1Ab* *bar*	Resistance to lepidopteran insects Glufosinate tolerance	USA 2012	Bayer Crop Science (including fully and partly owned companies)
T304-40	*cry1Ab* *bar*	Resistance to lepidopteran insects Glufosinate tolerance	USA 2011	Bayer Crop Science (including fully and partly owned companies)
MON15985 × MON1445	*cry1Ac* *cry2Ab2* *cp4 epsps*	Resistance to lepidopteran insects Glyphosate tolerance	Australia 2002, Costa Rica 2009	Roundup Ready™ Bollgard II™ cotton
COT102 × COT67B × MON88913	*cry1Ab* *vip3A(a)* *cp4 epsps*	Resistance to lepidopteran insects Glyphosate tolerance	Costa Rica 2009	VIPCOT™ Roundup Ready Flex™ cotton
MON-88913-8 × MON-15985-7	*cry1Ac* *cry2Ab2* *cp4 epsps*	Confers resistance to lepidopteran insects Glyphosate tolerance	Mexico 2006, Australia 2006, Japan 2006, South Africa 2007, Columbia 2007, Costa Rica 2009, Brazil 2012, Paraguay 2017	Roundup Ready™ Flex™ Bollgard II™ cotton
3006-210-23 × 281-24-236 × MON1445	*cp4 epsps* *cry1F* *cry1Ac, bar*	Glyphosate tolerance Resistance to lepidopteran insects Glufosinate tolerance	Japan (2006) Mexico (2005) Costa Rica (2009)	WideStrike™ Roundup Ready™ cotton Monsanto Company and Dow AgroSciences LLC

(continued)

Table 16.7 (continued)

Event name	Gene	Trait	Country and year of approval	Trade name/ developer
3006-210-23 × 281-24-236 × MON88913	*cp4 epsps* *cry1F,* *cry1Ac* *bar*	Glyphosate tolerance Resistance to lepidopteran insects Glufosinate tolerance	Japan (2006), Costa Rica (2009)	WideStrike™ Roundup Ready Flex™ cotton
LLCotton25 × MON15985	*cry1Ac* *cry2Ab2* *bar*	Confers resistance to lepidopteran insects Glufosinate tolerance	Australia 2006	Fibermax™ LibertyLink™ Bollgard II™
COT102 × COT67B × MON88913	*cry1Ab* *vip3A(a)* *cp4 epsps*	Resistance to lepidopteran insects Glyphosate tolerance	Costa Rica (2009)	VIPCOT™ Roundup Ready Flex™ cotton Syngenta and Monsanto Co.
T304-40 × GHB119	*cry2Ae* *cry1Ab* *Bar*	Resistance to lepidopteran insects Glufosinate tolerance	Brazil 2011 USA 2011 Argentina 2014	TwinLink™
GHB614 × T304-40 × GHB119	*2mepsps* *Bar, cry2Ae* *cry1Ab*	Glyphosate and glufosinate tolerance Resistance to lepidopteran insects	Brazil 2012	GlyTol™ × TwinLink™
GHB614 × LLCotton25 × MON15985	*cry1Ac* *cry*2Ab2 *Bar* *2mepsps*	Resistance to lepidopteran insects Glufosinate tolerance Glyphosate tolerance	Japan 2011	Bayer Crop Science
COT102 × MON15985 × MON88913	*cry1Ac,* *cry2Ab2* *vip3A(a)* *cp4 epsps*	Resistance to lepidopteran insects Glyphosate tolerance	Australia 2014 Brazil 2016	Bollgard® III × Roundup Ready™ Flex™

(continued)

Table 16.7 (continued)

Event name	Gene	Trait	Country and year of approval	Trade name/ developer
MON88701 × MON88913 × MON15985	*cry1Ac*, *cry2Ab2* *cp4 epsps* *bar*, *dmo*	Resistance to lepidopteran insects, Glyphosate tolerance Glufosinate tolerance Dicamba tolerance	Japan 2015 (for type I use only)	Monsanto Company (including fully and partly owned companies)
DAS-24236-5 × DAS-21Ø23-5 × SYN-IR1Ø2-7 × DAS-81910-7	*cry1F*, *cry1Ac* *vip3A(a)* *pat*	Resistance to lepidopteran insects Glufosinate tolerance	Brazil 2019, Japan 2016 (food and feed)	Dow AgroSciences LLC
DAS-21Ø23-5 × DAS-24236-5 × MON-88913-8 × SYN-IR1Ø2-7 × DAS-81910-7	*cry1F*, *cry1Ac* *vip3A(a)* *cp4 epsps* *pat*	Resistance to lepidopteran insects Glyphosate tolerance Glufosinate tolerance	Japan 2015, Mexico 2016, South Korea 2016–2017 (food and feed)	Dow AgroSciences LLC
COT102 × MON15985 × MON88913 × MON88701	*cry1Ac*, *cry2Ab2*, *vip3A(a)* *cp4 epsps* *Bar*, *dmo*	Resistance to lepidopteran insects Glyphosate tolerance Glufosinate tolerance Dicamba tolerance	Australia 2016 Brazil 2018	Monsanto Company (including fully and partly owned companies)
GHB614 × T304-40 × GHB119 × COT102	*2mepsps* *Bar*, *cry2Ae* *cry1Ab* *vip3A(a)*	Glyphosate and glufosinate tolerance Resistance to lepidopteran insects	Australia 2016 Brazil 2017	GlyTol™ × TwinLink™ × VIPCOT™ cotton
GHB614 × T304-40 × GHB119	*2mepsps* *Bar*, *cry2Ae* *cry1Ab* *vip3A(a)*	Glyphosate and glufosinate tolerance Resistance to lepidopteran insects	Approved for food (South Korea) and feed (Taiwan 2016) Feed (South Korea 2015)	GlyTol™ × TwinLink™ × VIPCOT™ cotton

(continued)

Table 16.7 (continued)

Event name	Gene	Trait	Country and year of approval	Trade name/ developer
3006-210-23 × 281-24-236 × MON88913 × COT102	*cry1F, cry1Ac vip3A(a) cp4 epsps pat*	Resistance to lepidopteran insects Glyphosate tolerance Glufosinate tolerance	Japan 2013, South Korea 2014–2015 (food and feed), Mexico 2014 (food)	WideStrike™ × Roundup Ready Flex™ × VIPCOT™ cotton
GHB614 × LLCotton25 × MON15985	*2mepsps Bar, cry1Ac, cry2Ab2*	Resistance to lepidopteran insects Glyphosate tolerance Glufosinate tolerance	Japan 2010, South Korea 2013 and 2011 (food and feed), Mexico (food) 2010, Taiwan (food) 2015	Bayer Crop Science (including fully and partly owned companies)
T304-40 × GHB119 × COT102	*cry2Ae cry1Ab, vip3A(a), bar*	Resistance to lepidopteran insects Glufosinate tolerance	Brazil 2018	Bayer Crop Science (including fully and partly owned companies)
GHB811 × T304-40 × GHB119 × COT102	*cry2Ae cry1Ab, vip3A(a), Bar, hppdPF W336*	Resistance to lepidopteran insects Glufosinate tolerance Isoxaflutole tolerance	Brazil 2019	BASF and Bayer Crop Science (including fully and partly owned companies)
Low seed gossypol				
TAM66274		Low seed gossypol	USA 2018	Texas A&M AgriLife Research University

Source: ISAAA's GM Approval Database. *http://www.isaaa.org/gmapprovaldatabase/*

16.10.1 Brazil

Brazil cultivates cotton over six provinces of which Motto Grasso cultivates the largest area under cotton. With an area of 1.6 Mha and a production of 2.77 million tonnes (ICAC Cotton Data book, 2019), Brazil in 2018 grew genetically modified cotton over 94% of the area. BRS-430 B2RF, BRS-432 B2RF, and BRS-433 B2RF, all with Bollgard II Roundup Ready Flex technology (Monsanto), were released by Embrapa/Bahia foundation, in addition to the foreign cultivars.

GM cotton events commercialized in Brazil:

Name	Event	Year of release	Target pest/herbicide
Lepidopteran pest tolerance	MON 531	2005	Bollworm tolerance
Roundup-resistant cotton	MON 1445	2008	Roundup Ready
LibertyLink cotton	LL cotton 25	2008	Glufosinate
Bollgard II cotton	MON 15985	2009	Enhanced spectrum of lepidopteran pest control
Bollgard RRF	MON 531 × MON 1445	2009	Lepidopteran and Roundup Ready resistant
GlyTol	GHB614	2010	Tolerant to glyphosate
TwinLink	T304-40 × GHB119	2011	Resistant to lepidopteran and tolerant to glufosinate
Glyphosate tolerance	MON 88913	2011	Roundup resistant
GlyTol × TwinLink	GHB614 × T304-40 × GHB119	2012	Tolerance to glyphosate and glufosinate ammonium, insect resistance
GlyTol × LibertyLink	GHB614 × LLCotton25	2012	Tolerance to glufosinate
Two-gene Bt cotton and glyphosate	MON15985 × MON88913	2012	Resistance to some lepidopteran and glyphosate tolerance

16.10.2 India

Six events have been approved for commercial cultivation in India, since the first approval of a GM event in the country in 2002. These events are Mon531, Mon15985, JK event 1, BNLA-601, *cry1C*Metahelix event, and the Nath event of which MON15985 is most widely cultivated. While almost all countries cultivate GM cotton varieties, Bt cotton was introduced as hybrids in India of which the Monsanto's events have been consistently the most popular. MON531 was soon replaced by hybrids containing MON15985. More than 95% of the area under cotton is cultivated with Bt cotton hybrids with MON15985 event. Unauthorized cultivation of GM cotton is often seen—illegal MON531 was being cultivated between 2002 and 2007, and currently unapproved cultivation of Roundup Ready Flex cotton is reported. Despite resistance to *cry* toxins detected in the pink bollworm since 2008 to single-gene Bt cotton and to the dual genes since 2010 and increase in the use of insecticides against sucking pests, the area under GM cotton has not substantially declined. It must be mentioned here that the single- and dual-gene BT cotton hybrids are still effective against *Helicoverpa*. An attempt is being made to cultivate short-duration, single-gene Bt varieties, by the public sector, and compact Bollgard II hybrids (by the private sector) under high-density planting, as an approach to overcome yield stagnation.

16.10.3 Pakistan

Cotton contributes 7.8% toward value addition in agriculture and fulfils 55% of country's domestic cooking oil requirements (Nazli et al. 2010). Only six GM events, two in cotton and four in maize, are approved in Pakistan. In 2005, Pakistan Atomic Energy Commission (PAEC) commercialized four varieties of Bt cotton exhibiting insect resistance (IR), i.e., IR-CIM-443, IR-CIM-448 (approved as IR-NIBGE-3701), IR-NIBGE-2 (approved as IR-NIBGE-1524), and IR-NIBGE-901 (Arshad et al. 2018). Punjab Seeds Council (PSC) approved 40 varieties, and by 2016, an additional 50 Bt cotton varieties were commercialized. At present, 96% of the total cotton production in Pakistan is Bt cotton which is planted on a total area of three million hectares. These varieties are backcross of Mon531 event which carries *cry1Ac* gene from Monsanto. At present more than 80 Bt cotton varieties having the single gene, i.e., *cry1Ac* (MON531 event), are available to farmers, and a couple of varieties having double genes, i.e., *cry1Ac* + *cry2Abcry2Ab* (CEMB-2 event), are in the pipeline.

Factors limiting seed cotton production in Pakistan are poor germination and limited availability of certified seed, reliance on single-gene event that has decreased efficacy against bollworms, increasing incidence and damage due to emerging pests, and narrow genetic base of the cotton germplasm from which varieties have been bred. The government, this year, has procured enough quantity of certified bioengineered seed of the latest cotton varieties that will increase farmers' choice to plant improved cultivars with the availability of free certified seed for 100,000 acres, through balloting. Arrangements for timely supply of fertilizer and pesticides were made. Regional trials with Bt varieties are recommended to identify the best performing ones with enhanced pest tolerance, leaf curl virus disease tolerance, and yield, for site-specific or region-specific recommendations (Karar et al. 2020).

16.10.4 Australia

Cotton is cultivated over 52,000 Ha in New South Wales and the rest of the 82,000 Ha is found in Queensland. Cotton production in Australia stands at 114,000 tonnes. All the 113 cotton varieties released for commercial cultivation in Australia are from CSIRO. Australia cultivates GM cotton with insecticide-resistant and herbicide-tolerant traits. GM cotton was first cultivated in 1996—Ingard 1 with *cry1Ac* was followed by Bollgard II and WideStrike. The herbicide-tolerant traits are Roundup Ready and Roundup Ready Flex, tolerant to glyphosate, and LibertyLink cotton, tolerant to glufosinate ammonium. With all their cultivars being GM, Australian cotton growers have put in place integrated insect pest management and integrated resistance management of weeds. Fall armyworm is being monitored having recently entered Australia. It is likely to remain within tropical areas away from most of the Australian cotton-growing regions, thereby having minimal impact on cotton production. However, pest management programs have been developed for its management should it infest biotech cotton (Biki and Flake 2020). Climate change is being experienced severely in Australia and with increasing degree days; cotton crop is seen to display rank vegetative growth that may necessitate changes in the doses of growth regulators used in canopy management. The stewardship provided by Australian cotton is facilitated across the cotton-growing regions through different committees, such as the insecticidal transgenic technical panel, insecticide technical panel, herbicide-tolerant crops technical panel, and in-season troubleshooting technical committee, with members whose responsibilities are assigned for a given region.

16.10.5 China

China is the largest cotton producer in the world and is also one of the largest pesticide users, overtaking the USA (FAOSTAT 2017). Almost 30–40% of pesticides, of which 40% are belonging to the extremely or highly hazardous, as categorized by the WHO are used on cotton (FAOSTAT 2013).

Introduction of genetically modified crops in China began with insect-resistant cotton, expressing *cry1Ac*, in 1997. From an area of 5.7%, the total planting area reached more than 3.7 mHa with 96% adoption in 10 years. The area planted with cotton decreased to 2.9 MHa in 2016–2017 due to structural changes in policies governing the sector. And*cry1Ac*, *cry1Ab1Ac*, and *cry1A + CpTI* events are cultivated in China (Li et al. 2017).

The practice of applying excessive amounts of highly toxic pesticides has continued even after the adoption of Bt cotton (Qiao et al. 2017; Pemsl et al. 2005; Yang et al. 2005a, b). Risk averseness and poor knowledge on IPM by cotton farmers, upsurge of secondary pests, and poor Bt seed quality are the suggested causes for increasing pesticide use on GM cotton in China.

New approaches to GM cotton involve pyramiding of RNA interference along with *cry* toxins as a resistant management tool (Li et al. 2020). A National Scientific and Technological Innovation Plan issued by the government proposed to strengthen research on GM crops and promote the industrialization of new varieties of Bt cotton, Bt corn, and herbicide-tolerant (HT) soybeans over the next 5 years (Deng et al. 2019).

16.10.6 Burkina Faso

South Africa introduced Bt cotton in 1997. While it performed well in the initial years of its commercialization, the technology was soon found unsustainable. Introduction of GM cotton in Makhathini (South Africa) disrupted, indirectly, easy access to institutional credit and a guaranteed market. Long-term impacts and benefits of the technology were not fully appreciated, and the data generated covered certain categories of farmers who benefit from these technologies over others, and these technologies disrupted farming systems. Burkina Faso followed South Africa to emerge as one of the leading adopters of agricultural biotechnology in sub-Saharan Africa (Vitale and Greenplate 2014). Genetically modified (GM) cotton, Monsanto's Bollgard© II, was legalized by the government of Burkina Faso in 2007 and was commercially introduced in 2009 (Vitale and Greenplate 2014). The success of Bt cotton in terms of yield gain and insecticide use reduction over conventional cotton was expected to impact the adoption of GM technology in Western and sub-Saharan Africa. Bt cotton outperformed conventional cotton by up to 2.2% in terms of yield. However, fiber quality was reduced in the upper half mean length and fiber strength by −1.45 to −2.09 mm and −19.7 to −40.57 kNm/Kg, respectively, while micronaire, maturity, short fiber index, reflectance, and yellowness were similar between Bt and conventional cotton. The reputation of cotton fiber from Burkina Faso was affected in the international market, and a penalty was promulgated on all export sales of cotton produced from 2010 (Fok 2016). In Burkina Faso, the decision to phase out Bt cotton was made by the cotton companies and not cotton farmers. "The higher yield of Bt cotton meant more income for farmers while the lower ginning ratio and shorter staple length meant less fiber, and of a lower quality, for cotton companies to sell. The case of Bt cotton in Burkina Faso exposed the conflicting interests within the cotton value chain, underlining how GM crops can produce different outcomes for different stakeholders" (Dowd-Uribe and Schnurr 2016). With this experience, steps such as introgression of traits into local varieties with backcrossing being made at the site, increasing the number of backcrosses, are now being put in place to secure the future road map for GM cotton in Burkina Faso.

Three regions cultivate cotton in Burkina Faso—Sofitex, Faso Cotton, and Socoma. Of the 615,000 ha, Sofitex cultivates 520,000 ha and cotton production in Burkina Faso is 170,000 tonnes. Three varieties are popularly cultivated in Burkina Faso that were released 6–24 years ago of which FK37, a 20-year-old variety, occupies 94% of the area. Fiber length of these varieties ranged from 28.1 to 30 mm, and fiber strength from 29.1 to 30.7 g/tex with a micronaire of 4–4.2.

16.10.7 USA

Thirty-two events have been reported in the USA, with traits responsible for insect resistance, herbicide tolerance, or both insecticide resistance and herbicide tolerance, according to ISAAA. Bollgard II (Mon15985), Bollgard, and VIPCOT regulate resistance in cotton plants to a wide range of lepidopteran pests. Of the herbicide-tolerant toxins, BXN (tolerance to oxynil herbicides), GlyTol (resistance to glyphosate), LibertyLink cotton (tolerance to glufosinate), and Roundup Ready and Roundup Ready Flex cotton (tolerance to glyphosate) have been listed under the ISAAA website. Two insecticide-resistant and herbicide-tolerant traits are registered—TwinLink cotton (glufosinate+ lepidopteran pest tolerance) and BXN plus Bollgard cotton (oxynil herbicide and lepidopteran pest tolerance) to combat weed and insect problems, simultaneously.

16.11 Economic Benefits of BT Cotton

There is a big debate still going on regarding the economic benefits derived due to the adoption of Bt cotton across the globe. Early studies proved that the major benefits from Bt cotton include effective control of bollworms leading to significant yield increase, drastic reduction in chemical sprays, and substantial increase in net profit to farmers (Manjunath 2007; Sadashivappa 2008; Subramanian and Qaim 2009; Rahman et al. 2015). The numbers of pesticide sprays and expenditure on insecticides decreased substantially, and higher yields were realized by the Bt cotton adopters (Dev and Rao 2007). The farmers in major cotton-growing states in India benefitted significantly from adopting Bt technology through higher profitability mainly due to reduced pest control costs and higher yields, though there was considerable variation in key variables like yield, cost, pesticide use, etc. (Ashok et al. 2012). In India the Bt cotton adoption rate has increased tangible socioeconomic benefits for small farm holders. Further, living standard of poor and small farm holders has increased by 18% by adopting Bt cotton; thus this technology contributes to positive socioeconomic development (Kathage and Qaim 2012). During the period 2002–2015, the total benefit gained due to the adoption of Bt cotton has been estimated to be of 220 billion with 85% share accruing to producers and 15% to the private seed companies/marketing firms (Ramasundaram et al. 2014). Brookes and Barfoot (2020) reviewed the global, socioeconomic, and environmental impacts of GM crops since its introduction for commercial cultivation from 1996 to 2018. Farmers around the world have adopted GM crops and continued to use the technology for their production system. The direct global farm income through GM in 2018 was $18.95 billion. Cotton sector significantly gained higher yields and lowers the production cost since 1996. Farm income level in 2018 was increased by $4.57 billion. Cumulatively global GM cotton farmers have benefitted as additionally of 65.8 billion since 1996 (Brookes and Barfoot 2020). The huge income gain for developing countries farmers has been through insect-resistant GM cotton and herbicide-tolerant soybean (Clive 2014). But some of the researchers attribute the increase in the yield during the Bt era to increased fertilizer use and the introduction of new insecticides.

16.12 Future Outlook

Technologies evolve concurrently with scientific advances. No technology is perfect. Every technology presents advantages and disadvantages. Because genetic changes are inheritable, GM technologies have the potential of offering unique advantages that many other nongenetic technologies do not. However, these heritable genetic changes may have the propensity to influence biodiversity in a significant manner. Therefore, GM technologies have attracted more attention of environmental activists, and the need for biosafety has been emphasized more with

GM products than with any other technologies. GM crops should not be seen as silver bullets. They are generally developed for one or two traits that can indeed have a strong impact on crop production and crop protection. Impact of any technology depends on stewardship and its methods of deployment. The 25 years of global GM cotton experience taught several lessons. GM technology has been readily accepted by majority of farmers because of the ease in adoption; GM technology is not a silver bullet; GM technology is not invincible, because insects display strong potential to develop resistance to Bt toxins and weeds develop resistance to herbicides, because of which the technology is rendered unsustainable. Experience also highlights the importance of preemptive strategies that must be developed and complied with, so as to delay resistance in insects and weeds and to conserve efficacy of the GM traits for the longest possible time. Experience in India shows that restricting the Bt technology to hybrid cotton varieties did not benefit rainfed regions, because of the suboptimal performance of input-intensive hybrids in rainfed conditions, thereby resulting in poor yields, for example, as in predominantly rainfed states such as Maharashtra.

Across the globe, except China and Uzbekistan, all other countries have adopted GM cotton products that were developed by multinational or transnational companies. The two main GM technologies, namely, herbicide-tolerant (HT) cotton and insect-resistant (IR) cotton, have been adopted in more than 75% of the global cotton acreage. A global analysis shows that *Bt* cotton technology has been effective so far in protecting cotton against the *Heliothine* species, except a few isolated cases of field resistance in the USA, China, and Pakistan. However, pink bollworm resistance to Bt cotton highlights the need for stringent compliance with insect resistance management (IRM) strategies. More and more genes are being continuously added in pyramids over the recent years to enhance the efficacy, increase the spectrum of efficacy, and decrease resistance risk. However upgrading a technology with new genes requires investment, which is recovered back eventually by the technology developers from farmers in the form of trait fee or technology fee or royalties. With continuous pyramiding of genes, GM products are progressively becoming expensive. With progressive exposure of the pyramided GM products to the ecology and environment, their efficacy is declining over time, which further necessitates gene pyramiding. It remains to be seen as to how long this cyclic pattern would continue to be technically feasible and economically acceptable especially by smallholder farmers who comprise more than 95% of the global cotton farmers.

Thus far Asia and African countries have been cultivating Bt cotton with either a single gene or with two genes. The two technologies that are being used in other countries and which India, Pakistan, and Africa (except South Africa) do not have are the "three-gene-based Bt cotton" and "herbicide-tolerant (HT) cotton." The three-gene-based *Bt* cotton neither kills the "Bt-resistant pink bollworms of India" nor any other insects that are not killed by the two-gene Bt cotton. Herbicide-tolerant cotton only facilitates weed control with herbicide spray; there is no evidence anywhere to show that the introduction of these technologies can help India to increase yields or reduce pesticide usage (Kranthi 2020).

What is the future outlook? What kind of GM products will be available to cause a breakthrough in Asia and Africa to increase yields? Will smallholder farmers be able to afford the more expensive multigene GM pyramid products? Have smallholder farmers in Asia been enticed into a technology trap that forces them progressively to adopt newer and more expensive technologies without allowing them any alternative options to move out? Smallholder farms obtain low yields in India, Pakistan, and Africa. While crop protection is just one facet of crop production, the main question remains: are there any GM traits available anywhere in the world that can increase yields or help to increase cotton yields?

A large section of the farming community in developing and developed countries has experienced the advantages and disadvantages of GM cotton over the past 25 years. With the existing technologies, a breakthrough in yields hasn't been apparent, at least until date. With new options of gene editing technologies such as CRISPR/Cas or RNA interference (RNAi) on the anvil, there is new hope that there could be newer technologies with better selectivity, with positive impacts on yields, and with least negative impacts on biosafety and biodiversity.

References

Adair D, Irwin R (2008) A practical guide to containment: plant biosafety in research greenhouses, 2nd edn. Information Systems for Biotechnology, Virginia Tech, p 76. http://www.isb.vt.edu/containment-guide.aspx

Ali MI, Luttrell RG (2007) Susceptibility of bollworm and tobacco budworm (*Lepidoptera*: *Noctuidae*) to *Cry2AbCry2Ab2* insecticidal protein. *J Econ Entomol*. https://doi.org/10.1093/jee/100.3.921

Ali MI, Luttrell RG, Abel C (2007) Monitoring Bt susceptibilities in *Helicoverpa zea and Heliothis virescens*: results of 2006 studies. In: Proceedings of the 2007 beltwide cotton conferences, pp 1062–1072

Arshad M, Zafar Y, Asad S (2013) Silicon carbide whisker-mediated transformation of cotton (*Gossypium hirsutum* L.). In: Zhang B (ed) Transgenic cotton: methods and protocols, methods in molecular biology, vol 958. https://doi.org/10.1007/978-1-62703-212-4_7

Arshad M, Khan RR, Aslam A, Akbar W (2018) Transgenic Bt cotton: effects on target and non-target insect diversity. In: Past, present and future trends in cotton breeding. pp 155–174

Ashok KR, Uma K, Prahadeeswaran M, Jeyanthi H (2012) Economic and environmental impact of Bt cotton in India. *Indian J Agric Econ* 67(3):405–428

Astwood JD, Leach JN, Fuchs RL (1996) Stability of food allergens to digestion *in vitro*. *Nat Biotechnol* 14:1269–1273. https://doi.org/10.1038/nbt1096-1269

Banks MS, Evans PK (1976) A comparison of the isolation and culture of mesophyll protoplasts from several Nicotiana species and their hybrids. *Plant Sci Lett* 7:409–416

Barker RF, Idler KB, Thompson DV, Kemp JD (1983) Nucleotide sequence of the T-DNA region from the *Agrobacterium tumefaciens* octopine Ti-plasmid pTi15955. *Plant Mol Biol* 2(6):335–350

Beckie HJ (2006) Herbicide-resistant weeds: management tactics and practices. *Weed Technol*. https://doi.org/10.1614/wt-05-084r1.1

Beekman M, Ratnieks FLW (2000) Long-term foraging by the honeybee *Apis mellifera* L. *Funct Ecol* 14(4):490–496

Beura K, Rakshit A (2011) Effect of Bt cotton on nutrient dynamics under varied soil type. *Ital J Agron* 6:25–28

Biki Z, Flake L (2020). https://apps.fas.usda.gov/newgainapi/api/Report/DownloadReportByFileName?fileName=Cotton%20and%20Products%20Annual_Canberra_Australia_04-01-2020

Bird L (2015) 2015-16 end of season resistance monitoring report conventional insecticide testing. http://hdl.handle.net/1/4428

Blanco CA, Andow DA, Abel CA et al (2009) *Bacillus thuringiensis Cry1Ac* resistance frequency in tobacco budworm (*Lepidoptera: Noctuidae*). *J Econ Entomol*. https://doi.org/10.1603/029.102.0149

Brod FCA, Van Dijk JP, Voorhuijzen MM et al (2014) A high-throughput method for GMO multi-detection using a microfluidic dynamic array. In: Analytical and bioanalytical chemistry. Springer, pp 1397–1410

Brookes G, Barfoot P (2020) GM crops: global socio-economic and environmental impacts 1996-2018. PG Economics Ltd, Dorchester. https://www.pgeconomics.co.uk/

Brower LP, Taylor OR, Williams EH, Slayback DA, Zubieta RR, Ramirez MI (2012) Decline of monarch butterflies overwintering in Mexico: is the migratory phenomenon at risk? *Insect Conserv Divers* 5:95–100. https://doi.org/10.1111/j.1752-4598.2011.00142.x

Burkitbayeva S, Qaim M, Swinnen J (2016) A black (white) hole in the global spread of GM cotton. *Trends Biotechnol* 34:260–263

Carzoli AK, Aboobucker SI, Sandall LL, Lübberstedt TT, Suza WP (2018) Risks and opportunities of GM crops: *Bt* maize example. *Glob Food Sec* 19:84–91

CFIA (2000) Directive Dir2000-07: conducting confined research field trials of plants with novel traits in Canada. Canadian Food Inspection Agency (CFIA), Ottawa, Canada. http://www.inspection.gc.ca/plants/plants-with-novel-traits/applicants/directive-dir2000–07/eng/1304474667559/1304474738697

CFIA (2007) Containment standards for facilities handling plant pests, 1st edn. Canadian Food Inspection Agency (CFIA), Ottawa. http://www.inspection.gc.ca/plants/plant-protection/biocontainment/containment-standards/eng/1412353866032/1412354048442

Chaperon S, Sauve S (2007) Toxicity interaction of metals (Ag, Cu, Hg, Zn) to urease and dehydrogenase activities in soils. *Soil Biol Biochem* 39:2329–2338

Choudhary B, Gaur K (2010) Bt Cotton in India: A Country Profile. ISAAA Series of Biotech Crop Profiles. ISAAA: Ithaca, NY

Christie PJ, Gordon JE (2014) The *Agrobacterium* Ti plasmids. *Microbiol Spectr* 2(6):295–313

Claire H (2005) Gene stacking in transgenic plants—the challenge for 21st century plant biotechnology. Plant Biotechnol J 3:141–155. https://doi.org/10.1111/j.1467-7652.2004.00113.x. © 2005 Blackwell Publishing Ltd

Clive J (1996) Global Review of the field testing and commercialization of transgenic plants: 1986-1995. In: The international service for the acquisition of agri-biotech applications

Clive J (2014) Global status of commercialized biotech/GM crops: ISAAA brief no. 49. ISAAA, Ithaca

Codex Alimentarius (2003) Codex Alinorm 03/34: Joint FAO/WHO Food Standard Programme, Codex Alimentarius Commission, Twenty-Fifth Session, Rome, Italy 30 June-5 July, 2003. Appendix III, Guideline for the conduct of food safety assessment of foods derived from recombinant-DNA plants, and Appendix IV, Annex on the assessment of possible allergenicity. Rome, Codex Alimentarius Commission, pp 47–60

Cottenet G, Blancpain C, Sonnard V, Chuah PF (2013) Development and validation of a multiplex real-time PCR method to simultaneously detect 47 targets for the identification of genetically modified organisms. *Anal Bioanal Chem* 405:6831–6844. https://doi.org/10.1007/s00216-013-7125-5

Cousins YL, Lyon BR, Llewellyn DJ (1991) Transformation of an Australian cotton cultivar: prospects for cotton through genetic engineering. *Aust J Plant Physiol* 18:481–494

Cunningham JP, Zalucki MP (2014) Understanding Heliothine (*Lepidoptera: Heliothinae*) pests: what is a host plant? *J Econ Entomol* 107:881–896

Dai SH, Zheng P, Marmey P, Zhang SP, Tian WZ, Chen SY, Beachy RN, Fauquet C (2001) Comparative analysis of transgenic rice plants obtained by *Agrobacterium*-mediated transformation and particle bombardment. *Mol Breed* 7:25–33

Dandan Z, Yutao X, Wenbo C et al (2019) Field monitoring of *Helicoverpa armigera* (Lepidoptera: Noctuidae) *Cry1Ac* insecticidal protein resistance in China (2005–2017). *Pest Manag Sci*. https://doi.org/10.1002/ps.5175

Daniell H (2002) Molecular strategies for gene containment in transgenic crops. *Nat Biotechnol* 20(6):581–586

DBT (1998) Revised guidelines for research in transgenic plants & guidelines for toxicity and allergenicity evaluation of transgenic seeds, plants and plant parts. Department of Biotechnology, Government of India

DBT (2008) Guidelines and standard operating procedures for confined field trials of regulated genetically engineered (GE) plants. http://dbtindia.gov.in/sites/default/files/uploadfiles/Guidelinesand_SOPs_for_confined_field_trials_of_regulated_GE_Plants%2C2008.pdf

De Framond AJ, Barton KA, Chilton MD (1983) Mini-Ti: a new vector strategy for plant genetic engineering. *Biotechnology* 5:262–269

Debode F, Hulin J, Charloteaux B et al (2019) Detection and identification of transgenic events by next generation sequencing combined with enrichment technologies. *Sci Rep* 9:1–9. https://doi.org/10.1038/s41598-019-51668-x

Deng H, Hu R, Pray C, Jin Y (2019) Perception and attitude toward GM technology among agribusiness managers in China as producers and as consumers. *Sustainability* 11(5):1342

Dev MS, Rao NC (2007 Nov) Socio-economic impact of Bt cotton. Monograph No. 3. Hyderabad: Centre for Economic and Social Studies (CESS)

Dhurua S, Gujar GT (2011) Field-evolved resistance to Bt toxin *Cry1Ac* in the pink bollworm, *Pectinophora gossypiella* (Saunders) (*Lepidoptera: Gelechiidae*), from India. *Pest Manag Sci*. https://doi.org/10.1002/ps.2127

Dinon AZ, Prins TW, Van Dijk JP et al (2011) Development and validation of real-time PCR screening methods for detection of *cry1A*.105 and *Cry2Ab2* genes in genetically modified organisms. *Anal Bioanal Chem* 400:1433–1442. https://doi.org/10.1007/s00216-011-4875-9

Dowd-Uribe B, Schnurr MA (2016) Briefing: Burkina Faso's reversal on genetically modified cotton and the implications for Africa. *Afr Aff* 115(458):161–172

Dubelman S, Ayden BR, Bader BM, Brown CR, Jiang C, Vlachos D (2005) *Cry1Ab* protein does not persist in soil after three years of sustained Bt corn use. *Environ Entomol* 34:915–921

EC (2001) Directive 2001/18/EC of the European Parliament and of the Council of 12 March 2001 on the deliberate release into the environment of genetically modified organisms and repealing Council Directive 90/220/EEC—Commission Declaration. 32001L0018. *Off J* L106:0001–0039

Ellstrand NC (2003) Current knowledge of gene flow in plants. *Philos Trans R Soc B Biol Sci* 358:1163–1170

EPA (2000) Mammalian toxicity assessment guidelines for protein plant pesticides. FIFRA Scientific Advisory Panel. Environmental Protection Agency, Arlington

EU (2009) Directive 2009/41/EC of the European Parliament and of the Council of 6 May 2009 on the contained use of genetically modified micro-organisms. Off J L125:75–97. http://eur-lex.europa.eu/legal-content/EN/ALL/?uri=CELEX:32009L0041

FAO/WHO (2001) Evaluation of the allergenicity of genetically modified foods. Report of a Joint FAO/WHO Expert Consultation on the Allergenicity of Foods Derived from Biotechnology, Rome

FAOSTAT (2013) Production-Crops. www.fao.org/faostat/en/#data/QC. Accessed 1 Aug 2013

FAOSTAT (2017) http://www.fao.org/faostat/en/#data/QC

Fedoroff NV, Battisti DS, Beachy RN, Cooper PJM, Fischhoff DA, Hodges CN, Knauf VC, Lobell D, Mazur BJ, Molden D, Reynolds MP, Ronald PC, Rosegrant MW, Sanchez PA, Vonshak A, Zhu JK (2010) Radically rethinking agriculture for the 21st century. *Science* 327(5967):833–834

Ferreira LHPL, Molina JC, Brasil C, Andrade G (2003) Evaluation of *Bacillus thuringiensis* bioinsecticidal protein effects on soil microorganism. *Plant Soil* 256:161–168
Finer JJ (1988) Plant regeneration from somatic embryogenic suspension cultures of cotton. *Gossypium hirsutum* L. *Plant Cell Rep* 7:399–402
Firoozabady E, DeBoer DL (1993) Plant regeneration via somatic embryogenesis in many cultivars of cotton (*Gossypium hirsutum* L.). *In Vitro Cell Dev Biol* 29:166–173
Fok M (2016) Impacts du coton-Bt sur les bilans financiers des sociétéscotonnières et des paysans au Burkina Faso. *Cahiers Agric* 25(3):35001
Fourie D, Van denBerg J, du Plessis H (2017) Evaluation of the susceptibility status of spiny bollworm *Earias biplaga* (Walker) (Lepidoptera: Noctuidae) to Bt cotton in South Africa. *Afr Entomol*. https://doi.org/10.4001/003.025.0254
Fraiture MA, Herman P, Taverniers I et al (2015) Current and new approaches in GMO detection: challenges and solutions. *Biomed Res Int* 2015:1–22. https://doi.org/10.1155/2015/392872
Garcia-Alonso M, Hendley P, Bigler F, Mayeregger E, Parker R, Rubinstein C, Satorre E, Solari F, McLean M (2014) Transportability of confined field trial data for environmental risk assessment of genetically engineered plants: a conceptual framework. *Transgenic Res* 23(6):1025–1041
GEAC (2015a) Confined field trials of genetically engineered (GE) plants, Phase II Capacity building project on biosafety. http://www.geacindia.gov.in/resource-documents/134Confined FieldTrials_of_GE_Plants.pdf
GEAC (2015b) Monitoring of confined field trials of regulated genetically engineered (GE) plants: monitoring manual. http://www.geacindia.gov.in/resource-documents/03-Monitoring ConfinedFieldTrials_of_Regulated_Genetically_Engineered_Plants_Monitoring_Manual.pdf
Gelvin SB (2005) Agricultural biotechnology: gene exchange by design. *Nature* 433(7026):583–584
Gianessi L, Silvers C, Sankula S, Carpenter J (2002). Plant biotechnology: current and potential impact for improving pest management. In: U.S. agriculture: an analysis of 40 case studies (executive summary). National Center for Food and Agricultural Policy, Washington, DC. Available on the World Wide Web: http://www.ncfap.org/40CaseStudies/NCFAB%20 Exec%20Sum.pdf
Guo B, Guo Y, Hong H, Qiu LJ (2016) Identification of genomic insertion and flanking sequence of G2-EPSPS and GAT transgenes in soybean using whole genome sequencing method. *Front Plant Sci* 7:1009. https://doi.org/10.3389/fpls.2016.01009
Hare PD, Chua NH (2002) Excision of selectable marker genes from transgenic plants. *Nat Biotechnol* 20(6):575–580
Harrigan GG, Lundry D, Drury S, Berman K, Riordan SG, Nemeth MA, Ridley WP, Glenn KC (2010) Natural variation in crop composition and the impact of transgenesis. *Nat Biotechnol* 28(5):402–404
Head G, Surber JB, Watson JA, Martin JW, Duan JJ (2002) No detection of *Cry1Ac* protein in soil after multiple years of transgenic Bt cotton (Bollgard) use. *Environ Entomol* 31:30–36
Heap I, Duke SO (2018) Overview of glyphosate-resistant weeds worldwide. *Pest Manag Sci* 74(5):1040–1049
Hernandez-Garcia CM, Finer JJ (2014) Identification and validation of promoters and cis-acting regulatory elements. *Plant Sci* 217(1):109–119
Herrera-Estrella L (1983) Transfer and expression of foreign genes in plants. Ph.D. Thesis, Laboratory of Genetics, Genetics University, Belgium
Heuberger S, Ellers-Kirk C, Tabashnik BE, Carrière Y (2010) Pollen and seed-mediated transgene flow in commercial cotton seed production fields. *PLoS One* 5(11):e14128. https://doi.org/10.1371/journal.pone.0014128
Hoa TT, Al-Babili S, Schaub P, Potrykus I, Beyer P (2003) Golden Indica and Japonica rice lines amenable to deregulation. *Plant Physiol* 133(1):161–169
Holst-Jensen A, Bertheau Y, de Loose M et al (2012) Detecting un-authorized genetically modified organisms (GMOs) and derived materials. *Biotechnol Adv* 30:1318–1335. https://doi.org/10.1016/j.biotechadv.2012.01.024
Howe GA, Jander G (2008) Plant immunity to insect herbivores. *Annu Rev Plant Biol* 59(1):41–66

Huang JK, Hu RF, Rozelle S, Qiao FB, Pray CE (2002a) Transgenic varieties and productivity of smallholder cotton farmers in China. *Aust J Agric Resour Econ* 46:367–387
Huang J, Rozelle S, Pray C, Wang Q (2002b) Plant biotechnology in China. *Science* 295(5555): 674–676
Huang J, Raifa H, Rozelle S, Pray C (2005) Insect-resistant GM rice in farmers fields: assessing productivity and health effects in China. *Science* 308:688–690
Hui SW (1995) Effects of pulse length and strength on electroporation efficiency. In: Nickoloff JA (ed) Methods in molecular biology. Plant cell electroporation and electrofusion protocols. Humana Press Inc., Totowa, pp 29–40
Hutchison WD, Burkness EC, Mitchell PD et al (2010) Area wide suppression of European corn borer with Bt maize reaps savings to non-Bt maize growers. *Science* 330:222–225. https://doi.org/10.1126/science.1190242
Icoz I, Saxena D, Andow DA, Zwahlen C, Stotzky G (2008) Microbial populations and enzyme activities in soil in situ under transgenic corn expressing *Cry* proteins from *Bacillus thuringiensis*. *J Environ Qual* 37:647–662
Inagaki S, Henry IM, Lieberman MC, Comai L (2015) High-throughput analysis of T-DNA location and structure using sequence capture. *PLoS One* 10(10):e0139672. https://doi.org/10.1371/journal.pone.0139672
ISAAA (2018) Global status of commercialized biotech/GM crops in 2018: biotech crops continue to help meet the challenges of increased population and climate change
Isenring R (2010) Pesticides and the loss of biodiversity. Pesticide Action Network, Europe. Development House 56e64 Leonard Street London EC2A 4LT
Ismael YR, Bennet R, Morse S (2001) Farm level impact of Bt cotton in South Africa. *Biotechnol Dev Monit* 48:15–19
James RR, Difazio SP, Brunner AM, Strauss SH (1998) Environmental effects of genetically engineered woody biomass crops. *Biomass Bioenergy* 14(4):403–414
Jefferson RA, Kavanagh TA, Bevan MW (1987) GUS fusions: beta-glucuronidase as a sensitive and versatile gene fusion marker in higher plants. *EMBO J* 6:3901–3907
Jones DD, Maryanski JH (1991) Safety considerations in the evaluation of transgenic plants for human food. In: Levin MA, Strauss HS (eds) Risk assessment in genetic engineering. McGraw-Hill, New York
Jones HD, Doherty A, Wu H (2005) Review of methodologies and a protocol for the *Agrobacterium*-mediated transformation of wheat. *Plant Methods* 1–5. https://doi.org/10.1186/1746-4811-1-5
Kamle M, Kumar P, Patra JK et al (2017) Current perspectives on genetically modified crops and detection methods. *3 Biotech* 7:219. https://doi.org/10.1007/s13205-017-0809-3
Kapur M, Bhatiya R, Pandey G, Pandey J, Paul D, Jain RK (2010) A case study for assessment of microbial community dynamics in genetically modified Bt cotton crop field. *Curr Microbiol* 61:118–124
Karar H, Bashir MA, Haider M, Haider N, Khan KA, Ghramh HA et al (2020) Pest susceptibility, yield and fiber traits of transgenic cotton cultivars in Multan, Pakistan. *PLoS One* 15(7):e0236340. https://doi.org/10.1371/journal.pone.0236340
Kathage J, Qaim M (2012) Economic impacts and impact dynamics of Bt (*Bacillus thuringiensis*) cotton in India. *Proc Natl Acad Sci U S A* 109:11652–11656. https://doi.org/10.1073/pnas.1203647109
Kofer W, Eibl C, Steinmüller K (1998) PEG-mediated plastid transformation in higher plants. *In Vitro Cell Dev Biol Plant* 34:303–309. https://doi.org/10.1007/BF02822739
Kousar S, Qaim M (2011) Impact of Bt cotton on pesticide poisoning in smallholder agriculture: a panel data analysis. *Ecol Econ* 11:2105–2113
Kramer MG, Redenbaugh K (1994) Commercialization of a tomato with an antisense polygalacturonase gene: the FLAVR SAVR™ tomato story. *Euphytica* 79(3):293–297
Kranthi KR (2016) Fertilizers gave high yields; Bt only Provided Cover. In: Cotton Statistics and News, No.39. http://www.cicr.org.in/pdf/pop_art/Fertilizers_and_Bt.pdf
Kranthi KR (2020) Cotton in India: long term trends and way forward. *ICAC Recorder* 38(1):3–40

Kranthi KR, Stone GD (2020) Long-term impacts of Bt Cotton in India. *Nat Plants* 6:188–196
Krishna VV, Qaim M (2012) Bt cotton and sustainability of pesticide reductions in India. *Agric Syst* 107:47–55
Kukanur VS, Singh TVK, Kranthi KR, Andow DA (2018) *Cry1Ac* resistance allele frequency in field populations of *Helicoverpa armigera* (Hübner) collected in Telangana and Andhra Pradesh, India. *Crop Prot*. https://doi.org/10.1016/j.cropro.2018.01.008
Lazzeri PA, Brettschneider R, Lührs R et al (1991) Stable transformation of barley via PEG induced direct DNA uptake into protoplasts. *Theor Appl Genet* 81(4):437–444
Li Y, Lu Y, Hallerman EM, Peng Y, Wu K (2017) Commercial use and governance of Bt cotton in China. In: Adenle AA, Morris EJ, Murphy DJ (eds) Genetically modified organisms in developing countries. Cambridge University Press, Cambridge, pp 225–235
Li P, Xue Y, Shi J et al (2018a) The response of dominant and rare taxa for fungal diversity within different root environments to the cultivation of Bt and conventional cotton varieties. *Microbiome* 184(6):1–15
Li Y, Wang NN, Wang Y et al (2018b) The cotton *XLIM* protein (*GhXLIM* 6) is required for fiber development via maintaining dynamic F-actin cytoskeleton and modulating cellulose biosynthesis. *Plant J* 96:1269–1282. https://doi.org/10.1111/tpj.14108
Li Y, Hallerman EM, Wu K, Peng Y (2020) Insect-resistant genetically engineered crops in China: development, application, and prospects for use. *Annu Rev Entomol* 65:273–292
Liu W, Liu X, Liu C et al (2020) Development of a sensitive monoclonal antibody-based sandwich ELISA to detect Vip3Aa in genetically modified crops. *Biotechnol Lett* 42:1467–1478. https://doi.org/10.1007/s10529-020-02854-9
Lu Y, Wu K, Jiang Y et al (2012) Widespread adoption of Bt cotton and insecticide decrease promotes bio-control services. *Nature* 487(7407):362–365
Lübeck M (2002) Detection of genetically modified plants: methods to sample and analyse GMO content in plants and plant products. Ministry of Environment, The Danish Forest and Nature Agency, Copenhagen. http://www.sns.dk/erhvogadm/bioteck/detection.htm
Luttrell RG, Wan L, Knighten K (1999) Variation in susceptibility of noctuid (Lepidoptera) larvae attacking cotton and soybean to purified endotoxin proteins and commercial formulations of *Bacillus thuringiensis*. *J Econ Entomol*. https://doi.org/10.1093/jee/92.1.21
Manjunath T (2007) Q&A on Bt-Cotton in India, All India Crop Biotechnology Association B-5. New Delhi
Marchetti MF (2014) The effects of a glyphosate-based herbicide (Roundup®) and temperature on the foraging of the wolf spider *Pardosa Milvina* (*Araneae: Lycosidae*). Master thesis, Department of Zoology, Faculty of Miami University
McCabe D, Christou P (1993) Direct DNA transfer using electric discharge particle acceleration (ACCELL™ technology). *Plant Cell Tissue Org Cult* 33:227–236
Merkle SA, Parrott WA, Flinn BS (1995) Morphogenic aspects of somatic embryogenesis. In: Thorpe TA (ed) In vitro embryogenesis in plants. Kluwer Academic Publishers, Dordrecht, pp 155–203
Mohan KS, Ravi KC, Suresh PJ et al (2016) Field resistance to the *Bacillus thuringiensis* protein *Cry1Ac* expressed in Bollgard® hybrid cotton in pink bollworm, *Pectinophora gossypiella* (Saunders), populations in India. *Pest Manag Sci* 72:738–746. https://doi.org/10.1002/ps.4047
Nadal A, Coll A, La Paz JL et al (2006) A new PCR-CGE (size and colour) method for simultaneous detection of genetically modified maize events. *Electrophoresis* 27:3879–3888. https://doi.org/10.1002/elps.200600124
Naik VCB, Kumbhare S, Kranthi S et al (2018) Field-evolved resistance of pink bollworm, *Pectinophora gossypiella* (Saunders) (*Lepidoptera: Gelechiidae*), to transgenic *Bacillus thuringiensis* (Bt) cotton expressing crystal *1Ac* (*Cry1Ac*) and *Cry2AbCry2Ab* in India. *Pest Manag Sci*. https://doi.org/10.1002/ps.5038
Nakai S, Hoshikawa K, Shimono A et al (2015) Transportability of confined field trial data from cultivation to import countries for environmental risk assessment of genetically modified crops. *Transgenic Res* 24:929–944

Nakayama T, Soma M, Rahmutula D et al (2001) Isolation of the 5'-flanking region of genes by thermal asymmetric interlaced polymerase chain reaction. *Med Sci Monit* 7(3):345–349

NAS (2016) Genetically engineered crops: experiences and prospects. The National Academies Press, 500 Fifth Street NW Washington, DC 20001

Nazli H, Sarker R, Meilke KD, Orden D (2010) Economic performance of Bt cotton varieties in Pakistan (No. 320-2016-10300)

NRC (2000) Genetically modified pest-protected plants: science and regulation. National Academy Press, Washington, DC

OECD (1986) Recombinant DNA safety considerations. Organisation for Economic Co-operation and Development (OECD), Paris, France. http://www.oecd.org/sti/biotech/40986855.pdf

OECD (1992) Safety considerations for biotechnology: scale up of crop plants. Paris, France

OECD (1998) Harmonised integrated hazard classification for human health and environmental effects of chemical substances as endorsed by the 28th Joint Meeting of the Chemicals Committee and the Working Party on Chemicals in November 1998, Part II, p.11 http://webnet1.oecd.org/oecd/pages/home/displaygeneral/0,3380,EN-documents521-14-no-24-no-0,FF.html

OECD (2000a) Guidance document on acute oral toxicity. Environmental Health and Safety Monograph Series on Testing and Assessment No. 24

OECD (2000b) Guidance document on the recognition, assessment and use of clinical signs as humane endpoints for experimental animals used in safety evaluation. Environmental Health and Safety Monograph Series on Testing and Assessment No. 19

OECD (2002) Report of the OECD workshop on the nutritional assessment of novel foods and feeds. Organization of Economic Cooperation and Development (OECD), Paris, France

OGTR (2001) PC3 Plant Facility Version 1.1. Office of the Gene Technology Regulator (OGTR), Department of Health, Australian Government, Canberra, Australia. http://www.ogtr.gov.au/internet/ogtr/publishing.nsf/content/PC3-4/$FILE/PC3PLv1-1.pdf

OGTR (2013) PC2 Plant Facility Version 3.2. Office of the Gene Technology Regulator (OGTR), Department of Health, Australian Government, Canberra, Australia. http://www.ogtr.gov.au/internet/ogtr/publishing.nsf/content/PC2-4/$FILE/PC2PLV3.2.pdf

Palm CJ, Seidler RJ, Schaller DL (1996) Persistence in soil of transgenic plant produced *Bacillus thuringiensis* var. kurstaki δ-endotoxin. *Can J Microbiol* 42:1258–1262

Paul J, Hennig B (2019) New world map of genetically modified organisms agriculture: North and South America=85%. *Acres Aust* 101:59–60

Pauwels K, DeKeersmaecker SC, DeSchrijver A et al (2015) Next-generation sequencing as a tool for the molecular characterization and risk assessment of genetically modified plants: add value or not. *Trends Food Sci Technol* 45:319–326

Pemsl D, Waibel H, Gutierrez AP (2005) Why do some Bt-cotton farmers in China continue to use high levels of pesticides? *Int J Agric Sustain* 3:44–56

Perlak FJ, Deaton RW, Armstrong TA et al (1990) Insect resistant cotton plants. *Nat Biotechnol* 8:939–943

Pray CE, Huang J, Hu R et al (2002) Five years of Bt cotton in China—the benefits continue. *Plant J* 31:423–430

Prins TW, van Dijk JP, Beenen HG et al (2008) Optimized padlock probe ligation and microarray detection of multiple (non-authorized) GMOs in a single reaction. *BMC Genomics* 9:584. https://doi.org/10.1186/1471-2164-9-584

Prins TW, van Dijk JP, van Hoef AMA et al (2010) Towards a multiplex cereal traceability tool using padlock probe ligation on genomic DNA. *Food Chem* 118:966–973. https://doi.org/10.1016/j.foodchem.2008.10.085

Purcell JP, Perlak FJ (2004) Global impact of insect resistant (Bt) cotton. *AgBioforum* 7(1&2):27–30

Qiao FB, Huang JK, Wang SK, Qiang LI (2017) The impact of Bt cotton adoption on the stability of pesticide use. *J Integr Agric* 16(10):2346–2356

Rahman M, Zaman M, Shaheen T, Irem S, Zafar Y (2015) Safe use of *Cry* genes in genetically modified crops. *Environ Chem Lett* 13:239–249

Rajeswari S, Muthuramu S, Chandirakala R et al (2010) Callus induction, somatic embryogenesis and plant regeneration in cotton (*Gossypium hirsutum* L.). *Electron J Plant Breed* 1(4):1186–1190

Ramasundaram P, Suresh A, Samuel J et al (2014) Welfare gains from application of first generation biotechnology in Indian agriculture: the case of Bt cotton. *Agric Econ Res Rev* 27:73–82

Ramaswami B, Pray CE, Lalitha N (2012) The spread of illegal transgenic cotton varieties in India: biosafety regulation, monopoly, and enforcement. *World Dev* 40:177–188. https://doi.org/10.1016/j.worlddev.2011.04.007

Randhawa G, Chhabra R, Bhoge RK et al (2015) Visual and real-time event-specific loop-mediated isothermal amplification based detection assays for Bt cotton events MON531 and MON15985. *J AOAC Int* 98:1207. https://doi.org/10.5740/jaoacint.14-269

Randhawa G, Singh M, Sood P (2016) DNA-based methods for detection of genetically modified events in food and supply chain. *Curr Sci* 110:1000–1009. https://doi.org/10.18520/cs/v110/i6/1000-1009

Rangan TS (1993) Regeneration of cotton. USA patent no. 5,244,802

Rangan TS, Rajasekaran K (1996) Regeneration of cotton plants in suspension culture. USA patent 5,583,036, 10 Dec 1996

Rashid B, Saleem Z, Husnain T et al (2008) Transformation and inheritance of Bt genes in *Gossypium hirsutum*. *J Plant Biol* 51:248–254. https://doi.org/10.1007/BF03036123

Raybould A (2007) Ecological versus ecotoxicological methods for assessing the environmental risks of transgenic crops. *Plant Sci* 173(6):589–602

Ribeiro TP, Arraes FBM, Lourenço-Tessutti IT et al (2017) Transgenic cotton expressing Cry10Aa toxin confers high resistance to the cotton boll weevil. *Plant Biotechnol J* 15:997–1009. https://doi.org/10.1111/pbi.12694

Ricroch AE (2012) Assessment of GE food safety using '-omics' techniques and long-term animal feeding studies. *New Biotechnol*. https://doi.org/10.1016/j.nbt.2012.12.001

Roberts RJ (2005) Classic perspective: how restriction enzymes became the workhorses of molecular biology. *Proc Natl Acad Sci* 102(17):5905–5908

Romeis J, Bartsch D, Bigler F et al (2008) Assessment of risk of insect-resistant transgenic crops to non-target arthropods. *Nat Biotechnol* 26:203–208

Rüdelsheim P (2015) Containment management for research and development of GM crops. *Collect Biosaf Rev* 9:40–71

Ruebelt MC, Leimgruber NK, Lipp M et al (2006) Application of two-dimensional gel electrophoresis to interrogate alterations in the proteome of genetically modified crops. 1. Assessing analytical validation. *J Agric Food Chem* 54:2154–2161. https://doi.org/10.1021/jf0523566

Ryffel GU (2014) Transgene flow: facts, speculations and possible countermeasures. *GM Crops Food* 5:249–258

Sadashivappa P (2008) Socio-economic impacts of Bt cotton adoption in India. *AgBioforum* 18(2):193–208

Sanford JC, Klein TM, Wolf ED et al (1987) Delivery of substances into cells and tissues using a particle bombardment process. *Part Sci Technol* 5:27–37

Sarkar B, Patra AK, Purakayastha TJ et al (2009) Assessment of biological and biochemical indicators in soil under transgenic Bt and non-Bt cotton crop in a sub-tropical environment. *Environ Monit Assess* 156:595–604

Saxena D, Stotzky G (2001) *Bacillus thuringiensis* (Bt) toxin released from root exudates and biomass of Bt corn has no apparent effect on earthworms, nematodes, protozoa, bacteria, and fungi in soil. *Soil Biol Biochem* 33:1225–1230. https://doi.org/10.1016/S0038-0717(01)00027-X

SCBD (2000) Cartagena protocol on biosafety to the convention on biological diversity: text and annexes. Secretariat of the Convention on Biological Diversity (SCBD), Montreal

Shang H, Liu C, Zhang C et al (2009) Progress in mechanisms of cotton somatic embryogenesis. *Acta Bot Boreali-Occiden Sin* 29:637–642

Shen RF, Cai H, Gong WH (2006) Transgenic Bt cotton has no apparent effect on enzymatic activities or functional diversity of microbial communities in rhizosphere soil. *Plant Soil* 285: 149–159. https://doi.org/10.1007/s11104-006-9000-z

Shiva V (1993) Monocultures of the mind. Zed Books Ltd, London

Shoemaker RC, Couche LJ, Galbraith DW (1986) Characterization of somatic embryogenesis and plant regeneration in cotton (*Gossypium hirsutum* L.). *Plant Cell Rep* 5:178–181. https://doi.org/10.1007/BF00269112

Siddiqui HA, Asif M, Asad S et al (2019) Development and evaluation of double gene transgenic cotton lines expressing Cry toxins for protection against chewing insect pests. *Sci Rep* 9:1–7. https://doi.org/10.1038/s41598-019-48188-z

Sinebo W, Maredia K (2016) Innovative farmers and regulatory gatekeepers: genetically modified crops regulation and adoption in developing countries. *GM Crops Food* 7:1–11. https://doi.org/10.1080/21645698.2016.1151989

Singh RJ, Ahlawat IPS, Singh S (2013) Effects of transgenic Bt cotton on soil fertility and biology under field conditions in subtropical inceptisol. *Environ Monit Assess* 185:485–495. https://doi.org/10.1007/s10661-012-2569-1

Singh M, Pal D, Sood P, Randhawa G (2020) Construct-specific loop-mediated isothermal amplification: rapid detection of genetically modified crops with insect resistance or herbicide tolerance. *J AOAC Int*. https://doi.org/10.1093/JAOACINT/QSAA043

Sivamani E, Delong RK, Qu R (2009) Protamine-mediated DNA coating remarkably improves bombardment transformation efficiency in plant cells. *Plant Cell Rep* 28:213–221. https://doi.org/10.1007/s00299-008-0636-4

Smets G, Rüdelsheim (2018) Global trends and developments in confined field trials with genetically modified plants. *Collect Biosaf Rev* 10:53–77

Smith, H.O. and Welcox, K.W. (1970) A restriction enzyme from Hemophilus influenzae: I. Purification and general properties. 51(2): 379–391

Smith JL, Farhan Y, Schaafsma AW (2019) Practical resistance of Ostrinia nubilalis (*Lepidoptera: Crambidae*) to *Cry1FBacillus thuringiensis* maize discovered in Nova Scotia, Canada. *Sci Rep* 9:1–10. https://doi.org/10.1038/s41598-019-54263-2

Southgate EM, Davey MR, Power JB, Marchant R (1995) Factors affecting the genetic engineering of plants by microprojectile bombardment. *Biotechnol Adv* 13:631–651. https://doi.org/10.1016/0734-9750(95)02008-X

Spalinskas R, Van den Bulcke M, Milcamps A (2013) Efficient retrieval of recombinant sequences of GM plants by Cauliflower Mosaic Virus 35S promoter-based bidirectional LT-RADE. *Eur Food Res Technol* 237:1025–1031. https://doi.org/10.1007/s00217-013-2078-7

Subramanian A, Qaim M (2009) Village-wide effects of agricultural biotechnology: the case of Bt cotton in India. *World Dev* 37:256–267. https://doi.org/10.1016/j.worlddev.2008.03.010

Sun CX, Chen LJ, Wu ZJ et al (2007) Soil persistence of *Bacillus thuringiensis* (Bt) toxin from transgenic Bt cotton tissues and its effect on soil enzyme activities. *Biol Fertil Soils* 43:617–620. https://doi.org/10.1007/s00374-006-0158-6

Sun R, TIian R, Ma D et al (2018) Comparative transcriptome study provides insights into acquisition of embryogenic ability in upland cotton during somatic embryogenesis. *J Cott Res* 1:9. https://doi.org/10.1186/s42397-018-0010-1

Sweet JB, Shepperson R (1997) The impact of genetically modified herbicide tolerant oilseed rape in UK. *Acta Hortic* 459:225–234

Tabashnik BE (1994) Evolution of resistance to *Bacillus thuringiensis*. *Annu Rev Entomol* 39(39):47–49. https://doi.org/10.1146/annurev.ento.39.1.47

Tabashnik BE (2008) Delaying insect resistance to transgenic crops. *Proc Natl Acad Sci* 105:19029–19030. https://doi.org/10.1073/pnas.0810763106

Tabashnik BE, Carrière Y (2017) Surge in insect resistance to transgenic crops and prospects for sustainability. *Nat Biotechnol* 35:926. https://doi.org/10.1038/nbt.3974

Tabashnik BE, Carrière Y (2019) Global patterns of resistance to Bt crops highlighting pink bollworm in the United States, China, and India. *J Econ Entomol* 112:2513–2523. https://doi.org/10.1093/JEE/TOZ173

Tabashnik BE, Carrière Y, Dennehy TJ et al (2003) Insect resistance to transgenic Bt crops: lessons from the laboratory and field. *J Econ Entomol* 96:1031–1038. https://doi.org/10.1093/jee/96.4.1031

Tabashnik BE, Gassmann AJ, Crowder DW, Carrière Y (2008) Insect resistance to Bt crops: evidence versus theory. *Nat Biotechnol* 26:199–202. https://doi.org/10.1038/nbt1382
Tabashnik BE, Van Rensburg JBJ, Carríre Y (2009) Field-evolved insect resistance to Bt crops: definition, theory, and data. *J Econ Entomol* 102:2011–2025. https://doi.org/10.1603/029.102.0601
Tabashnik BE, Brévault T, Carrière Y (2013) Insect resistance to Bt crops: lessons from the first billion acres. *Nat Biotechnol* 31:510–521. https://doi.org/10.1038/nbt.2597
Tabashnik BE, Carrière Y, Brewer M (2020) Evaluating cross-resistance between vip and cry toxins of *Bacillus thuringiensis*. *J Econ Entomol* 113:553–561. https://doi.org/10.1093/jee/toz308
Tappeser B, Reichenbecher W, Teichmann H (2014) Agronomic and environmental aspects of the cultivation of genetically modified herbicide-resistant plants. A BfN-FOEN-EAA-Joint paper
Tizaoui K, Kchouk ME (2012) Genetic approaches for studying transgene inheritance and genetic recombination in three successive generations of transformed tobacco. *Genet Mol Biol* 35:640–649. https://doi.org/10.1590/S1415-47572012000400015
Touraev A, Stoger E, Voronin V, Heberle-Bors E (1997) Plant male germ line transformation. *Plant J* 12:949–956
Trolinder NL, Goodin J (1987) Somatic embryogenesis and plant regeneration in cotton (*Gossypium hirsutum* L.). *Plant Cell Rep* 6:231–234. https://doi.org/10.1007/BF00268487
USDA APHIS (2011) Biotechnology regulatory services user guide. Notification. Animal and Plant Health Inspection Service (APHIS), United States Department of Agriculture (USDA), Riverdale
Vaeck M, Reynaerts A, Höfte H et al (1987) Transgenic plants protected from insect attack. *Nature* 328:33–37. https://doi.org/10.1038/328033a0
Van Montagu M, Zambryski P (2013) Agrobacterium and Ti plasmids. In: Maloy S, Hughes K (eds) Brenner's encyclopedia of genetics, vol 1, 2nd edn. Elsevier Science, pp 55–57
Veettil PC, Krishna VV, Qaim M (2014) Bt cotton and ecosystem impacts of pesticide reductions, Global Food Discussion Papers, No. 41, Georg-August-University Göttingen, Research Training Group (RTG) 1666—Global Food, Göttingen
Veettil PC, Krishna VV, Qaim M (2016) Ecosystem impacts of pesticide reductions through Bt cotton adoption. *Aust J Agric Resour Econ* 61:115–134. https://doi.org/10.1111/1467-8489.12171
Velmourougane K, Sahu A (2013) Impact of transgenic cottons expressing *Cry1Ac* on soil biological attributes. *Plant Soil Environ* 59:108–114. https://doi.org/10.17221/616/2012-PSE
Verma SR (2013) ISRN biotechnology. 2013:820671, 11 p. https://doi.org/10.5402/2013/820671
Vitale J, Greenplate J (2014) The role of biotechnology in sustainable agriculture of the twenty-first century: the commercial introduction of Bollgard II in Burkina Faso. In: Convergence of food security, energy security and sustainable agriculture. Springer, Berlin, Heidelberg, pp 239–293
Vogelstein B, Kinzler KW (1999) Digital PCR. *Proc Natl Acad Sci* 96:9236–9241. https://doi.org/10.1073/pnas.96.16.9236
Walsh T, James B, Chakroun M et al (2018) Isolating, characterising and identifying a *Cry1Ac* resistance mutation in field populations of Helicoverpa punctigera. *Sci Rep* 8:1–9. https://doi.org/10.1038/s41598-018-21012-w
Wan P, Xu D, Cong S et al (2017) Hybridizing transgenic Bt cotton with non-Bt cotton counters resistance in pink bollworm. *Proc Natl Acad Sci U S A* 114:5413–5418. https://doi.org/10.1073/pnas.1700396114
Wang XB, Jiang LX, Wei L et al (2010) Integration and insertion site of *EPSPs* gene on the soybean genome in genetically modified glyphosate-resistant soybean. *Acta Agron Sin* 36:365–375. https://doi.org/10.3724/sp.j.1006.2010.00365
Wang M, Zhang B, Wang Q (2013) Cotton transformation via pollen tube pathway. *Methods Mol Biol* 958:71–77. https://doi.org/10.1007/978-1-62703-212-4-6
Wang L, Wang X, Jin X et al (2015) Comparative proteomics of Bt-transgenic and non-transgenic cotton leaves. *Proteome Sci* 13:15. https://doi.org/10.1186/s12953-015-0071-8
Wrinn KM, Evans SC, Rypstra AL (2012) Predator cues and an herbicide affect activity and emigration in an agrobiont wolf spider. *Chemosphere* 87:390–3966
Xanthopoulos FP, Kechagia UE (2000) Natural crossing in cotton (*Gossypium hirsutum* L.). *Aust J Agric Res* 51:979–983. https://doi.org/10.1071/AR00026

Yang L, Chen J, Huang C et al (2005a) Validation of a cotton-specific gene, Sad1, used as an endogenous reference gene in qualitative and real-time quantitative PCR detection of transgenic cottons. *Plant Cell Rep* 24:237–245. https://doi.org/10.1007/s00299-005-0929-9
Yang P, Iles M, Su Y, Jolliffe F (2005b) Farmers' knowledge, perceptions and practices in transgenic Bt cotton in small producer systems in northern China. *Crop Prot* 24:229–239
Yang L, Wang C, Holst-Jensen A et al (2013) Characterization of GM events by insert knowledge adapted re-sequencing approaches. *Sci Rep*. https://doi.org/10.1038/srep02839
Yang F, Santiago González JC, Williams J et al (2019) Occurrence and ear damage of Helicoverpa zea on transgenic *Bacillus thuringiensis* maize in the field in Texas, U.S. and its susceptibility to Vip3A protein. *Toxins (Basel)* 11:1–13. https://doi.org/10.3390/toxins11020102
Yasin S, Asghar HN, Ahmad F et al (2016) Impact of Bt-cotton on soil microbiological and biochemical attributes. *Plant Prod Sci* 19:458–467. https://doi.org/10.1080/1343943X.2016.1185637
Yohannes T, Woldesemayate T (2017) Biotechnology and biodiversity conservation: are they conflicting? *Int J Curr Res* 9(8):55427–55433
Yu-Jia Z, Pei-Lei F, Liang L et al (2020) Development and detection of genetically modified crops. *E3S Web Conf* 145:2019–2021. https://doi.org/10.1051/e3sconf/202014501013
Zaman M, Mirza MS, Irem S, Zafar Y, Rahman M (2015) A temporal expression of *Cry1Ac* protein in cotton plant and its impact on soil health. *Int J Agric Biol* 17:280–288
Zhang BH, Feng R, Liu F, Wang Q (2001) High frequency somatic embryogenesis and plant regeneration of an elite Chinese cotton variety. *Bot Bull Acad Sin* 42:9–16
Zhang BH, Pan XP, Guo TL et al (2005) Measuring gene flow in the cultivation of transgenic cotton (*Gossypium hirsutum* L.). *Mol Biotechnol* 31:11–20. https://doi.org/10.1385/MB:31:1:011
Zhang C, Yu S, Fan S et al (2011a) Inheritance of somatic embryogenesis using leaf petioles as explants in upland cotton. *Euphytica* 181:55–63. https://doi.org/10.1007/s10681-011-0380-7
Zhang M, Feng M, Xiao L, Song X, Ding G, Yang W (2019) Persistence of *Cry1Ac* protein from Transgenic Bt cotton cultivation and residue returning in field and its effect on functional diversity of soil microbial communities. *Pedosphere* 29(1):114–122
Zhao Z, Chen Y, Xu W, Ma M (2013) Surface plasmon resonance detection of transgenic *Cry1Ac* cotton (*Gossypium* spp.). *J Agric Food Chem* 61:2964–2969. https://doi.org/10.1021/jf3050439
Zhou G, Weng J, Zeng Y et al (1983) Introduction of exogenous DNA into cotton embryos. *Methods Enzymol* 29:433–481. https://doi.org/10.1016/0076-6879(83)01032-0

Websites

http://bch.cbd.int/protocol/background/
http://dbtindia.gov.in/guidelines-biosafety
http://geacindia.gov.in/capacity-building-resources.aspx
http://geacindia.gov.in/india-bch.aspx
http://gmo-crl.jrc.ec.europa.eu/gmomethods/
http://www.cicr.org.in/techbank/bt_express.pdf
http://www.cicr.org.in/tech_bank/Bt_kit.pdf
http://www.isaaa.org/gmapprovaldatabase/crop/default.asp?CropID=7&Crop=Cotton
http://www.ucsusa.org/superweeds
http://www.weedscience.org/Pages/crop.aspx
https://www.canr.msu.edu/news/superweeds-secondary-pests-lack-of-biodiversity-are-frequent-gmo-concerns

Chapter 17
Cotton Production Beyond 2030

Don Keim, M. Rafiq Chaudhry, Sandhya Kranthi, and Dean Ethridge

17.1 Introduction

Cotton production practices have been continuously changing due to better understanding of the cotton plant behavior, response, and needs coupled with composite efforts to improve lint yield and fiber quality. Thus, changes have occurred in all aspects of production technology including input quantity and their timing. For a long time, development of varieties remained the predominant focus of bringing improvement in production, quality, and suitability to the growing environment. In order to exploit the best of the cotton plant, it needs to be fed properly and protected against vagaries of insect pests, diseases, and weather stress. Over the last about a century, some of the major aspects of producing cotton that have changed drastically are application of fertilizers, use of pesticides and consequences, variety development, machine measurement of fiber quality characteristics, and the latest being the use of biotechnology applications. Agriculture is nearing rapidly to utilization of drone technology as a component of the precision agriculture. Drone could be used for spraying cotton. Additions to the list will continue to increase, while improvement/perfections in the current practices will not stop. How all these developments will impact cotton production practices beyond 2030 is the subject matter of this chapter.

D. Keim
Stoneville, MS, USA

M. R. Chaudhry (✉)
Springfield, VA, USA

S. Kranthi
International Cotton Advisory Committee, Washington, DC, USA

D. Ethridge
Lubbock, TX, USA

Mehboob-ur-Rahman et al. (eds.), *Cotton Precision Breeding*,
https://doi.org/10.1007/978-3-030-64504-5_17

17.2 Cotton Breeding

The state of the art in plant breeding has advanced dramatically in the last 20 years. This is primarily a result of advances in molecular marker technology and data science. The adoption of these and other tools in breeding programs have shown to both (a) enhance genetic gain and (b) reduce the selection cycle time. Advances in data collection automation and analysis have also enabled more efficient programs. Applied breeding efforts in cotton, whether in private- or public-funded organizations, have generally lagged behind other major field crops such as corn, wheat, and soybean. Consequently, adoption of new technologies in cotton programs has generally been slower. But, as these technologies are adapted to cotton breeding, large advances in breeding efficiency have been realized.

The breeding efficiency in cotton is much lower (higher cost per unit of genetic gain) than most field crops due to inherent nature of the cotton plant. In comparison to other field crops, the cotton plant is indeterminate, having sequential blooming and relatively long growing cycles. Cotton is a perennial woody bush, which has been adapted to and commercially grown as an annual field crop. Other characteristics of the crop have added to the high "breeding cost" (Keim 2007). These include requirements for ginning, seed delinting, hand harvesting in early breeding stages, hand self-pollination, and costly fiber evaluation. Because of this high cost, adopting any new technologies, such as genomic selection (GS), that can enable fewer progeny generations to be grown, harvested, and processed will have a major effect on breeding efficiency. Below are described how some of these new breeding technologies are being implemented in some large breeding programs. In doing so, a description will emerge of what an advanced applied cotton breeding program could be like by 2030. Discussion will be limited to varietal breeding, as this is a characteristic of most applied cotton breeding programs worldwide.

17.3 Breeding Goals

The goals established in any breeding programs will dictate which tools will be useful. Fiber productivity (yield of marketable lint per unit area) is typically the primary breeding goal. This includes fiber quality as a critical objective when considering the "marketability" of the end product. Fiber length, strength, and fineness tend to have negative associations with fiber yield. Thus, neglect of these traits in a program can lead to produce high-yield varieties having poor marketable fiber. Other critical goals include adaptability to a range of environments, stress tolerance, and disease and insect resistance. The use of biotechnology in the form of transgenic traits has long been a critical breeding goal in many programs worldwide. Primary transgenic traits commercialized to date have included lepidopteron insect resistance and herbicide tolerance. New traits such as sucking pest tolerance are showing promise (Akbar et al. 2019). Additionally, as pests overcome the resistance traits, a search for new transgenic traits is continuing.

17.4 New Breeding Tools

Tools available to plant breeders have advanced dramatically. Until recently, breeders were limited by the resources necessary to select plants solely on a phenotypic basis. Once initial breeding development and selection were accomplished, costly multiyear wide-scale testing was needed to identify the most productive and most widely adapted genotypes for a given area. In addition, extensive fiber quality evaluation was needed to ensure the released genotypes meet the market needs. Although advanced-stage phenotypic evaluation will continue to be necessary, any tools available to reduce the needed time and effort will be of great advantage.

Off-cycle generation advance. Although this has long been an integral part of breeding in several crops, use of off-season nurseries in cotton, until recently, has been quite limited. Some applied programs are beginning to adopt strategies where early generations (from crossing up to F_4) can be advanced quickly using greenhouses and off-season nurseries. In a traditional breeding program, this can have the advantage of reducing the breeding cycle by up to one-third. Additionally, use of off-cycle generation advance is highly advantageous in marker-assisted backcross programs. Backcross development cycles can be reduced by up to two-thirds.

Because genetic gain is a function of time, use of off-cycle nurseries can be of great advantage. However, large resource commitments are necessary to take full advantage. Development and maintenance of off-season nurseries and greenhouse facilities can be cost prohibitive for some programs.

Advances in phenotyping. Performance evaluation remains a critical part of cotton breeding. Productivity, performance stability across years and environments, maturity, stress response, and fiber quality are characteristics essential to successful varietal development. In recent years, mechanized planting, harvest, and processing have enabled commercial breeding programs to reduce the labor input in their testing programs. Thus, breeders have been able to either expand their program testing footprint or reduce their overall testing costs. The former approach has the advantage of more thorough evaluation of genotypes. Such an approach could reduce the testing cycle by a year or more.

High-throughput phenotyping technology has advanced dramatically in recent years. A recent demonstration using LiDAR (light detection and ranging/light imaging, detection, and ranging) in assessing cotton plant growth was conducted in Georgia, USA (Sun et al. 2018). This technology has the capability of rapid assessment of plant growth characteristics related to productivity on large numbers of genotypes.

Data analysis and management. These new phenotyping technologies have the capability assessing large sets of genotypes but require the acquisition and analysis of a huge amount of data. Statistical and computational methods incorporating GxE effects have advanced dramatically (Perez-Rodriquez et al. 2015).

17.5 Marker-Assisted Breeding Tools

Molecular marker technology has had major breakthroughs in recent years, dramatically reducing the cost per data point, as well as increasing the usefulness of molecular data. This has enabled the use of new breeding approaches in a cost-effective manner.

Marker-assisted selection (MAS). Marker-assisted selection is quite useful in selection for simply inherited traits in cotton but requires the prior establishment of marker-trait associations (QTLs). A particular usefulness of the tool has been in the selection for traits that are very difficult to phenotype accurately. It has also proven beneficial where phenotyping requires destructive sampling of live plants. For example, selection for nematode resistance, both for root knot and reniform nematode, in cotton has been made using DNA markers. To accurately determine whether a plant is resistant to nematode, the plant must be grown in nematode-infected soil. After the plants are infected, the roots must be carefully removed, carried to the lab, and washed carefully. Finally, physical nematode counts must be made. Once QTLs have been identified, MAS for nematode resistance has made breeding for these traits a feasible endeavor. Public and private US breeding programs are currently using MAS for these traits.

Marker-assisted backcrossing (MABC). MABC has the primary advantage of reducing the number of backcrossing cycles needed to essentially recover the recurrent parent genotype. It uses the traditional backcross method to transfer alleles at one or a few loci from a donor parent to an elite variety (recurrent parent). Markers are utilized not only to select the trait of interest but also to recover genotype of the recurrent parent quickly. The trait of interest must have associated a strong link with a small group of markers. To recover the recurrent parent, a set of genome-wide markers that differentiate the two parent genotypes are utilized. Included are markers close to the allele of interest (flanking markers) that are utilized to minimize the carryover donor segments located near the target locus. This allows more complete recovery of the recurrent parent genotype, as well as the elimination of any potential linkage drag.

In cotton, MABC is widely utilized in transferring transgenic traits to elite breeding lines. It is increasingly used to introduce "native" traits that are rare in the elite germplasm breeding pools of a program. Often programs have several traits to transfer using MABC into elite germplasm. The plant numbers needed to genotype and the complexity of the breeding schemes increase dramatically when more than three loci are transferred from a donor parent.

Genomic selection (GS). Genomic selection is being evaluated and initiated in some applied cotton breeding programs (Gapare et al. 2018). A review of GS in plant breeding is presented in Trends in Plant Science (Crossa et al. 2017). The GS uses a set of genome-wide markers to predict the breeding value (BV) of individuals. Genomic selection is enabled by establishing a training population where individuals are both genotyped and phenotyped. The association of performance data and marker data is used to "train" the model used in BV prediction (GP). Individuals from breeding populations are genotyped, and a genomic estimated breeding value (GEBV) is established for each, for use in selection. The accuracy of the GS model is assessed by correlating actual phenotypic values with GEBVs in additional populations.

Genomic selection can reduce the breeding cycle time in two primary ways. First, an initial generation of field phenotyping can be eliminated by reliance on selection solely based on GEBV. Additionally, individuals with high breeding values can be recombined earlier, before multiyear performance testing establishes their true value.

GS is particularly useful for complex traits, such as lint yield and some fiber traits that have many small genetic effects. However, at the same time, MAS can be incorporated for simply inherited traits, such as disease resistance, where marker-trait associations have been previously established. Several schemes can be implemented using combinations of phenotypic selection, MAS, and pedigree information. The cost and time factors in the genotyping are key considerations in the planning of the breeding schemes.

A problem with initiating a GS program is the high start-up costs related to the needed technology, as well as costs associated with development and maintenance of a large training population. Historical phenotype data could be used to initiate GS, reducing the time and cost of maintaining an initial training population (Gapare et al. 2018). This would allow realization of the benefit GS, prior to incorporating contemporary data and recalibrating the model. Breeding goals also change over time due to changing environmental factors, changes in production systems, and changes in biotic and abiotic stresses. This requires the ongoing introduction of new germplasm into the base populations. These changes will necessitate a continual retraining of the model to maintain validity of the GP.

17.6 Cotton Breeding in 2030

Cotton breeding program structures in the future will depend primarily on the resources available. Resources from governmental and nongovernmental (NGO) institutions are now and will continue to be limited. Several NGOs strongly support breeding in major food crops with good reason. In cotton, strong support will come primarily from private industries that will potentially economically benefit directly from the products developed by their associated breeding programs. This direct benefit will come in the form of direct seed sales, sales of associated products (agrochemicals), and/or market share advantages. Direct economic benefit could also come for the end user of fiber and seed.

Ideally, an active commercial breeding program will utilize a combination of breeding schemes utilizing the new tools mentioned above. Because many of these tools are well developed in other crops, it would be a matter of transferring to and further adapting this technology to cotton.

Genomic selection will become the major component of commercial breeding programs. It will be further complemented by the intensive use of off-season nurseries and greenhouses and thus will shorten breeding cycles; both together will enhance the breeding efficiency. GS will be coupled with MAS for common but essential QTLs, which will enhance the efficiencies of the marker technologies.

Because of market demands, transgenic traits will continue an important part of most breeding programs. However, these traits will change or be added to over time. Thus, it will be necessary to uncouple transgenic breeding from breeding for yield and other agronomic traits. Thus, transgenic and other rare "native traits will continue to be added to MABC breeding schemes.

Performance evaluation across years and locations will continue to be an essential part of varietal development program. New tools in the growing, harvesting, processing, data acquisition, and analysis will increase the efficiencies of evaluation leading to final varietal selection. Several breeding programs will not have all the resources needed to implement all these technologies. However, certain technologies can be integrated partially in a breeding program with minimal costs. Examples of breeding schemes utilizing GS in limited programs are well presented in wheat (Bassi et al. 2016). Outsourcing of genotyping is now readily available and is continuing with reductions in cost per data point. Publicly available cotton marker data is also readily available (Yu et al. 2014).

17.7 Cotton Breeding Beyond 2030

Two particular technologies are described that are not yet developed in cotton. However, they have potential in dramatically advancing cotton breeding far into the future.

Doubled haploids (DH). Use of doubled haploids has proved an important tool for breeding few crops (Dwivedia et al. 2015). Haploids are produced and doubled, making the genotype fully homozygous. This takes place in two generations than that of the conventional breeding scheme that may require several years to achieve the acceptable level of homozygosity. Use of doubled haploids in a GS or MABC breeding schemes could substantially reduce the breeding development time.

Semigamy was discovered in *Gossypium* several decades ago (Turcotte and Feaster 1963). The semigametic strain of *G. barbadense* when used as a female parent in crossing produced 1% androgenetic haploids in F_1. Further development in the frequency of haploid production and in chromosome doubling could make doubled haploids a useful tool in cotton breeding.

Apomixis-asexual reproduction of F_1 hybrids. Hybrid cotton had been a commercial reality in India for several decades. Heterosis for fiber yield has been extensively demonstrated. However, hybrid seed production is costly as it is accomplished by hand emasculation and hand pollination of the female parent. This type of hybrid seed production is cost prohibitive in most countries. Any method whereby asexual reproduction of F_1 hybrids occurred could dramatically reduce seed production costs while exploiting the inherent heterosis in the crop. Although proposed, no successful demonstration of large-scale asexual reproduction has been demonstrated in cotton or any other crop plant. A recent study in rice establishes the feasibility of asexual reproduction in crops and could enable the maintenance of hybrids clonally through seed propagation (Khanday et al. 2019).

17.8 Physiological Challenges

The cotton plant has a huge yield potential that has been partially utilized. The physiological aspirations that have a direct bearing on the plant's performance, either in terms of higher yield or better quality, have yet to be materialized. The biggest loss in yield comes from fruit shedding whether it is a tiny invisible bud or a more visible fruiting form. The struggle to avoid this loss is ongoing with only a partial success. A multidimensional approach to this problem will continue and even get more attention.

There are two main theories about shedding of buds (invisible and visible) allude to an imbalance between hormones and the supply of carbohydrates. The imbalance between auxin and growth-retarding antiauxin hormones prevents the plant from producing flower buds. The carbohydrate supply to the existing fruiting load impedes formation of more fruiting forms. There are yet "plant training" commercial practices, wherein the plant is directed to enhance physiological processes by eliminating older/inefficient leaves; biotechnology can induce a characteristic or characteristics whereby the active life of the leaves is extended or the leaves are made to carry out photosynthesis more efficiently. There could be more physiological ways to handle or minimize lost fruiting points on the plant. Other innovative approaches, in particular biotechnological, will become more practical to save physiological losses in terms of empty nodes on the plant.

Cotton is a C3 plant and photorespires about 30% of the photosynthetic rate. During the process of photosynthesis, carbohydrates are formed from water and CO_2 absorbed from the air. The C4 plants have a potential to utilize all carbohydrates formed during photosynthesis and thus have a higher growth rate. The cotton plant is unable to utilize all the carbohydrates and tend either to burn or release into the atmosphere a good percentage of the carbohydrates in different forms. This is what makes cotton to grow slower and produce yield lower than its potential. Efforts have been made in the past to reduce photorespiration through artificial means to increase yield. Just prior to the introduction of transgenic cotton, CO_2 enrichment and methanol application were tried. Methanol application increased water use efficiency, increased leaf surface area and leaf thickness, increased fruiting points, and improved growth rate (ICAC 1994). However, artificially changing the microclimate of the plant does not seem to be a feasible solution for various reasons. The need is still there to lower photorespiration rate, and researchers beyond 2030 will still be struggling to deal with the loss of carbohydrates during photosynthesis.

17.9 Input Use

Almost 3% of the world cotton area does not get any fertilizer application. Two reasons for this are (1) fertilizer applications have no impact on yield and (2) inability to buy inputs. Organic fertilization is used on a limited scale in few countries. Wherever fertilizer is used, nitrogen application is a must over phosphorus and potassium because of the immobility of phosphorus and lack of potassium impact on yield under suboptimal yield levels. Soil depletion under such circumstances will ultimately require farmers to use fertilizer under all circumstances in the next 15–20 years. Nitrogen will continue to be required at current levels, while P and K applications will depend on cropping systems and soil types. Overuse of nitrogen, resulting in excessive vegetative growth, is not yet a common problem, but is expected to increase with the overuse of nitrogen in an effort to further increase yields. Breeders will have to focus on earlier entry by the plant into the reproductive phase in order to further increase yields. More efficient use of nitrogen by the cotton plant will continue to be targeted.

17.10 Irrigation

Scarcity and/or inaccessibility of water preclude complete irrigation of cotton. Almost 40% of the world cotton area is produced under rainfed conditions, in spite of the fact that yields under rainfed conditions are only 40–45% of the comparable irrigated fields. This is why 40% of rainfed area accounts for less than 30% of world production. The lower yield potential and higher yield variability generally dictates reduced costly inputs for rainfed cotton. Water will continue to be limited and will have to be shared with competing crops, in particular with food crops. Conserving the existing available water and using it more efficiently by minimizing losses will receive increasing levels of attention in the future. The current trend to shift from flood irrigation to row/furrow irrigation will continue. The best option will be to supply only as much water as needed or is taken up by the plant. Where feasible, subsurface or above-surface drip irrigation will likely be used to supply only as much water as needed by the plant. However, this will be limited by the large capital investments required to install a system. Biotechnological approaches in breeding offer hope for the development of "drought-tolerant" cottons offering superior performance under rainfed conditions.

17.11 Cotton Pest Management

Insect pests, nematodes, weeds, and diseases cause serious economic losses and deteriorate fiber quality. Since the early 1990s, new pesticide chemistries and biotech cotton have dramatically impacted plant protection in cotton, which in turn contributed directly to increases in yield and production of cotton. Newer strategies to increase seed cotton production will mandate responsible use of biotech approaches through clustered regularly interspaced short palindromic repeats (CRISPR/cas9), ribonucleic acid interference (RNAi), and gene pyramiding. Mechanization, robotics, artificial intelligence, nanotechnology, nano-formulations of agrochemicals, and information technology mainly through the use of mobile applications (apps) will significantly impact plant protection in cotton beyond 2030. These technologies may go hand in hand with the conventional and traditional methods of plant protection that are currently being adopted, not necessarily replacing them.

17.11.1 Protection Technologies of Today

Promising, effective, and "greener" technologies have been made available to cotton growers in many cotton-growing countries in the last 30 years. Biotech cotton has replaced conventional varieties on over 80% of the world cotton area (Kranthi 2018). Novel insecticides, mainly seed treatment formulations of neonicotinoids, have significantly contributed to the control of sap-sucking pests, thereby enabling the bollworm-resistant biotech cotton varieties to realize their genetic potential to a greater extent. However, these technologies have not only been effective but also expensive, thus leading to increased cost of production.

17.11.2 Biotech Cotton

Genetic engineering in cotton leads to the development of biotech cotton varieties, through gene transfer or gene editing methods (RNA interference and CRISPR/cas9 technologies). To date, transgenics generated through gene transfer have played a major role in cotton production where farmers relied heavily on seed industry for obtaining genetically modified (GM) or biotech cottonseeds. A total of 510 genetic-transformation events for improving 40 traits across 30 crops have been reported; of these 7 GM traits have been most popularly commercialized across 44 countries. Of these, two traits have been exploited in cotton, for insect resistance (IR) against lepidopteran and hemipteran pests and herbicide tolerance (HT) against 2,4D, dicamba, glyphosate, glufosinate, isoxaflutole, oxynil, and sulfonylurea herbicides (http://www.isaaa.org/gmapprovaldatabase/eventslist/default.asp). The IR, HT, and IR + HT cottons have been generally associated with target pest suppression (Wu et al. 2008; Carpenter 2010; Lu et al. 2012; Wang and Fok 2018). Though biotech cotton assists only in pest management, it is also perceived as a yield enhancement technology (Thirtle et al. 2003; Qaim and Matuschke 2005; Cattaneo et al. 2006; Traxler et al. 2013). However, these generalizations are debatable in some countries.

Development of resistance in target insects to biotech cotton was anticipated due to widespread and prolonged cultivation, particularly under poor stewardship.

Practical resistance was reported in 16 cases, while no decrease in susceptibility was noticed in 17 cases when data from 16 countries, to 5 toxins expressed in Bt crops, for populations of 9 species of lepidopteran pests was analyzed (Tabashnik and Carrière 2017). The pink bollworm resistance to Cry1Ac (Dhurua and Gujar 2011) and to *Cry1Ac + Cry2Ab* (Naik et al. 2018) in India adds to the list of insect pests resistant to Bt cotton. The pink bollworm is the second insect pest to have developed resistance to Bt cotton, the first being the cotton bollworm, *Helicoverpa zea* (Tabashnik et al. 2013).

Cotton diseases have also been pursued with the help of genetic engineering; however, till-to-date disease-resistant cotton has not been commercialized; however, success in lab-scale experiments was claimed. A few examples of disease-resistant biotech cotton are as follows: D4E1, a synthetic peptide expressed in cotton, imparted tolerance to black root rot caused by the fungal pathogen, *Thielaviopsis basicola* (Rajasekaran et al. 2005); an *endochitinase* gene expressed from *Trichoderma virens* enhanced fungal resistance in biotech cotton (Emani et al. 2003); a plant defensin gene *NaD1* expressed in cotton conferred field resistance to *Fusarium oxysporum* and *Verticillium dahliae* (Gaspar et al. 2014); a non-expresser of pathogenic related gene 1 (*npr1*) from *Arabidopsis*, overexpressed in cotton, imparted resistance to *V. dahliae*, *F. oxysporum*, *Rhizoctonia solani*, *Alternaria alternata*, and reniform nematode (Parkhi et al. 2010). Leaf curl-tolerant biotech cotton with antisense against the pre-coat protein *av2* gene has also been reported (Sohrab et al. 2014).

17.11.3 Conventional Technologies

Application of pesticides has emerged as the integral part of the integrated pest management strategies across the world. Window-based, temporally selective pesticide-use strategies were developed to conserve native natural enemies and have been adopted in some cotton-growing countries (Kranthi et al. 2002; Wilson et al. 2004; Dhawan et al. 2009; Silvie et al. 2013). Novel and highly selective pesticide chemistries with low ecotoxicological profiles are recommended in specified temporal windows when the pest exceeds the economic threshold level (ETL). Such pesticides are often expensive and need to be used prudently to ensure economic and environmental sustainability. Integrated pest management strategies across the world incorporate the use of all pest control measures that favor eco-friendly and economically viable crop protection. Apart from the use of synthetic pesticides, biopesticides such as *Bacillus thuringiensis* (*Bt*), nuclear polyhedrosis viruses (NPV), entomopathogenic fungi, neem-based insecticides, and exploitation of natural enemies such as parasites and predators have constituted the main conventional methods of cotton pest management. Adoption of cultural methods such as closed season for the pink bollworm (Watson et al. 1978), use of food sprays to attract native natural enemies (Mensah 2002), use of pheromone lures and traps for monitoring, mass capture, and mating confusants (Lance et al. 2016; Carriere et al. 2017) have also been popular in cotton IPM programs. Eradication of the pink bollworm in the USA is a landmark achievement, where pheromones and lab-reared sterile moths were used on a massive in undative scale in demarcated areas (Simmons et al. 2011; Morrison et al. 2012; Tabashnik et al. 2012).

While no till, conservation tillage, extensive herbicide use, and mechanized weeding have been popular for weed control for many years in developed countries; hand weeding is still adopted in small-scale production systems. Integrated pest management strategies have limitations. Pest-resistant varieties are not very common; novel eco-friendly insecticides are expensive; quality control of biopesticides is a challenge; and availability of biopesticides and biocontrol organisms for augmentative or innundative releases is limited. Moreover, cotton growers, in countries such as China, India, Pakistan, and many African countries, have small-scale holdings where a community approach that would be more effective in pest management is not very feasible and is rarely adopted. Also farmers need guidance for a proactive implementation of IPM strategies, particularly to prevent pest outbreaks and for effective and sustainable pest control. Advances in improving the efficacy of conventional pest management technologies through the use of novel pesticide application technologies, nano-formulations, electrostatic sprays, solar-powered sprayers, food sprays, novel pheromone formulations, and conservation biological control techniques have been made, but more work is needed.

17.11.4 Technologies for the 2030s

The future of cotton pest management would depend on how well production practices are integrated with varietal improvement, genetic engineering, and IPM strategies. Creation of biotech cotton varieties using novel genes, RNA interference, and CRISPR/cas9 technologies fall in the domain of advances in genetic engineering. Transgenics involve introduction of genes across genera. RNAi involves modulating gene expression. CRISPR/cas9 involves gene editing. Biotech cottons using *cry* toxin genes and herbicide tolerance genes are well known throughout the world, as cotton was the first biotech crop approved for commercial cultivation in 1995 in the USA. While *Bt* cotton has been very effective in controlling the target insects, stewardship of the transgenic technology to suit local conditions is of paramount importance to utilize the technology to its fullest potential before resistance sets in. Failure to do so has caused yield stagnation and resistance development as in India or a ban as in Burkina Faso, due to various reasons.

17.11.5 CRISPR/cas9

CRISPR/cas9 is an in vivo genome editing technology in a biologist's tool kit that modifies DNA sequences in a genome to turn on or off specific genes in cells and organisms with relative ease and can also help raise transgene-free genetically modified crops (Soda et al. 2018). Insects from five orders (Diptera, Orthoptera,

Coleoptera, Hymenoptera, and Lepidoptera) have been subjected to CRISPR/cas9 study, and *Drosophila* spp. have been extensively studied (Sun et al. 2017). CRISPR/cas9 deployed for targeted mutagenesis in cotton (Chen et al. 2017) was used to design unique and broad-spectrum control of cotton leaf curl disease-associated begomovirus complex and its associated satellite molecules (Ali et al. 2016; Iqbal et al. 2016; Khatodia et al. 2017). Using CRISPR/cas9 as an efficient genome editing tool, HaCad was confirmed as a functional receptor of *Cry1Ac* in *Helicoverpa armigera* (Wang et al. 2016) and also to demonstrate that HaABCA2 played a key role in mediating toxicity of both *Cry2Aa* and *Cry2Ab* in *H. armigera* but remained unchanged with *Cry1Ac* (Wang et al. 2017). CRISPR/cas9 was used to characterize the pheromone-binding protein (PBP1) gene in *H. armigera* to provide the first in vivo evidence that HaPBP1 played an important role in perception of female sex pheromone (Ye et al. 2017). The *or16* gene when knocked out and delivered through plasmid mRNA provided novel strategies to destroy insect pest mating (Chang et al. 2017). While almost all of the plant genome editing is yet to reach a stage of commercialization, in light of the research findings, it is expected that the 2030s would see successful CRISPR/cas9-generated biotech cotton in fields. Most of these strategies could also be meant to target pests resistant to biotech cotton such as Cry toxin-resistant insect pests and herbicide-tolerant weeds.

The rising world population is bound to increase pressure on available arable land, thereby accentuating the need for new technologies that can help to produce more food and fibers from the limited land with less agrochemical inputs. Genetic engineering has the potential to increase yields and reduce input usage by introducing novel traits and modifying existing traits for genetic improvement of crops. Needless to emphasize, the deployment of biotech crops including cotton would be subjected to the condition of proven biosafety of the transgenes and CRISPR/cas9 technologies. An institutional setup for monitoring the release, adoption, and effects of new biotech technologies in crops must be envisioned before its adoption. Biosafety guidelines for biotech crops are available in most countries. Biotech cotton has been cultivated for almost 15–25 years in many developed and developing countries. It would be useful if the experiences were shared and common biosafety-regulatory guidelines developed so as to enable developing and least developed countries to devise their own biosafety-regulatory guidelines based on the common framework by incorporating the local requirements.

17.11.6 RNA Interference (RNAi)

RNAi can generate new crop quality traits or provide protection against insects, nematodes, and pathogens without introducing new proteins into food and feed products (Auer and Frederick 2009). RNAi is an evolutionarily conserved mechanism that, in primitive organisms, protects the genome from viruses and regulates gene expression during development (Dykxhoorn and Lieberman 2005) through gene silencing. The antisense (guide strand) of short dsRNA is incorporated into an RNA-induced silencing complex that can either suppress protein expression or direct the degradation of mRNAs containing homologous sequences. Elucidation of RNAi won Andrew Fire and Craig Mello a Nobel Prize in 2006. RNAi-based control of insect pests of cotton has been targeted by gene silencing of *Cyp6AE14*, *GST*, *trypsin*, *chymotrypsin*, and *JhAMT* through oral feeding or through the development of biotech cotton (Asokan et al. 2011). Cyp6Ae14 is an oxygenation enzyme expressed in the midgut of *Helicoverpa armigera* that enables larvae to tolerate gossypol. When larvae were fed dsRNA specific to Cyp6AE14 through transgenic plants, the level of cyp6ae14 transcripts in the gut decreased, and larval growth was retarded due to gossypol toxicity (Mao et al. 2011a, b). Gong and co-workers (2014) demonstrated that insect-resistant transgenic plants could be obtained through plant RNAi-mediated silencing of insect *carboxylesterase* genes. Expression of dsRNA from ECR (ecdysone receptor complex) of the cotton bollworm in tobacco improved its pest resistance (Zhu et al. 2012). RNAi was explored for the control of whitefly, *Bemisia tabaci*, for which dsRNA and siRNA were synthesized for five different genes and evaluated using a simplified oral route of feeding that caused 29–97% mortality after 6 days of feeding. Phloem-specific expression of dsRNA of PPLq and VATPase in transgenic plants for protection against whiteflies has been generated and is currently being evaluated in trials (Upadhyay et al. 2011). The dsRNA microinjection of chitin synthesis in *Anthonomus grandis* female moths resulted in normal oviposition of unviable eggs and malformed live larvae that were unable to develop on artificial diet (Firmino et al. 2013). RNAi was used against a molt regulating transcription factor gene HaHR3 of *H. armigera* through transgenic plants to disrupt its development (Xiong et al. 2013). The dsRNA of the gene governing the expression of 3-hydroxy-3-methyl-glutaryl-CoA reductase (*hmgr* gene) that catalyzes an enzymatic reaction in the JH (juvenile hormone) biosynthesis pathway in bollworms was expressed in cotton. When *H. armigera* larvae fed on these transgenics, transcription levels of HMGR were reduced unto 81%, and expression of vitellogenin was reduced by 76% reduction that resulted in impaired larval growth and development. Odorant-binding proteins (AgOBP2) in *A. gossypii* play a crucial role in host location, selection, and oviposition. Silencing it through RNA interference would play a crucial role in *A. gossypii* control (Rebijith et al. 2016). However, RNAi-mediated knockdown of expression of polygalcturonase (*pg1*) mRNA in adult salivary glands of *Lygus lineolaris* was not very successful in impacting feeding (Walker and Allen 2010).

New approaches to gene silencing involve generation of biotech crops, which is expensive, oral feeding of the dsRNA that is impractical in fields or through the virus-induced gene silencing (VIGS). VIGS is a method where a recombinant virus is used as a silencing vector to target host plant genes by triggering the antiviral RNA silencing pathway with little induction of viral disease symptoms in the host plant. Its silencing effect is transient since genetic material does not integrate in the plant genome.

17.11.7 Gene Pyramiding

Trait pyramiding and gene pyramiding are instrumental in extending the trait value, to enhance efficacy and to extend the durability of the trait. Gene pyramiding for cotton crop protection is adopted to ensure redundant killing of the target insect pests that are not controlled by one toxin which will be killed by the other. Pyramiding toxins is particularly effective when resistance is recessive. The chances of the target pest being resistant to both toxins are rare especially if populations of the pest have not been exposed to any of the toxins (Brévault et al. 2013). Factors such as a decline in toxin expression as the plant ages (Greenplate 1999; Kranthi et al. 2005; Knight et al. 2013) and asymmetric cross resistance (Wei et al. 2015) negatively impact the effects of gene pyramiding. Introducing *Bt* genes as pyramids is fairly common and has been proven effective thus far. Cowpea trypsin inhibitor genes expressed along with *Cry1Ac* in biotech cotton enhanced resistance to *Helicoverpa armigera* though it did not lead to increased mortality of the pest in the field (Cui et al. 2011). Four methods of gene pyramiding are hybrid stacking (where each of the parent involved in the cross carries transgenes of interest, e.g., Roundup Ready Flex Bollgard II), co-transformation (different constructs carrying desired transgenes are introduced simultaneously into the plant cell), multigene cassette transformation (two or more transgenes are present within the same construct), and retransformation (where a transgene is introduced into a transgenic plant, e.g., generation of Bollgard II). Gene pyramiding for pest control is the immediate future for transgenics where multiple genes for insect pest resistance would be expressed along with multiple genes for herbicide tolerance and/or disease resistance. In fact 95% of transgenic cotton being cultivated in India carries two genes for pest resistance, and more than 80% of the cotton in the USA carries stacked genes for lepidopteran and herbicide tolerance. For example, the new biotech cotton called Bollgard® 3 XtendFlex® technology has two traits and six trait-specific functional genes with three genes (*cry1Ac* + *cry2Ab* + *vip3Aa*) for insect resistance and three genes, *epsps, bar*, and *demethylase*, for tolerance to the herbicides glyphosate, glufosinate, and dicamba, respectively. Recently, biotech cotton, carrying a new event, was developed to express a modified Bt *Cry51Aa2* protein conferring tolerance to hemipteran insects. Genes for hemipteran, lepidopteran pest tolerance and herbicide tolerance have together been stacked in cotton that is under evaluation.

Stewardship of the stacked gene technology is of utmost importance. Caution must be exercised to ensure that populations of the pest have not been exposed individually to genes in the stack prior to the introduction of stacked product. Introduction of the two-gene (*cry1Ac* + *cry2Ab*) biotech cotton after 3 years of cultivation of *cry1Ac*-based biotech cotton led to a sequential exposure of the worms to the toxins, as a result of which both the genes became ineffective to the pink bollworm within 10 years of cotton cultivation. However, though Bt cotton is still effective on the polyphagous pest *H. armigera*, modeling results show that the pest may soon develop resistance to the two genes sooner because of the sequential exposure. Also, transgenes deployed in biotech cotton must be homozygous. Currently almost all the biotech F_1 hybrid plants produce bolls that have a small proportion of non-*Bt* seeds in the bolls, which allow the survival of a few larvae, thereby permitting compromising bollworm protection.

Gene pyramiding using the RNAi approach along with conventional transgene expression could be the next-generation approach to develop durable biotech cotton. Pyramiding of RNAi with *Bt* cotton was done by silencing *juvenile hormone methyltransferase* and the JH-binding protein that are crucial for JH (juvenile hormone) synthesis and transport of JH to the organs, respectively, for effective control of *H. armigera* (Yu et al. 2016). Results from modeling experiments demonstrate that combining RNAi protection with one or more Bt toxins can delay the evolution of resistance but the gain in durability depends on the percentage of the refuge (Ni et al. 2017).

17.11.8 Introduction of New Pesticides

Significant changes in cotton crop protection have occurred through introduction of new chemistries. Novel insecticides with new modes of action and new formulations were introduced and registered for use on cotton in the past 20 years. Insecticides based on their mode of action have been classified into 32 groups of which insecticides belonging to 14 groups are permitted for use on cotton by the Insecticide Resistance Action Committee. Many of the new insecticides are highly specific to the targeted site in the insect and are necessary to be used in milli- and micro-quantities of the active ingredient per hectare. New formulations have also been introduced, thus facilitating the handling of lower volumes of pesticides. Typically, large containers of formulated products that once occupied market shelves are currently being replaced with less than 100 mL containers for one-time use in small-scale cotton production systems, which not only prevents insecticide misuse but also makes it more cost-effective for the end user. Novel formulations also include nano-formulations that facilitate increased solubility of the active ingredient, thereby enhancing its bioavailability ensuring targeted delivery, controlled release, and protection against degradation (Hazra 2017).

Old chemistries are being replaced with new molecules, especially of those active ingredients whose patents are set to expire and would be phased out beginning 2020. Also insecticides whose negative impact on human health or environment was hitherto unknown may also be replaced. Glyphosate as a possible carcinogen and its fate as the most popular weedicide are being debated (Andreotti et al. 2018; Berezow 2018). Pesticides that are being phased out are usually more inexpensive than the new active ingredients that are being registered. The new molecules meet stringent regulatory requirement, are effective, are patent protected, and are therefore expensive.

Novel technologies to enhance the efficacy of biocontrol agents in the field may also become popular. Evidence of food sprays (to attract native natural enemies) and behavior-modifying compounds (e.g., magnet) that provide long-range attraction of

Helicoverpa moths through plant volatiles have been demonstrated to be effective in field trials. A new product called Sero X was developed based on short-range, non-volatile compounds and was found to be effective against *Helicoverpa*, whiteflies, mirids, and other homopterans. The product is in the advanced stage of commercialization (Mensah et al. 2013). Noctovis is a tank-mixed formulation designed to target both sexes of multiple species of noctuid moths. It is a blend of plant-based volatile kairomones and has been demonstrated to reduce insecticide inputs by more than 98% in soybean in Brazil (Borges et al. 2015). Noctovis may be recommended for cotton in the days to come. Specialized pheromone and lure application technology (SPLAT) comprises of food-safe, organic, and inert ingredients, has a thixotropic property, and facilitates fast and simple application, either manually or mechanically, thus producing a series of environmentally hardy, rainfast point sources affixed firmly to substrates. It can carry active ingredients responsible for mating disruption; attract and kill, push and pull strategies; and more. Such advancements may have a positive impact on cotton crop protection in the next decade.

17.11.9 Mechanization and Robotics

Advances in pesticide application technology have been made with electrostatic sprays, nano-formulations, battery-operated solar sprays, and sensor-based drone sprays. Digital technology is currently being used to perform agricultural tasks and facilitating data-driven decision-making. Robotics in agriculture is in the development stage. Use of sensors is becoming popular in some countries particularly for detecting nutrient anomalies, insect pest, and disease damage. Drones are used for survey and scouting in cotton fields through automated systems using a range of sensors to assess crop health matrixed with global positioning system (GPS). This would ensure timeliness of operation and would also be cost-effective especially when trained manpower is a constraint. The use of drones could help to measure crop fields, determine the best strategy for more profitable planting, gather data which then could be used to determine the fertilizers or inputs required for certain stages of the crop, collect aerial images, detect diseases and pests, forecast yields, and provide recommendations for water management, among other uses. Xiongkui et al. (2017) reported experiments with drones also in cotton fields of Southeast Asia. Agriculture is moving into an area of digital farming (Korres et al. 2019), and this is applicable to cotton as well. Information on cotton crop protection that was being disseminated through conventional methods such as posters and farmer fairs is now also being disseminated in real time, using the mobile technology.

Personal digital assistants (PDA) and handheld GPS have become important in cotton-growing countries, especially where farmers use complementary remote sensing, plant mapping and grid information of soil nutrient status, and infield decision support system such as COMAN impacted adoption of site-specific management (Walton et al. 2010).

Digitized weed maps for weed monitoring through artificial intelligence (AI) and to carry out real-time weed detection are mounted on field spraying machines resulting in weed-activated sprays. A real-time precision robotic weed control system was developed and tested in commercial cotton fields where it was capable of distinguishing grasslike weeds from cotton plants and applying a chemical spray only to targeted weeds while traveling at a continuous speed of 0.45 m/s. The robot consisted of a real-time machine vision system, a controlled illumination chamber, and a precision chemical applicator. In commercial cotton fields, the system correctly sprayed 89% of the weeds while correctly identifying and not spraying 79% of the cotton plants while traveling at 0.45 m/s (Lamm et al. 2002). Such technologies have not yet been adopted widely, and it remains to be seen if digital technology can make a difference in cotton crop protection.

17.11.10 Harmonization of Integrated Pest Management and Regulatory Guidelines for Biotech Cotton

Every cotton-growing country has developed and implemented sound integrated pest management (IPM) in cotton, particularly after 1990 (Kogan 1998). In fact, IPM was the key to pest management till the introduction of biotech cotton. Integrated pest management now needs to be harmonized taking into consideration the IRAC (Insecticide Resistance Action Committee), WHO (World Health Organization), and IOBC (International Organization for Biological Control) ratings of insecticides and their safety to human beings and natural enemies.

Some countries devised excellent stewardship for biotech cotton that included IPM strategies. Stewardship was compromised in other countries where poor attention was paid to pest management leading to problems with sustainability of biotech cotton. Unviable refuge recommendation and its poor adoption is often an example of poor stewardship. Monopolization of biotech cottonseed trade led to unfair trade practices such as the use of fake unauthorized seed with poor plant protection. Patent laws on biotech cotton were unclear that led to the exploitation of small-scale producers in developing countries.

Guidelines governing the strategies of integrated pest management need to be revised in light of large-scale adoption of biotech cotton that has led to a change in the pest scenario. New diseases and pests and development of pest resistance to biotech crops are the two factors that are most likely to emerge as serious concerns in the next decade and beyond. The threat of the fall armyworm looms large on crops including cotton. Emphasis must be laid on the use of stochastic models to determine the timeline of likely probability of resistance in insect pests and weeds based on the existing data, in each country. Patent laws must be upheld across countries, and clean, transparent, and harmonized systems must be established to prevent economic, social, and ecological exploitation that leads to the failure of technologies in the field.

17.12 Picking of Cotton

According to the study conducted by the International Cotton Advisory Committee (ICAC) and published in October (2017), 62% of cotton produced was picked by hand and 38% with machines. The ICAC data on cost of production have shown that machine picking is less expensive compared to manual picking, though the data is limited to picking operations and does not analyze the subsequent effects on quality and costs involved in handling cotton loaded with trash. As a share of the total cost of production, picking cost, at 13%, ranks just behind fertilizer costs, which accounts for 14% (ICAC 2016). Scarcity and cost of labor are pushing more countries toward machine picking. However, cotton harvesting machines are much easier to use under large-scale commercial production. Factors limiting the use of mechanical harvesters include small field size, unsuitable morphology of varieties, row spacing, lack of defoliation, and lack of requisite cleaning equipment at gins. Machine picking has the biggest disadvantage of huge picking losses, which is a direct loss in yield.

Countries that have yet not adopted machine picking need smaller machines that can accommodate differing row spacing, can be operated in a manner similar to small spray machines, and require few if any additional costs in allied operations. Ideally, the harvesters would collect a minimal amount of vegetative and trash contamination. Perhaps advancing robotics will enable small harvesters to pick boll-by-boll without grabbing dry or green leaves.

Three characteristics that create demand for small-scale picking machines are advancing industrialization, improvements in social structures, and urbanization. Developing countries that are increasingly exhibiting these characteristics will increasingly pursue mechanical harvesters. But the size will not compare to the harvesters used in the advanced cotton-producing countries. The objective will be to increase harvesting capabilities from 25 to 30 kg per person per day with hand picking to a few hundreds of kilograms per person per day. The need for such small-scale mechanical cotton harvesters is generally accepted, and development work is ongoing. It is likely that commercialization of such harvesters will be evident by 2030.

17.13 Engineering of Ginning

Cotton is produced for fiber, though seed itself has a tremendous value and still underutilized and much less explored. Lint fiber is only about 35–40% of the total seed cotton production. The only two processes of separating lint from seed are roller ginning and saw ginning, and both were developed when hand picking was the only procedure of picking cotton. Saw ginning which was considered to be an improvement over roller ginning for higher efficiency was the need of time. The saw gin developed by Eli Whitney and patented in 1794 is truly the last development in the process of removing fiber from the seed coat. The inabilities to gin huge quantities of seed cotton had become a constraint for expanding production. Saw ginning efficiency revolutionized cotton production in the world. The two processes of ginning, saw ginning and roller ginning, are a harsh way of removing fibers from the seed. The roller ginning that has gone through many changes, as has been saw ginning, removes fibers by pulling fibers against the stationary seeds. The engineering of saw ginning is a mix of pulling and beating actions. There is no doubt that the roller ginning has a gentler process of pulling fibers over saw ginning but it has some limitations. Saw ginning is not recommended for long and extra-long fiber because fiber breakage degrades the length distribution.

Without going into the impacts of the two methods of ginning, the focus here continues on actual method of removing fibers from the seed coat. Almost 2/3 of cotton produced in the world is ginned on saw gins and 1/3 on roller gins. It is so because no other better options are available. Efforts have been continuously made to find better mechanical means of removing fibers from the seed without sacrificing efficiency and preservation of quality but without any success. The target is to make the pulling and beating actions as gentler as possible without any physical damage to the state of fiber achieved in production. In this regard the two processes that provided hope for a better ginning were Cage ginning and Templeton ginning. The two processes airstream pressure (Cage ginning) and improved double roller mechanism with leather and knife (Templeton ginning) were close to be commercialized, but it never happened due to various reasons. Currently, there are no new technologies, not even in the offing, that have potential to replace neither roller ginning nor saw ginning.

Cotton ginning beyond 2030 will in no way be different from today. What has changed in ginning is the type of gin stands used and the amount and type of handing and cleaning equipment used in a particular ginning system. How softer and longer fibers you may produce, you have to treat them at the mercy of saws and rollers. The monitoring and control of ginning processes have changed and will continue to be improved without any change in the harsher process of removing fibers from seeds. The future thrust will continue to be the focus of exploring new methods of separating lint and seed with these targets:

- Develop a ginning method that reduces fiber damage and maintains fiber length properties.
- Avoid fiber breakage and improve length uniformity data.
- Improve ginning efficiency and lower cost of ginning.

17.14 Spinners Needs, Quality Traits, and Cotton Classing

Cotton classing is a unique manifestation of time-honored systems for sorting and grading that are done on commodities to improve efficiency of a marketing system. All such systems are justified when the important properties—that determine the use value, or "quality," of the commodity or product—are not easily and readily identified by the buyers. Methods for sorting and grading may be descriptive and/or quantitative. Cotton certainly qualifies as a candidate for a *sophisticated* classification system, because efficient marketing requires the critical properties of cotton fibers to be specified with a degree of *precision*. Such precision is especially challenging when using only *thousands* of fibers to represent *billions* of fibers with multiple properties having nontrivial distributions across the individual fibers. The *distributional characteristics* of the fibers make cotton classification unusually difficult.

Sophistication notwithstanding, it must be that cotton classing is a *commercial* task—not a *research* task. The procedures used and the results communicated by cotton classing should be predicated on research results to the extent possible. But both procedures used and results produced from cotton classing are substantially constrained by *economic* and *time* considerations. The value of information must be commensurate with the cost of information; a research project cannot be funded for each bale of cotton sold. Furthermore, the natural flow process of cotton marketing does not allow much time for intensive evaluation of properties.

The cotton fiber properties needed by yarn spinners are generally well known throughout the industry; however, substantial confusion still exists within the global industry about actual fiber properties versus the limitations of commercial measurements of these properties. Such confusion emerged with the very first commercial instrument measurement of cotton more than 50 years ago, the micronaire. Commercial interest in micronaire measurements was driven by a desire of spinners to get information on the *maturity* of the cotton fibers. But an education process still continues regarding the micronaire's ambivalent measurement of the *combination* of genetic *fineness* versus *maturity* of the fibers. Fiber fineness (characterized by the linear density or the circumference of individual fibers) is determined genetically, while fiber maturity (characterized by the amount of cellulose deposition in individual fibers) is determined by the interaction between genetics and environmental factors.[1] The micronaire measurement indicates the *specific surface area* of a compressed bundle of cotton fibers. As such, it does not provide distinct estimates of either fineness or maturity, but a conflated estimate of the combination.

This distinction between the fundamental fiber properties and the measurement of these properties is important when trying to predict future progress in (1) achieving genetic improvements in cotton fiber quality and (2) commercially useful measurements of improved fiber quality. From the standpoint of fiber technology, the three iconic fiber properties that are the benchmarks for superior fiber quality are fiber *length*, *strength*, and *fineness*. Thus, these are the three defining properties for premium cotton fibers provided by the "extra-long staple," or ELS, types of cotton. These include the Pima, Giza, and Sea Island types, which all belong to a tetraploid, *G. barbadense* species and are the longest, strongest, and finest cotton fibers available. Nevertheless, more detailed information is needed for efficient purchasing and handling by yarn spinners.

[1]This interaction is in the domain of phenomics, while fiber fineness is in the domain of genomics.

17.14.1 Current State of Commercial Measurement Technology

The current paradigm for providing commercially useful measurements is the HVI (High-Volume Instrument) system, which was developed starting in the late 1960s in the USA and became the basis for classification of US-produced cotton in the early 1990s. The HVI machine used by the Agricultural Marketing Service of the US Department of Agriculture[2] is comprised of a cluster of computer-controlled instruments that provide measurements for length (L), micronaire (Mic), strength (Str), length uniformity index (UI), leaf grade, color Rd, color +b, color grade, and trash percent area.[3] In addition, human classers provide identification of any extraneous matter among the fibers.

Strictly speaking, measurements for leaf grade, trash, and extraneous matter are not inherent cotton fiber properties; nevertheless, these are important quality issues for yarn spinners and weavers. Indeed, cotton fibers that are baled and shipped with little or no contamination are always in demand by the global spinning industry. The large-scale growing and machine-harvested cotton production sectors, as in the USA, Australia, and Brazil, have reputations for low contaminations, which give these countries a competitive advantage in global cotton sales.

The color measurements are used in various ways. The most basic use is to assess deterioration of the fibers from weather or other postharvest damage. An abnormal "grayness" of the fibers (based on the +b reading) indicates weather or water damage that has degraded fiber strength and partially removed the natural finishes of waxes and pectins on the fiber surfaces. These finishes are necessary for efficient processing through opening, cleaning carding, drawing, and spinning.

Another use of color measurements is for blending to achieve more efficient dyeing. Even "white" cottons exhibit different shades of white. Healthy cotton fibers may have differing "yellowness" readings (the Rd measurement) based on inherent properties. If these differ significantly, then blended fibers are supposed to bleach more heavily to bring the colors close enough together to achieve a consistent shade in dyeing. High levels of leaf and other vegetative contaminations may also result in staining, spotting, and off-color fibers (affecting both color measurements and the color grade).

The HVI measurements for L, Mic, Str, and UI relate to key fiber properties that are targeted—or should be targeted—by all cotton plant breeders. Improvement in the fiber properties that are related to such measurements as these will continue to preoccupy everyone focused on continuous improvements in cotton fibers as an industrial raw material in textile manufacturing. By implication, new or improved measurements related to these properties should be the future focus of measurement research to increase the use value of cotton fibers.

---

[2]AMS, USDA, uses Uster HVI systems, made by Uster Technologies, Inc. in Charlotte, North Carolina, USA. A comparable system is offered by Premier Evolvics in Coimbatore, India.

[3]See "Cotton Classification: Understanding the Data," www.ams.usda.gov/cotton/UnderstandtheData.

17.14.2 Limitations on Commercial Measurements of Fiber Properties

While there have been substantial improvements in sensory precision, greatly enhanced computational capacities, useful algorithmic enhancements, and time-saving automation of the HVI measurements, there have been few fundamental changes made in commercial HVI machines since the USDA began using these for cotton classification. All measurements except the UI are measures of central tendencies; i.e., these do not give information on the distribution of the fiber properties. The UI relates to the length distribution but is a very rudimentary indicator, being expressed as a simple ratio of the mean length of fibers in a sampled fiber bundle divided by the upper-half mean length of that fiber bundle, expressed as a percentage.[4]

The overriding reason for the lack of more sophisticated fiber measurements is revealed by the measuring system's name: *high-volume* instrument. Any additional measurement provided to the marketing system must not add more than a few seconds to the total time required to run the fiber sample through the HVI. There is an unavoidable inverse relationship between the sophistication of a measurement and the time required to obtain it, which inevitably increases the cost of measurement. Participants in the marketing system are notoriously reluctant to bear the added cost.

The spinning also requires some additional information about the extent of neps and short fiber content (SFC)[5] in cotton fiber before buying. This response reflects production managers' observed problems and conventional thinking regarding the yarns made in the spinning plants. Both manifestations of these quality problems are magnified by broken fibers, which always result to some degree from mechanical action on the fibers—whether at harvesting, ginning, or during the opening and cleaning processes in yarn spinning plants. Therefore, the *propensity* of fibers to break under mechanical stress is the underlying metric. To complicate matters, this propensity toward breakage is impacted by a complex of other fiber properties. For example, weaker fibers, whether due to genetic reasons or to fiber immaturity, are much more susceptible to breakage. This helps explain why yarn spinners are vitally interested in measurements of both fiber strength and micronaire.

The foregoing explanation illustrates the prevalence of autocorrelation among the critical fiber properties. The length distribution of cotton fibers received by yarn spinners reflects both (1) unnecessary mechanical breakage and (2) the impacts of any faults in other fiber properties. Thus, the genetic staple length and distribution, the maturity, the strength, the elongation, etc., all contribute to the length distribu-

---

[4]The upper-half mean length (UHML) measurement from the HVI is analogous to the traditional staple length of fibers when hand classing is used. Both of these measurements are intended to identify the "dominant long fibers" in the length distribution of the fiber sample.

[5]Neps are hopelessly entangled masses of fibers that protrude from the yarns and fabrics, degrading both appearance and touch. Short fiber content (SFC) is a traditional benchmark based on the share of the fibers that are one-half inch or shorter in length. The short lengths do not incorporate well into the yarns, degrading the strength and appearance. Problems caused by SFC include low yarn strength, unevenness, thick places, thin places, and neps.

tion confronted by yarn spinners. It is well known, however, that the more *uniform* is the length distribution of the cotton fibers received by spinners, the less problems they will have with neps and short fibers, the more efficient will be the spinning process, and the higher quality will be the yarns.

Pre-spinning, measurement-based indicators of the potential for making neps and short fibers are feasible, but the measurements are interdependent with the measurement techniques, which must control the stress put on the fibers prior to and during measurement. While useful measurements may be done, rapid measurement is not currently feasible. The most common commercial laboratory instrument for assessing neps and SFC is the AFIS (Advanced Fiber Information System) instrument made by Uster Technologies. This instrument separates the fibers from a technician-prepared sliver and takes electro-optical measurements of individual fibers. Sample preparation requires operator consistency to achieve sample consistency, and each sample requires several minutes for measurement. In addition to neps and SFC, measurements are produced for length, fineness, maturity, trash, and dust. More advanced spinning plants around the world have used an AFIS instrument on a limited basis to screen fibers either before these are subjected to spinning process or to monitor performance of the processing machinery over time. Use for monitoring in textile mills has decreased in recent years, however, as other in-line measurement technologies are increasingly used by spinners to monitor neps during the spinning process.

The spinning industry's fixation on SFC is explained by the fact that, until recent history, it was impossible to obtain more precision in measurement of the *distributions* of the cotton fiber lengths. Arduous hand extraction and arrangement of fibers from single seeds demonstrated long ago that cotton fibers exhibit a natural frequency distribution that mimics a bell curve, with a relatively small percentage of shorter fibers, a predominant percentage of medium-to-long fibers, and then trailing off to a smaller percentage of longer fibers. With commercial handling of the cotton, the tail of the distribution with short fibers typically is elevated; i.e., there is a larger percentage of very short fibers than there is of very long fibers. This result is due to the breakage of fibers from the mechanical stresses required in commercial handling. Thus, the observed length distribution of any cotton sample is determined by a combination of natural biological distribution and fiber breakage.

The observed length distribution of large quantities of cotton is also complicated by variations across the cotton boll positions on the plants. Given the indeterminate fruiting behavior of cotton, different bolls will be of different ages, which brings the likelihood of different maturity levels, different strength levels, different staple lengths, etc. All of these differences would be "captured" in the length distributions of these fibers.

Cotton spinners have always known about the importance of reducing short fibers for producing high-quality yarns. Combing machines were developed to remove most of the short fibers, with the machine settings targeting the one-half inch threshold of fiber lengths for removal. Commercial combing of cotton generally removes 15–25% of the fibers by weight. The waste fibers, called noils, are clean fibers that find other uses, but the value of these noils is significantly reduced. The combination of a costly mechanical combing process and the loss of valuable fibers makes combed cotton textiles expensive. But eliminating the short fibers elevates the quality of textiles made from yarns to a new level. Quality characteristics—dye uptake, color fastness, softness, shedding, and durability—are all notably improved.

17.14.3 The Fourth Fundamental Fiber Property: Length Uniformity

Looking to the future, the three iconic fiber properties of length, strength, and fineness may become four properties with the addition of length uniformity. Measuring length uniformity requires measuring the entire distribution of fiber lengths in a cotton sample. The ideal length distribution would be a uniform distribution; i.e., all fibers should have the same length. This is impossible, but some cotton varieties produce fibers that are genetically more uniform than others and are more resistant to breakage. Thus, weak fibers will break more often when mechanically stressed. Fibers that are immature will be weak. Therefore, weakness that is due either to genetics or to immaturity will be manifested in distorted length distributions. HVI measurements of both strength and micronaire are "bundle" or "central-tendency" measurements, but fiber breakage occurs from stresses applied to *individual* fibers and will be manifested in the length distribution.

The use of SFC as an important fiber property by yarn spinners is due to the historical infeasibility of commercial measurement of the length distribution of cotton fibers. The half-inch benchmark used for SFC is arbitrary; it represents a heretofore practical compromise between fiber wastage and yarn quality. But the entire frequency distribution of fiber length has important implications for both spinning performance and yarn quality. Combining advanced sensor technology with computer technology has made measurement of the entire length distribution more feasible. While it cannot be accomplished at a speed or a cost level required for the HVI, it can be done for more limited purposes. For example, it can be achieved when developing new cotton varieties. Although the HVI cannot detect the difference, varieties with more uniform fiber length distributions will definitely exhibit superior performance.

Looking to the future, it seems inevitable that the length distribution of cottons will command a greater focus by both cotton plant breeders and by cotton spinners. To better understand why, some explanation of the alternative spinning technologies is needed.

17.14.4 Three Major Spinning Technologies and Their Fiber Requirements

The three major spinning technologies are, in order of dominance:

1. Ring spinning
2. Rotor spinning
3. Air-jet spinning

Ring spinning is the oldest technology among these three. It also yields the slowest throughput, requires the greatest amount of labor, and has the highest cost per unit of yarn produced. However, it produces the highest quality of textiles, with strength, appearance, and "feel" (called "hand" in the industry) that sets the standard for comparison. It is also the least capital-intensive technology and requires the least technical expertise to keep the spinning machines functioning satisfactorily. Given the persistent migration of textile manufacturing toward less-developed, cheap-labor, low-skilled countries and the policy of the governments of these countries to provide large numbers of jobs, the labor requirements are not a definitive barrier to the continuing use of ring spinning.

Rotor spinning became a commercial technology during the 1970s and during the 1980s and 1990s captured a significant share of the total spinning capacity, based on its productivity in making larger-sized yarns. It is more capital- and energy-intensive than ring spinning but produces about ten times more throughput, requires only a fraction of the labor, and is the lowest cost per unit of yarn produced of the three technologies. However, its advantages are available only for large- and medium-size yarns, and both its strength and hand are inferior to the other two technologies. Trained technicians are required to get good performance from rotor spinning machines. Use of rotor spinning generally increases its share of spinning capacity as a country develops and labor costs become a major factor in the per-unit cost of yarn production.

Air-jet spinning also became commercial technology during the 1970s, not long after rotor spinning. It is the most capital- and energy-intensive of the three technologies, costs somewhat more per unit of yarn production than does rotor spinning, but produces the highest throughput (1.5–2 times that of rotor spinning). Air-jet yarns more closely approximate the appearance and hand of ring-spun yarns than does the rotor spinning. The comparative advantage of this technology is maximum for making small yarns, and it cannot adequately spin larger yarn sizes. Air-jet spinning is also the most demanding of technical expertise and requires precise maintenance to deliver consistent performance. These factors have limited its market penetration.

Of these three technologies, only ring spinning can commercially produce a full range of yarn sizes. Approximate share of global capacity of ring spinning is between 70 and 75%. Rotor spinning accounts for between 20 and 25%, while the share of air-jet spinning is not more than 5%. These shares are changing slowly if at all and are not expected to be significantly different a decade from now.

Using the four basic measurements of cotton fibers—length, length uniformity, strength, and fineness—these may be ranked in order to their importance for each of the three existing spinning technologies. It must be emphasized that all four of these properties are critically important to yarn spinners; thus, if any one of these is unacceptable, spinning performance and yarn quality will be unacceptable. Nevertheless, the three spinning systems exhibit different *sensitivities* to each of these properties, which allow a useful 1-through-4 ranking, as shown in Table 17.1.

Length uniformity is ranked third for ring spinning, fourth for rotor spinning, but first for air-jet spinning. The importance of length uniformity explains why air-jet spinning of cotton is most often done in a blend with synthetic fibers, for which length uniformity is more controllable. It also explains the fact that combed cotton is most often used for air-jet yarns.

Table 17.1 Ranking importance of four fiber properties for three spinning systems

Rank	Ring spinning	Rotor spinning	Air-jet spinning
1	Length	Strength	Length uniformity
2	Strength	Fineness	Length
3	Length uniformity	Length	Fineness
4	Fineness	Length uniformity	Strength

Superior length uniformity is critical for enabling cotton fibers to penetrate further into air-jet spinning. Furthermore, length uniformity belongs among the top four cotton fiber properties for all commercial spinning technologies. For both ring spinning and air-jet spinning, cottons with a shorter staple length but a higher length uniformity can outperform cottons with a longer staple length but lower length uniformity. Within limits, these systems can be set to accommodate different staple lengths, and then superior length uniformity will produce superior yarns. Rotor spinning is the most tolerant of a lack of length uniformity, but superior uniformity results in superior rotor-spun yarns.

17.14.5 Status and Potential of Length Uniformity Measurements

As a matter of historical focus on commercial measurement of important cotton fiber properties, length uniformity was neglected, owing to difficulty in its measurement. Fiber length, strength, and fineness are more amenable to useful measurements of central tendencies, while length uniformity is overtly a *distributional* property. As previously emphasized, the ratio of mean length to upper-half mean length that is currently provided by the HVI, denoted as the uniformity index (UI), does not begin to capture the complexities of fiber length distribution.[6] Nevertheless, this measurement has increasingly become a factor in determining the market prices of cotton in recent years, which testifies that informed participants in the commercial spinning industry are increasingly aware that they need information on length uniformity.

Ironically, dominance of the HVI measurements for marketing cotton has inhibited a proper focus on length uniformity in the research and development of improved cotton varieties. This is understandable, since the HVI was a giant step in measurement technology and cotton breeders naturally tended to utilize the low-

[6] For example, a simple ratio of mean length to upper-half mean length does not inform about fiber distributions below the mean length, yet this information is important for spinning performance and yarn quality. Moreover, the UI lacks precision, meaning that the interval of uncertainty around the point estimate of the UI is quite wide. Thus, it may be that a UI estimate for a single cotton sample of 80 is not significantly different from an estimate of 81. See "Cotton Classification: Understanding the Data," www.ams.usda.gov/cotton/UnderstandtheData.

cost data it provided to guide development of cotton fibers that classed well on the HVI (and which sold for higher prices in the market). It is regrettable, however, because slower and more costly measurement technologies are available to reveal superior length uniformity and neither the slowness nor the cost is prohibitive within the context of a plant breeding program. Furthermore, research suggests that length uniformity can be selected within cotton breeding programs if it can be adequately measured.[7]

Looking to the future, a combination of computer power, advancing sensor technologies, and sophisticated algorithmic programs has potential to enable high-volume length distribution measurements that provide more detailed, precise, and accurate guidance on spinning performance and yarn quality. Incorporation of these measurements into existing HVI technology would be one of the most significant advancements in commercial measurement technology since the inception of the HVI. Furthermore, this would influence cotton breeding programs and encourage increased utilization of more sophisticated, lower-volume measurement technologies that could provide better guidance toward achieving genetic improvements in the uniformity of the fiber length distributions.

17.15 Enhancing Cotton Utilization

Fiber is the primary reason cotton is produced, but the seed is a significant byproduct that has potential for further value-added developments. For every kilogram of cotton fiber produced, about 1.4 kg of cottonseed is produced. Furthermore, cottonseed is a unique oilseed that yields premium oil, along with valuable meal, linters, and hulls. When *G. hirsutum* cottonseed is processed through state-of-the-art oil mills, approximate yields are as follows:

- Oil—16%
- Meal—46%
- Linters—8%
- Hulls—27%
- Waste—3%

These percentages are somewhat dependent on varieties, but differences between species are more significant. The *G. barbadense* varieties have minimum amounts of "fuzz" and produce minimum amounts of linters. *G. hirsutum* varieties typically require two operations to remove linters, usually called first cut linters and second cut linters. Linters are almost 100% cellulose and as such are valuable for a multitude of chemical and nonchemical commercial uses.

[7] Research has indicated that genetic components significantly impact within-plant variability of fiber length and maturity. See "Evaluating the within-plant variability of cotton fiber length and maturity," by A. Ayele, B. Kelly, and E.F. Hequet, *Agronomy Journal*. 110(1) 47–55.

Unprocessed, whole cottonseed is also a premium ingredient for feeds to dairy cattle as well as for other livestock. In recent years, the value of cottonseed has accounted for 15–20% of the total revenue to farmers from cotton production. Dowd and Wakelyn (2010) have extensively discussed the current and future utilization of cottonseed products. They have also discussed enhanced uses of cottonseed oil including some of the hindrances to make it more valuable. Whole cottonseed, as well as cottonseed meal and flour consumption by both animals and humans, has generally been limited by the trace presence of gossypols—polyphenolic disesquiterpene compounds that are seen as dark brown to blackish dots on the plant parts. The plant body having gossypols also transmits gossypols to the seed. The presence of gossypols in the seed carries toxicity and prohibits livestock producers limit the quantity of cottonseed fed to animals. Gossypols are desirable for defense against insects but they are also transmitted to the seed. Nonruminant animals such as poultry cannot handle much gossypol before toxicity signs develop. Cattles have the ability to detoxify but to a certain extent. On the other hand, gossypol-free seed kernels can be used in combination with wheat and corn flours to enrich the protein content of these products. Breakthrough genetic engineering research (Rathore et al. 2008) has successfully produced cotton with gossypol eliminated from the seeds; therefore, a research barrier to widespread human consumption of cottonseed flour has been broken, and it is possible that commercial varieties that have gossypol-free cottonseeds would be in combination with wheat and corn flours, in order to enrich the protein content of these products.

The production of baled cotton fibers at gins results in waste byproducts that are underutilized in most of the world. These waste byproducts include linters, burrs, and sticks. Significant research and development has been carried out at the USDA Gin Lab in Lubbock, Texas, USA, to capture, process, and exploit these waste products in various ways. Examples include processing as compost and for hydro-mulch suitable for land filling and cattle feed. The benefits are twofold: (1) additional revenue is generated and (2) waste disposal costs and pollution are eliminated. The large-scale, technology-oriented ginning sectors are already acting to utilize these waste products, and these actions will increase and spread in upcoming years.

The harvest index as measured in terms of dry mass of seed cotton to total dry mass of the cotton plant is only 0.33, and the index will vary greatly depending on growing conditions and production practices. Not counting sugarcane, which has a different harvesting, cotton's harvest index is only 65% of wheat, rice, and maize. Bulk of the dry matter left in the cotton field is cotton sticks that are not properly utilized. Mulching back cotton stalks into the soil is a genuine return to the soil but cotton stalks can find better uses. One such use that has been explored to some extent is making particle boards from cotton stalks. The International Cotton Advisory Committee with funding from the Common Fund for Commodities undertook a pilot project in India aiming to produce hardboards. The work done by local researchers involved chipping stalks to an appropriate mesh of 1.5–2.0 cm, mixing chips with urea formaldehyde or phenol formaldehyde, and preparing coarse and fine layer mats. A coarse layer mat and two fine chip mats, on either side of the coarse mat, were pressed with a hydraulic pressing having heated platens. The boards thus made under standard conditions of resin concentration, pressure and temperature, etc. produced an acceptable quality of board. This aspect of the raw materials produced by the cotton plant will get higher attention beyond 2030.

References

Akbar W, Gowda A, Ahrens JE, Stelzer JW, Brown RS, Bollman SL, Greenplate JT, Gore J, Catchot A, Lorenz LG, Stewart SD, Kerns DL, Greene JK, Toews MD, Herbert DA, Resig DD, Sword GA, Ellsworth PC, Godfrey LD, Clark TL (2019) First transgenic trait for control of plant bugs and thrips in cotton. *Pest Manag Sci* 75(3):867–877

Ali Z, Ali S, Tashkandi M, Zaidi SS, Mahfouz MM (2016) CRISPR/Cas9-mediated immunity to Gemini viruses: differential interference and evasion. *Sci Rep* 6:26912

Andreotti G, Koutros S, Hofmann JN, Sandler DP, Lubin JH, Lynch CF, Lerro CC, De Roos AJ, Parks CG, Alavanja MC, Silverman DT, Freeman LEB (2018) Glyphosate use and cancer incidence in the agricultural health study. *J Natl Cancer Inst* 110(5):509–516. https://doi.org/10.1093/jnci/djx233

Armijo CB, Whitelock DP, Thomas JW, Hughs SE, Gillum MN (2017) Roller ginning. *J Cotton Sci* 21:199–209

Asokan R, Chandra GS, Manamohan M, Krishna Kumar NK (2011) Effect of diet delivered various concentrations of double-stranded RNA in silencing a midgut and a non-midgut gene of *Helicoverpa armigera*. *Bull Entomol Res* 103:555–563

Auer C, Frederick R (2009) Crop improvement using small RNAs: application and predictive ecological risk assessment. *Trends Biotechnol* 27(11):644–651

Bassi FM, Bentley AR, Charmet G, Ortiz R, Crossa J (2016) Breeding schemes for the implementation of genomic selection in wheat (Triticum spp.). *Plant Sci* 242:23–36

Berezow A (2018) California's glyphosate judgement—emotion, bad science and Greed Win the Day. *Outlooks Pest Manag* 29(5):204–205

Borges R, Oliveira Da Silva R, Bernardi C, Urrutia W, MafraNeto A (2015) Noctovi, an effective food based attractant for lepidopteran pests. Poster presented at Entomological Society of America

Brévault T, Heuberger S, Zhang M, Ellers-Kirk C, Ni X, Masson L, Li X, Tabashnik BE, Carrière Y (2013) Potential shortfall of pyramided transgenic cotton for insect resistance management. *Proc Natl Acad Sci* 110(15):5806–5811

Carpenter JE (2010) Peer-reviewed surveys indicate positive impact of commercialized GM crops. *Nat Biotechnol* 28:319–321

Carriere Y, Antilla L, Liesner L, Tabashnik BE (2017) Large-scale evaluation of association between pheromone trap captures and cotton boll infestation for pink bollworm (Lepidoptera: Gelechiidae). *J Econ Entomol* 110(3):1345–1350

Cattaneo MG, Christine Y, Schmidt C, Huang C, Rahman M, Olson C, Kirk CE, Orr BJ, Marsh SE, Antilla L, Dutilleul P, Carrière Y (2006) Farm-scale evaluation of the impacts of transgenic cotton on biodiversity, pesticide use, and yield. *Proc Nat Acad Sci* 103(20):7571–7576. https://doi.org/10.1073/pnas.0508312103

Chang H, Liu Y, Ai D, Jiang X, Dong S, Wang G (2017) A pheromone antagonist regulates optimal mating time in the moth *Helicoverpa armigera*. *Curr Biol* 27(11):1610–1615

Chen X, Lu X, Shu N, Wang S, Wang J, Wang D, Guo L, Ye W (2017) Targeted mutagenesis in cotton (*Gossypium hirsutum* L.) using the CRISPR/Cas9 system. *Sci Rep* 7:44304

Crossa J, Perez-Rodriguez P, Cuevas J, Montesinos-Lopez O, Jarquin D, de los Campos G, Burgueno J, Gonzalez-Camacho JM, Perez-Elizalde S, Beyene Y, Dreisigacker S, Singh R, Zhang X, Gowda M, Roorkiwal M, Rutkoski J, Varshney RK (2017) Genomic selection in plant breeding: methods, models, and perspectives. *Trends Plant Sci* 22(11):961–975

Cui J, Luo J, Van Der Werf W, Ma Y, Xia J (2011) Effect of pyramiding Bt and CpTI genes on resistance of cotton to *Helicoverpa armigera* (Lepidoptera: Noctuidae) under laboratory and field conditions. *J Econ Entomol* 104:673–684

Dhawan AK, Singh S, Kumar S (2009) Integrated Pest Management (IPM) helps reduce pesticide load in cotton. J Agric Sci Technol 11:599–611

Dhurua S, Gujar GT (2011) Field-evolved resistance to Bt toxin Cry1Ac in the pink bollworm, *Pectinophora gossypiella* (Saunders) (Lepidoptera: Gelechiidae) from India. *Pest Manag Sci* 67:898–903

Dowd, M. K. and P. J. Wakelyn. 2010. Cottonseed: current and future utilization. In the Cotton: technology for the 21st century. Phillip J. Wakelyn and M. Rafiq Chaudhry, International Cotton Advisory Committee, Washington, DC

Dwivedia SL, Britt AB, Tripathi L, Sharma S, Upadhyaya HD, Ortiz R (2015) Haploids: constraints and opportunities in plant breeding. *Biotechnol Adv* 33(6 Part 1):812–829

Dykxhoorn DM, Lieberman J (2005) The silent revolution: RNA interference as basic biology, research tool, and therapeutic. *Ann Rev of Med* 56(1):401–423

Emani C, Garcia JM, Lopata-Finch E, Pozo MJ, Uribe P, Kim DJ, Kumar GS, Cook DR, Kenerley CM, Rathore KS (2003) Enhanced fungal resistance in transgenic cotton expressing an endochitinase gene from *Trichoderma virens*. *Plant Biotechnol J* 1:321–336

Firmino AAP, Fonseca FCDA, de Macedo LLD, Coelho RR, Antonino de Souza JD Jr, Togawa RC (2013) Transcriptome analysis in Cotton Boll Weevil (*Anthonomus grandis*) and RNA interference in insect pests. *PLoS One* 8(12):e85079. https://doi.org/10.1371/journal.pone.0085079

Gapare W, Liu S, Conaty W, Zhu Q, Gillespie V, Llewellyn D, Stiller W, Wilson I (2018) Historical datasets support genomic selection models for the prediction of cotton fiber quality phenotypes across multiple environments. *G3 (Bethesda)* 8(5):1721–1732

Gaspar YM, McKenna JA, McGinness BS, Hinch J, Poon S, Connelly AA, Anderson MA, Heath RL (2014) Field resistance to *Fusarium oxysporum* and *Verticillium dahliae* in transgenic cotton expressing the plant defensin *NaD1*. *J Exp Bot* 65(6):1541–1550. https://doi.org/10.1093/jxb/eru021

Gong Y-H, Yu XR, Shang QL, Shi XY, Gao XW (2014) Oral Delivery Mediated RNA Interference of a carboxylesterase gene results in reduced resistance to organophosphorus insecticides in the cotton aphid, *Aphis gossypii* Glover. *PLoS One* 9(8):e102823. https://doi.org/10.1371/journal.pone.0102823

Greenplate JT (1999) Quantification of *Bacillus thuringiensis* insect control protein Cry1Ac over time in Bollgard cotton fruit and terminals. *J Econ Entomol* 92:1377–1383

Hazra DK (2017) Nano-formulations: high definition liquid engineering of pesticide for advanced crop protection in agriculture. *Adv Plant Agric Res* 6(3):1–2

Hughs SE, Parnell CB Jr, Wakelyn PJ (2010) Cotton harvesting and ginning in the 21st century. In: Wakelyn PJ, Chaudhry MR (eds) Cotton: technology for the 21st century. International Cotton Advisory Committee, Washington, DC

Iqbal Z, Sattar MN, Shafiq M (2016) CRISPR/Cas9: a tool to circumscribe cotton leaf curl disease. *Front Plant Sci.* https://doi.org/10.3389/fpls.2016.00475

Keim DL (2007) Industry highlights: cotton breeding. In: Principles of plant genetics and breeding. Blackwell, Malden, p 569

Khanday I, Skinner D, Yang B, Mercier R, Sundaresan V (2019) A male-expressed rice embryogenic trigger redirected for asexual propagation through seeds. *Nature* 565(7737):91–95

Khatodia S, Bhatotia K, Tuteja N (2017) Development of CRISPR/Cas9 mediated virus resistance in agriculturally important crops. *Bioengineered* 8(3):274–279. https://doi.org/10.1080/21655979.2017.1297347

Knight K, Head G, Rogers J (2013) Season-long expression of Cry1Ac and Cry2Ab proteins in Bollgard II cotton in Australia. *Crop Prot* 44:50e58

Kogan M (1998) Integrated pest management: historical perspectives and contemporary developments. *Annu Rev Entomol* 43(1):243–270

Korres NE, Burgos NR, Travlos I, Vurro M, Gitsopoulos TK, Varanasi VK, Duke SO, Kudsk P, Brabham C, Rouse CE, Salas-Perez R (2019) New directions for integrated weed management: modern technologies, tools and knowledge discovery. *Adv Agron* 155:243

Kranthi KR (2018) Biotech cotton-relevance for Africa, *ICAC Recorder* 29–34
Kranthi KR, Russell D, Wanjari R, Kherde M, Munje S, Lavhe N, Armes N (2002) In-season changes in resistance to insecticides in *Helicoverpa armigera* (Lepidoptera: Noctuidae) in India. *J Econ Entomol* 95(1):134–142. https://doi.org/10.1603/0022-0493-95.1.134
Kranthi KR, Naidu S, Dhawad CS, Tatwawadi A, Mate K, Patil E, Bharose AA, Behere GT, Wadaskar RM, Kranthi S (2005) Temporal and intra-plant variability of Cry1Ac expression in Bt-cotton and its influence on the survival of the cotton bollworm, *Helicoverpa armigera* (Hübner) (Noctuidae: Lepidoptera). *Curr Sci* 89:291–298
Lamm RD, Slaughter DC, Giles DK (2002) Precision weed control system for cotton. *Trans ASAE* 45(1):231
Lance DR, Leonard DS, Mastro VC, Walters ML (2016) Mating disruption as a suppression tactic in programs targeting regulated lepidopteran pests in US. *J Chem Ecol* 42(6):590–601
Lu Y, Wu K, Jiang Y, Guo Y, Desneux N (2012) Widespread adoption of Bt cotton and insecticide decrease promotes biocontrol services. *Nature* 487:362–365
Mao YB, Tao XY, Xue XY, Wang LJ, Chen XY (2011a) Silencing a cotton bollworm P450 monooxygenase gene by plant-mediated RNAi impairs larval tolerance of gossypol. *Nat Biotechnol* 25:1307–1313
Mao Y, Tao X, Xue X, Wang L, Chen X (2011b) Cotton plants expressing CYP6AE14 double-stranded RNA show enhanced resistance to bollworms. *Transgenic Res* 20:665–673
Mensah RK (2002) Development of an integrated pest management programme for cotton. Part 2: integration of a lucerne/cotton interplant system, food supplement sprays with biological and synthetic insecticides. *Int J Pest Manag* 48:95–105
Mensah RK, Peter AF, Gregg C, Alice B, Del Socorro P, Christopher BJ, Moore D, Hawes JAC, Watts N (2013) Integrated pest management in cotton: exploiting behaviour-modifying (semiochemical) compounds for managing cotton pests. *Crop Pasture Sci* 64:763–773
Morrison NI, Simmons GS, Fu G, O'Connell S, Walker AS, Dafa'alla T, Walters M, Claus J, Tang G, Jin L, Marubbi T (2012) Engineered repressible lethality for controlling the pink bollworm, a lepidopteran pest of cotton. *PLoS One* 7(12):e50922
Naik VC, Kumbhare S, Kranthi S, Satija U, Kranthi KR (2018) Field-evolved resistance of pink bollworm, *Pectinophora gossypiella* (Saunders) (Lepidoptera: Gelechiidae), to transgenic *Bacillus thuringiensis* (Bt) cotton expressing crystal 1Ac (Cry1Ac) and Cry2Ab in India. *Pest Manag Sci* 74:2544–2554. https://doi.org/10.1002/ps.5038
Ni M, Ma W, Wang X, Gao M, Dai Y, Wei X, Zhang L, Peng Y, Chen S, Ding L, Tia Y, Li J, Wang H, Wang X, Xu G, Guo W, Yang Y, Wu Y, Heuberger S, Tabashnik BE, Zhang T, Zhu Z, Ni M et al (2017) Next-generation transgenic cotton: pyramiding RNAi and Bt counters insect resistance. *Plant Biotechnol J* 15:1204–1213
Parkhi V, Kumar V, Campbell LM, Bell AA, Shah J, Rathore KS (2010) Resistance against various fungal pathogens and reniform nematode in transgenic cotton plants expressing Arabidopsis NPR1. *Trans Res* 19:959–975
Perez-Rodriquez P, Crossa J, Bondalapati K, De Meyer G, Pita F, de los Campos G (2015) A pedigree-based reaction norm model for prediction of cotton yield in multienvironment trials. *Crop Sci* 55:1143–1151
Qaim M, Matuschke I (2005) Impacts of genetically modified crops in developing countries: a survey. *Q J Int Agric* 44:207–227
Rajasekaran K, Cary JW, Jaynes JM, Cleveland TE (2005) Disease resistance conferred by the expression of a gene encoding a synthetic peptide in transgenic cotton (*Gossypium hirsutum* L.) plants. *Plant Biotechnol J* 3:545–554
Rathore KS, Sunilkumar G, Stipanovic RD, Wedegaertner TC (2008) RNAi-mediated, selective and substantial reduction in gossypol levels from cottonseed to enhance its food and feed value. In: Proceedings of world cotton conference, September 10–14, Lubbock
Rebijith KB, Asokan R, Hande HR, Kumar NKK, Krishna V, Vinutha J, Bakthavatsalam N (2016) RNA interference of odorant-binding protein 2 (OBP2) of the cotton aphid, *Aphis gossypii* (Glover), resulted in altered electrophysiological responses. *Appl Biochem Biotechnol* 178(2):251–266

Silvie PJ, Renou A, Vodounnon S, Bonni G, Adegnika MO, Héma O, Prudent P, Sorèze J, Ochou GO, Togola M, Badiane D, Ndour A, Akantetou PK, Ayeva B, Brévault T (2013) Threshold-based interventions for cotton pest control in West Africa: what's up 10 years later? *Crop Prot* 43:157–165

Simmons GS, McKemey AR, Morrison NI, O'Connell S, Tabashnik BE, Claus J, Fu G, Tang G, Sledge M, Walker AS, Phillips CE (2011) Field performance of a genetically engineered strain of pink bollworm. *PLoS One* 6(1):e24110

Soda N, Verma L, Giri J (2018) CRISPR-Cas9 based plant genome editing: significance, opportunities and recent advances. *Plant Physiol Biochem* 131:2–13

Sohrab SS, Azhar EI, Kamal MA, Bhattacharya PS, Rana D (2014) Genetic variability of Cotton leaf curl beta satellite in Northern India. *Saudi J Biol Sci* 21(6):626–631

Sun D, Guo Z, Liu Y, Zhang Y (2017) Progress and prospects of CRISPR/Cas systems in insects and other arthropods. *Front Physiol* 8:608

Sun S, Li C, Paterson AH, Jiang Y, Xu R, Robertson JS, Snider JL, Chee PW (2018) In-field high throughput phenotyping and cotton plant growth analysis using LiDAR. *Front Plant Sci* 9:1–17

Tabashnik BE, Carrière Y (2017) Surge in insect resistance to transgenic crops and prospects for sustainability. *Nat Biotechnol* 35:926–935

Tabashnik BE, Morin S, Unnithan GC, Yelich AJ, Ellers-Kirk C, Harpold VS, Sisterson MS, Ellsworth PC, Dennehy TJ, Antilla L, Liesner L (2012) Sustained susceptibility of pink bollworm to *Bt* cotton in the United States. *GM Crops Food* 3(3):194–200

Tabashnik BE, Brévault T, Carrière Y (2013) Insect resistance to Bt crops: lessons from the first billion acres. *Nat Biotechnol* 31:510–521

Technical Information Section (1994 Sept) Methanol use in cotton. THE ICAC RECORDER, vol XII, no. 3. International Cotton Advisory Committee, Washington, DC

Technical Information Section (2016 Oct) Cost of production of cotton. International Cotton Advisory Committee, Washington, DC

Technical Information Section (2017 Oct) Cotton production practices. International Cotton Advisory Committee, Washington, DC

Thirtle C, Beyers L, Ismael Y, Piesse J (2003) Can GM-technologies help the poor? The impact of Bt cotton in Makhathini Flats, KwaZulu-Natal. *World Dev* 31:717–732

Traxler G, Godoy-Avila S, Falck-Zepeda J, Espinoza-Arellano J (2013) Transgenic cotton in Mexico: economic and environmental impacts. In: Kalaitzandonakes N (ed) Economic and environmental impacts of first generation biotechnologies. Kluwer Academic Publishers, New York, pp 183–202

Turcotte EL, Feaster CV (1963) Haploids: high frequency production from single-embryo seeds in a line of Pima cotton. *Science* 140:1407–1408

Upadhyay SK, Chandrashekar K, Thakur N, Verma PC, Borgio JF, Singh PK, Tuli R (2011) RNA interference for the control of whiteflies (*Bemisia tabaci*) by oral route. *J Biosci* 36:153–161

Walker WB, Allen ML (2010) Expression and RNA interference of salivary polygalacturonase genes in the tarnished plant bug, *Lygus lineolaris*. *J Insect Sci* 10(1):173. https://doi.org/10.1673/031.010.14133

Walton J, Larson J, Roberts R, Lambert D, English B, Larkin S, Marra MC, Martin SW, Paxton KW, Reeves J (2010) Factors influencing farmer adoption of portable computers for site-specific management: a case study for cotton production. *J Agric Appl Econ* 42(2):193–209. https://doi.org/10.1017/S1074070800003400

Wang G, Fok M (2018) Managing pests after 15 years of Bt cotton: farmers' practices, performance and opinions in Northern China. *Crop Prot* 110:251–260

Wang J, Zhang H, Wang H, Zhao S, Zuo Y, Yang Y et al (2016) Functional validation of cadherin as a receptor of Bt toxin Cry1Ac in *Helicoverpa armigera* utilizing the CRISPR/Cas9 system. *Insect Biochem Mol Biol* 76:11–17. https://doi.org/10.1016/j.ibmb.2016.06.008

Wang J, Zhang H, Wang H, Zhao S, Zuo Y, Yang Y, Wu Y (2017) Functional validation of cadherin as a receptor of Bt toxin Cry1Ac in *Helicoverpa armigera* utilizing the CRISPR/Cas9 system. *Insect Biochem Mol Biol* 87:147–153

Wei J, Guo Y, Liang G, Wu K, Zhang J, Tabashnik BE, Li X (2015) Cross-resistance and interactions between Bt toxins Cry1Ac and Cry2Ab against the cotton bollworm. *Sci Rep* 5:7714

Wilson LJ, Mensah RK, Fitt GP (2004) Implementing integrated pest management in Australian cotton. In: Insect pest management. pp 97–118

Wu KM, Lu YH, Feng HQ, Jiang YY, Zhao JZ (2008) *Suppression* of cotton bollworm in multiple crops in China in areas with Bt toxin-containing cotton. *Science* 321:1676–1678

Xiong Y, Zeng H, Zhang Y, Xu D, Qiu D (2013) Silencing the HaHR3 gene by transgenic plant-mediated RNAi to disrupt *Helicoverpa armigera* development. *Int Biol Sci* 9(4):370–381. https://doi.org/10.7150/ijbs.5929

Xiongkui H, Bonds J, Herbst A, Langenakens J (2017) Recent development of unmanned aerial vehicle for plant protection in East Asia. *Int J Agric Biol Eng* 10(3):18–30

Ye ZF, Liu XL, Han Q, Liao H, Dong XT, Zhu GGH, Dong SL (2017) Functional characterization of PBP1 gene in Helicoverpa armigera (Lepidoptera: Noctuidae) by using the CRISPR/Cas9 system. *Sci Rep* 7:8470. https://doi.org/10.1038/s41598-017-08769-2

Yu J, Jung S, Cheng C, Ficklin SP, Lee T, Zheng P, Jones D (2014) CottonGen: a genomics, genetics and breeding database for cotton research. *Nucleic Acids Res* 42:D1229–D1236

Yu XD, Liu ZC, Huang SL, Chen ZQ, Sun YW, Duan PF, Ma YZ, Xia LQ (2016) RNAi-mediated plant protection against aphids. *Pest Manag Sci* 72:1090–1098

Zhu J-Q, Liu S, Ma Y, Zhang JQ, Qi HS, Wei ZJ (2012) Improvement of pest resistance in transgenic tobacco plants expressing dsRNA of an insect-associated gene *EcR*. *PLoS One* 7(6):e38572. https://doi.org/10.1371/journal.pone.0038572

Correction to: Evolution and Diversity of the Cotton Genome

Guanjing Hu, Corrinne E. Grover, Josef Jareczek, Daojun Yuan, Yating Dong, Emma Miller, Justin L. Conover, and Jonathan F. Wendel

Correction to: Chapter 2 in: M. U. Rahman et al. (eds.), *Cotton Precision Breeding*, https://doi.org/10.1007/978-3-030-64504-5_2

The original version of this chapter was revised to include a co-author. The correct version of the list of authors is as follows:

Guanjing Hu, Corrinne E. Grover, Josef Jareczek, Daojun Yuan, Yating Dong, Emma Miller, Justin L. Conover, Jonathan F. Wendel

The updated version of this chapter can be found at
https://doi.org/10.1007/978-3-030-64504-5_2

M. U. Rahman et al. (eds.), *Cotton Precision Breeding*,
https://doi.org/10.1007/978-3-030-64504-5_18

Index

Mehboob-ur-Rahman et al. (eds.), *Cotton Precision Breeding*,
https://doi.org/10.1007/978-3-030-64504-5

D

G

T

The manufacturer's authorised representative in the EU is Springer Nature Customer Service Centre GmbH, Europaplatz 3, 69115 Heidelberg, Germany. If you have any concerns regarding our products, please contact ProductSafety@springernature.com

Printed and bound by CPI Group (UK) Ltd, Croydon, CR0 4YY
28/09/2026
02224067-0003